T0251103

Computer Methods in Structural Analysis

BOOKS ON STRUCTURES PUBLISHED BY E & FN SPON

Analysis of Concrete Structures by Fracture Mechanics
Edited by L. Elfgren and S.P. Shah

Behaviour and Design of Steel Structures
N.S. Trahair and M.A. Bradford

Bridge Deck Behaviour
E.C. Hambly

Computer Analysis of Skeletal Structures
C.T.F. Ross and T. Johns

Developments in Structural Engineering
Edited by B.H.V. Topping

The Plastic Methods of Structural Analysis
B.G. Neal

Reinforced Concrete Designers Handbook
C.E. Reynolds and J.C. Steedman

Structural Analysis using Virtual Work
F. Thompson and G.G. Haywood

Structural Analysis – A unified classical and matrix approach
A. Ghali and A.M. Neville

Vibration of Structures – Applications in civil engineering design
J.W. Smith

For more information about these and other titles published by us,
please contact:

The Promotion Department, E & FN Spon, 2–6 Boundary Row,
London, SE1 8HN

Computer Methods in Structural Analysis

J.L. Meek

Associate Professor, Civil Engineering Department, University of
Queensland, Australia

CRC Press is an imprint of the
Taylor & Francis Group, an **informa** business

CRC Press
Taylor & Francis Group
6000 Broken Sound Parkway NW, Suite 300
Boca Raton, FL 33487-2742

First issued in hardback 2017

© 1991 J.L. Meek
CRC Press is an imprint of Taylor & Francis Group, an Informa business

No claim to original U.S. Government works

ISBN-13: 978-0-4191-5440-2 (pbk)
ISBN-13: 978-1-1384-7038-5 (hbk)

This book contains information obtained from authentic and highly regarded sources. Reasonable efforts have been made to publish reliable data and information, but the author and publisher cannot assume responsibility for the validity of all materials or the consequences of their use. The authors and publishers have attempted to trace the copyright holders of all material reproduced in this publication and apologize to copyright holders if permission to publish in this form has not been obtained. If any copyright material has not been acknowledged please write and let us know so we may rectify in any future reprint.

Except as permitted under U.S. Copyright Law, no part of this book may be reprinted, reproduced, transmitted, or utilized in any form by any electronic, mechanical, or other means, now known or hereafter invented, including photocopying, microfilming, and recording, or in any information storage or retrieval system, without written permission from the publishers.

Trademark Notice: Product or corporate names may be trademarks or registered trademarks, and are used only for identification and explanation without intent to infringe.

Visit the Taylor & Francis Web site at
http://www.taylorandfrancis.com

and the CRC Press Web site
http://www.crcpress.com

British Library Cataloguing in Publication Data
Meek, J.L.
Computer methods in structural analysis.
I. Title
624.10285

ISBN0-419-15440-X

Library of Congress Cataloguing-in-Publication Data
Available

CONTENTS

CHAPTER 6   THE DIRECT STIFFNESS METHOD — NON-LINEAR ANALYSIS   236

Preface

This book grew out of experience gained in three decades of development of matrix structural analysis, firstly as a research tool, then as post graduate lecture material and finally as an integral part of the undergraduate curriculum. The basic building blocks of a unified theory for structural analysis are to be found in matrix and linear algebra. It is with the use of matrix theory that the duality of the concepts of equilibrium and compatibility can be brought together via the principles of virtual displacements and virtual forces and expressed as the contregredient law. For many of these basic ideas the author is indebted to Dr K. Liversley of Cambridge University, Mr J.C. DeC Henderson formerly of Imperial College and the late Mr R.S. Jenkins of Ove Arup and Partners. The spirit of the book is to look firstly at these underlying principles of mechanics, developing them as a coherent and unified theory which can then be used in the approximate solution of the partial differential equations of solid mechanics, as applied to various structural forms such as skeletal frames, plates, shells and the solid continuum.

Because the line element structure (truss, frame and cable) plays an important role in the teaching of elementary structural mechanics, this form has received a reasonable amount of attention (Chapters 4, 5 and 6). The plan for the book then is to start with an introduction to vector, tensor and matrix notations in Chapter 1, together with a discussion of the interpolation theory which forms a basic part of the finite element method. In Chapter 2, the fundamental theorems necessary to apply the principles of virtual displacements and forces to both discrete and continuous structures are fully developed. It is found convenient to use the ubiquitous Gauss divergence theorem for the latter case. Material constitutive laws are discussed and the incremental elasto-plastic constitutive equations developed so that the reader will be able to apply the theory to a variety of material models currently available.

Chapter 3 introduces the finite element method in a number of its forms, and the use of the contregredient law in the development of the finite element stiffness matrices explored for the displacement, hybrid and equilibrium types of element. With nearly 40 000 papers now written on the finite element method this Chapter must by its very briefness be one of highly distilled material.

Chapter 4 deals with the force method of analysis in abbreviated form. Although the stiffness method has gained ascendency in use in the numerous computer software packages commercially available for structural analysis, the principles of statics, deflection calculation etc., are still taught in the first instance via the force method. Thus, there is still rationale for an understanding of the force method and it is still valid to observe that computer software for both the force and the displacement methods can be made to look identical at the user interface. Only for large scale analyses does the displacement method have significant advantages.

In Chapter 5 the displacement method is detailed and in this context attention is focused on the direct stiffness method. It must be realized that the direct stiffness method is only the name given to the process in which the various element stiffness properties are firstly expressed in the common global coordinate system and then assembled via unit matrix transformations which reflect the structure connectivity or topology.

In Chapter 6 some attention has been give to the calculation of elastic critical loads, not only because this is a useful exercise, but also because it now proves to be a rather minor yet elegant extension of the simple linear theory. A treatment has been given in this Chapter of the geometric non-linear analysis, both for line elements and membrane shell structures, because these forms have an important application to the shape finding, analysis and design, of tension structures.

Chapter 7 deals with the finite element analysis of the solid continuum which of course can be divided into the planar situations (plane stress, plane strain and axisymmetry), and the three-dimensional stress state. A feature in this Chapter is the discussion of the natural mode method of Professor J.H. Argyris which succinctly separates the rigid-body motion and straining modes of an element. The St Venant's torsion problem is analysed by the use of the cross section warping displacement function, and the means given for location the shear centre of those cross sections under St Venant's torsion stress.

Chapter 8 introduces a detailed discussion of the small deflection plate bending analysis. Because of its importance in understanding the bending action associated with shell behaviour many element types are given, including the flat facet elements and the degenerated isoparametric and heretosis elements. Comparisons of element computational efficiency and accuracy are given. The DKL (Discrete Kirchoff with Loof nodes) is introduced as a possible contender for the inclusion of plate elements in shell analysis via flat facet elements.

The analysis of shells is given in Chapter 9. With over 4000 papers now published on shell theory it becomes necessary to give a brief review of the literature. This is followed by a description of flat facet a type elements (plane stress plus plate element). Again the DKL element is highlighted for some advantages it appears to possess for moment connections in box type structures as well as in general shell analysis. It is similar to the Morley and semi-Loof type elements in that it has the rotational nodal variables embedded in the element sides rather than at the apex nodes. In looking at curved shell elements one has to make some compromise between computational efficiency and accuracy on the one hand and practical use on the other. It would appear that those elements which require high order derivatives in their nodal variables have limited application. Thus the decision was made to give details of the isoparametric shell element (Ahmad and Irons) and to give a reasonable discussion therein of the problems of membrane and shear locking problems and the means to avoid these troubles.

In studying the book the suggested order is to first read Chapters 1–3, and if the intention then is to pursue finite element analysis, to move to Chapters 5, 7, 8 and 9, in that order. For a lower level of study one may choose parts of Chapters 1 and 2 and then move to Chapter 4 and 5 to obtain a knowledge of matrix structural analysis as applied to line element (skeletal) structures. It is strongly recommended that some of the material of Chapter 6 is included to extend the student's capabilities to the calculation of frame buckling loads.

The author would like to thank the many people who have contributed to the book. Firstly, to Professor R.W. Clough for his introduction to the writer matrix structural analysis at UC (Berkeley) in the spring of 1958 and for the three decades of inspirational research which has come from UC (Berkeley) under the leadership of Professors Popov, Scordelis, Wilson, Powell and many others. To Professor Carlos Fellipa, now at Boulder, CO, USA, whose insight into finite element analysis has always been at the frontiers of knowledge and an inspiration to the writer. One cannot write a preface such as this without referring to the forty years of leadership provided by Professor J.H. Argyris with his marvellously simple natural mode technique, and to the quiet courage of the late Professor Bruce Irons who gave his all to the cause.

My thanks too, to the many students who have contributed through their research efforts: Drs W. Tranberg, H.S. Tan, K.Y. Tan and S. Loganathan, and to W. Lin and H. Goa for assisting with proof reading of the manuscript.

J.L. Meek, Brisbane, Australia

CHAPTER 1

Mathematical preliminaries

1.1 INTRODUCTION

In writing this book the intention is to assist the student in the learning process for structural and finite element analysis, in which he must master, not only the numerical techniques available, but also the mathematical skills necessary for the efficient description of the physics of the problem at hand. As such it is a teaching book. The format, then, is first to present a discussion on mathematical preliminaries in which the concepts of vectors, tensors and matrices are introduced. Which notation should be used? Vector, tensor or matrix? The answer is not simple, and the three notations are certainly not mutually exclusive. Following a discussion of the merits of the various notations, the reader is introduced to the ubiquitous Gauss' divergence theorem. This theorem forms the basis of most descriptions of the integral of the rate of change of a variable over a region to its values on the surface of the region. In structural mechanics the relationship has been discovered independently and given such names as principle, of virtual displacements, principle of virtual forces, or simply virtual work. The advantage of the more general formulation by Gauss is in its adaptability to a variety of other physical situations where the concepts of static equilibrium are not available. In structural mechanics it is possible to extend the concept of virtual displacements (forces) to the contragredient principle which succinctly describes the relationship between statically equivalent force systems and their compatible displacements. The contragredient principle can be loosely described as a reflective principle (for the purposes of memory only). That is, if the force systems (P, Q) are connected through the relationship,

$$\{P\} = [B]\{Q\}$$

then the corresponding compatible displacements (p, q) are connected as follows:

$$\{q\} = [B]^T \{p\}$$

Without the ability to apply both Gauss' celebrated theorem and the contragredient principle, the student's capabilities of analysis are seriously restricted. The finite element method can be classified as a means for the approximate solution of the partial differential equations of mathematical physics. It requires the subdivision of the region under consideration into a number of geometrically definable domains. In each domain, the unknown variables (for which the approximation is desired), are expressed in terms of certain values (either generalized coordinates or nodal values), by functions of the coordinate system. A study of this problem and the choice of nodal values rather than generalized coordinates lead to the definition and construction of orthogonal interpolation functions. The chapter concludes with a discussion of coordinate systems and the various element domains (isoparametric and triangular) and the calculation of function derivatives in, and integrals over, the domain. In subsequent chapters, the various physical situations encountered in both line structures and continuum mechanics are introduced and their finite element approximations discussed.

An attempt is made to give a fairly complete description of a wide selection of finite element types. In the first instance, attention is focused on small displacement, linear elasticity. In general the emphasis will be on the displacement element formulation with compatible displacements at the element interfaces. It will be shown that this approach may also be formulated as a Galerkin method of weighted residuals (again a useful concept in non–structural applications). The discussion will, however, by no means be restricted only to displacement models, and sections will be reserved for discussions of the natural mode technique, developed by Argyris [1], and the hybrid stress elements of Pian [2]. It may be pertinent at this stage in the book to draw the reader's attention to the work, *Energy Theorems and Structural Analysis* by Argyris and Kelsey [3], published in 1959. The material in this work belies its title, and the work actually contains a very modern discussion of the matrix methods of structural analysis. Because the calculation of elastic critical loads is merely an extension of the linear theory to include the effects of axial forces when the equilibrium equations are written in the deformed rather than the initial position, it is natural that the book includes a chapter on critical load (or eigenvalue) calculation and the corresponding mode shapes. This is a break in tradition from classical texts in that the computer has made the calculation of the linear eigenvalue problem for the first critical load a simple extension of the first–order theory. Because the effectiveness of the finite element method lies in part in its successful application to non–linear analysis problems of both a material and geometric nature, the text concludes with some of the techniques currently available for these analyses. In particular a discussion is given of net and membrane structures and shape finding techniques and of the non–linear analysis of structures which may present rather mischievous behaviour in their load deflection relationships. The solution to the transient heat flow problem and the related elasto–plastic thermal stresses will be given as an example of the non–linear material behaviour.

1.2 NOTATION: VECTOR – MATRIX – INDEX

1.2.1 Vector notation

The concept of vectors arises from the geometrical representation of forces, displacements and their time derivatives and from the parallelogram law of addition of such quantities. A vector is defined as a quantity in space which has direction as well as magnitude, and it will be denoted by $\vec{F}$, $\vec{R}$, etc. A point P in space is defined by its position vector $\vec{r}$, as in Fig. 1.1(a). In Fig. 1.1(b), the vector addition of $\vec{A}$ and $\vec{B}$ is defined as

$$\vec{C} = \vec{A} + \vec{B} \tag{1.1}$$

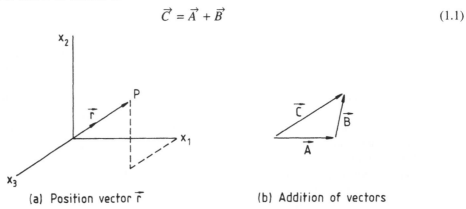

(a) Position vector $\vec{r}$ (b) Addition of vectors

Fig. 1.1   Vector components - vector addition.

It is soon found, however, that further operations are necessary for the satisfactory manipulation of vector quantities; the two most commonly used are the dot and cross products. The first produces a scalar and the second a vector.

Dot product

$$\vec{A} \cdot \vec{B} = |A||B|\cos\theta \qquad (1.2)$$

Cross product

$$\vec{A} \times \vec{B} = |A||B|\sin\theta\,\hat{n} \qquad (1.3)$$

The sense of $\hat{n}$ in Fig. 1.2(b) is given by the right–hand screw rule from $\vec{A}$ to $\vec{B}$. The use of these two operators is illustrated in the examples which follow. Firstly, consider the position vector $\vec{r}$ given as the vector sum of its three components in the directions of the Cartesian coordinates (x_1, x_2, x_3), base unit vectors $(\hat{e}_1, \hat{e}_2, \hat{e}_3)$.

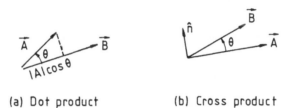

(a) Dot product   (b) Cross product

Fig 1.2   Vector operations.

That is,

$$\vec{r} = x_1\hat{e}_1 + x_2\hat{e}_2 + x_3\hat{e}_3 \qquad (1.4)$$

The i th component of $\vec{r}$ is given by

$$x_i = \vec{r} \cdot \hat{e}_i \qquad (1.5)$$

The magnitude of the vector $\vec{F}$ is given by

$$|F| = (\vec{F} \cdot \vec{F})^{\frac{1}{2}} \qquad (1.6)$$

Consider now, the force vector $\vec{F}$ acting through the point P, position vector $\hat{r}$. It is required to calculate the moment vector $\vec{M}_0$ of $\vec{P}$ about O. From Fig. 1.3(a), the magnitude of this moment is given by

$$|M_0| = |F|p \qquad (1.7)$$

where p is the perpendicular distance from O to the line of $\vec{F}$. The moment $\vec{M}_0$ is perpendicular to the plane of $\vec{F}$ and $\vec{r}$ and in the sense shown in Fig. 1.3(a). Thus,

$$|\vec{M}_0| = \vec{r} \times \vec{F} = |F||r|\sin\theta\,\hat{n} = |F|p\hat{n} \qquad (1.8)$$

A seemingly unrelated problem is shown in Fig. 1.3(b). The point P ($\vec{r}$) is part of a rigid body that contains O. The rigid body is given an infinitesimal rotation $\vec{\beta}$ about O. It is required to calculate the displacement vector $\vec{u}$ of P due to this rotation. From Fig. 1.3(b) it is seen that $\vec{u}$ is perpendicular to the plane of $\vec{\beta}$ and $\vec{r}$ and has magnitude,

$$|u| = |\beta|p = |\beta||r|\sin\theta$$

That is,

$$\vec{u} = \vec{\beta} \times \vec{r} \tag{1.9}$$

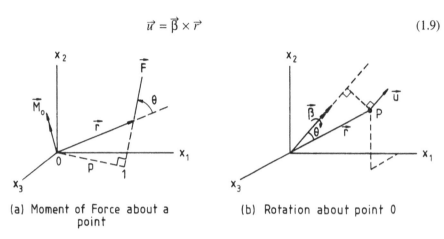

(a) Moment of Force about a point

(b) Rotation about point 0

Fig. 1.3   Moment and rotation vectors.

The similarity between eqns (1.8) and (1.9) is noted. This will be highlighted later as a simple example of the contragredient principle. The use of vector operators has a considerable advantage in the development of theory. However, for computational purposes it is perhaps better to work in terms of the components of the vectors in their Cartesian coordinate system. That is,

$$\vec{A} = A_1\hat{e}_1 + A_2\hat{e}_2 + A_3\hat{e}_3, \text{ etc.} \tag{1.10}$$

Then the dot product is given,

$$\vec{A} \cdot \vec{B} = (A_1\hat{e}_1 + A_2\hat{e}_2 + A_3\hat{e}_3) \cdot (B_1\hat{e}_1 + B_2\hat{e}_2 + B_3\hat{e}_3) = A_1B_1 + A_2B_2 + A_3B_3 \tag{1.11}$$

Since

$$\hat{e}_i \cdot \hat{e}_j = 0 \text{ for } i \neq j,$$

$$= 1 \text{ for } i = j \tag{1.12}$$

The cross product of two vectors is

$$\vec{C} = \vec{A} \times \vec{B} = (A_1\hat{e}_1 + A_2\hat{e}_2 + A_3\hat{e}_3) \times (B_1\hat{e}_1 + B_2\hat{e}_2 + B_3\hat{e}_3)$$

$$= (A_2B_3 - A_3B_2)\hat{e}_1 + (A_3B_1 - A_1B_3)\hat{e}_2 + (A_1B_2 - A_2B_1)\hat{e}_3 \tag{1.13}$$

since

$$\vec{e}_2 \times \vec{e}_3 = \vec{e}_1 \text{ and } \vec{e}_1 \times \vec{e}_1 = 0, \text{ etc.} \tag{1.14}$$

The expression for $\vec{C}$ is conveniently remembered as the value of the determinant with rows $(\hat{e}, \vec{A}, \vec{B})$ as follows:

$$\vec{C} = \vec{A} \times \vec{B} = \begin{vmatrix} \hat{e}_1 & \hat{e}_2 & \hat{e}_3 \\ A_1 & A_2 & A_3 \\ B_1 & B_2 & B_3 \end{vmatrix} \tag{1.15}$$

Some useful geometric results can be deduced with the aid of the dot and cross products. Thus, the area of the triangle shown in Fig. 1.4(a) can be calculated from the magnitude of the cross product of vectors of direction and magnitude of two of its sides for example, $\vec{AB}$ and $\vec{AC}$. That is,

$$\text{Area of triangle ABC} = \frac{1}{2} |\vec{AB} \times \vec{AC}| = \frac{1}{2} |\vec{BC} \times \vec{BA}| = \frac{1}{2} |\vec{CA} \times \vec{CB}| \tag{1.16}$$

From Fig. 1.4(b), the volume of the parallelopiped is calculated as

Volume = perpendicular distance EF × area of base $(ABDC)$

$$= \vec{AE} \cdot \hat{n} |\vec{AB} \times \vec{AC}|$$

That is,

$$\text{Volume} = \vec{AE} \cdot (\vec{AB} \times \vec{AC}) \tag{1.17}$$

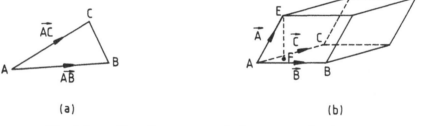

(a)   (b)

Fig. 1.4   Use of vectors to calculate areas and volumes.

From eqn (1.15),

$$\text{Volume} = \begin{vmatrix} A_1 & A_1 & A_3 \\ B_1 & B_2 & B_3 \\ C_1 & C_2 & C_3 \end{vmatrix} \tag{1.18}$$

The expression for the volume is thus the scalar triple product of the three vectors $\vec{AE}$, $\vec{AB}$, $\vec{AC}$ forming the sides of the parallelopiped.

One of the important relationships associated with vectors is the law of transformation of components from one Cartesian set to a second, obtained from the first by a rotation of axes. The two sets of axes are shown in Fig. 1.5.

The vector $\vec{A}$ may equally well be represented in either coordinate system as,

$$\vec{A} = A_1\,\hat{e}_1 + A_2\,\hat{e}_2 + A_3\,\hat{e}_3 = A'_1\,\hat{e}'_1 + A'_2\,\hat{e}'_2 + A'_3\,\hat{e}'_3 \qquad (1.19)$$

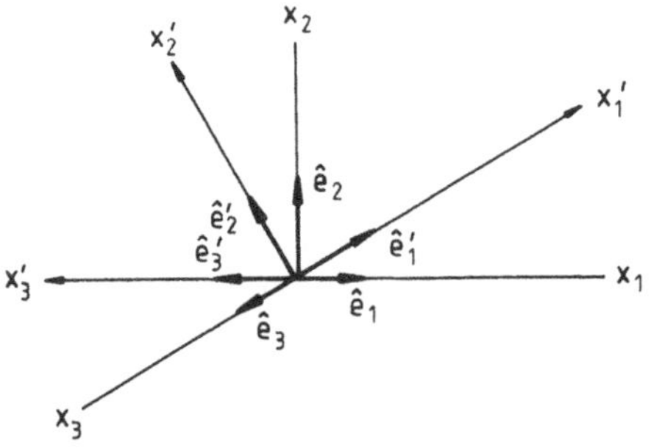

Rotation of co-ordinate axis

Fig. 1.5   Base vectors – original and rotated coordinate axes.

Suppose now that $\hat{e}'_i$ has components (l_{i1}, l_{i2}, l_{i3}) in the unprimed coordinate set. Since $\hat{e}'_i$ is a unit vector, its components are, in fact, the direction cosines of the x'_i–axis x relative to the x_i–axes. Thence,

$$\hat{e}'_i = l_{i1}\hat{e}_1 + l_{i2}\hat{e}_2 + l_{i3}\hat{e}_3 \quad \text{where } i = 1, 2 \text{ or } 3 \qquad (1.20)$$

and

$$|\hat{e}'| = \hat{e}'_i \cdot \hat{e}'_i = l_{i1}^2 + l_{i2}^2 + l_{i3}^2 = 1 \qquad (1.21)$$

and

$$\hat{e}'_i \cdot \hat{e}'_j = 0$$

Substitution in eqn (1.19) gives

$$\vec{A} = A'_1(l_{11}\hat{e}_1 + l_{12}\hat{e}_2 + l_{13}\hat{e}_3) + A'_2(l_{21}\hat{e}_1 + l_{22}\hat{e}_2 + l_{23}\hat{e}_3) + A'_3(l_{31}\hat{e}_1 + l_{32}\hat{e}_2 + l_{33}\hat{e}_3)$$

Then,

$$A_i = \vec{A} \cdot \vec{e_i} = A'_1 l_{1i} + A'_2 l_{2i} + A'_3 l_{3i} \qquad (1.22)$$

Now, the components of $\hat{e}_i$ in the primed coordinate set are again the direction cosines of the $\hat{e}_i$ set with respect to x'_1, x'_2, x'_3. It is seen then that

$$\hat{e}_i = l_{1i}\hat{e}'_1 + l_{2i}\hat{e}'_2 + l_{3i}\hat{e}'_3 \qquad (1.23)$$

Hence,

$$A'_i = \vec{A} \cdot \hat{e}'_i = A_1 l_{i1} + A_2 l_{i2} + A_3 l_{i3} \qquad (1.24)$$

The transformations in eqns (1.22) and (1.24) could be taken as the definition of a vector. That is, a physical quantity, Cartesian components (A_1, A_2, A_3), is a vector if its components obey these rules of transformation for rotation of coordinate axes. The vector notation introduced thus far proves to be very useful. However, because its notation is based on operators rather than components, it tends to be analysis rather than computation oriented.

Also when physical quantities such as stresses, strains and inertias are considered, their components form matrices, and not vectors, and thus do not obey the simple rules of vector transformation. The vector operators may be extended to include these new quantities and their operators (see, for example, [4]), but herein it is considered preferable to change to either matrix or index notation.

1.2.2 Matrix notation and theory

Matrix theory recognizes the fact that the components of the vector $\vec{A}$ form either the column matrix $\{A\}$ or the row matrix $<A>$. Thus,

$$\{A\} = \left\{ \begin{array}{c} A_1 \\ A_2 \\ A_3 \end{array} \right\} \tag{1.25}$$

and

$$<A> = [A_1\ A_2\ A_3]$$

These are, of course, only special cases of the general $m \times n$ matrix, $[A]$

$$[A] = \left[\begin{array}{cccc} A_{11} & A_{12} & \cdot & A_{1n} \\ A_{21} & A_{22} & \cdot & A_{2n} \\ & \cdot & \cdot & \cdot \\ & \cdot & \cdot & \cdot \\ & \cdot & \cdot & \cdot \\ A_{m1} & A_{m2} & \cdot & A_{mn} \end{array} \right] \tag{1.26}$$

Now the rules must be established for matrix operations. The elementary operations are addition, subtraction and multiplication. Certain rules for compatibility of matrices for these operations must be observed. Thus, for addition or subtraction of two matrices, $[A]$, $[B]$, their number of rows and columns must be the same; that is,

$$[C] = [A] \pm [B] \tag{1.27}$$

This implies term–by–term operation, so that

$$c_{ij} = a_{ij} \pm b_{ij} \tag{1.28}$$

The multiplication rule is the sum of row into column multiplication of corresponding terms. Thus,

$$[C] = [A][B] \tag{1.29}$$

where by definition

$$C_{ij} = A_{i1}B_{1j} + A_{i2}B_{2j} + \cdots + A_{ip}B_{pj} \tag{1.30}$$

Of course, the above notation is clumsy if the order of the matrices must be written along with each operation. In general, then, it will be assumed that the compatibility rule for the matrices is satisfied *a priori* by the conditions of the problem.

A further useful matrix operation is that of transposition. Thus, $[B]$ the transpose of $[A]$, is written,

$$[B] = [A]^T \tag{1.31}$$

where

$$B_{ij} = A_{ji}$$

Thus, the row and column vectors are given by $<A> = \{A\}^T$, and vice versa. It is seen that square matrices may possess symmetric or antisymmetric properties, defined as follows:

Symmetric property

$$[A]^T = [A] \tag{1.32a}$$

Antisymmetric property

$$[A]^T = -[A] \tag{1.32b}$$

The diagonal terms of an antisymmetric matrix are zero. Any square matrix $[A]$ may be decomposed into the sum of a symmetric matrix $[A_S]$ and an antisymmetric matrix $[A_A]$. Thus, write

$$[A] = [A_S] + [A_A]$$

Then, from eqn (1.32),

$$[A]^T = [A_S]^T + [A_A]^T = [A_S] - [A_A]$$

Addition and subtraction of these equations yields

$$[A_S] = \tfrac{1}{2}\{[A] + [A]^T\} \tag{1.33a}$$

$$[A_A] = \tfrac{1}{2}\{[A] - [A]^T\} \tag{1.33b}$$

From the definition of the dot product of two vectors, eqn (1.11), it is seen that

$$\vec{A} \cdot \vec{B} = \{A\}^T \{B\} \tag{1.34}$$

Thence the magnitude of the vector $\vec{A}$ is given by

$$|A| = (\{A\}^T \{A\})^{1/2} \tag{1.35}$$

It is now possible to write the transformation of components of a vector due to coordinate axes rotation in matrix form. Thus, from eqn (1.24),

$$A'_i = [l_{i1} \ l_{i2} \ l_{i3}] \begin{Bmatrix} A_1 \\ A_2 \\ A_3 \end{Bmatrix} \tag{1.36}$$

For values of $i = 1, 2, 3$, the direction cosines l_{i1}, etc. form the rows of a matrix $[L]$, such that

$$[L] = \begin{bmatrix} l_{11} & l_{12} & l_{13} \\ l_{21} & l_{22} & l_{23} \\ l_{31} & l_{32} & l_{33} \end{bmatrix} \tag{1.37}$$

Using eqn (1.37),

$$\{A\}' = [L]\{A\} \tag{1.38}$$

An advantage of the matrix notation becomes apparent in the fact that all components are now treated simultaneously. The nature of the transformation is revealed further when the same substitution is made in eqn (1.22), in which case,

$$\{A\} = [L]^T \{A\}' \tag{1.39}$$

Evidently, there is a need for a further matrix operation, namely the inverse of a square matrix, denoted by $[A]^{-1}$, such that

$$[A]^{-1}[A] = [I] \tag{1.40}$$

In eqn (1.40), $[I]$ has the usual definition of the unit matrix, with unit diagonal terms and zeros elsewhere. From eqns (1.38) and (1.39) it is seen that, for the transformation to be reversible,

$$[L]^T = [L]^{-1} \tag{1.41}$$

The matrix $[L]$ whose transpose is also its inverse is defined as an orthogonal matrix. The condition under which the inverse of a square matrix exists is that its determinant should be non–zero. If the matrix is of order n, then this is expressed *that the rank of the matrix should be n* ([5] gives for further details of the properties of the inverse). It is now worthwhile to examine the expressions for the moment of a force about a point and the rigid–body rotation about the origin in matrix notation. Again, further properties of these transformations are revealed. Eqn (1.8) results in a moment vector $\vec{M}_0$, $\{M_0\}$ whose components can be extracted from either eqn (1.13) or (1.15). That is,

$$\begin{Bmatrix} M_{01} \\ M_{02} \\ M_{03} \end{Bmatrix} = \begin{bmatrix} 0 & -x_3 & x_2 \\ x_3 & 0 & -x_1 \\ -x_2 & x_1 & 0 \end{bmatrix} \begin{Bmatrix} F_1 \\ F_2 \\ F_3 \end{Bmatrix}_P \tag{1.42}$$

The subscript P denotes that the force acts through the point P, coordinates $<x_1 \; x_2 \; x_3>$. Write eqn (1.42) as

$$\{M_0\} = [T]\{F_P\} \tag{1.43}$$

The matrix $[T]$ is skew symmetric. Examine now the effect of the rigid–body rotation $\vec{\beta}$ on the displacements at P, as given by eqns (1.9) and (1.15). Substitution of eqn (1.9) into eqn (1.15) gives

$$\begin{Bmatrix} u_1 \\ u_2 \\ u_3 \end{Bmatrix}_P = \begin{bmatrix} 0 & x_3 & -x_2 \\ -x_3 & 0 & x_1 \\ x_2 & -x_1 & 0 \end{bmatrix} \begin{Bmatrix} \beta_{01} \\ \beta_{02} \\ \beta_{03} \end{Bmatrix} \tag{1.44}$$

From eqns (1.42) and (1.43), this equation is written

$$\{u_P\} = [T]^T \{\beta_0\} \tag{1.45}$$

The contragredient nature of the transformation eqns (1.43) and (1.45) is thus revealed, i.e. the second transformation is the transpose of the first. The matrix notation of eqns (1.38) and (1.39) may be extended to tensor quantities. The most elementary example of such a quantity is the matrix formed by the post–multiplication of a position vector $\{r\}$ by its transpose. Thus, define the matrix $[\Delta Z]$

$$[\Delta Z] = \{r\}\{r\}^T = \begin{Bmatrix} x_1 \\ x_2 \\ x_3 \end{Bmatrix} <x_1 \ x_2 \ x_3> = \begin{bmatrix} x_1^2 & x_1 x_2 & x_1 x_3 \\ x_1 x_2 & x_2^2 & x_2 x_3 \\ x_1 x_3 & x_2 x_3 & x_3^2 \end{bmatrix} \tag{1.46}$$

The quantity $[\Delta Z']$ is defined in the rotated coordinate set,

$$[\Delta Z]' = \{r\}'\{r\}'^T \tag{1.47}$$

However, substitution of eqn (1.38) gives

$$[\Delta Z]' = [L]\{r\}\{r\}^T[L]^T = [L][\Delta Z][L]^T \tag{1.48}$$

The transformation law for the matrix $[\Delta Z]$ has been established. The notation $[\Delta Z]$ is used because if eqns (1.46) and (1.47) are multiplied by the elementary mass $m dV$ and integrated over the volume of the body, the resulting matrix $[Z]$ is the inertia matrix of the body. In two dimensions the same theory applies to the transformation of the matrix forming the second moment of area. The transformation in eqn (1.48) may be used as a definition of a tensor quantity. Alternatively, returning to eqn (1.24), it is seen that a vector associates a scalar with each direction in space by means of an expression which is linear and homogeneous in the direction cosines. Generalizing this definition, a tensor associates a vector with each direction in space by means of an expression which is linear and homogeneous in the direction cosines. That is, a tensor T is specified in the $(x_1 \ x_2 \ x_3)$ coordinate set by vectors, (T_1, T_2, T_3).

The vector $T^{(\alpha)}$ associated with direction (α), $(l_{\alpha 1}, l_{\alpha 2}, l_{\alpha 3})$ is given

$$T^{(\alpha)} = T_1 l_{\alpha 1} + T_2 l_{\alpha 2} + T_3 l_{\alpha 3} \tag{1.49}$$

This definition will be expanded in section 1.2.3 to give the same result as in eqn (1.48). Given the tensor $[T]$, it is possible to find its components $[T_0]$ by a suitable rotation of axes such that $[T_0]$ is a diagonal matrix ($T_{0ij} = 0$, for $i \neq j$). That is,

$$[L][T][L]^T = [T_0] ; \quad \text{or} \quad [T][L]^T = [L]^T[T_0]$$

Examine the multiplication of the first column of $[L]^T$, so that

$$[T]\begin{Bmatrix} l_{11} \\ l_{12} \\ l_{13} \end{Bmatrix} = T_{011}\begin{Bmatrix} l_{11} \\ l_{12} \\ l_{13} \end{Bmatrix} \qquad (1.50)$$

Thus, for the ith column of $[L]^T$,

$$\{[T] - T_{0ii}[I]\}\begin{Bmatrix} l_{i1} \\ l_{i2} \\ l_{i3} \end{Bmatrix} = 0 \qquad (1.51)$$

A solution to these equations exists if

$$|[T] - T_{0ii}[I]| = 0 \qquad (1.52)$$

Expanding this equation, the three values of T_{0ii} can be obtained, and substitution in eqn (1.50) gives the corresponding directions. This theory may be used to calculate principal inertias and principal axes of beam cross sections, and of course principal stresses or strains and their directions. It should be realized from the above illustrations that matrix theory forms a very useful basis for the manipulation of the components of vector and tensor quantities. Not only this, but also some useful properties, such as orthogonality of transformations, the contragredient nature of moment and rotation transformations, and the calculation of principal values and directions are revealed. Matrix theory is, in a sense, however, based on a command notation. Its algebra depends on the definition of operations on the individual terms of the matrices. The idea of matrix algebra as a command structure can be conveyed into a computer program, both within the program structure itself and in the user instruction language. An example of such a matrix command language is CAL78, developed by Prof. E.L. Wilson at the University of California, Berkeley, [6].

The obvious commands are

LOAD A 6 6	Load a matrix A(6,6)
DELETE A	Delete a matrix A
MULT A B C	Multiply two matrices A, B and store the result in C
INVERT A	Invert a square matrix A previously defined

1.2.3 Index notation

The index notation chooses to refer to matrices and vectors by their general term, with the indices ranging over the dimensions of the matrix or vector. This range will be either 3 or 2, depending on whether the problem is in three– or two–dimensional space. Thus, for vectors the various equivalent notations are

vector	$\vec{A}$
matrix	$\{A\}$
index	A_i
matrix	$[A] = A_{ij}$

The use of the index notation is greatly enhanced by the Einstein summation convention. This simple idea states that a repeated index in an expression implies a sum over the range of the index. For example, the expression $A_i B_i$ for two three–dimensional vectors $\{A\}$, $\{B\}$ is

$$A_i B_i = A_1 B_1 + A_2 B_2 + A_3 B_3$$

If the vector components are in the Cartesian coordinate axes, the result obtained is simply the dot product of the two vectors. The multiplication of two matrices $[A]$ and $[B]$ is now given as

$$C_{ij} = A_{ik} B_{kj} \tag{1.53}$$

The repeated index k implies that the number of columns of $[A]$ must equal the number of rows of $[B]$. K is seen that the index notation in a fashion, carries its operation definitions in the expressions themselves. This is often a considerable benefit. The transformation rule for vector components under rotation of axes [see eqns (1.38) and (1.39)] is now given

$$A'_i = l_{ij} A_j \; ; \quad \text{and} \quad A_i = l_{ji} A'_j \tag{1.54}$$

In the index notation, two special quantities, the Kronecker delta (δ_{ij}) and the permutation factor (e_{ijk}), must be defined for use in the various operations on vectors and matrices.

The Kronecker delta is defined

$$\delta_{ij} = 1 \; \text{ if } i = j$$

$$= 0 \; \text{ if } i \neq j \tag{1.55}$$

The Kronecker delta has the properties of the unit matrix $[I]$.

The permutation factor (e_{ijk}) is defined as follows:

$$e_{ijk} = +1 \text{ if } i, j, k \text{ is an even permutation of 1, 2, 3} \tag{1.56a}$$

$$e_{ijk} = -1 \text{ if } i, j, k \text{ is an odd permutation of 1, 2, 3} \tag{1.56b}$$

$$e_{ijk} = 0 \text{ if any values are repeated} \tag{1.56c}$$

From the above definitions it is seen that

$$e_{123} = e_{231} = e_{312} = 1$$

$$e_{321} = e_{213} = e_{132} = -1$$

and $e_{112} = 0$, etc.

The uses of these two operations will be illustrated in the examples that follow.

Example 1.1 *Replacement of an index*

$$\delta_{ik} A_{ij} = \delta_{1k} A_{1j} + \delta_{2k} A_{2j} + \delta_{3k} A_{3j} = A_{kj}$$

Example 1.2 *Repeated index in Kronecker delta*

$$\delta_{ii} = \delta_{11} + \delta_{22} + \delta_{33} = 3$$

Example 1.3 *Repeated indices in permutation factor*

$$e_{ijk}\, e_{ilm} = \delta_{jl}\, \delta_{km} - \delta_{jm}\, \delta_{kl}$$

$$e_{ijk}\, e_{ijk} = \delta_{jj}\, \delta_{kk} - \delta_{jk}\, \delta_{kj} = 9 - 3 = 6$$

$$e_{ijk}\, e_{ijm} = \delta_{jj}\, \delta_{km} - \delta_{jm}\, \delta_{km} = 3\delta_{km} - \delta_{km} = 2\delta_{km}$$

The examples given previously in vector and matrix notations are now rephrased in the index notation. A matrix [A] is symmetric if

$$A_{ji} = A_{ij}$$

and antisymmetric if

$$A_{ji} = -A_{ij}$$

For a matrix A_{ij}, the symmetric and antisymmetric components may be written $A_{(ij)}$ and $A_{[ij]}$, respectively. Then,

$$A_{(ij)} = \tfrac{1}{2}(A_{ij} + A_{ji}) \tag{1.57a}$$

$$A_{[ij]} = \tfrac{1}{2}(A_{ij} - A_{ji}) \tag{1.57b}$$

The magnitude of the vector A_i is

$$|A| = (A_i A_i)^{\frac{1}{2}} \tag{1.58}$$

To obtain the cross product of two vectors, the permutation factor is used. From eqn (1.13), it is easily deduced that if $\vec{D} = \vec{B} \times \vec{C}$, then

$$d_i = e_{ijk}\, b_j\, c_k \tag{1.59}$$

The scalar triple product of three vectors $\vec{A}$, $\vec{B}$, $\vec{C}$ is given from eqns (1.17) and (1.59)

$$S = a_i d_i = e_{ijk}\, a_i\, b_j\, c_k \tag{1.60}$$

It is easily shown that S is the value of the determinant

$$\begin{vmatrix} \{A_i\}^T \\ \{B_i\}^T \\ \{C_i\}^T \end{vmatrix}$$

Examine now the expressions for the moment of a force about a point and the deflections due to a rotation about that point, in index notation.

Moment of a force P about origin

Using the definition given in eqns (1.8) and (1.59),

$$M_{oi} = e_{ijk}\, r_j\, F_k \tag{1.61}$$

Define the matrix as follows:

$$T_{ik} = e_{ijk} r_j \tag{1.62}$$

It is seen from the properties of e_{ijk}, that T_{ik} is skew symmetric in $\{i\text{-}k\}$. Then eqn (1.61) is written

$$M_{0i} = T_{ik} F_k \tag{1.63}$$

Displacement at P due to a rotation at the origin

From eqn (1.9),

$$u_{Pi} = e_{ijk} \beta_j r_k \tag{1.64}$$

Interchange dummy indices j and k so that

$$u_{Pi} = e_{ikj} r_j \beta_k \tag{1.65}$$

Then, because e_{ikj} is skew symmetric in k and j,

$$u_{Pi} = -e_{ijk} r_j \beta_k \tag{1.66}$$

It is seen now that

$$u_{Pi} = T_{ki} \beta_k \tag{1.67}$$

The coefficient matrix in eqn (1.67) is simply the transpose of that in eqn (1.63). This is, of course, the identical result obtained in eqn (1.45). Now, however, the eqn (1.62) gives a convenient means for the definition of the $[T]$ matrix.

1.2.4   Transformation of tensor components

The index equivalent to eqn (1.48) is simply

$$\Delta Z'_{ij} = l_{ik} \Delta Z_{km} l_{jm} = l_{ik} l_{jm} \Delta Z_{km} \tag{1.68}$$

Again, as with vectors and matrices, it would be possible to extend the discussion of the index notation with further examples. However, the purpose here is one of comparison of the notations in common use. It can be seen that the index notation is particularly powerful. Operations expressed by equations are defined within the index notation, and thus the equations become transparent. This feature of the index notation can be useful when translating an expression into the equivalent computer coding. For example, in eqns (1.48) and (1.68) two matrix multiplications are involved. However, eqn (1.68) allows the term $\Delta Z'_{ij}$ to be calculated by using loops on the indices k and m. The whole matrix $[\Delta Z]$ is calculated by allowing i and j to range over their respective values. Unfortunately, this line of reasoning may produce very inefficient coding, as is easily seen by writing down the expression for the multiplication of several matrices in index notation and performing the operation in a single set of nested do loops.

1.3 THE GAUSS DIVERGENCE THEOREM

The Gauss divergence theorem is introduced at this early stage because of the fundamental role that it plays, not only as a basis of the virtual displacement (force) principles, but also in deriving theorems such as those of Green and Stoke with their well known application in continuum mechanics.

Consider the function A, bounded and differentiable in the region V, as shown in Fig. 1.6. It is required to express the integral of $\dfrac{\partial A}{\partial x_i}$ over the volume in terms of a surface integral of A. Consider firstly $\dfrac{\partial A}{\partial x_1}$, and calculate I, such that

$$I = \iiint \frac{\partial A}{\partial x_1} dx_1 dx_2 dx_3 \tag{1.69}$$

Take the elementary prism shown in Fig. 1.6(a), of cross section dx_2, dx_3, cutting the surface of V at L and R where the unit outward normals are $\hat{n}^L$ and $\hat{n}^R$ respectively. Firstly, it is required to calculate the relationship between the area ds and the cross–section of the prism $<dx_2, dx_3>$. From Fig. 1.6(b), it is seen that, for dx_2, dx_3 infinitesimally small, the vectors $\vec{dl}$, $\vec{\delta l}$ are defined

$$\vec{dl} = <dx_1\hat{e}_1, -dx_2\hat{e}_2, 0> \tag{1.70}$$

$$\vec{\delta l} = <\delta x_1\hat{e}_1, 0, dx_3\hat{e}_3> \tag{1.71}$$

Furthermore,

$$|ds|\hat{n}^R = \vec{\delta l} \times \vec{dl} = \begin{vmatrix} \hat{e}_1 & \hat{e}_2 & \hat{e}_3 \\ \delta x_1 & 0 & dx_3 \\ dx_1 & -dx_2 & 0 \end{vmatrix} \tag{1.72}$$

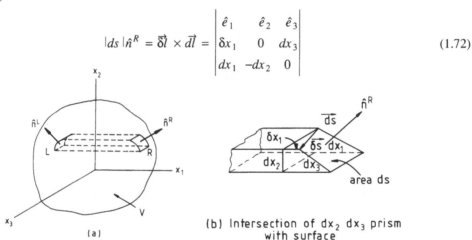

(a)

(b) Intersection of dx_2 dx_3 prism with surface

Fig. 1.6   Volume integration – Gauss' theorem.

By using eqn (1.60), or, alternatively, expanding the determinant and taking the dot product, it is shown that

$$|ds|\hat{n}^R \cdot \hat{e}_1 = dx_2\ dx_3 \tag{1.73}$$

By similar reasoning,

$$|ds|\hat{n}^L \cdot \hat{e}_1 = -dx_2\ dx_3 \tag{1.74}$$

Integrate the function $\dfrac{\partial A}{\partial x_1}$ from L to R, so that,

$$\int_L^R \frac{\partial A}{\partial x_1} dx_1\ dx_2\ dx_3 = (A_R - A_L)dx_2\ dx_3 \tag{1.75}$$

For all such prisms,

$$\iiint_V \frac{\partial A}{\partial x_1} dx_1 \, dx_2 \, dx_3 = \sum (A_r - A_L) dx_2 \, dx_3 \tag{1.76}$$

However, from eqns (1.73) and (1.74),

$$A_R \, dx_2 \, dx_3 = A_R \, \hat{n}^R \cdot \hat{e}_1 ds \quad \text{and} \quad A_L \, dx_2 \, dx_3 = -A_L \, \hat{n}^L \cdot \hat{e}_1 ds$$

Hence the summation on the right–hand side of eqn (1.76) may be replaced by the surface integral,

$$\int_S A\hat{n} \cdot e_1 \, ds = \int_S An_1 \, ds \tag{1.77}$$

Substitution in eqn (1.76) gives the result,

$$\iiint_V \frac{\partial A}{\partial x_1} dx_1 \, dx_2 \, dx_3 = \int_S An_1 \, ds \tag{1.78}$$

Similar expressions are obtained for $\dfrac{\partial A}{\partial x_2}, \dfrac{\partial A}{\partial x_3}$, so that, in general for all terms,

$$\iiint_V \frac{\partial A}{\partial x_i} dx_1 \, dx_2 \, dx_3 = \int_S An_i \, ds \tag{1.79}$$

Suppose now, that A is a component of the vector $\vec{F}$, so that

$$A = F_i$$

and use the notation to denote derivatives with respect to the independent variables,

$$\frac{\partial A}{\partial x_j} = A_{,j} = F_{i,j} \tag{1.80}$$

By making $i = j$, the divergence of the vector $\vec{F}$ written as $divF$ is obtained:

$$F_{i,i} = \frac{\partial F_1}{\partial x_1} + \frac{\partial F_2}{\partial x_3} + \frac{\partial F_3}{\partial x_2} = div\vec{F} \tag{1.81}$$

Substitution in eqn (1.79) gives

$$\int_V F_{i,i} = \int_S F_i n_i \, ds \tag{1.82}$$

In vector notation,

$$\int_V div\vec{F} \, dv = \int_S F \cdot \hat{n} \, ds \tag{1.83}$$

If matrix notation is used, it becomes necessary to define the differential operator $[D]$ such that,

$$[D]^T = \{ \frac{\partial}{\partial x_1} \quad \frac{\partial}{\partial x_2} \quad \frac{\partial}{\partial x_3} \}^T \tag{1.84}$$

Then Gauss' theorem becomes

$$\int_V \{D\}^T \{F\} \, dV = \int_S \{F\}^T \{n\} \, ds \tag{1.85}$$

Example 1.4 *Green's theorem*

Green's theorem proves to be a simple and useful extension of Gauss' theorem. Thus, let U, V be scalar point functions such that

$$\nabla V = V_{,i} \, e_i \tag{i}$$

Then it is easily shown that

$$div U \, \nabla V = \nabla U \cdot \nabla V + U \nabla^2 V \tag{ii}$$

Apply Gauss' theorem eqn (1.83) to eqn (ii), so that

$$\int_S U \nabla V \cdot n dS = \int_V (\nabla U \cdot \nabla V + U \nabla^2 V) dV$$

or,

$$\int_V U \nabla^2 V dV = \int_S (U \nabla V \cdot n) dS - \int_V (\nabla U \cdot \nabla V) dV \tag{iii}$$

This expression is useful in the application of weighted residuals to either Poisson's or Laplace's equations (Chapter 3, section 5).

1.4 INTERPOLATION

1.4.1 Element domains

A basic step in the finite element method is the subdivision of the region under consideration into a number of domains (or finite elements). In each domain the geometry, the displacements and the stress values may be interpolated in terms of either nodal values or certain generalized values which finally may be expressed in terms of nodal values. Usually the nodal values will be displacements and their first derivatives for displacement models, although in some cases second order derivatives may be required. It suffices to say that the integrals in eqns (1.82) or (1.85) become the summation of separate integrals over the various domains on the one hand, and of the surface integrals, on the other. Typical finite element families for linear, two–dimensional and three–dimensional space are shown in Fig. 1.7(a), (b), (c). For each element type with curved sides, there is a straight–sided equivalent. The common straight elements are

linear space	:	straight line
two dimensions	:	triangle
	:	quadrilateral
three dimensions	:	tetrahedron
	:	polyhedron

Because the sides of curved elements are defined in terms of nodes within sides, the shape of a side must be consistent with the order of approximation available. For example, a quadratic shape requires one internal node in a side; a cubic shape, two per side.

The planar element types are shown in Figs. 1.7(a), and again with quadratic and cubic shapes in Fig. 1.8. Returning to Fig. 1.7(b), we see that, in the non–dimensional (ξ,η,ζ) space, every element of a particular classification has the same size, being contained in unit space. The problem of interpolation can be considered as a one–to–one mapping of functions at each (ξ,η,ζ) point to the corresponding point in (x_1, x_2, x_3) space. Integrals and derivatives in (x_1, x_2, x_3) space must be related to interpolation functions given in (ξ,η,ζ) space.

1.4.2 Polynomial interpolation

The space P_n of all polynomials of degree $\leq n$ in the Euclidean space E_S contains the independent functions,

$$m = \frac{(n + s)!}{n!s!}$$

For example, let $s = 1,\quad n = 1,\quad m = 2! = 2.$ That is, for a line element,

$$f(x) = A + Bx$$

Again, let $s = 2$ and $n = 2$; this is a quadratic function in two–dimensional space.

$$m = \frac{(4!)}{(2!\ 2!)} = 6$$

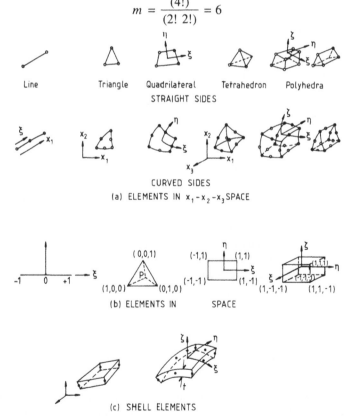

(a) ELEMENTS IN $x_1 - x_2 - x_3$ SPACE

(b) ELEMENTS IN   SPACE

(c) SHELL ELEMENTS

Fig. 1.7   Finite element types – global and interpolation space.

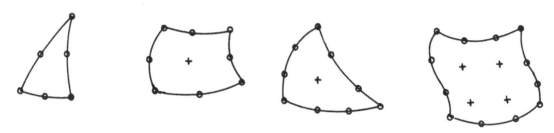

quadratic
one intermediate node per side
(optional internal node for quadrilateral)

cubic
one and four internal nodes for
triangle and quadrilateral, respectively

Fig. 1.8   Planar elements.

Thence,

$$f(x_1, x_2) = A + Bx_1 + Cx_2 + Dx_1^2 + Ex_1x_2 + Fx_2^2$$

Call P_n a complete polynomial space of order n and then P_n contains all $P_r (r \leq n)$ as linear subspaces. Let $X = (x_1, x_2, \cdots, x_s)$ denote a point in a bounded, simply connected domain D of Euclidean space E_S. Consider the m–dimensional vector space Γ of functions $f(x)$ defined and differentiable in D and spanned by a basis,

$$\{\phi\}^T = <\phi_1(X)\ \phi_2(X)\ \cdots\ \phi_m(X)>$$

By definition, any $f(X)$ contained in Γ can be expressed as a linear function of $\{\phi_2\}$ as follows:

$$f(X) = a_1\phi_1 + a_2\phi_2 + \cdots + a_m\phi_m$$

or,

$$f(X) = \{a\}^T\{\phi(X)\} = \{\phi(X)\}^T\{a\} \qquad (1.86)$$

In eqn (1.86), $\{a\}^T = <a_1a_2 \cdots a_m>$ are the generalized coordinates of $f(X)$ in the basis ϕ.

Consider now a configuration (N) of m distinct points of D,

$$(N) : (X_1, X_2, \cdots, X_m),$$

at which $f(X)$ takes the values $f_i = f(X_i)$. Then let

$$\{f\}^T = <f_1f_2 \cdots f_m> \qquad (1.87)$$

The value f_i is obtained by evaluating $f(X)$ at the node value X_i. Then, from eqn (1.86),

$$
\begin{Bmatrix} f_1 \\ f_2 \\ \cdot \\ \cdot \\ \cdot \\ f_m \end{Bmatrix} =
\begin{bmatrix}
\phi_1(x_1) & \phi_2(x_1) & \cdots & \phi_m(x_1) \\
\phi_1(x_2) & \phi_2(x_2) & \cdots & \phi_m(x_2) \\
\cdot & \cdot & \cdots & \cdot \\
\cdot & \cdot & \cdots & \cdot \\
\cdot & \cdot & \cdots & \cdot \\
\cdot & \cdot & \cdots & \cdot \\
\phi_1(x_m) & \phi_2(x_m) & \cdots & \phi_m(x_m)
\end{bmatrix}
\begin{Bmatrix} a_1 \\ a_2 \\ \cdot \\ \cdot \\ \cdot \\ a_m \end{Bmatrix}
\tag{1.88}
$$

The coefficient matrix on the right–hand side of eqn (1.88) will be defined such that

$$[A] = A_{ij} = \phi_i(X_j) \tag{1.89}$$

Then eqn (1.88) is written

$$\{f\} = [A]^T \{a\} \tag{1.90}$$

If the determinant $|A| \neq 0$, it is possible to obtain the generalized coordinates $\{a\}$, since then

$$\{a\} = ([A]^T)^{-1} \{f\} \tag{1.91}$$

An orthogonal basis, $\{\Phi\}^T = <\Phi_1(X)\ \Phi_2(X) \cdots \Phi_m(X)>$, is defined as one for which the nodal parameters f_i are also the generalized coordinates a_i. That is,

$$\{f(X)\} = \{\phi(X)\}^T \{a\} = \{\Phi(X)\}^T \{f\} \tag{1.92}$$

Let $[\bar{A}]$ be the matrix constructed from $\{\phi(x)\}$ in the same way as $[A]$; that is,

$$\bar{A}_{ij} = [\Phi_i(X_j)]$$

Then, from eqn (1.92),

$$\{f\} = [\bar{A}]^T \{f\}$$

Hence $[\bar{A}] = [I]$ and $\phi_i(X)_j = \delta_{ij}$. It is now possible to construct the basis $\{\Phi\}$ if $\{\phi\}$ is given by substitution for $\{a\}$ from eqn (1.91) in eqn (1.92). That is,

$$\{\phi(X)\}^T ([A]^T)^{-1} \{f\} = \{\Phi(X)\}^T \{f\}$$

Since the values of $\{f\}$ are arbitrary, it follows that

$$\{\Phi(X)\} = ([A]^T)^{-1} \{\phi(X)\} \tag{1.93}$$

It should be noted that the nodal values $\{f\}$ of the function can contain not only the function values but also values of its first and second derivatives, the only requirement being that $|A| \neq 0$.

1.4.3 Examples of the construction of orthogonal base systems

Example 1.5 *Linear space, quadratic function*

Consider the quadratic function in one dimension for which f is known at $x = 0$, $x = 1/2$, and $x = 1$.

In this case,

$$m = \frac{(n+s)!}{n!s!} = \frac{3!}{2!} = 3$$

Thence take

$$\{\phi\}^T = \{1 \ x \ x^2\}^T$$

$$f(x) = \{a\}^T \begin{bmatrix} 1 \\ x \\ x^2 \end{bmatrix}$$

Construct $[A]$ by substituting for $x = 0, \frac{1}{2}, 1$, so that

$$[A] = \begin{bmatrix} 1 & 1 & 0 \\ 1 & \frac{1}{2} & \frac{1}{4} \\ 1 & 1 & 1 \end{bmatrix} \quad \text{and} \quad [A]^{-1} = \begin{bmatrix} 1 & 0 & 0 \\ -3 & 4 & -1 \\ 2 & -4 & 2 \end{bmatrix}$$

Substitution in eqn (1.93) gives

$$\{\Phi(X)\} = \begin{bmatrix} 1 & -3 & 2 \\ 0 & 4 & -4 \\ 0 & -1 & 2 \end{bmatrix} \begin{bmatrix} 1 \\ x \\ x^2 \end{bmatrix} = \begin{bmatrix} 1 - 3x + x^2 \\ 4x(1-x) \\ x(2x-1) \end{bmatrix}$$

Example 1.6 *Linear space, quadratic function*

In this example the quadratic function is required for the region $X = 0, 1$, with the function values being given as

$$x = 0, f_0, \frac{\partial f}{\partial x_0}; x = 1, f_1$$

Again, choose the polynomial $\{\phi\}$, as in example 1.5. Then,

$$f(X) = \{a\}^T \begin{bmatrix} 1 \\ x \\ x^2 \end{bmatrix}; \quad \frac{\partial f}{\partial x} = \{a\}^T \begin{bmatrix} 0 \\ 1 \\ 2x \end{bmatrix}$$

In this case,

$$[A] = \begin{bmatrix} 1 & 0 & 0 \\ 0 & 1 & 0 \\ 1 & 1 & 1 \end{bmatrix} \quad \text{and} \quad [A]^{-1} = \begin{bmatrix} 1 & 0 & 0 \\ 0 & 1 & 0 \\ -1 & -1 & 1 \end{bmatrix}$$

Again, from eqn (1.93),

$$\Phi(X) = \begin{bmatrix} 1 & 0 & -1 \\ 0 & 1 & -1 \\ 0 & 0 & 1 \end{bmatrix} \begin{bmatrix} 1 \\ x \\ x^2 \end{bmatrix} = \begin{bmatrix} 1 - x^2 \\ x(1-x) \\ x^2 \end{bmatrix}$$

Example 1.7 *Linear space, cubic function with end derivatives*

The line element shown in Fig. 1.9 has a cubic function, interpolated in terms of derivatives at the ends of the element. Because it has been assumed that the function values are zero at the ends, only two cubic functions are required. Choose homogeneous polynomials for the $\{\phi\}$ functions. That is,

$$\{\phi\} = \begin{Bmatrix} \zeta_1^2 \zeta_2 \\ \zeta_1 \zeta_2^2 \end{Bmatrix} \quad \text{with} \quad \zeta_1 + \zeta_2 = 1, \quad \zeta_1 = 1 - \frac{x}{l}, \quad \zeta_2 = \frac{x}{l}$$

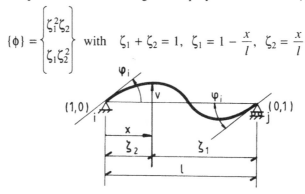

Cubic function interpolation

Fig. 1.9   Beam displacements–cubic interpolation using end slopes.

Then,

$$v = \{\phi\}^T \{a\}$$

The derivative with respect to x is

$$v' = \frac{\partial v}{\partial x} = \frac{\partial v}{\partial \zeta_1} \frac{\partial \zeta_1}{\partial x} + \frac{\partial v}{\partial \zeta_2} \frac{\partial \zeta_2}{\partial x} = \frac{1}{l}\left\{ -\frac{\partial v}{\partial \zeta_1} + \frac{\partial v}{\partial \zeta_2} \right\}$$

Then,

$$v' = \frac{1}{l}[\zeta_1(\zeta_1 - 2\zeta_2) \; \zeta_2(2\zeta_1 - \zeta_2)] \begin{Bmatrix} a_1 \\ a_2 \end{Bmatrix}$$

Substitution for ζ_1, ζ_2 values at i and j yields:

$$[A] = \frac{1}{l}\begin{bmatrix} 1 & 0 \\ 0 & -1 \end{bmatrix} ; \quad \text{and hence,} \quad [A]^{-1} = l\begin{bmatrix} 1 & 0 \\ 0 & -1 \end{bmatrix}$$

Thence, substitution in eqn (1.93) yields

$$\{\Phi\} = l\begin{bmatrix} -\zeta_1^2 \zeta_2 \\ \zeta_1 \zeta_2^2 \end{bmatrix}$$

As mentioned in section 1.4.1, it is advantageous to use interpolation functionsin the non–dimensional domains as given in Fig. 1.7(b), and to transform integrals and derivatives into (x_1, x_2, x_3) space. Interpolation may use either nodal values alone, or nodal values and nodal derivatives.

1.4.4 Common interpolation functions for line, quadrilateral and cubic elements

The line element interpolation functions for linear, quadratic and cubic interpolation are given in Fig. 1.10(a), (b), (c). Equally spaced node points have been used. If $f(\xi)$ is the actual function, $\tilde{f}(\xi)$ is its approximation.

(a) Linear

$$\tilde{f} = \{f_1 \, f_2\}^T \begin{bmatrix} \dfrac{1}{2}(1 - \xi) \\ \dfrac{1}{2}(1 + \xi) \end{bmatrix} = \{f\}^T \{N_1\}$$

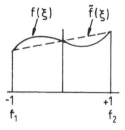

(b) Quadratic

$$\tilde{f} = \{f_1 \, f_2 \, f_3\}^T \begin{bmatrix} -\dfrac{1}{2}\xi(1 - \xi) \\ (1 - \xi^2) \\ \dfrac{1}{2}\xi(1 + \xi) \end{bmatrix} = \{f\}^T \{N_2\}$$

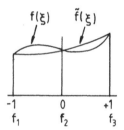

(c) Cubic

$$\tilde{f} = \{f_1 \, f_2 \, f_3 \, f_4\}^T \begin{bmatrix} \dfrac{1}{16}(1 - \xi)(3\xi + 1)(3\xi - 1) \\ \dfrac{9}{16}(1 - 3\xi)(1 - \xi)(1 + \xi) \\ \dfrac{9}{16}(1 + 3\xi)(1 - \xi)(1 + \xi) \\ \dfrac{1}{16}(1 + \xi)(3\xi + 1)(3\xi - 1) \end{bmatrix} = \{f\}^T \{N_3\}$$

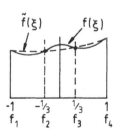

Fig. 1.10   Linear interpolation functions.

To obtain the interpolation functions in (ξ, η) two–dimensional space, the simplest approach is to multiply together the corresponding shape functions in each of the two directions.

(a) Linear

$[\Phi_{ij}] = [N_1][N_1]^T$

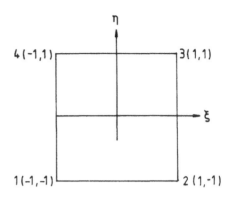

(b) Quadratic

$[\Phi_{ij}] = [N_2][N_2]^T$

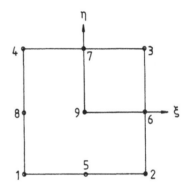

(c) Cubic

$[\Phi_{ij}] = [N_3][N_3]^T$

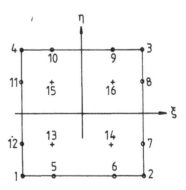

Fig. 1.11   Two–dimensional quadrilateral interpolation.

In Fig. 1.11(a), for example,

$$\Phi_{11} = \frac{1}{4}(1 - \xi)(1 - \eta) , \quad \Phi_{21} = \frac{1}{4}(1 + \xi)(1 - \eta) , \text{ etc.}$$

In Fig. 1.11(b),

$$\Phi_{11} = \frac{1}{4}\xi\eta(1 - \xi)(1 - \eta) , \quad \Phi_{31} = -\frac{1}{4}\xi\eta(1 + \xi)(1 - \eta) , \text{ etc.}$$

In Fig. 1.11(c),

$$\Phi_{11} = \frac{1}{256}(1 - \xi)(3\xi + 1)(3\xi - 1)(1 - \eta)(3\eta + 1)(3\eta - 1)$$

and

$$\Phi_{43} = \frac{9}{256}(1 + \xi)(3\xi + 1)(3\xi - 1)(1 - 3\eta)(1 - \eta)(1 + \eta) , \text{ etc.}$$

It is seen that interpolation functions obtained in this way contain internal node values, except in the simple case of the combination of linear functions. They are the Lagrange interpolation functions. The polynomials so obtained are incomplete. However, along lines of ξ or η constant, the other variable follows the linear, quadratic or cubic variation as required for compatibility between elements. The internal nodes may be condensed out before the 'stiffness' matrix is assembled. The static condensation algorithm to achieve this effect will be discussed in Chapter 5. The use of the Lagrange elements has some advantages, not the least of which is the ease with which they can be incorporated in automatic mesh generation routines. The identical process may be used to construct Lagrange interpolation functions for the cube in (ξ , η , ζ) space. The linear and quadratic elements are shown in Fig. 1.12. The cubic element contains 64 nodes and is not shown because it, probably too complicated in its connectivity for any practical application. Because the polynomials used in the interpolation are incomplete, the possibility exists for choosing functions which satisfy the necessary conditions for inter–element compatibility on their boundaries, but contain no internal nodes.

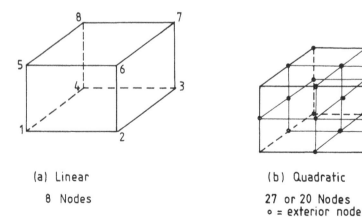

(a) Linear

8 Nodes

(b) Quadratic

27 or 20 Nodes
○ = exterior node
+ = interior node

Fig. 1.12   Three–dimensional interpolation.

The interpolation functions thus formed have been named 'serendipity functions', and the general class of element with curved sides in (x_1, x_2, x_3), space 'isoparametric elements' [7]. The serendipity family for two dimensions (quadratic and cubic) is given in Fig. 1.13.

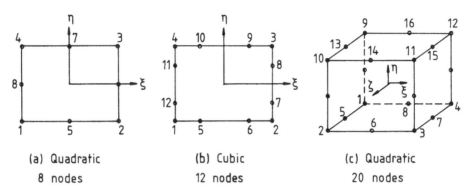

(a) Quadratic	(b) Cubic	(c) Quadratic
8 nodes	12 nodes	20 nodes

Fig. 1.13   Isoparametric interpolation.

In all cases,

$$\tilde{f} = [N_1 N_2 \cdots] \begin{Bmatrix} f_1 \\ f_2 \\ \cdot \\ \cdot \\ \cdot \end{Bmatrix}$$

For the 8– node element in Fig. 1.13(a),

$$4N_i = (1 + \xi_i \xi)(1 + \eta_i \eta)(\xi_i \xi + \eta_i \eta - 1) , \quad \text{for } i = 1 \text{ to } 4;$$

$$2N_i = (1 + \eta_i \eta)(1 - \xi^2) , \quad \text{for } i = 5, 7;$$

$$2N_i = (1 + \xi_i \xi)(1 - \eta^2) , \quad \text{for } i = 6, 8. \tag{1.94}$$

In each case, (ξ_i, η_i) signifies the (ξ, η) coordinates of node i.

For the 12– node element in Fig. 1.13(b):

$$32N_i = (1 + \xi_i \xi)(1 + \eta_i \eta)[- 10 + 9(\xi^2 + \eta^2)] \tag{1.95a}$$

for nodes $\xi_i = \pm 1 \quad \eta_i = \pm 1;$

$$32N_i = 9(1 + \xi_i \xi)(1 - \eta^2)(1 + 9\eta_i \eta) \tag{1.95b}$$

for nodes $\xi_i = \pm 1, \eta_i = \pm 1/3;$

$$32N_i = 9(1 + \eta_i \eta)(1 - \xi^2)(1 + 9\xi_i \xi) \tag{1.95c}$$

for nodes $\xi_i = \pm 1/3, \eta_i = \pm 1.$

For the 20– node solid brick:

    corner nodes, (numbers 1, 2, 3, 4, 10, 11, 12, 13):

$$8N_i = (1 + \xi_i\xi)(1 + \eta_i\eta)(1 - \zeta_i\zeta)(\xi_i\xi + \eta_i\eta + \zeta_i\zeta - 2); \tag{1.96a}$$

    mid–side nodes, (numbers 5, 7, 13, 15):

$$8N_i = (1 - \xi^2)(1 + \eta_i\eta)(1 + \zeta_i\zeta); \tag{1.96b}$$

    mid–side nodes, (numbers 17, 18, 19, 20):

$$8N_i = (1 + \xi_i\xi)(1 + \eta_i\eta)(1 - \zeta^2) \tag{1.96c}$$

1.4.5   Differentiation and integration using interpolation polynomials

1.4.5.1   Differentiation

The interpolation functions are used to obtain the approximate value of a function ($\tilde{f}$) at the point (ξ_i) in the interpolation space, given the nodal values (f). In the use of the finite element method in the (x_i) domain, derivatives and integrals are required in this domain. It is necessary to relate these quantities to interpolated values expressed in the ξ domain. Firstly, it is observed that the interpolation functions (ξ space Fig. 1.7(b), may be used to map the x space, as in Fig. 1.7(a). Use is made of the following definitions, given for generality in the three–dimensional space. For two dimensions, simply drop the third dimension.

$$\{\tilde{x}\} = \begin{Bmatrix} \tilde{x}_1 \\ \tilde{x}_2 \\ \tilde{x}_3 \end{Bmatrix} \tag{1.97a}$$

$$[X] = \begin{bmatrix} x_{11} & x_{21} & x_{31} \\ x_{12} & x_{22} & x_{32} \\ . & . & . \\ x_{1N} & x_{2N} & x_{3N} \end{bmatrix} \tag{1.97b}$$

    Note that $\{\tilde{x}\}$ is the current coordinate point, and $[X]$ is the matrix of nodal coordinates, where the element has N node points. Then the mapping of ξ space into x space is expressed as

$$\{\tilde{x}\}^T = \{\phi\}^T[X] \tag{1.98}$$

    This mapping of ($\tilde{x}$, $\tilde{y}$) space into ($\tilde{\xi}$, $\tilde{\eta}$) space is shown in Fig. 1.14. Then the derivatives of the function f are given:

    ξ space

$$\{f_{,\xi}\} = \begin{Bmatrix} f_{,\xi_1} \\ f_{,\xi_2} \\ f_{,\xi_3} \end{Bmatrix} \tag{1.99a}$$

x space,

$$\{f_{,x}\} = \begin{Bmatrix} f_{,x_1} \\ f_{,x_2} \\ f_{,x_3} \end{Bmatrix} \tag{1.99b}$$

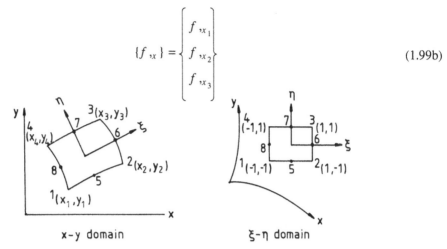

x-y domain   ξ-η domain

Fig. 1.14   Mapping in $X-Y$ and $\xi-\eta$ space.

Then, from eqns (1.98) and (1.99),

$$\{\tilde{x}_{,\xi}\}^T = \{\phi_{,\xi}\}^T[X] \tag{1.100}$$

The chain rule of differentiation is used to express ξ derivatives, in terms of x_i,

$$\frac{\partial f}{\partial \xi_i} = \frac{\partial f}{\partial x_j}\frac{\partial x_j}{\partial \xi_i} \tag{1.101}$$

In matrix notation for all components,

$$\{f_{,\xi}\} = [\tilde{x}_{,\xi}]^T\{f_{,x}\} \tag{1.102}$$

Substituting from eqn (1.100), and noting that [ ] brackets now define a matrix,

$$\{f_{,\xi}\} = [\phi_{,\xi}]^T[X]\{f_{,x}\} \tag{1.103}$$

Define the matrix $[J]$ such that

$$[J] = [\phi_{,\xi}]^T[X] \tag{1.104}$$

Then the x derivatives of f are given by the inverse transformation to eqn (1.103),

$$\{f_{,x}\} = [J]^{-1}\{f_{,\xi}\} \tag{1.105}$$

Suppose, for example, f is equal to u_i, the ith component of the displacement vector at a point Then,

$$\{u_{i,x}\} = [J]^{-1}\{u_{i,\xi}\} \tag{1.106}$$

1.4.5.2   Integration

The integration of the function $f(x_i)$ in the x domain is calculated by transforming the integral $f(\xi_i)$ in the ξ domain. Thus,

$$I = \int_{vol} f(x_1, x_2, x_3) dx_1 \, dx_2 \, dx_3 \tag{1.107a}$$

$$I = \int_{-1}^{+1} \int_{-1}^{+1} \int_{-1}^{+1} f(\xi_1 \, \xi_2 \, \xi_3) dV \tag{1.107b}$$

It is necessary to express dV in ξ space and to obtain the relationship between the coordinate systems. The parametric lines, with increments $d\xi_1$, $d\xi_2$, $d\xi_3$ are shown in Fig. 1.15.

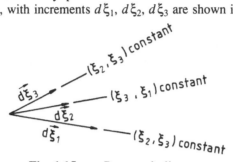

Fig. 1.15   Parametric lines.

The vector,

$$d\vec{\xi}_1 = \left\{ \frac{\partial \tilde{x}_1}{\partial \xi_1}, \frac{\partial \tilde{x}_2}{\partial \xi_1}, \frac{\partial \tilde{x}_3}{\partial \xi_1} \right\} d\xi_1 = \{\tilde{x}_{,\xi_1}\}^T d\xi_1 \tag{1.108}$$

with similar expressions for $d\vec{\xi}_2$ and $d\vec{\xi}_3$. Then, from eqn (1.17),

$$dV = |d\vec{\xi}_3 \cdot (d\vec{\xi}_1 \times d\vec{\xi}_2)| = |d\vec{\xi}_1 \cdot (d\vec{\xi}_2 \times d\vec{\xi}_3)| \tag{1.109}$$

That is,

$$dV = \begin{vmatrix} \dfrac{\partial \tilde{x}_1}{\partial \xi_1} & \dfrac{\partial \tilde{x}_2}{\partial \xi_1} & \dfrac{\partial \tilde{x}_3}{\partial \xi_1} \\[2mm] \dfrac{\partial \tilde{x}_1}{\partial \xi_2} & \dfrac{\partial \tilde{x}_2}{\partial \xi_2} & \dfrac{\partial \tilde{x}_3}{\partial \xi_2} \\[2mm] \dfrac{\partial \tilde{x}_1}{\partial \xi_3} & \dfrac{\partial \tilde{x}_2}{\partial \xi_3} & \dfrac{\partial \tilde{x}_3}{\partial \xi_3} \end{vmatrix} d\xi_1 d\xi_2 d\xi_3 = |[\tilde{x}_{,\xi}]^T| d\xi_1 d\xi_2 d\xi_3 = \det J \, d\xi_1 d\xi_2 d\xi_3 \tag{1.110}$$

where the notation, $\det J = |J|$, is the determinant of the matrix $[J]$. By similar reasoning for the two–dimensional case, the infinitesimal area is given by

$$dA = |d\vec{\xi}_1 \times d\vec{\xi}_2| = |[\tilde{x}_{,\xi}]^T| d\xi_1 d\xi_2 = \det J \, d\xi_1 d\xi_2 \tag{1.111}$$

Thus, for three dimensions, from eqn (1.110),

$$I_3 = \int_{-1}^{+1} \int_{-1}^{+1} \int_{-1}^{+1} f(\xi_1, \xi_2, \xi_3) \det J \, d\xi_1 d\xi_2 d\xi_3 \tag{1.112}$$

and for two dimensions, from eqn (1.111),

$$I_2 = \int_{-1}^{+1}\int_{-1}^{+1} f(\xi_1,\xi_2)\det J \; d\xi_1 d\xi_2 \tag{1.113}$$

Finally, for one dimension:

$$I_1 = \int_{-1}^{+1} f(\xi_1)|\frac{\partial \tilde{x}_1}{\partial \xi_1}|d\xi_1 \tag{1.114}$$

In most cases det J varies from point to point within the domain so that I must be calculated numerically. Any numerical integration scheme may be used; however, Gauss integration appears to be not only the most popular but also the most powerful. Thus, the approximate values of I_3, I_2 are expressed

$$I_3 = \sum w_i w_j w_k (f \; \det J)_{ijk} \tag{1.115}$$

$$I_2 = \sum w_i w_j (f \; \det J)_{ij} \tag{1.116}$$

In both cases, summation is implied on the repeated indices in this case over the range of the integration order. The notation $(f \; \det J)_{ijk}$ means the value calculated at the integration point $(i - j - k)$, and w_i, w_j, w_k are the corresponding weight factors. The ξ_i coordinates and the weight factors for integration orders 1 to 4 are given in Table 1.1.

Table 1.1

Numerical integration			
n	i	ξ_i	w_i
1	I	0	2
2	I	$-1/\sqrt{3}$	$+1$
	II	$+1/\sqrt{3}$	$+1$
3	I	$-\sqrt{0.6}$	5/9
	II	0	8/9
	III	$+\sqrt{0.6}$	5/9

Numerical integration			
n	i	ξ_i	w_i
4	I	$-\dfrac{\sqrt{3+\sqrt{4.8}}}{\sqrt{7}}$	$\dfrac{1}{2}-\dfrac{\sqrt{30}}{36}$
	II	$-\dfrac{\sqrt{3-\sqrt{4.8}}}{\sqrt{7}}$	$\dfrac{1}{2}+\dfrac{\sqrt{30}}{36}$
	III	$\dfrac{\sqrt{3-\sqrt{4.8}}}{\sqrt{7}}$	$\dfrac{1}{2}+\dfrac{\sqrt{30}}{36}$
	IV	$\dfrac{\sqrt{3+\sqrt{4.8}}}{\sqrt{7}}$	$\dfrac{1}{2}-\dfrac{\sqrt{30}}{36}$

In Table 1.2 the polynomial functions x^3, x^4, x^5 have been integrated over the range of x, 0 to 10, using the Gauss approximation of orders 1 to 4. The exact values are given for comparison.

Table 1.2

	Comparison of Gauss integration with exact values		
n	x^3	x^4	x^5
1	78.125	195.312	488.281
2	176.384	654.945	2387.281
3	158.365	627.997	2604.167
4	156.592	624.999	2604.167
exact	156.250	625.000	2604.167

Example 1.8 *Moments of inertia*

It is required to calculate the first and second moments of area of a quadrilateral about the coordinate axes. The quadrilateral is shown in Fig. 1.16 and is defined by apex coordinates, (x_{1i}, x_{2i}), $i = 1, 2, 3, 4$. In this case, $[X]$, given in eqn (1.97), is

$$[X] = \begin{bmatrix} x_{11} & x_{21} \\ x_{12} & x_{22} \\ x_{13} & x_{23} \\ x_{14} & x_{24} \end{bmatrix}$$

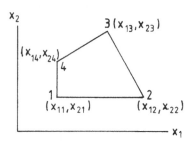

Fig 1.16   Arbitrary quadrilateral coordinates.

The interpolation polynomial is simply $\{\phi_1\}$, given in Fig. 1.11(a); that is,

$$\{\phi_1\} = \frac{1}{4} \begin{bmatrix} (1 - \xi_1)(1 - \xi_2) \\ (1 + \xi_1)(1 - \xi_2) \\ (1 + \xi_1)(1 + \xi_2) \\ (1 - \xi_1)(1 + \xi_2) \end{bmatrix}$$

The first moment of area of the quadrilateral is calculated as follows. For the quadrilateral, $f(\xi_1, \xi_2)$ is given by

$$f(x_1, x_2) = [\tilde{x}_1 \ \tilde{x}_2] = \{\tilde{x}\}^T = [\phi_1]^T [X]$$

Hence, from eqn (1.113),

$$I_1 = \int\limits_{-1}^{+1}\int\limits_{-1}^{+1} [\phi_1] \det J \ d\xi_1 d\xi_2 \ [X]$$

The area of the quadrilateral is simply

$$A = \int\limits_{-1}^{+1}\int\limits_{-1}^{+1} \det J \ d\xi_1 d\xi_2$$

Hence, the centroid of the quadrilateral is located at

$$[\bar{x} \ \ \bar{y}] = \frac{I_1}{A}$$

For the special case of a rectangle, it is easily shown from eqn (1.104) that $\det J$ is constant over the rectangle, and thence

$$I_1 = [1 \ 1 \ 1 \ 1][X] \det J \quad \text{and,} \quad \text{area} = A = 4 \det J$$

Hence,

$$[\bar{x} \ \ \bar{y}] = \frac{1}{4}[1 \ 1 \ 1 \ 1][X]$$

The second moment of area of the quadrilateral is calculated as follows: The second moment of area has for its function, $f(x_1, x_2)$

$$f(x_1, x_2) = \begin{bmatrix} \tilde{x}_1 \\ \tilde{x}_2 \end{bmatrix} [\tilde{x}_1 \ \tilde{x}_2] = [X]^T [\phi_1][\phi_1]^T [X]$$

The integral is given by

$$I_2 = [X]^T \int\limits_{-1}^{+1}\int\limits_{-1}^{+1} [\phi_1][\phi_1]^T \det J \ d\xi_1 d\xi_2 [X]$$

Again, for the rectangle, it is easily shown from eqn (1.114) that

$$\det J = \left\| \begin{bmatrix} \frac{1}{4}\{(x_{12} - x_{11}) + (x_{13} - x_{14})\} & \frac{1}{4}\{(x_{22} - x_{21}) + (x_{23} - x_{24})\} \\ \frac{1}{4}\{(x_{14} - x_{11}) + (x_{13} - x_{12})\} & \frac{1}{4}\{(x_{24} - x_{21}) + (x_{23} - x_{22})\} \end{bmatrix} \right\|$$

Thus, $\det J$ is constant, and, in this case,

$$\int\limits_{-1}^{+1}\int\limits_{-1}^{+1} [\phi_1][\phi_1]^T d\xi_1 d\xi_2 = \frac{1}{9}\begin{bmatrix} 4 & 2 & 1 & 2 \\ 2 & 4 & 2 & 1 \\ 1 & 2 & 4 & 2 \\ 2 & 1 & 2 & 4 \end{bmatrix}$$

This gives the value of I_2 as

$$I_2 = [X]^T \frac{1}{9} \begin{bmatrix} 4 & 2 & 1 & 2 \\ 2 & 4 & 2 & 1 \\ 1 & 2 & 4 & 2 \\ 2 & 1 & 2 & 4 \end{bmatrix} [X] \det J$$

For the special case when the origin is at the centre of the rectangle and the sides are parallel with the coordinate axes so that the corner nodes are $(\pm\frac{a}{2},\pm\frac{b}{2})$, $\det J = \frac{ab}{4}$. Then,

$$I_2 = \frac{1}{4}\frac{1}{9} \begin{bmatrix} -a & -a & a & a \\ -b & b & b & -b \end{bmatrix} \begin{bmatrix} 4 & 2 & 1 & 2 \\ 2 & 4 & 2 & 1 \\ 1 & 2 & 4 & 2 \\ 2 & 1 & 2 & 4 \end{bmatrix} \begin{bmatrix} -a & -b \\ -a & b \\ a & b \\ a & -b \end{bmatrix} \frac{ab}{4} = \frac{ab}{12} \begin{bmatrix} a^2 & 0 \\ 0 & b^2 \end{bmatrix}$$

Thus, I_2 is simply the principal axes tensor for the second moment of area matrix.

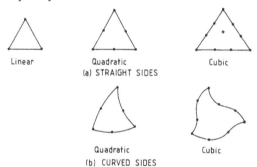

Linear   Quadratic   Cubic

(a) STRAIGHT SIDES

Quadratic   Cubic

(b) CURVED SIDES

Fig. 1.17   Triangular domains. Straight and curved sides.

1.4.6   The triangular domain – coordinate systems

The triangular domain with various configurations of node points is shown in Fig. 1.17(a), (b). The elements with curved sides are treated like the isoparametric elements in section 1.4.5. There are three types of coordinate systems used in the description of triangular element properties. These are:

1.   The global coordinates of the total system. Input of data to the computer program is generally in this system, as is the calculation of nodal displacements. There may also be a local global system associated with each triangle and referenced to one of the triangle apices as origin.

2.   The local coordinate systems associated with each side. There are three such sets. These coordinates are particularly useful in the expression of derivatives normal to the sides of the triangle.

3.   The area or natural coordinates. These are useful for general polynomial interpolation over the triangle. Derivatives and integrals of polynomials are readily calculated in this system.

It is seen that the coordinate systems (1) and (2) express the Cartesian dimensions of the triangle, whereas those in (3) will be shown to be non–dimensional in the range (0, 1).

1.4.6.1   Global coordinates

A plane triangle 1–2–3 or (i, j, k) lies in the x_1, x_2 plane of the global coordinate system. The apex coordinates are (x_{1i}, x_{2i}), $i = 1, 2, 3$. The global dimensions of the triangle are taken parallel to the coordinate axes and are denoted a_i, b_i, $i = 1, 2, 3$. Thus, for example, (a_1, b_1) are the global dimensions of side 1. From Fig. 1.18 it is seen that

$$a_i = x_{1k} - x_{1j} \quad \text{and} \quad b_i = x_{j2} - x_{k2} \tag{1.117}$$

The indices i, j, k are cyclic permutations of 1, 2, 3. Thus, for example,

$$a_1 = x_{13} - x_{12}; \quad b_1 = x_{22} - x_{23}, \text{ etc.}$$

The sequence for a_i is anticlockwise and that for b_i clockwise around the triangle. It is seen that the sums of the a and the b values are equal to zero. That is,

$$\sum a_i = \sum b_i = 0 \tag{1.118}$$

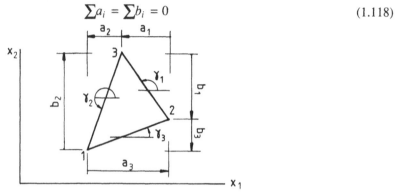

Fig. 1.18   Global dimensions of triangle.

The area of the triangle is easily calculated as half the magnitude of the cross product of two adjacent side vectors taken in anticlockwise sense. These side vectors are

$$\vec{s_i} = \{a_i, -b_i\} \tag{1.119}$$

So that

$$2 \text{ area} = |\vec{s_i} \times \vec{s_j}| = (a_j b_i - a_i b_j) \tag{1.120}$$

where i, j, k are cyclic permutations of 1, 2, 3. It is seen that the global dimensions describe the physical shape of the triangle as well as its orientation in space.

1.4.6.2   Local coordinates

There are three local or side coordinate systems (s_i, n_i), $i = 1, 2, 3$ (see Fig. 1.19a), with the s_i axis along the side i in an anti clockwise sense, and with the origin at node j. The n_i axis, taken as the inwards normal at node j, forms a right–handed coordinate set. In these coordinate axes, the coordinates of the apices of the triangle are expressed in terms of the side lengths l_i, the heights h_i, and the projections of the apex i on side i, d_i. The quantities (l_i, h_i, d_i) are called the intrinsic dimensions of the triangle and describe the physical shape of the triangle in the ith local set, but not its global disposition. The intrinsic dimensions are shown in Fig. 1.19(b).

From Fig. 1.19(b), it is seen that in coordinate system (s_i, n_i) the apex nodes have coordinates.

node	s_i, n_i coordinates
i	(d_i, h_i)
j	$(0, 0)$
k	$(l_i, 0)$

Here again, i, j, k have the cyclic permutations on 1, 2, 3.

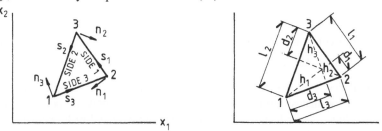

(a) Local coordinate axes   (b) Intrinsic dimensions

Fig. 1.19   Local coordinates and intrinsic dimensions of triangle.

The area of the triangle is expressed by

$$2A = l_i h_i \tag{1.121}$$

It is seen that

$$l_i = \frac{1}{l_i}[a_i \ -b_i]\begin{bmatrix} a_i \\ -b_i \end{bmatrix} = \frac{1}{l_i}\vec{s_i} \cdot \vec{s_i} \tag{1.122}$$

and

$$d_i = \frac{1}{l_i}[a_i \ -b_i]\begin{bmatrix} -a_k \\ b_k \end{bmatrix} \tag{1.123}$$

1.4.6.3   Area or natural coordinates

Let P be an interior point of the triangle (Fig. 1.20). Then A_i is the area of the subtriangle, apex P, base side l_i (nodes $j-k$).

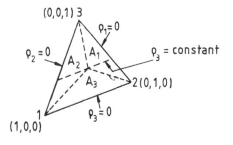

Natural or area coordinates

Fig. 1.20   Natural or area coordinates.

It is seen that the sum of the subareas is the area of the triangle A, so that

$$A_1 + A_2 + A_3 = A$$

The three area coordinates of P are defined by

$$\zeta_i = \frac{A_i}{A}, \quad i = 1, 2, 3 \tag{1.124}$$

and hence,

$$\zeta_1 + \zeta_2 + \zeta_3 = 1 \tag{1.125}$$

The triangle has now become non–dimensionalized since the apices of the triangle have coordinates,

Apex 1: (1,0,0)   Apex 2: (0,1,0)   Apex 3: (0,0,1)

From Fig. 1.20 it is seen that side i has $\zeta_i = 0$ and the line $\zeta_i = constant$ is parallel with the side s_i. The natural coordinates will be used to construct the interpolation polynomials for functions whose nodal values (function, first or second derivatives) are assumed to be known. In the development of the finite element theory it will be necessary to derive not only the relationships between the various coordinate systems but also those between the derivatives of the function in the various systems. The derivatives of functions may be calculated as:

(1) global derivatives with respect to the x_i coordinate axes;
(2) side derivatives with respect to the s_i axis;
(3) normal derivatives with respect to the n_i axis.

1.4.6.4   The relationships between the coordinate systems

(1)   *Global and local coordinates*

These may given in trigonometric form by using the γ_i angles shown in Fig. 1.18, although the alternative form using the global dimensions is often more convenient. That is,

$$\begin{bmatrix} s_i \\ n_i \end{bmatrix} = \begin{bmatrix} \cos \gamma_i & \sin \gamma_i \\ -\sin \gamma_i & \cos \gamma_i \end{bmatrix} \begin{bmatrix} x_1 - x_{1j} \\ x_2 - x_{2j} \end{bmatrix} \tag{1.126}$$

and

$$\begin{bmatrix} s_i \\ n_i \end{bmatrix} = \frac{1}{l_i} \begin{bmatrix} a_i & -b_i \\ b_i & a_i \end{bmatrix} \begin{bmatrix} x_1 - x_{1j} \\ x_2 - x_{2j} \end{bmatrix} \tag{1.127}$$

Again, i, j, k have cyclic permutations on 1, 2, 3. The inverse transformation from local to global coordinates is obtained by solving eqn (1.127) for x_1, x_2; that is,

$$\begin{bmatrix} x_1 \\ x_2 \end{bmatrix} = \frac{1}{l_i} \begin{bmatrix} a_i & b_i \\ -b_i & a_i \end{bmatrix} \begin{bmatrix} s_i \\ n_i \end{bmatrix} + \begin{bmatrix} x_{j1} \\ x_{j2} \end{bmatrix} \tag{1.128}$$

(2)   *Global and natural coordinates*

From Figs. 1.18 and 1.20, it is seen that the x_i coordinates are obtained from the ζ_i coordinates by the transformation,

$$\begin{bmatrix} 1 \\ x_1 \\ x_2 \end{bmatrix} = \begin{bmatrix} 1 & 1 & 1 \\ x_{11} & x_{12} & x_{13} \\ x_{21} & x_{22} & x_{23} \end{bmatrix} \begin{bmatrix} \zeta_1 \\ \zeta_2 \\ \zeta_3 \end{bmatrix} \tag{1.129}$$

On solving eqn (1.129) for ζ_1, ζ_2, ζ_3 and using eqn (1.117), it is seen that the inverse transformation is given by

$$\begin{bmatrix} \zeta_1 \\ \zeta_2 \\ \zeta_3 \end{bmatrix} = \frac{1}{2A} \begin{bmatrix} 2A_{23} & b_1 & a_1 \\ 2A_{31} & b_2 & a_2 \\ 2A_{12} & b_3 & a_3 \end{bmatrix} \begin{bmatrix} 1 \\ x_1 \\ x_2 \end{bmatrix} \tag{1.130}$$

The area $2A$ is calculated from eqn (1.120). The term A_{ij} in eqn (1.130) is the area subtended by the apices i-j and the origin of the coordinates. Thus, for example,

$$A_{ij} = x_{1i}x_{2j} - x_{1j}x_{2i} \tag{1.131}$$

Since $\zeta_i = constant$ is the equation of a line parallel to side i, so also is

$$2A_{jk} + b_i x_1 + a_i x_2 = constant$$

In particular, the equation for side i of the triangle is

$$2A_{jk} + b_i x_1 + a_i x_2 = 0$$

(3)   *Local and natural coordinates*

In this case, from Figs 1.19 and 1.20,

$$\begin{bmatrix} 1 \\ s_i \\ n_i \end{bmatrix} = \begin{bmatrix} 1 & 1 & 1 \\ 0 & l_i & d_i \\ 0 & 0 & h_i \end{bmatrix} \begin{bmatrix} \zeta_j \\ \zeta_k \\ \zeta_i \end{bmatrix} \tag{1.132}$$

On solving eqn (1.132), for the ζ values, it is seen that

$$\begin{bmatrix} \zeta_j \\ \zeta_k \\ \zeta_i \end{bmatrix} = \frac{1}{2A} \begin{bmatrix} 2A & -h_i & (d_i - l_i) \\ 0 & h_i & -d_i \\ 0 & 0 & l_i \end{bmatrix} \begin{bmatrix} 1 \\ s_i \\ n_i \end{bmatrix} \tag{1.133}$$

1.4.6.5   Calculation of derivatives

From eqns (1.129) and (1.130), it is seen that

$$\frac{\partial x_i}{\partial \zeta_j} = x_{ji}$$

and

$$\frac{\partial \zeta_i}{\partial x_1} = \frac{b_i}{2A} \quad \text{and} \quad \frac{\partial \zeta_i}{\partial x_2} = \frac{a_i}{2A} \tag{1.134}$$

The ζ derivatives with respect to s_i and n_i are defined by the matrices $[D_{ij}]$ and $[C_{ij}]$ such that

$$[D_{ij}] = \left[\frac{\partial \zeta_i}{\partial s_j} \right] = \begin{bmatrix} 0 & \dfrac{1}{l_2} & -\dfrac{1}{l_3} \\[2ex] -\dfrac{1}{l_1} & 0 & \dfrac{1}{l_3} \\[2ex] \dfrac{1}{l_1} & -\dfrac{1}{l_2} & 0 \end{bmatrix}$$

and

$$[C_{ij}] = \left[\frac{\partial \zeta_i}{\partial n_j} \right] = \frac{1}{2A} \begin{bmatrix} l_1 & -d_2 & (d_3 - l_3) \\[1ex] (d_1 - l_1) & l_2 & -d_3 \\[1ex] -d_1 & (d_2 - l_2) & l_3 \end{bmatrix}$$

From the above values, derivatives of a function $f(\zeta_1, \zeta_2, \zeta_3)$ with respect to local or global Cartesian coordinates can be obtained by the chain rule of differentiation. Thus,

$$\frac{\partial f}{\partial s_i} = \frac{1}{l_i} \left(\frac{\partial f}{\partial \zeta_k} - \frac{\partial f}{\partial \zeta_j} \right) \tag{1.135}$$

and

$$\frac{\partial f}{\partial n_i} = \frac{1}{2A} \left[\frac{\partial f}{\partial \zeta_i} l_i + \frac{\partial f}{\partial \zeta_j} (d_i - l_i) - \frac{\partial f}{\partial \zeta_k} d_i \right] \tag{1.136}$$

Note in eqns (1.135) and (1.136) there is no sum on repeated indices.

$$\frac{\partial f}{\partial x_1} = \frac{1}{2A} b_i \frac{\partial f}{\partial \zeta_i} \quad \text{and} \quad \frac{\partial f}{\partial x_2} = \frac{1}{2A} a_i \frac{\partial f}{\partial \zeta_i} \tag{1.137}$$

In eqns (1.136) and (1.137) sum in the usual way on the repeated index i, whereas in eqns (1.134) and (1.135) no sum applies.

1.4.7 Polynomial interpolation and function integration for plane triangles

1.4.7.1 Polynomial interpolation

The triangular domain is attractive to use because of the relative ease with which complete polynomials may be constructed. A complete polynomial P_n in two dimensions is spanned by coordinate functions (section 1.4.2),

$$\{\phi_i\}^T = (1, x_1, x_2, x_1^2, x_1 x_2, x_2^2, \cdots x_2^n)$$

and has dimension,

$$m = \frac{(n + 2)(n + 1)}{2} \tag{i}$$

Any polynomial $P_n(x_1 x_2)$ of order n may also be expressed as a polynomial $P_r(\zeta), r \geq n$ in triangular coordinates. Such a representation is not unique since each term may contain arbitrary factors $(\zeta_1 + \zeta_2 + \zeta_3)^q \equiv 0$.

When all such factors have been removed, $P_r(\zeta) \equiv P_n(\zeta)$ has degree n and is said to be irreducible. The number of polynomial terms that can be formed in P_n from the three ζ_i variables is given by,

$$r = \frac{(n+3)!}{n!3!} = \frac{1}{6}(n+3)(n+2)(n+1)$$ (ii)

However, in the Euclidean space, $s = 2$ that is, in the $x_1 x_2$ plane, the number of independent functions is given by eqn (i). Hence from (i) and (ii) the number of dependent functions is

$$r - m = \frac{1}{6}(n+2)(n+1)n$$ (iii)

It is found in the triangle that for $n \geq 3$ a complete irreducible polynomial $P_n(\zeta)$ cannot be specified uniquely by boundary values. This is because terms of the type $\zeta_1^p \zeta_2^q \zeta_3^r$, with $p, q, r \geq 1$ occur and these belong to the subspace of polynomials of order $\leq n$ which have zero values on all sides. Similarly, for $n \geq 6$ a complete polynomial $P_n(\zeta)$ cannot be uniquely determined for values of P_n and the normal gradient $\partial P_n / \partial n$ on the boundary. In this case the arbitrary terms $\zeta_1^p \zeta_2^q \zeta_3^r$, with $p, q, r \geq 2$ belonging to the subspace of all polynomials of order $\leq n$, may be added, whose values and normal derivatives are zero on the sides. Consider now the three important cases of linear, quadratic and cubic interpolation of a function over the triangle. The orthogonal interpolation functions will be obtained together with the requisite number of nodal values.

Case 1. Linear function

In this case $n = 1$ and $s = 2$, so that

$$m = \frac{(1+2)!}{1!2!} = 3$$

That is, three nodes are required. A suitable choice for these will be the vertices (1, 2, 3), as in Fig. 1.21(a). The nodal configuration for the cubic interpolation is not unique, and other nodal configurations exist than that shown in Fig. 1.21(c). For example,

1. $f, f_{,x_1}, f_{,x_2}$ at nodes 1, 2, 3 and f at the centroid.

2. $f, f_{,x_1}, f_{,x_2}$ at nodes 1, 2, 3 and $f_{,n_1}$ at node 4.

For the linear interpolation function, $\tilde{f}(x_1, x_2)$ may be expressed

$$\tilde{f}(x_1, x_2) = \alpha_1 + \alpha_2 x_1 + \alpha_3 x_2$$ (1.138)

The α coordinates are obtained from eqn (1.138) by substituting for x_1 and x_2 at the nodes 1, 2 and 3.

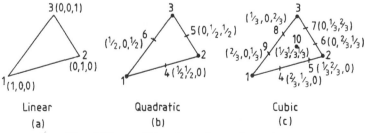

Linear
(a)

Quadratic
(b)

Cubic
(c)

Fig. 1.21   Triangle node configurations.

Then the function values at the nodes are given:

$$
\begin{bmatrix} f_1 \\ f_2 \\ f_3 \end{bmatrix} = \begin{bmatrix} 1 & x_{11} & x_{21} \\ 1 & x_{12} & x_{22} \\ 1 & x_{13} & x_{23} \end{bmatrix} \begin{bmatrix} \alpha_1 \\ \alpha_2 \\ \alpha_3 \end{bmatrix}
\tag{1.139}
$$

It is seen that the matrix in eqn (1.139) is simply the transpose of that in eqn (1.129) giving the transformation from ζ_i to x_i coordinates. However, the concept of orthogonal interpolation functions has been introduced expressly to avoid the inversion needed in eqn (1.139) to obtain the α_i values, even though in this case the inverse may be obtained by inspection. If $\tilde{f}$ is expressed as a function of the ζ coordinates, then

$$
\tilde{f} = \alpha_1 \zeta_1 + \alpha_2 \zeta_2 + \alpha_3 \zeta_3
\tag{1.140}
$$

When nodal coordinates are substituted in eqn (1.140), it is found that

$$
\begin{bmatrix} f_1 \\ f_2 \\ f_3 \end{bmatrix} = \begin{bmatrix} 1 & 0 & 0 \\ 0 & 1 & 0 \\ 0 & 0 & 1 \end{bmatrix} \begin{bmatrix} \alpha_1 \\ \alpha_2 \\ \alpha_3 \end{bmatrix}
$$

That is, the interpolation polynomial $\{\phi_1\}^T = <\zeta_1 \ \zeta_2 \ \zeta_3>$ is the orthogonal linear interpolation set. Then,

$$
\tilde{f} = [\zeta_1 \ \zeta_2 \ \zeta_3] \begin{bmatrix} f_1 \\ f_2 \\ f_3 \end{bmatrix} = \{\phi_1\}^T \{f\} = \{f\}^T \{\phi_1\}
\tag{1.141}
$$

Case 2. Quadratic interpolation functions

In this case $n = 2$ and $s = 2$; hence

$$
m = \frac{(2+2)!}{2!2!} = 6
$$

Thus, six nodal values are required for the complete polynomial, and in the Cartesian coordinates, $\tilde{f}$ is expressed as

$$
\tilde{f}(x_1, x_2) = \alpha_1 + \alpha_2 x_1 + \alpha_3 x_2 + \alpha_4 x_1^2 + \alpha_5 x_1 x_2 + \alpha_6 x_2^2
$$

Again the α_i values may be obtained by substituting (x_1, x_2) values at the six nodes. The nodes 4, 5, 6 will be taken at the midpoints of the sides. The quadratic polynomial in ζ_i coordinates is given by the homogeneous functions,

$$
\{\phi\}^T = <\zeta_1^2 \ \zeta_2^2 \ \zeta_3^2 \ \zeta_2 \zeta_3 \ \zeta_3 \zeta_1 \ \zeta_1 \zeta_2>
\tag{1.142}
$$

Now $\{\phi\}$ is not the orthogonal interpolation polynomial. The theory given in section 1.4.2 is now used to construct $\{\Phi\}$, the orthogonal basis. The orthogonal basis is constructed by application of eqn (1.93).

Thus, substituting nodal coordinates in $\{\phi\}$,

$$[A]^T = \begin{bmatrix} 1 & 0 & 0 & 0 & \frac{1}{4} & \frac{1}{4} \\ 0 & 1 & 0 & \frac{1}{4} & 0 & \frac{1}{4} \\ 0 & 0 & 1 & \frac{1}{4} & \frac{1}{4} & 0 \\ 0 & 0 & 0 & \frac{1}{4} & 0 & 0 \\ 0 & 0 & 0 & 0 & \frac{1}{4} & 0 \\ 0 & 0 & 0 & 0 & 0 & \frac{1}{4} \end{bmatrix}$$

This matrix may be inverted by partitioning into four (3×3) submatrices. Thence,

$$[A^{-1}]^T = \begin{bmatrix} 1 & 0 & 0 & 0 & -1 & -1 \\ 0 & 1 & 0 & -1 & 0 & -1 \\ 0 & 0 & 1 & -1 & -1 & 0 \\ 0 & 0 & 0 & 4 & 0 & 0 \\ 0 & 0 & 0 & 0 & 4 & 0 \\ 0 & 0 & 0 & 0 & 0 & 4 \end{bmatrix}$$

Then, using eqn (1.93),

$$[A^{-1}]^T \{\phi\} = \{\Phi_2\} = \begin{bmatrix} \zeta_1^2 - \zeta_3\zeta_1 - \zeta_1\zeta_2 \\ \zeta_2^2 - \zeta_2\zeta_3 - \zeta_1\zeta_2 \\ \zeta_3^2 - \zeta_2\zeta_3 - \zeta_3\zeta_1 \\ 4\zeta_2\zeta_3 \\ 4\zeta_3\zeta_1 \\ 4\zeta_1\zeta_2 \end{bmatrix} = \begin{bmatrix} \zeta_1(2\zeta_1 - 1) \\ \zeta_2(2\zeta_2 - 1) \\ \zeta_3(2\zeta_3 - 1) \\ 4\zeta_2\zeta_3 \\ 4\zeta_3\zeta_1 \\ 4\zeta_1\zeta_2 \end{bmatrix} \qquad (1.143)$$

In eqn (1.143) use has been made of relationships such as

$$\zeta_1^2 - \zeta_1(\zeta_2 + \zeta_3) = \zeta_1(2\zeta_1 - 1)$$

Case 3. Cubic interpolation polynomial

For $n = 3$, and $s = 2$, the number of independent terms is

$$m = \frac{(3 + 2)!}{3!2!} = 10$$

It has been shown that it is not possible to define the complete third–degree ζ polynomial in terms of values on the boundary because of the bubble function, $\zeta_1\zeta_2\zeta_3 = 0$, on all sides. To define this function, a node (10 in Fig. 1.21(c)) must be used. It is convenient to use the centroid, coordinates 1/3, 1/3, 1/3. The derivation of the orthogonal interpolation functions proceeds as outlined in case 2, and the result is given in eqn (1.144).

$$\{\Phi_3\} = \frac{1}{2}\begin{bmatrix} \zeta_1(3\zeta_1-1)(3\zeta_1-2) \\ \zeta_2(3\zeta_2-1)(3\zeta_2-2) \\ \zeta_3(3\zeta_3-1)(3\zeta_3-2) \\ 9\zeta_2\zeta_3(3\zeta_2-1) \\ 9\zeta_2\zeta_3(3\zeta_3-1) \\ 9\zeta_3\zeta_1(3\zeta_3-1) \\ 9\zeta_3\zeta_1(3\zeta_1-1) \\ 9\zeta_1\zeta_2(3\zeta_1-1) \\ 9\zeta_1\zeta_2(3\zeta_2-1) \\ 54\zeta_1\zeta_2\zeta_3 \end{bmatrix} \tag{1.144}$$

An alternative $\{\Phi_3\}$ polynomial, $\{\overline{\Phi}_3\}$, can be obtained, for the function given by $(f, f_{,x_1}, f_{,x_2})$ at nodes 1, 2, and 3 and $f_{,n}$ at node 4. In this case it can be shown that

$$\{\overline{\Phi}_3\} = \begin{bmatrix} \zeta_3^2(3 - 2\zeta_3) \\ \zeta_3^2(b_1\zeta_2 - b_2\zeta_1) \\ \zeta_3^2(a_1\zeta_2 - a_2\zeta_1) \\ \zeta_1^2(3 - 2\zeta_1) + 6\mu_3\zeta_1\zeta_2\zeta_3) \\ \zeta_1^2(b_2\zeta_3 - b_3\zeta_2) - (b_3\mu_3 - b_1)\zeta_1\zeta_2\zeta_3 \\ \zeta_1^2(a_2\zeta_3 - a_3\zeta_2) - (a_3\mu_3 - a_1)\zeta_1\zeta_2\zeta_3 \\ \zeta_2^2(3 - 2\zeta_2) + 6\lambda_3\zeta_1\zeta_2\zeta_3 \\ \zeta_2^2(b_3\zeta_1 - b_1\zeta_3) - (b_2 - b_3\lambda_3)\zeta_1\zeta_2\zeta_3 \\ \zeta_2^2(a_3\zeta_1 - a_1\zeta_3) - (a_2 - a_3\lambda_3)\zeta_1\zeta_2\zeta_3 \\ 4h_3\zeta_1\zeta_2\zeta_3 \end{bmatrix} \tag{1.145}$$

In eqn (1.145), from eqn (1.123),

$$\lambda_3 = \frac{d_3}{l_3} = \frac{1}{l_3^2}[a_3 - b_3]\begin{bmatrix} -a_2 \\ b_2 \end{bmatrix} \tag{1.146a}$$

$$\mu_3 = 1 - \lambda_3 = \frac{1}{l_3^2}[a_3 - b_3]\begin{bmatrix} -a_1 \\ b_1 \end{bmatrix} \tag{1.146b}$$

1.4.7.2 Integration of ζ functions over the area of the triangle

The use of the ζ functions for interpolation leads to integrals of the form,

$$I_2 = \int_{area} \zeta_1^a \, \zeta_2^b \, \zeta_3^c \, dA \tag{1.147}$$

By making the substitution, $\zeta_3^c = (1 - \zeta_1 - \zeta_2)^c$, this integral is always reducible to an integral in ζ_1 and ζ_2. It is now required to express dA in terms of $d\zeta_1 \, d\zeta_2$ and the coordinate transformations. The infinitesimal triangle $i' \, j' \, k'$ in Fig. 1.22 is bounded by the vectors $\vec{ds_i}, \vec{ds_j}, \vec{ds_k}$, such that from eqn (1.132),

$$|\vec{ds_i}| = l_i \, d\zeta_k \tag{1.148}$$

Now the components of $\vec{ds_i}$ are

$$\vec{ds_i} = \begin{Bmatrix} \dfrac{\partial \tilde{x}_1}{\partial s_i} \\[2mm] \dfrac{\partial \tilde{x}_2}{\partial s_i} \end{Bmatrix} ds_i = \begin{Bmatrix} \dfrac{\partial \tilde{x}_1}{\partial s_i} \\[2mm] \dfrac{\partial \tilde{x}_2}{\partial s_i} \end{Bmatrix} l_i \, d\zeta_k$$

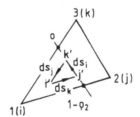

Fig. 1.22   Infinitesimal triangle.

That is, from eqn (1.134),

$$\vec{ds_i} = \frac{1}{l_i} \begin{bmatrix} \dfrac{\partial \tilde{x}_1}{\partial \zeta_k} - \dfrac{\partial \tilde{x}_1}{\partial \zeta_j} \\[3mm] \dfrac{\partial \tilde{x}_2}{\partial \zeta_k} - \dfrac{\partial \tilde{x}_2}{\partial \zeta_j} \end{bmatrix} l_i \, d\zeta_k = \vec{a}_{jk} \, d\zeta_k, \text{ no sum on } k \tag{1.149}$$

Thence,

$$\vec{ds}_2 = \vec{a}_{31}d\zeta_1 ; \quad \vec{ds}_3 = \vec{a}_{12}d\zeta_2$$

and the infinitesimal area is given by

$$dA = |\vec{ds}_2 \times \vec{ds}_3| = |\vec{a}_{31} \times \vec{a}_{12}|d\zeta_1 d\zeta_2 \tag{1.150}$$

Finally,

$$dA = \begin{vmatrix} a_2 & a_3 \\ -b_2 & -b_3 \end{vmatrix} d\zeta_1 d\zeta_2 = 2A d\zeta_1 d\zeta_2 \tag{1.151}$$

Substitution in eqn (1.147) gives the integrals as

$$I_2 = 2A \int_0^1 \int_0^{1-\zeta_1} \zeta_1^a \zeta_2^b (1 - \zeta_1 - \zeta_2)^c \, d\zeta_2 \, d\zeta_1$$

Substitute $t = \dfrac{\zeta_2}{1 - \zeta_1}$; then,

$$I_2 = 2A \int_0^1 \zeta_1^a (1 - \zeta_1)^{b+c+1} d\zeta_1 \int_0^1 t^b (1 - t)^c \, dt$$

Each integral on the right–hand side of this equation is in the form of a beta function where,

$$B(z,w) = \int_0^1 t^{z-1}(1 - t)^{w-1} dt = \frac{\Gamma(z)\Gamma(w)}{\Gamma(z + w)}$$

and $\Gamma(n + 1) = n!$ for integers $n \geq 0$. Thence,

$$\int_A \zeta_1^a \zeta_2^b \zeta_3^c \, dA = 2A \frac{\Gamma(a + 1)\Gamma(b + 1)\Gamma(c + 1)}{\Gamma(a + b + c + 3)}$$

so that

$$I_2 = 2A \frac{a!b!c!}{(a + b + c + 2)!} \tag{1.152}$$

Fig. 1.23   Linear and tetrahedron elements.

In a similar way the integrals may be calculated for linear and three-dimensional spaces (Fig. 1.23) to be

$$I_1 = \int_L \zeta_1^a \zeta_2^b \, dl = \frac{La!b!}{(a + b + 1)!} \tag{1.153}$$

and for the tetrahedron,

$$I_3 = \int_V \zeta_1^a \zeta_2^b \zeta_3^c \zeta_4^d \, dV = \frac{6Va!b!c!d!}{(a+b+c+d+3)!} \tag{1.154}$$

Example 1.9   *Sample integrals of ζ functions over the triangle*

(1)   Linear function, ζ_1.

$$I_2 = \int_A \zeta_1 dA = 2A\frac{1!}{3!} = \frac{A}{3}$$

This integral gives the volume of the prism with the triangle as base and of height unity, as shown in Fig. 1.24(a).

(2)   Quadratic function, ζ_1^2.

$$I_2 = \int_A \zeta_1^2 dA = 2A\frac{2!}{4!} = \frac{A}{6}$$

This function is shown in Fig. 1.24(b).

Fig. 1.24   Examples of polynomial integration over triangle.

The use of eqns (1.152), and (1.154) makes the application of the ζ functions particularly powerful.

Example 1.10   *First and second moments of area of triangle about the coordinate axes*

This is a problem similar to that solved in example 1.8 for the quadrilateral. The results are simple because of the unique geometric properties of the triangle.

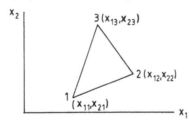

Fig. 1.25   Triangular coordinates.

The triangle is shown in Fig. 1.25, and the matrix $[X]$ is given

$$[X] = \begin{bmatrix} x_{11} & x_{21} \\ x_{12} & x_{22} \\ x_{13} & x_{23} \end{bmatrix}$$

The interpolation polynomial is simply

$$\{\Phi_1\}^T = \{\zeta_1 \ \zeta_2 \ \zeta_3\}$$

(1) First moment of area of the triangle about the coordinate axes

In this case $f(x_1 \ x_2)$ is given by

$$f(x_1 \ x_2) = [\tilde{x}_1 \ \tilde{x}_2] = \{\tilde{x}\}^T = [\Phi_1]^T [X]$$

Hence, from eqn (1.151),

$$I_2 = 2A \iiint [\zeta_1 \ \zeta_2 \ \zeta_3] d\zeta_1 d\zeta_2 \ d\zeta_3 [X] = \frac{A}{3}[1 \ 1 \ 1][X]$$

The area of the triangle is given by

$$\text{area} = 2A \iiint d\zeta_1 \ d\zeta_2 \ d\zeta_3 = A$$

Hence, the coordinates of the centroid are

$$[\bar{x} \ \bar{y}] = \frac{I_2}{A} = \frac{1}{3}[1 \ 1 \ 1][X] = \left[\frac{x_{11} + x_{12} + x_{13}}{3} \quad \frac{x_{21} + x_{22} + x_{23}}{3} \right]$$

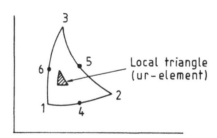

Fig. 1.26   Six–node, curved–sided triangular domain.

(2) Second moment of the area of the triangle about the coordinate axes

The function for the second moment of the area is

$$f(x_1 \ x_2) = \begin{bmatrix} \tilde{x}_1 \\ \tilde{x}_2 \end{bmatrix} [\tilde{x}_1 \ \tilde{x}_2] = \{\tilde{x}\}\{\tilde{x}\}^T = [X]^T[\Phi_1][\Phi_1]^T[X]$$

Again, because this is a function of $\zeta_1 \ \zeta_2 \ \zeta_3$, from eqn (1.152),

$$I_2 = 2A [X]^T \iiint [\Phi_1][\Phi_1]^T d\zeta_1 \ d\zeta_2 \ d\zeta_3 [X]$$

so that

$$I_2 = \frac{A}{3}[X]^T \begin{bmatrix} 2 & 1 & 1 \\ 1 & 2 & 1 \\ 1 & 1 & 2 \end{bmatrix}[X] \tag{1.155}$$

1.4.8   The isoparametric triangle

The isoparametric, or curved–sided, triangle is shown in Fig. 1.17(b). It is of sufficient importance to devote this section to the derivation of its properties; that is, the $(x_1 \, x_2)$ derivatives, and integrals of functions over the triangle area when the interpolation functions are given in ζ space as for the plane triangle. The curved–sided triangle differs from the plane triangle in that the dimensions of the infinitesimal triangle vary from point to point, and so integration must be performed numerically. In this case the quadratic interpolation polynomial is given by eqn (1.143), so that

$$\{\tilde{x}\}^T = [\Phi_2]^T [X] \tag{1.156}$$

In eqn (1.156),

$$[X] = \begin{bmatrix} x_{11} & x_{21} \\ x_{12} & x_{22} \\ x_{13} & x_{23} \\ x_{14} & x_{24} \\ x_{15} & x_{25} \\ x_{16} & x_{26} \end{bmatrix} \tag{1.157}$$

The local triangle

At any point within a curved triangle an infinitesimal local triangle (Fig. 1.25) can be constructed with side vectors given by the expression,

$$\vec{a}_{ij} = \left\{ -\frac{\partial \tilde{x}}{\partial \zeta_i} + \frac{\partial \tilde{x}}{\partial \zeta_j} \right\} \tag{1.158}$$

This is a process identical to that in eqn (1.149). Then, as for the plane triangle, the area of the infinitesimal triangle is given by

$$2 \, area = d\Omega = |\vec{a}_{31} \times \vec{a}_{12}| d\zeta_1 \, d\zeta_2, \text{ etc.} \tag{1.159}$$

Define a matrix $[J_A]$ such that

$$[J_A] = [e_3 \mid [\frac{\partial \tilde{x}}{\partial \zeta}]^T] = [e_3 \mid \frac{\partial [\Phi_2]^T}{\partial \zeta}[X]] \tag{1.160}$$

The matrix $[J_A]$ is 3×3 and if eqns (1.159) and (1.160) are expanded it is proven that

$$d\Omega = \det J_A \, d\zeta_i \, d\zeta_j \tag{1.161}$$

This is, of course, the corresponding expression to that obtained for the plane sided triangle where $\det J_A = 2A$ The derivatives of the function $f(x_1, x_2) \to f(\zeta_1 \, \zeta_2 \, \zeta_3)$ are obtained by the chain rule as,

$$\{\tilde{f}_{,x}\} = \begin{bmatrix} \dfrac{\partial \zeta_1}{\partial x_1} & \dfrac{\partial \zeta_2}{\partial x_1} & \dfrac{\partial \zeta_3}{\partial x_1} \\[2mm] \dfrac{\partial \zeta_1}{\partial x_2} & \dfrac{\partial \zeta_2}{\partial x_2} & \dfrac{\partial \zeta_3}{\partial x_2} \end{bmatrix} \{\tilde{f}_{,\zeta}\} \tag{1.162}$$

However, $\{\tilde{f}\}$ is interpolated in terms of its nodal values by

$$\tilde{f} = [\Phi_2]^T \{f\} \quad \text{and} \quad \{\tilde{f}_{,\zeta}\} = [\Phi_{2,\zeta}]^T \{f\} \tag{1.163}$$

where,

$$[\Phi_{2,\zeta}]^T = \begin{bmatrix} \dfrac{\partial \Phi_2^T}{\partial \zeta_1} \\ \dfrac{\partial \Phi_2^T}{\partial \zeta_2} \\ \dfrac{\partial \Phi_2^T}{\partial \zeta_3} \end{bmatrix} = \begin{bmatrix} (4\zeta_1 - 1) & 0 & 0 & 0 & 4\zeta_3 & 4\zeta_2 \\ 0 & (4\zeta_2 - 1) & 0 & 4\zeta_3 & 0 & 4\zeta_1 \\ 0 & 0 & (4\zeta_3 - 1) & 4\zeta_2 & 4\zeta_1 & 0 \end{bmatrix} \equiv [D] \tag{1.164}$$

Substitute eqn (1.164) in eqn (1.162) so that

$$\{\tilde{f}_{,x}\} = [L_R][D]\{f\} \tag{1.165}$$

As yet, the terms of $[L_R]$, in eqn (1.165) are unknown. To calculate $[L_R]$ proceed as follows:

$$\frac{\partial x_\nu}{\partial x_\mu} = \frac{\partial x_\nu}{\partial \zeta_1}\frac{\partial \zeta_1}{\partial x_\mu} + \frac{\partial x_\nu}{\partial \zeta_2}\frac{\partial \zeta_2}{\partial x_\mu} + \frac{\partial x_\nu}{\partial \zeta_3}\frac{\partial \zeta_3}{\partial x_\mu} = \delta_{\nu\mu} \tag{1.166}$$

where $x_\nu = x_1$ or x_2. Also,

$$\frac{\partial \zeta_1}{\partial x_\mu} + \frac{\partial \zeta_2}{\partial x_\mu} + \frac{\partial \zeta_3}{\partial x_\mu} = 0 \quad \text{since,} \quad \zeta_1 + \zeta_2 + \zeta_3 = 1 \tag{1.167}$$

Thence, from eqns (1.166) and (1.167),

$$\begin{bmatrix} 1 & 1 & 1 \\ \dfrac{\partial x_1}{\partial \zeta_1} & \dfrac{\partial x_1}{\partial \zeta_2} & \dfrac{\partial x_1}{\partial \zeta_3} \\ \dfrac{\partial x_2}{\partial \zeta_1} & \dfrac{\partial x_2}{\partial \zeta_2} & \dfrac{\partial x_2}{\partial \zeta_3} \end{bmatrix} \begin{bmatrix} \dfrac{\partial \zeta_1}{\partial x_1} & \dfrac{\partial \zeta_1}{\partial x_2} \\ \dfrac{\partial \zeta_2}{\partial x_1} & \dfrac{\partial \zeta_2}{\partial x_2} \\ \dfrac{\partial \zeta_3}{\partial x_1} & \dfrac{\partial \zeta_3}{\partial x_2} \end{bmatrix} = \begin{bmatrix} 0 & 0 \\ 1 & 0 \\ 0 & 1 \end{bmatrix} \tag{1.168}$$

However, this is simply the equation,

$$[J_A]^T [L_R]^T = \begin{bmatrix} 0 & 0 \\ 1 & 0 \\ 0 & 1 \end{bmatrix} \tag{1.169}$$

It is easily shown that

$$\begin{bmatrix} \cdot \\ L_R \end{bmatrix} = [J_A]^{-1} \tag{1.170}$$

This gives the means to calculate $[L_R]$, and from eqn (1.165), derivatives of the function $\{f\}$ may be approximated at any point in the triangle.

Integration of functions over the curved triangle area

The function $\tilde{f}(x_1 x_2) \rightarrow f(\zeta_1 \zeta_2 \zeta_3)$ is calculated as

$$I_2 \approx w_i (f \; detJ_A)_i \tag{1.171}$$

The integration points are given in Table 1.3 for the seven–point integration scheme. See [8] for other integration formulae.

Table 1.3

Point	Coordinates			weight	
	Numerical integration over triangle				
a	α	α	α	0.225	
b	β_1	α_1	β_1	0.13239415	$\alpha = 0.33333333$
c	β_1	β_1	α_1	0.13239415	$\alpha_1 = 0.05971587$
d	α_1	β_1	β_1	0.13239415	$\beta_1 = 0.47014206$
e	β_2	β_2	α_2	0.12593918	$\alpha_2 = 0.60695079$
f	α_2	β_2	β_2	0.12593918	$\beta_2 = 0.19652460$
g	β_2	α_2	β_2	0.12593918	

CHAPTER 2

Basis of solid mechanics–fundamental theorems

2.1 STRAIN

2.1.1 Introduction

In the application of forces or displacements to a body, the effects are observed by the measurement of displacements (or strains). To predict the behaviour, tests must be carried out on specimens of the material to determine not only the elastic constants of the material but also the laws for its behaviour in the inelastic range. The determination of material properties can be an exacting task, particularly if they are required over the whole range of temperatures up to the softening point of the material. Such information is, of course, vitally necessary in thermal stress analysis, e.g. in the calculation of residual welding stresses where extremely high temperatures occur. Since all problems in the inelastic range may be treated as being composed of linear steps, it is essential first to study elastic (or linear) behaviour. Linear elastic behaviour is characterized by a linear relationship between forces and the corresponding displacements. This study leads to the definitions of strain and stress and the relationships (or constitutive laws), between these two quantities. Strain is the more natural phenomenon to study first because it is the observed quantity. With this in mind the concepts of strain will be introduced, followed by those of stress and the constitutive relationships which connect the two. In keeping with the spirit of the text the theory is presented in index notation, but where results are also useful in matrix form from the viewpoint of computational mechanics, both will be given.

2.1.2 Definition of strain

When forces are applied to a body or there is a change in temperature, deformations occur. These deformations are not necessarily elastic in the Hookean sense. That is, the stress–strain curve may not be a straight line and the deformations not recoverable. It is usual to assume, however, in the classical theories of elasticity and plasticity that the deformations are small. This requirement of small deformations ensures that the equations of equilibrium written in terms of the undistorted geometry will not be in serious error. In the theory developed in this section on strain, it is found that certain non-linear quadratic terms occur which together form a quadratic strain tensor. It is also discovered that the quadratic strain tensor is useful in the analysis of bodies whose displacements are large and in the calculation of the critical loads based on first–order theory. The displacements of a body, general point P in Fig. 2.1, are examined in the neighbourhood of P and the definitions of strain developed. Let P be the point x_i and Q a neighbouring point $(x_i + \Delta x_i)$. The vector Δx_i is assumed small by the mathematical concept of neighbouring point. When the body is deformed, let P move to P_D, defined by displacements u_i from P. The coordinates of P_D are then $x_i + u_i$. Then Q will be displaced to Q_D by this same amount plus the vector which gives the variation in u_i because of the distance Δx_i from P to Q. Notice that it will be implied that the displacement field and its first derivatives are continuous functions of x_i. Let δu_i be the variation in the displacement field from P to Q so that the coordinates of Q_D are $x_i + \Delta x_i + u_i + \delta u_i$.

The vector $P\vec{Q}$ is given by

$$P\vec{Q} = \vec{Q} - \vec{P} = (x_i + \Delta x_i) - x_i = \Delta x_i \qquad (2.1)$$

After straining, the vector $P_D\vec{Q}_D$ is given:

$$P_D\vec{Q}_D = \vec{Q}_D - \vec{P}_D = \Delta x_i + \delta u_i \qquad (2.2)$$

The difference between $P_D\vec{Q}_D$ and $P\vec{Q}$ is a measure of the deformation at P. Because P and Q are neighbouring points, and because of the continuity of u_i and its derivatives, the variations δu_i may be written as linear functions of Δx_i and the derivatives of the displacements at P. That is, for Q close to P,

$$\delta u_i = u_{i,j} \Delta x_j \qquad (2.3)$$

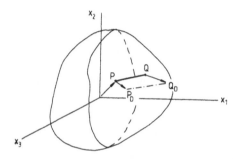

Fig. 2.1   Initial and final states of deformed body.

The tensor quantity $u_{i,j}$ may be decomposed into its symmetric and antisymmetric components by defining

$$\varepsilon_{ij} = \frac{1}{2}(u_{i,j} + u_{j,i}) \qquad (2.4)$$

and

$$\theta_{ij} = \frac{1}{2}(u_{i,j} - u_{j,i}) \qquad (2.5)$$

so that,

$$M_{ij} = u_{i,j} = \varepsilon_{ij} + \theta_{ij} \qquad (2.6)$$

Write eqn (2.3), using eqn (2.6), as

$$\delta u_i = M_{ij} \Delta x_j \qquad (2.7)$$

It is seen then that

$$P_D\vec{Q}_D = \Delta x_i + M_{ij} \Delta x_j = (\delta_{ij} + M_{ij})\Delta x_j \qquad (2.8)$$

The relationship between $P_D\vec{Q}_D$ and $P\vec{Q}$ is shown graphically in Fig. 2.2. It is seen that δu_i is a measure of the total strain and rotation at P. The direct strain at P along PQ will be measured by the difference in magnitudes of the vectors $P_D\vec{Q}_D$ and $P\vec{Q}$. Now,

$$|PQ|^2 = \Delta x_i \Delta x_i \qquad (2.9)$$

and

$$|P_D Q_D|^2 = (\delta_{ij} + M_{ij})\Delta x_j(\delta_{ik} + M_{ik})\Delta x_k$$

$$= (\delta_{ij}\delta_{ik} + \delta_{ij}M_{ik} + \delta_{ik}M_{ij} + M_{ij}M_{ik})\Delta x_j \Delta x_k \qquad (2.10)$$

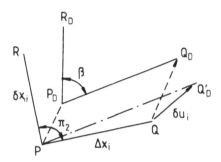

Fig. 2.2   Deformation at a point P in a body.

That is,

$$|P_D Q_D|^2 = (\delta_{jk} + M_{jk} + M_{kj} + M_{ij}M_{ik})\Delta x_j \Delta x_k$$

$$= \Delta x_k \Delta x_k + 2M_{jk}\Delta x_j \Delta x_k + M_{ij}M_{ik}\Delta x_j \Delta x_k \qquad (2.11)$$

Now Δx_i, the components of $\vec{PQ}$, may be expressed in terms of $|PQ|$ and the components of the unit vector t_i and the direction $\vec{PQ}$, that is,

$$\Delta x_i = \vec{PQ} = |PQ|\hat{t} = |PQ|t_i$$

So that

$$t_i = \frac{\Delta x_i}{|PQ|} \qquad (2.12)$$

Thence, substituting in eqn (2.11),

$$|P_D Q_D|^2 = |PQ|^2(1 + 2M_{jk}t_j t_k + M_{ij}M_{ik}t_j t_k) \qquad (2.13)$$

For small strains,

$$|P_D Q_D| = |PQ|(1 + M_{jk}t_j t_k + \tfrac{1}{2}M_{ij}M_{ik}t_j t_k) \qquad (2.14)$$

so that the strain in the direction PQ, ε_t is

$$\varepsilon_t = \frac{|P_D Q_D| - |PQ|}{|PQ|} = t_j t_k (M_{jk} + \tfrac{1}{2}M_{ij}M_{ik}) \qquad (2.15)$$

Now express M_{jk} in terms of its symmetric and antisymmetric components, and define η_{jk} such that

$$\eta_{jk} = \frac{1}{2}M_{ij}M_{ik} = \frac{1}{2}u_{i,j}u_{i,k} \qquad (2.16)$$

Then, from eqn (2.15),

$$\varepsilon_t = t_j t_k (\varepsilon_{jk} + \theta_{jk} + \eta_{jk}) \tag{2.17}$$

because θ_{jk} is antisymmetric $t_j t_k \theta_{jk} = 0$, representing rotation without strain. Then,

$$\varepsilon_t = t_j t_k (\varepsilon_{jk} + \eta_{jk}) \tag{2.18}$$

If quadratic terms are neglected, as is the case in small strain elasticity,

$$\varepsilon_t = t_j t_k \varepsilon_{jk} \tag{2.19}$$

The matrix equations equivalent to eqns (2.18) and (2.19) are

$$(2.18): \quad \varepsilon_t = \{t\}^T ([\varepsilon] + \frac{1}{2}[M]^T[M])\{t\} \tag{2.20}$$

$$(2.19): \quad \varepsilon_t = \{t\}^T [\varepsilon]\{t\} \tag{2.21}$$

Examine the terms of ε_{ij}. It is seen that

$$\varepsilon_{11} = u_{1,1}; \quad \varepsilon_{22} = u_{2,2}; \quad \varepsilon_{33} = u_{3,3}$$

are the direct strains in the directions of the coordinate axes. The terms such as $\varepsilon_{12} = \varepsilon_{21}$ are associated with the change in the angle between the x_1 and x_2 axes and are shear strains, as is explained below.

2.1.3 Shear strain at the point P

In Fig. 2.3 the displacements for P and Q have been plotted for the two–dimensional x_1, x_2 space, so that the only strains occurring are ε_{11}, ε_{22} and $\varepsilon_{12} = \varepsilon_{21}$. It will be seen from the figure that the change in the right angle from X_1OX_2 to $\bar{Q}'\bar{P}\bar{Q}$ in the deformed state is given by, change in right angle = $u_{1,2} + u_{2,1}$. However, this quantity is simply twice the strain ε_{12} defined in eqn (2.4), and leads to the definition of ε_{12} as 1/2 change in the right angle X_1OX_2. It is seen now that the shear strain at P, associated with the direction defined by $\hat{t}$, will be half the change in the angle between $\hat{t}$ and its normal $\hat{n}$. Let the components of $\vec{PR}$ be $\Delta x'_i$, so that

$$n_i = \frac{\Delta x'_i}{|PR|} \tag{2.22}$$

Then, because $\vec{PQ}$ and $\vec{PR}$ are orthogonal,

$$\vec{PQ} \cdot \vec{PR} = \Delta x_i \Delta x'_i = t_i n_i = 0 \tag{2.23}$$

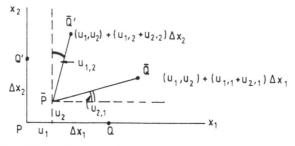

Fig. 2.3   Shear strain defined as change in right angle.

In the deformed state, the vector $P_D \vec{R}_D$ has components

$$P_D \vec{R}_D = (\delta_{ij} + M_{ij})\Delta x'_j$$

Furthermore, if β is the angle between the deformed vectors $P_D \vec{Q}_D$ and $P_D \vec{R}_D$, then

$$P_D \vec{Q}_D \cdot P_D \vec{R}_D = |P_D Q_D||P_D R_D|\cos \beta$$

However,

$$\cos \beta = \sin \left(\frac{\pi}{2} - \beta\right) \approx \frac{\pi}{2} - \beta \approx 2\varepsilon_{nt}$$

Therefore,

$$|P_D Q_D||P_D R_D|2\varepsilon_{nt} = (\delta_{ij} + M_{ij})\Delta x'_j \, (\delta_{ik} + M_{ik})\Delta x_k$$

$$= (\delta_{ij}\delta_{ik} + \delta_{ij}M_{ik} + \delta_{ik}M_{ij} + M_{ij}M_{ik})\Delta x'_j \Delta x_k$$

$$= (\delta_{jk} + M_{jk} + M_{kj} + (\text{higher order terms}))\Delta x'_j \Delta x_k$$

$$= \Delta x_j \Delta x'_j + (M_{jk} + M_{kj})\Delta x_k \Delta x'_k$$

The first term is zero (from eqn (2.23)), and, using eqns (2.12) and (2.23),

$$2\varepsilon_{nt} = (M_{jk} + M_{kj})t_k n_j \qquad (2.24)$$

Now express M_{jk} in terms of ε_{jk} and θ_{jk}, so that now,

$$2\varepsilon_{nt} = (\varepsilon_{jk} + \varepsilon_{kj} + \theta_{jk} + \theta_{kj})t_k n_j$$

Now because ε_{ij} is symmetric and θ_{ij} is antisymmetric, it follows that

$$\varepsilon_{nt} = \varepsilon_{jk} t_k n_j \qquad (2.25)$$

In matrix form,

$$\varepsilon_{nt} = \{n\}^T [\varepsilon]\{t\} \qquad (2.26)$$

2.1.4 Transformation of the strain tensor for rotation of axes

The transformation of the strain tensor ε_{ij} to components ε_{ij}' in the rotated coordinate set x'_i follows immediately from eqn (2.19), when it is recognized that the vector components t_i obey the usual transformation rule of eqn (1.54). That is,

$$t_i = l_{ji} t'_j \qquad (2.27)$$

Substitute in eqn (2.19), so that

$$\varepsilon_t = l_{mj} t'_m l_{nk} t'_n \varepsilon_{jk}$$

Rearranging terms,

$$\varepsilon_t = t'_m t'_n l_{mj} l_{nk} \varepsilon_{jk} \qquad (2.28)$$

However, ε_t may also be written directly in terms of ε'_{ij} and t'_i by following the same reasoning, leading to eqn (2.19), in which case,

$$\varepsilon_t = t'_m t'_n \varepsilon'_{mn} \qquad (2.29)$$

Because the direction of $\hat{t}$ is arbitrary it follows that

$$\varepsilon'_{mn} = l_{mj}\,l_{nk}\,\varepsilon_{jk}$$

and in matrix form,

$$\{\varepsilon\}' = [L][\varepsilon][L]^T \qquad (2.30)$$

Thus, the quantity ε_{ij} obeys the same law of transformation as the inertia tensor, eqns (1.48) and (1.68). The principal strain directions are defined as those directions for which the shear strains are zero. In this case, solutions of the equations

$$\varepsilon_{ij} X_j = \lambda X_i$$

yield the principal strains $\bar{\varepsilon}_{11}$, $\bar{\varepsilon}_{22}$, $\bar{\varepsilon}_{33}$ as the eigenvalues. The solutions for $\bar{X}_i$ corresponding to $\bar{\varepsilon}_{ii}$ will be the principal axes' direction cosines.

2.1.5   The strain quadratic

Associated with the strain ε_t (direction $\hat{t}$) define the position vector $\vec{R}$ whose components define the point P (R_i) such that

$$R_i = \varepsilon_t^{-\frac{1}{2}} t_i \qquad (2.31)$$

Then,

$$t_i = \varepsilon_t^{\frac{1}{2}} R_i \qquad (2.32)$$

Substitution of eqn (2.32) in eqn (2.19) leads to the quadratic function associated with the strain tensor ε. That is,

$$1 = R_j R_k \varepsilon_{jk} \qquad (2.33)$$

or in matrix notation,

$$1 = \{R\}^T [\varepsilon]\{R\} \qquad (2.34)$$

Expand the terms in eqn (2.34); thence,

$$\{R\}^T [\varepsilon]\{R\} = x_1^2 \varepsilon_{11} + x_2^2 \varepsilon_{22} + x_3^2 \varepsilon_{33} + 2x_1 x_2 \varepsilon_{12} + 2x_2 x_3 \varepsilon_{23} + 2x_3 x_1 \varepsilon_{31} = 1 \qquad (2.35)$$

A suitable rotation of axes reduces eqn (2.35) to

$$x_1^2\,\bar{\varepsilon}_{11} + x_2^2\,\bar{\varepsilon}_{22} + x_3^2\,\bar{\varepsilon}_{33} = 1 \qquad (2.36)$$

The properties of the ellipsoid in eqns (2.35) and (2.36) are discussed in further detail in section 2.3, in relation to the stress tensor.

2.2 STRESS

2.2.1   Definition of stress

When forces or displacements are applied to a body, internal forces are induced. Thus, across an internal surface dA shown in Fig. 2.4, the force vector $d\vec{F}$ is the influence of the right portion on the left. This force $d\vec{F}$ may be expressed in its components normal and tangential to the surface dA $(d\vec{F}_n, d\vec{F}_t)$.

The stress components acting are then given as

$$\sigma_n = \lim_{dA \to 0} \frac{dF_n}{dA}$$

and, for the shear stress,

$$\sigma_t = \lim_{dA \to 0} \frac{dF_t}{dA}$$

The shear stress component may be further decomposed into two orthogonal components, σ_{t_1} and σ_{t_2}.

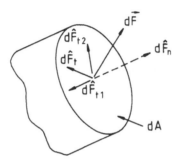

Fig. 2.4   Internal force components.

Thus, it is seen that on any internal section there are three stress components, one normal and two tangential to the section. When the stress components are written on the faces of an infinitesimal cube dx_1, dx_2, dx_3, it is found that there are nine components, as shown in Fig. 2.5. It is noted that a stress component has two indices; the first index indicates the direction of the normal to the plane on which it acts, and the second refers to the direction of the stress component. A further convention is required; that positive stress acts on a positive face in the positive coordinate sense or on a negative face in a negative sense. This is the convention for direct stress, tension positive.

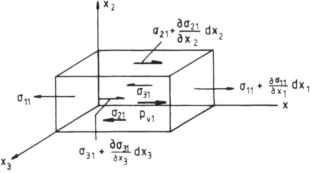

Fig. 2.5   Stress components–infinitesimal element.

The stress tensor σ_{ij} is defined as

$$\sigma_{ij} = \begin{bmatrix} \sigma_{11} & \sigma_{12} & \sigma_{13} \\ \sigma_{21} & \sigma_{22} & \sigma_{23} \\ \sigma_{31} & \sigma_{32} & \sigma_{33} \end{bmatrix} \tag{2.37}$$

The fundamental problem associated with the stress tensor is, given its components in one coordinate set, find the components in a second set, obtained by a rotation of axes. This is of course the same problem that has been solved already in section 2.1.2 for the strain components. The stress matrix is symmetric, since by taking moments about axes through the centre of the cube, it is easily shown that

$$\sigma_{12} = \sigma_{21}, \; \sigma_{13} = \sigma_{31}, \; \sigma_{23} = \sigma_{32}$$

or, in general,

$$\sigma_{ij} = \sigma_{ji} \tag{2.38}$$

In order to study the transformation of stress, isolate an infinitesimal prism bounded by the sides dx_1, dx_2, dx_3 and the face ds whose outwards normal is $\vec{n}$, as shown in Fig. 2.6. Let the force per unit area acting on ds be p_i in the coordinate direction x_i. The stress components are shown in Fig. 2.6 acting on the faces of the prism. It is then possible to write the equations of equilibrium in the directions of the coordinate axes. It has been shown in section 1.3, eqn (1.73) that

$$ds \; n_i = dx_j \, dx_k \tag{2.39}$$

That is, $dx_j \, dx_k$ is the projection of ds parallel with the x_i axis. Write the equation of equilibrium for forces acting in the x_1 direction. Then,

$$p_1 ds - (\sigma_{11} \, dx_2 \, dx_3 + \sigma_{21} \, dx_3 \, dx_1 + \sigma_{31} \, dx_1 \, dx_2) = 0$$

Substituting eqn (2.39) gives

$$p_1 - (\sigma_{11} n_1 + \sigma_{21} n_2 + \sigma_{31} n_3) = 0$$

For all components, this equation is written

$$p_i = \sigma_{ji} n_j \tag{2.40}$$

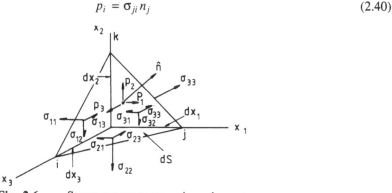

Fig. 2.6   Stress components–prism element.

2.2.2   Transformation of the stress tensor

In order to obtain the transformation for components of σ_{ij} for the rotation of axes, first express the components of the vectors (p_i, n_i) in terms of their components in the rotated axes, using eqn (1.54). That is,

$$l_{mi} p'_m = \sigma_{ji} l_{jn} n'_n$$

Multiply both sides of this equation by l_{ki}, noting $l_{mi} l_{ki} = \delta_{mk}$

$$\delta_{mk} p'_m = l_{ki} l_{jn} \sigma_{ji} n'_n$$

That is,

$$p'_k = l_{ki} l_{jn} \sigma_{ji} n'_n \qquad\qquad (2.41)$$

However, starting with the x'_i axes and the stress tensor σ'_{nk}, the expression for p'_k is

$$p'_k = \sigma'_{nk} n'_n$$

Since the direction n'_n is arbitrary, it follows that

$$\sigma'_{nk} = l_{ki} l_{jn} \sigma_{ji}$$

Since both σ'_{nk} and σ_{ji} are symmetric,

$$\sigma'_{kn} = l_{ki} l_{jn} \sigma_{ij} \qquad\qquad (2.42a)$$

Again, from eqn (2.30), it is seen that this transformation is identical in form to that for the strain tensor components. In matrix form eqn (2.42a) is written

$$[\sigma'] = [L][\sigma][L]^T \qquad\qquad (2.42b)$$

2.2.3   The stress quadratic

The normal stress σ_n on the plane i, j, k in Fig. 2.6 may be found by resolving the force components $p_i \, ds$ on the plane in the direction of the normal $\hat{n}$, and dividing by the area ds. That is,

$$\sigma_n = n_i p_i \qquad\qquad (2.43)$$

Substituting for p_i, from eqn (2.40),

$$\sigma_n = n_i n_j \sigma_{ji} = \{\hat{n}\}^T [\sigma]\{\hat{n}\} \qquad\qquad (2.44)$$

Consider the vector $\vec{R}$, whose components define the point P, coordinates (R_i) such that

$$R_i = \sigma_n^{-\frac{1}{2}} n_i$$

or,

$$n_i = \sigma_n^{\frac{1}{2}} R_i \qquad\qquad (2.45)$$

Substitute for n_i in eqn (2.44),

$$\sigma_n = R_i R_j \sigma_{ji} \sigma_n$$

Hence,

$$R_i R_j \sigma_{ji} = \{R\}^T [\sigma]\{R\} = 1 \qquad\qquad (2.46)$$

When eqn (2.46) is expanded in terms of σ_{11}, σ_{12}, etc. and x_i, the expression obtained is the equation to an ellipsoid. That is,

$$R_i R_j \sigma_{ji} = R_1^2 \sigma_{11} + R_2^2 \sigma_{22} + R_3^2 \sigma_{33} + 2R_1 R_2 \sigma_{12} + 2R_2 R_3 \sigma_{23} + 2R_3 R_1 \sigma_{31} = 1 \qquad (2.47)$$

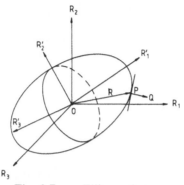

Fig. 2.7   Ellipse of stress.

If the vector $\overrightarrow{OP} = \vec{R}$ is plotted from the origin O (Fig. 2.7), P lies on the quadratic surface given by eqn (2.47). The principal axes of the quadratic surface are such that for these axes $\vec{R}$ is also normal to the surface. By a rotation of axes, $[\sigma]$ may be transformed to the principal axes for which the shear stresses are zero. The equation to the quadratic surface in the principal axes is given by

$$R_1^2 \bar{\sigma}_{11} + R_2^2 \bar{\sigma}_{22} + R_3^2 \bar{\sigma}_{33} = 1 \qquad (2.48)$$

Let l_{ij} be the rotation matrix for the transformation of R_i to the principal axes of the conic. Then the vector $\bar{R}_i$ in the principal axes is given by

$$\bar{R}_i = l_{ij} R_j$$

and the inverse transformations by

$$R_j = l_{mj} \bar{R}_m; \quad R_i = l_{ki} \bar{R}_k$$

Substitute in eqn (2.46) for the quadratic surface so that,

$$l_{ki} l_{mj} \sigma_{ji} \bar{R}_i \bar{R}_m = 1 \qquad (2.49)$$

From eqns (2.48) and (2.49), it is seen that for l_{ij} to be the required transformation matrix,

$$\bar{\sigma}_{km} = l_{ki} l_{mj} \sigma_{ji}$$

That is, the same transformation to transform the conic to principal axes also transforms the stress matrix to its principal axes. It is now of interest to obtain the means by which to calculate the l_{ij} matrix and the principal stress values. It has been shown in section 2.1.3 that the principal stresses will be given by the solutions for λ in the equation

$$\sigma_{ij} X_j = \lambda X_i \qquad (2.50)$$

and that the X_i are the principal directions.

The solution exists if the determinant,

$$|\sigma_{ij} - \delta_{ij}\lambda| = 0 \tag{2.51}$$

This determinant leads to the cubic equation in λ,

$$\lambda^3 - J_1\lambda^2 + J_2\lambda - J_3 = 0 \tag{2.52}$$

The coefficients J_1, J_2, J_3 are invariant under coordinate transformation (otherwise multiple solutions would exist for λ). On expansion of eqn (2.52) it is found that

$$J_1 = \sigma_{ii} \tag{2.53a}$$

$$J_2 = \begin{vmatrix} \sigma_{22} & \sigma_{23} \\ \sigma_{32} & \sigma_{33} \end{vmatrix} + \begin{vmatrix} \sigma_{11} & \sigma_{13} \\ \sigma_{31} & \sigma_{33} \end{vmatrix} + \begin{vmatrix} \sigma_{11} & \sigma_{12} \\ \sigma_{21} & \sigma_{22} \end{vmatrix}$$

$$= \sigma_{11}\sigma_{22} + \sigma_{22}\sigma_{33} + \sigma_{33}\sigma_{11} - \sigma_{12}^2 - \sigma_{23}^2 - \sigma_{31}^2 = \frac{1}{2}(\sigma_{ii}\sigma_{jj} - \sigma_{ij}\sigma_{ij}) \tag{2.53b}$$

$$J_3 = |\sigma| \tag{2.53c}$$

2.3 THE INERTIA TENSOR – INERTIA MATRIX

2.3.1   The definition of the inertia tensor

The tensor transformations already presented in section 1.2.2 are further developed and applied herein to the important problem of the calculation of the principal axes and principal moments of inertia of plane areas (that is, of beam cross sections). For this presentation it is convenient to use matrix notation. The terminology moment of inertia has been carried over from dynamics to describe the second moment of area, i.e. of the moment of inertia of a disc of unit density and thickness.

Consider the plane area shown in Fig. 2.8 lying in the $X_2 X_3$ plane whose normal coincides with the X_1 axis. For a prismatic member whose centroidal axis coincides with OX_1, the area will be the member cross–sectional area (Fig. 2.8). The moment of inertia (or second moment of area) about the axis OX_3 is defined by the integral,

$$J_{22} = \int_{area} x_2^2 \; dA$$

and, similarly,

$$J_{33} = \int_{area} x_3^2 \; dA$$

These terms alone are insufficient to define the quadratic functions of the area associated with the OX_2, OX_3 axes, and it is necessary to introduce the product of inertia defined by

$$J_{23} = J_{32} = \int_{area} x_2 x_3 \; dA$$

The four quantities so defined may be grouped in the (2×2) symmetric matrix defined as the moment of inertia matrix $[J]$.

That is,

$$[J] = \begin{bmatrix} J_{22} & J_{23} \\ J_{32} & J_{33} \end{bmatrix} \tag{2.54}$$

The construction of $[J]$ is, of course, conveniently defined, as in section 1.2, as the integral of the matrix product $\{Y\}\{Y\}^T$, where,

$$\{Y\} = \begin{bmatrix} x_2 \\ x_3 \end{bmatrix} \tag{2.55}$$

Then,

$$[J] = \int_{area} \{Y\}\{Y\}^T dA \tag{2.56}$$

2.3.2   Transformation of the components of the inertia matrix

The transformation law of the vectors $\vec{Y}$ and $\vec{Y}'$ for rotation of axes about OX_1 is known, and it is thus possible to determine the transformation of $[J]$. Now, from eqn (1.39),

$$\{Y\} = [L]^T \{Y'\} \tag{2.57}$$

Substitute in eqn (2.56),

$$[J] = [L]^T \int_{area} \{Y'\}\{Y'\}^T dA \, [L] \tag{2.58}$$

The matrix $[L]$ is taken outside the integral because it is constant. Alternatively, if $[J']$ is formed directly in the prime coordinate set,

$$[J'] = \int_{area} \{Y'\}\{Y'\}^T dA \tag{2.59}$$

Substitution in eqn (2.58) gives

$$[J] = [L]^T [J'][L] \tag{2.60}$$

The rule for transformation of $[J]$ components under rotation of axes about OX_1 has been established, and, of course, it is seen to be yet another example of tensor transformations. From the properties of $[L]$,

$$[J'] = [L][J][L]^T \tag{2.61}$$

In this particular case, $[L]$ may be written in terms of θ_{x_1},

$$[L] = \begin{bmatrix} \cos \theta_{x_1} & \sin \theta_{x_1} \\ -\sin \theta_{x_1} & \cos \theta_{x_1} \end{bmatrix} = \begin{bmatrix} c & s \\ -s & c \end{bmatrix}$$

where $c = \cos \theta_{x_1}$, and $s = \sin \theta_{x_1}$. Expansion of eqn (2.61) then gives, for example,

$$J'_{22} = J_{22} c^2 + 2J_{23} cs + J_{33} s^2 \tag{2.62a}$$

$$J'_{23} = (J_{33} - J_{22})\, sc + J_{23}\, (c^2 - s^2) \tag{2.62b}$$

These are the familiar expressions found in the elementary texts on strength of materials.

2.3.3 Principal axes – principal moments of inertia

The principal axes of the cross section are defined as those axes for which the product of inertia vanishes. They are also the axes for which the moments of inertia have maximum and minimum values. When the inertia matrix is expressed in principal axes it is a diagonal matrix; that is, $[\bar{J}]$ is of the form,

$$[\bar{J}] = \begin{bmatrix} \bar{J}_{22} & 0 \\ 0 & \bar{J}_{33} \end{bmatrix}$$

The question then is how to establish the transformation which reduces $[J]$ given in eqn (2.56) to diagonal form. This is achieved by using eqn (2.62b) and setting $J'_{23} = 0$. Then,

$$J'_{23} = 0 = (J_{33} - J_{22})\, sc + J_{23}\, (c^2 - s^2)$$

Thence,

$$\tan 2\theta = \frac{2J_{23}}{J_{22} - J_{33}} \tag{2.63}$$

The solution of eqn (2.63) gives two values of θ, 90° apart. It is then possible to calculate $\bar{J}_{22}$, $\bar{J}_{33}$ directly from eqn (2.61a). An interesting alternative approach is to use the properties of the characteristic equation,

$$[J]\{X\} = \lambda[I]\{X\}$$

or,

$$([J] - \lambda[I])\{X\} = 0 \tag{2.64}$$

These homogeneous equations may be solved only if the determinant of the coefficient matrix of $\{X\}$ is equal to zero; that is, if

$$|[J] - \lambda[I]| = 0 \tag{2.65}$$

This is the identical mathematical problem encountered in the calculation of principal stresses. Here in two dimensions it leads to the solution of the quadratic eqn (2.66) in λ. On expansion of eqn (2.65),

$$\lambda^2 - \lambda(J_{22} + J_{33}) - J_{23}^2 + J_{22}J_{33} = 0 \tag{2.66}$$

Hence,

$$\lambda_{1,2} = \frac{1}{2}[(J_{22} + J_{33}) \pm \sqrt{(J_{22} + J_{33})^2 + 4(J_{23}^2 - J_{22}J_{33})}] \tag{2.67}$$

Corresponding to the roots $\lambda_{1,2}$, eqn (2.64) will yield solutions $\{X_1\}$ and $\{X_2\}$, which are now shown to be orthogonal; that is,

$$\{X_1\}^T \{X_2\} = 0 \tag{2.68}$$

From eqn (2.64) it is seen that only one of the terms of $\{X\}$ is independent, so that solutions may be normalized, giving

$$\{X_1\}^T \{X_1\} = \{X_2\}^T \{X_2\} = 1 \tag{2.69}$$

Substitute λ_1, λ_2 in eqn (2.64),

$$[J]\{X_1\} = \lambda_1 [I]\{X_1\} \tag{2.70a}$$

and

$$[J]\{X_2\} = \lambda_2 [I]\{X_2\} \tag{2.70b}$$

Premultiplication of eqn (2.70a) by $\{X_2\}^T$, and of eqn (2.70b) by $\{X_1\}^T$ gives

$$\{X_2\}^T [J]\{X_1\} = \lambda_1 \{X_2\}^T \{X_1\} \tag{2.71a}$$

and

$$\{X_1\}^T [J]\{X_2\} = \lambda_2 \{X_1\}^T \{X_2\} \tag{2.71b}$$

Transpose both sides of this equation, and, remembering $[J]$ is symmetric, subtract from eqn (2.71a) to obtain,

$$0 = (\lambda_1 - \lambda_2)\{X_2\}^T \{X_1\} \tag{2.72}$$

If $\lambda_1 \neq \lambda_2$, it follows that $\{X_1\}$, $\{X_2\}$ are orthogonal since, from eqn (2.72),

$$\{X_2\}^T \{X_1\} = 0 \tag{2.73}$$

Now form the matrix $[Q]$ composed of $\{X_1\}$, $\{X_2\}$ as columns:

$$[Q] = [X_1 \ X_2] \tag{2.74}$$

Then,

$$[Q]^T [J][Q] = \begin{bmatrix} X_1^T \\ X_2^T \end{bmatrix} [J] [X_1 \ X_2] = \begin{bmatrix} X_1^T J X_1 & X_1^T J X_2 \\ X_2^T J X_1 & X_2^T J X_2 \end{bmatrix}$$

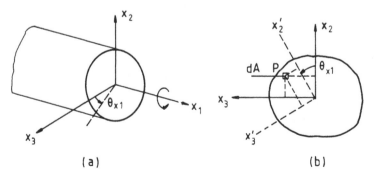

(a)   (b)

Fig. 2.8   Coordinates of member cross section.

Hence, from eqns (2.71) and (2.73),

$$[Q]^T[J][Q] = \begin{bmatrix} \lambda_1 & 0 \\ 0 & \lambda_2 \end{bmatrix} \tag{2.75}$$

It is seen that λ_1 and λ_2 are the principal moments of inertia and $[Q]$ is the required rotation matrix necessary to define the principal axes.

2.3.4   The parallel axes theorem

The discussion on the inertia tensor of plane areas is concluded with the proof of the parallel axes shift theorem. That is, given $[J]$ in the coordinate axes X_1, X_2 through the centroid of the area, determine the value of $[J']$ for a set of axes parallel to the first obtained by an origin shift:

$$\{C\} = \begin{Bmatrix} x_{2c} \\ x_{3c} \end{Bmatrix}$$

Then coordinates in the primed set are related to those in the first via the transformation,

$$\begin{bmatrix} x'_2 \\ x'_3 \end{bmatrix} = \begin{bmatrix} x_2 \\ x_3 \end{bmatrix} - \begin{bmatrix} x_{2c} \\ x_{3c} \end{bmatrix}$$

Then, with the notation used in eqn (2.56),

$$\{Y'\} = \{Y\} - \{C\} \tag{2.76}$$

For the axes through the new origin,

$$[J'] = \int_{area} \{Y'\}\{Y'\}^T dA \tag{2.77}$$

Substitute for $\{Y'\}$ from eqn (2.76) in this equation, so that

$$[J'] = \int \{Y - C\}\{Y - C\}^T dA$$

On multiplication,

$$[J'] = \int \{Y\}\{Y\}^T dA - \int \{Y\}dA\{C\}^T - C\int \{Y\}^T dA + \{C\}\int dA\{C\}^T \tag{2.78}$$

However, because the origin of the X_i coordinates is at the centroid,

$$\int \{Y\}dA = 0 \quad \text{and} \quad \int dA = A$$

the area of the cross section. Hence,

$$[J'] = [J] + \{C\}\{C\}^T A \tag{2.79}$$

Equating terms of the matrices on both sides of eqn (2.79) gives the usual expressions for the inertias about parallel axes; for example,

$$J'_{22} = J_{22} + Ax_{2c}^2 \tag{2.80}$$

2.4 STRESS EQUILIBRIUM

2.4.1   The equations of infinitesimal equilibrium

Having defined stress, as in section 2.2.1, it is necessary to determine the conditions of equilibrium which relate the stress components within a body to the body forces and at the surface to the externally applied pressures and tractions. The stress components may be expressed in any coordinate system, i.e. rectangular Cartesian, polar or general curvilinear. However, in the finite element analyses it is usual to use rectangular Cartesian coordinates to express stress and strain components. As set out in section 2.2.1, internal forces arise in a body when forces or displacements are applied which produce internal strain. These internal forces are described by the stress tensor σ_{ij}, with nine components given in eqn (2.37). In addition to the stress components acting on an infinitesimal element of the body, it will be assumed that body forces are also present and have intensity p_i per unit volume in the coordinate direction x_i. The stress components vary throughout the body, and it is assumed that these components and their first partial derivatives are continuous functions of the coordinates. That is, if on the face $dx_2\, dx_3$, the stress component is σ_{1i}, then at the distance dx_1 away from this face the component now has the value, $\sigma_{1i} + \dfrac{\partial \sigma_{1i}}{\partial x_1} dx_1$. In Fig. 2.9, the stress components acting on the infinitesimal cube $dx_1\, dx_2\, dx_3$ which have resultants in the x_1 direction are shown. The remaining components have been omitted for clarity. From Fig. 2.9 write down the equation of equilibrium in the x_1 direction; that is

$$\sum F_{x1} = 0 = -\sigma_{11}\, dx_3\, dx_2 + (\sigma_{11} + \sigma_{11,1}\, dx_1)dx_3\, dx_2 - \sigma_{21}\, dx_1\, dx_3 + \sigma_{21} + \sigma_{21,2}\, dx_2)dx_1\, dx_3$$

$$-\sigma_{31}\, dx_2\, dx_1 + (\sigma_{31} + \sigma_{31,3}\, dx_3)dx_2\, dx_1 + p_{v1}\, dx_1 dx_2\, dx_3$$

That is, on simplifying this equation,

$$\sigma_{11,1} + \sigma_{21,2} + \sigma_{31,3} + p_{v1} = 0 \tag{2.81}$$

In eqn (2.81), the notation $\dfrac{\partial \sigma_{11}}{\partial x_1} = \sigma_{11,1}$, adopted previously, has been used. When all three equations have been developed, it is readily shown that they may be expressed in the single equation,

$$\sigma_{ij,i} + p_{vj} = 0 \tag{2.82}$$

This is the equation of equilibrium in index notation. It is often convenient to express the six independent components of the stress tensor in vector form, and derive an equilibrium equation equivalent to eqn (2.82) in matrix form. Let $\{\sigma\}^T = <\sigma_{11}\ \sigma_{22}\ \sigma_{33}\ \sigma_{12}\ \sigma_{23}\ \sigma_{31}>$ and define the differential operator $[D]$,

$$[D] = \begin{bmatrix} \dfrac{\partial}{\partial x_1} & 0 & 0 & \dfrac{\partial}{\partial x_2} & 0 & \dfrac{\partial}{\partial x_3} \\[2mm] 0 & \dfrac{\partial}{\partial x_2} & 0 & \dfrac{\partial}{\partial x_1} & \dfrac{\partial}{\partial x_3} & 0 \\[2mm] 0 & 0 & \dfrac{\partial}{\partial x_3} & 0 & \dfrac{\partial}{\partial x_2} & \dfrac{\partial}{\partial x_1} \end{bmatrix} \tag{2.83}$$

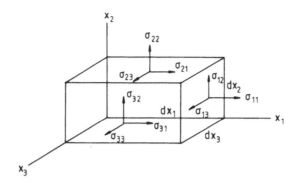

Fig. 2.9   Stress components x_1 direction only.

and

$$\{p_v\}^T = \{p_{v1}\ p_{v2}\ p_{v3}\}^T$$

In matrix form, the equations of infinitesimal equilibrium are written (noting $\sigma_{12} \equiv \sigma_{21}$ etc.)

$$[D]\{\sigma\} + \{p_v\} = 0 \tag{2.84}$$

It is interesting and also useful to define a strain vector $\{\varepsilon\}$ corresponding to σ, such that

$$\{\varepsilon\}^T = \{\varepsilon_{11}\ \varepsilon_{22}\ \varepsilon_{33}\ 2\varepsilon_{12}\ 2\varepsilon_{23}\ 2\varepsilon_{31}\}^T$$

Then the relationship between $\{\varepsilon\}$ and the displacements $\{u\}^T = <u_1\ u_2\ u_3>$ is given by

$$\{\varepsilon\} = [D]^T \{u\} \tag{2.85}$$

2.4.2   Stress boundary conditions

Conditions different from those expressed in eqn (2.82) or (2.85) exist on the boundary of the domain where internal forces must equilibrate externally applied boundary tractions $\{p_s\}$. The situation is shown in Fig. 2.10. The situation in Fig. 2.10 is similar to that shown in Fig. 2.6, except that now ds is an external face, whereas previously it was an internal face acted upon by internal forces. As in eqn (2.39),

$$ds\ n_i = dx_j\,dx_k \tag{2.86}$$

Summing the force components in the x_1 direction (and noting $p_v(1/3)dx_1\ dx_2\ dx_3$ is of a higher order than force components involving stresses), the equation of equilibrium is given by

$$-\sigma_{11}\ dx_2\ dx_3 - \sigma_{21}\ dx_3\ dx_1 - \sigma_{31}\ dx_1\ dx_2 + p_{s1}\ ds = 0$$

Using eqn (2.86),

$$\sigma_{11}n_1 + \sigma_{21}n_2 + \sigma_{31}n_3 = p_{s1}$$

For all such equations,

$$\sigma_{ij}\,n_i = p_{sj} \tag{2.87}$$

In matrix notation, the operator $[D]$ is again used, and $[(Dn)]$ is defined such that all terms operate on the same scalar n. Then stress boundary conditions are written as

$$[\sigma]^T\{n\} = \{p_s\} \tag{2.88}$$

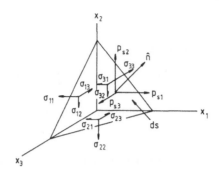

Fig. 2.10   Stress boundary conditions.

2.5 APPLICATION OF GAUSS' DIVERGENCE THEOREM

2.5.1   Principle of virtual displacements

Once the relationships between strain and displacements have been established, and the internal and surface stress equilibrium conditions derived, it is possible to obtain the relationships between external forces and displacements via Gauss' divergence theorem. Historically, the resulting expressions have been derived independently, using the principles of mechanics, and are known respectively, as the principle of virtual displacements and the principle of virtual forces. The term principle of virtual work is purposely avoided in this text because it is usually used in non–rigorous descriptions of either of the principles. To derive the principle of virtual displacements, proceed as follows. Let a body be in static equilibrium under the action of specified body and surface forces. The boundary S shall be assumed to consist of two parts, S_σ, and S_u.

Over S_σ the surface traction p_{si} is specified.
Over S_u the displacements u_i is specified.

Consider a set of arbitrary displacements $\bar{u}_i$ such that $\bar{u}_i$ vanishes on S_u and is arbitrary on S_σ. Internal compatibility conditions are also satisfied by $\bar{u}_i$. The body is in equilibrium under surface forces p_{si} and body forces p_{vi}. Then the work done for an arbitrary displacement δu_i, equal to $\bar{u}_i$, is

$$\delta W_e = \int\limits_v p_{vi}\bar{u}_i\,dv + \int\limits_{S_\sigma} p_{si}\bar{u}_i\,ds \tag{2.89}$$

The stress boundary conditions, from eqn (2.40), are

$$p_{si} = \sigma_{ji}n_j$$

Thence,

$$\int\limits_{S_\sigma} p_{si}\bar{u}_i\,ds = \int\limits_{S_\sigma} \sigma_{ji}n_j\bar{u}_i\,ds \tag{2.90}$$

Now use Gauss' divergence theorem, eqn (1.82), to express the right–hand side of eqn (2.90) as a volume integral; that is,

$$\int_{S_\sigma} (\sigma_{ji} \bar{u}_i) n_j \, ds = \int_v (\sigma_{ji} \bar{u}_i)_{,j} \, dv$$

Expanding the derivative on the right–hand side,

$$\int_{S_\sigma} (\sigma_{ji} \bar{u}_i) n_j \, ds = \int_v \sigma_{ji,j} \bar{u}_i \, dv + \int_v \sigma_{ji} \bar{u}_{i,j} \, dv \tag{2.91}$$

However, the equations of static equilibrium within the body are, from eqn (2.82),

$$\sigma_{ji,j} + p_{vi} = 0$$

so that the first integral on the right–hand side of eqn (2.91) is equal to $-\int_v p_{vi} \bar{u}_i \, dv$. Furthermore, strains are given

$$\bar{\varepsilon}_{ij} = \frac{1}{2}(\bar{u}_{i,j} + \bar{u}_{j,i})$$

and because σ_{ij} is symmetric,

$$\sigma_{ji} \bar{u}_{i,j} = \sigma_{ij} \frac{1}{2}(\bar{u}_{i,j} + \bar{u}_{j,i}) = \sigma_{ij} \bar{\varepsilon}_{ij}$$

With these substitutions, eqn (2.91) becomes

$$\int_{S_\sigma} \sigma_{ji} \bar{u}_i n_j \, ds = -\int_v p_{vi} \bar{u}_i \, dv + \int_v \sigma_{ij} \bar{\varepsilon}_{ij} \, dv \tag{2.92}$$

Substitution in eqn (2.89) gives

$$\delta W_e = \int_v p_{vi} \bar{u}_i \, dv + \int_{S_\sigma} p_{si} \bar{u}_i \, ds = \int_v \sigma_{ij} \bar{\varepsilon}_{ij} \, dv \tag{2.93}$$

However, the right–hand side of eqn (2.93) is simply the work done by the internal forces σ_{ij}, so the result follows as

$$\delta W_e = \delta W_I \tag{2.94}$$

This is the principle of virtual displacements. It is evidently an equilibrium principle in that a given set of forces and stresses $(\sigma_{ij}, p_{vi}, p_{si})$ is tested by the virtual displacements $(\bar{u}_i, \bar{\varepsilon}_{ij})$. Since

$$\delta W_I - \delta W_e = 0$$

The principle of virtual displacements may be expressed, that for equilibrium of the stress field σ_{ij} and the applied forces p_{vi}, p_{si},

$$\int_v \sigma_{ij} \bar{\varepsilon}_{ij} \, dv - \int_v p_{vi} \bar{u}_i \, dv - \int_{S_\sigma} p_{si} \bar{u}_i \, ds = 0 \tag{2.95}$$

2.5.2 Principle of virtual forces

The principle of virtual forces is established through reasoning similar that used to derive the principle of virtual displacements. Suppose a virtual stress field $\bar{\sigma}_{ij}$ exists, and that it is assumed that $\bar{p}_{si}$ is zero on S_σ where external forces are specified. The virtual stress field is assumed to be in equilibrium.

Apply the actual displacements to this system. By the same reasoning by which eqn (2.95) was obtained, it is found that

$$\int_v \bar{\sigma}_{ij} \varepsilon_{ij} \, dv - \int_v \bar{p}_{vi} u_i \, dv - \int_{S_u} \bar{p}_{si} u_i \, ds = 0 \qquad (2.96)$$

This principle evidently tests the compatibility of the given strain field (ε_{ij}, u_i) by applying the virtual force field. The method of calculating deflections called the 'dummy unit load method' is derived directly from the principle of virtual forces. To this end, suppose $\bar{p}_{vi} = 0$ and $\bar{p}_{si}$ is a unit load at a specified point. Then,

$$\int_{S_u} \bar{p}_{si} u_i \, ds = u_i$$

Hence, from eqn (2.96),

$$u_i = \int_v \bar{\sigma}_{ij} \varepsilon_{ij} \, dv \qquad (2.97)$$

The stress field $\bar{\sigma}_{ij}$ is, of course, in equilibrium with the load $\bar{p}_{si} = 1$. Using the matrix notation developed in section 2.4.1, the expressions become:

Principle of virtual displacements

$$\int_v \{\sigma\}^T \{\bar{\varepsilon}\} dv - \int_v \{p_v\}^T \{\bar{u}\} dv - \int_{S_\sigma} \{p_s\}^T \{\bar{u}\} ds = 0 \qquad (2.98)$$

Principle of virtual forces

$$\int_v \{\bar{\sigma}\}^T \{\varepsilon\} dv - \int_v \{\bar{p}_v\}^T \{u\} dv - \int_{S_u} \{\bar{p}_s\}^T \{u\} ds = 0 \qquad (2.99)$$

2.6 THE CONTRAGREDIENT PRINCIPLE

In other texts [9], the contragredient principle has been derived by using discrete force displacement systems and then generalizing these to the continuous domain. Herein the alternative approach has been used by deriving the principle via Gauss' divergence theorem and the associated virtual displacement and force principles. The first expression of the contragredient principle is obtained from eqn (2.98), in which it is assumed that one of the force systems (conveniently, $\{p_v\}$) is equal to zero. Furthermore, it is assumed that the virtual strain $\{\bar{\varepsilon}\}$ at any point can be expressed in terms of the surface displacement parameters $\{\bar{r}\}$ by the transformation,

$$\{\bar{\varepsilon}\} = [a]\{\bar{r}\} \qquad (2.100)$$

Finally, it is assumed that the surface virtual displacements $\{\bar{u}\}$ are expressed in terms of $\{\bar{r}\}$ and $\{p_s\}$, in terms of generalized forces $\{P_s\}$, by the expressions,

$$\{\bar{u}\} = [G]\{\bar{r}\} \qquad (2.101)$$

and

$$\{p_s\} = [L]\{P_s\} \qquad (2.102)$$

With these substitutions,

$$\int_v \{\sigma\}^T [a] \, dv \, \{\bar{r}\} = \{P_s\}^T \int_{S_\sigma} [L]^T [G] \, ds \, \{\bar{r}\} \tag{2.103}$$

Because $\{\bar{r}\}$ is arbitrary, it follows that,

$$\int_{S_\sigma} [G]^T [L] \, ds \, \{P_s\} = \int_v [a]^T \{\sigma\} \, dv \tag{2.104}$$

The more usual form of this equation is obtained by assuming that $\{P_s\}$ forces act at the $\{\bar{r}\}$ node points and in the same directions as the displacements. Then, $[L]$ and $[G]$ are both unit matrices, and the integral on the surface becomes a simple summation, to give

$$\{P_s\} = \int_v [a]^T \{\sigma\} \, dv \tag{2.105}$$

The eqns (2.100) and (2.105) form the contragredient principle, which states that if the strains $\{\bar{\varepsilon}\}$ within the body are related to the nodal displacements $\{\bar{r}\}$ by the transformation matrix $[a]$, i.e. eqn (2.100), then the equilibrium nodal forces corresponding to the internal stress field $\{\sigma\}$ are given by eqn (2.105). Again, it is noted that the direct application of the principle of virtual displacements leads to an expression for the equilibrium between internal and external forces. These equations are found to form the basis of the displacement method of structural analysis. The application of the principle of virtual forces leads to the alternative force method of structural analysis. The theory proceeds by assuming that the virtual stresses (eqn (2.99)), $\{\bar{\sigma}\}$, are related to the virtual node forces by the transformation,

$$\{\bar{\sigma}\} = [b]\{\bar{R}\} \tag{2.106}$$

Again, assume $\{\bar{p}_v\} \equiv 0$, and that as in eqns (2.101) and (2.102),

$$\{\bar{p}_s\} = [L]\{\bar{R}\} ; \quad \text{and} \quad \{u\} = [G]\{r\} \tag{2.107}$$

Then, substituting in eqn (2.99),

$$\{\bar{R}\}^T \int_v [b]^T \{\varepsilon\} \, dv = \{\bar{R}\}^T \int_{S_u} [L]^T [G] \, ds \, \{r\} \tag{2.108}$$

It follows, because $\{\bar{R}\}$ is arbitrary, that

$$\int_{S_u} [L]^T [G] \, ds \, \{r\} = \int_v [b]^T \{\varepsilon\} \, dV \tag{2.109}$$

Furthermore, if it is assumed that $\{\bar{R}\}$ consists of concentrated nodal forces and $\{r\}$ the corresponding displacements, so that $[L]$, $[G]$ are unit matrices and the integral over the surface is a simple summation, then

$$\{r\} = \int_v [b]^T \{\varepsilon\} \, dv \tag{2.110}$$

Eqns (2.106) and (2.110) give a second expression of the contragredient principle, which states that if eqn (2.106) relates the stress system $\{\bar{\sigma}\}$ to the nodal forces $\{\bar{R}\}$, then eqn (2.110) relates the corresponding nodal displacements $\{r\}$ to the internal strains $\{\varepsilon\}$ corresponding to $\{\bar{\sigma}\}$. It should be realized that these equations form a powerful means to calculate deflections and are the basis of the force method of analysis.

2.7 DISCRETE FORCE AND DISPLACEMENT SYSTEMS

2.7.1 Contragredient principle

The contragredient principle for discrete force and displacement systems is of equal importance and is now proved independently of the above theory. Thus, consider two statically equivalent force systems $\{Q\}$ and $\{P\}$ connected by the linear transformation,

$$\{Q\} = [B]\{P\} \qquad (2.111)$$

In certain circumstances this may be a reversible transformation, that is, $\{P\} = [B]^{-1}\{Q\}$. This is, however, an unnecessary restriction, and it will be assumed that in general there will be more terms in $\{Q\}$ than in $\{P\}$. This implies that there are solutions to $\{Q\}$ for which $\{P\}$ is zero. For example, in the truss shown in Fig. 2.11, there is a relationship connecting member forces $<F_1\ F_2\ F_3> = \{Q\}^T$ and $<R_x\ R_y> = \{P\}^T$. However, there will be solutions for which $<F_1\ F_2\ F_3>$ have zero external force components since the three member forces alone may satisfy the conditions of joint equilibrium, $\sum F_{x_1} = 0$, $\sum F_{x_2} = 0$. This corresponds, of course, to a tempera- ture change or lack of fit of one or more of the members. Associated with the force systems $\{Q\}$ and $\{P\}$ will be sets of displacements $\{q\}$ and $\{p\}$. These are compatible displacements, so that if the displacements in the directions of $\{Q\}$ are $\{q\}$, those in the directions $\{P\}$ are $\{p\}$. If two statically equivalent force systems are given compatible displacements, the work done in each sys- tem is the same (since both force and displacement systems have the same components when transformed to a common reference origin). That is,

$$\{P\}^T\{p\} = \{Q\}^T\{q\} \qquad (2.112)$$

However, from eqn (2.111),

$$\{Q\}^T = \{P\}^T[B]^T$$

Hence,

$$\{P\}^T\{p\} = \{P\}^T[B]^T\{q\} \qquad (2.113)$$

The concept that $\{P\}$ is arbitrary is now invoked, so it follows that,

$$\{p\} = [B]^T\{q\} \qquad (2.114)$$

Eqns (2.111) and (2.114) are equivalent to eqns (2.106) and (2.110) for discrete systems and are used to form the basis of the force method of structural analysis. The contragredient law appears in the second form in which $\{Q\}$ and $\{P\}$ are statically equivalent force systems, and the associated sets of displacements $\{q\}$ and $\{p\}$, are related by the linear transformation,

$$\{q\} = [C]\{p\} \qquad (2.115)$$

Then, for equality of work done, for compatible displacements,

$$\{P\}^T\{p\} = \{Q\}^T\{q\} = \{Q\}^T[C]\{p\} \qquad (2.116)$$

Again, invoking the arbitrariness of $\{p\}$, it follows that

$$\{P\} = [C]^T\{Q\} \qquad (2.117)$$

Under certain circumstances the transformation eqns (2.111) and (2.117) are the inverse of one another. This occurs when $\{Q\}$ and $\{P\}$ are vectors of the same dimension and $[B]$ is a non–singular square matrix. Then,

$$\{P\} = [B]^{-1}\{Q\} \tag{2.118}$$

and, hence,

$$[C]^T = [B]^{-1} ; \quad \text{or, alternatively,} \quad [B] = [C^T]^{-1} \tag{2.119}$$

It can be shown that the above reversible transformation occurs in the relationship between node and member forces and node and member displacements in statically determinate structures. The second expression of the contragredient principle in eqns (2.114) and (2.116) is equivalent to that of the eqns (2.100) and (2.105) and thus is based on the principle of virtual displacements. These equations form the basis of the displacement method of structural analysis. It is now seen that eqns (1.43) and (1.45) are simple expressions of the contragredient law, since $\{M_0\}$ and $\{\beta_0\}$ are corresponding force and displacement quantities at the origin, and $\{F_P\}$ and $\{u_P\}$ the corresponding force and displacement quantities at P. The general use of the contragredient principle is demonstrated here in its application to the calculation of deflections of structures and of joint equilibrium. The formal presentation of its application will be given in more detail in Chapters 4 and 5.

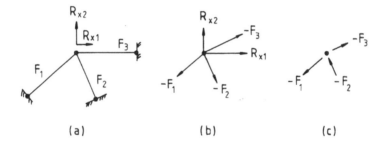

(a)   (b)   (c)

Fig. 2.11   Statically equivalent forces.

2.7.2   Deflections of structures

Suppose that for any structure, subdivided in some way into members, a force transformation exists relating member forces $\{S\}$ to node forces $\{R\}$; that is,

$$\{S\} = [b]\{R\} \tag{2.120}$$

Then the contragredient principle (eqn (2.113)) shows that the deflections at $\{R\}$ caused by member distortions $\{v\}$ are given simply by

$$\{r\} = [b]^T\{v\} \tag{2.121}$$

The distortions $\{v\}$ may be totally independent of $\{S\}$.

2.7.3 Joint equilibrium equations of structures

Suppose that for any structure again subdivided in some way into members, a displacement transformation exists relating member distortions, $\{v\}$, to node displacements, $\{r\}$, such that

$$\{v\} = [a]\{r\} \tag{2.122}$$

Then the joint node forces are again given by eqn (2.116)

$$\{R\} = [a]^T\{S\} \tag{2.123}$$

The use of this equation becomes apparent when it is written

$$\{R\} - [a]^T\{S\} = 0$$

It is seen immediately that if $\{R\}$ forces are defined in the global coordinate system, eqn (2.123) is an expression of the joint equations of equilibrium, in the global coordinates.

2.8 THE CONSTITUTIVE RELATIONSHIPS

2.8.1 Index notation

In the theory of elasticity developed herein, attention is focused on materials which are homogeneous and isotropic and exhibit a linear relationship between stress and strain. Beyond the yield point of the material the plastic behaviour will be characterized by the effective stress–strain curve determined by experiment. It can be shown [10] that the general constitutive equation of a material relating strain and stress, is of the form,

$$\varepsilon_{ij} = C_{ijlm}\sigma_{lm} \tag{2.124}$$

This equation in which thirty six elastic constants appear, can be simplified so that for the homogeneous isotropic material only two independent elastic constants are required. Various combinations of the constants can be used; for example,

$(E\,v)$ Young's modulus and Poisson's ratio;
$(G\,,K)$ shear modulus and bulk modulus;
(λ,v) Lame constants, being the commonly used sets.

Because only two constants are available, relationships must obviously exist between the various groups above. In index notation, the relationship between the strain and the stress tensors for the homogeneous isotropic material is given by

$$\varepsilon_{ij} = \frac{1+v}{E}\sigma_{ij} - \frac{v}{E}\sigma_{kk}\delta_{ij} \tag{2.125}$$

This is a flexibility relationship because it relates strain to stresses. The basic components in eqn (2.125) are of two types, namely:

Direct strain: $\varepsilon_{11} = \dfrac{1}{E}(\sigma_{11} - v\sigma_{22} - v\sigma_{33})$, etc.; shear strain: $\varepsilon_{12} = \dfrac{1+v}{E}\sigma_{12}$, etc.

To obtain the inverse, or stiffness relationship, proceed as follows. In eqn (2.125) put $i = j = k$, so that

$$\varepsilon_{kk} = \frac{1+v}{E}\sigma_{kk} - \frac{3v}{E}\sigma_{kk} = \frac{1-2v}{E}\sigma_{kk} \quad \text{or,} \quad \sigma_{kk} = \frac{E}{1-2v}\varepsilon_{kk} \tag{2.126}$$

Then, from eqn (2.125),

$$\varepsilon_{ij} + \frac{v}{1 - 2v}\varepsilon_{kk}\delta_{ij} = \frac{1 + v}{E}\sigma_{ij}$$

Finally,

$$\sigma_{ij} = E^*((1 - 2v)\varepsilon_{ij} + v\varepsilon_{kk}\delta_{ij}) \tag{2.127}$$

where

$$E^* = \frac{E}{(1 + v)(1 - 2v)}$$

The bulk modulus K can be obtained from eqn (2.126), since the volume change of a cube of material $dx_1\ dx_2\ dx_3$ is given by

$$dV = (1 + \varepsilon_{11})(1 + \varepsilon_{22})(1 + \varepsilon_{33})dx_1\ dx_2\ dx_3 - dx_1\ dx_2\ dx_3$$

so that

$$\frac{dV}{V} = \varepsilon_{11} + \varepsilon_{22} + \varepsilon_{33} = \varepsilon_{kk} \tag{2.128}$$

Now the mean stress is given by

$$p = \frac{\sigma_{kk}}{3} \quad \text{so that} \quad \frac{dV}{V} = \varepsilon_{kk} = \frac{3(1 - 2v)}{E}p \tag{2.129}$$

That is, the relationship between K, E and v is expressed

$$K = \frac{E}{3(1 - 2v)} \tag{2.130}$$

Example 2.1 *Relationship between E, v, and G*

The derivation of the relationship between the shear modulus (G), Young's modulus (E) and Poisson's ratio (v) is as follows. The stress state $\sigma_{11} = -\sigma_{22} = p$ is shown in Fig. 2.12(a). The stress state in coordinate axes at 45° to the global x_1, x_2 axes is obtained by using eqn (2.42) with the transformation matrix $[L]$,

$$[L] = \frac{1}{\sqrt{2}}\begin{bmatrix} 1 & 1 \\ -1 & 1 \end{bmatrix}$$

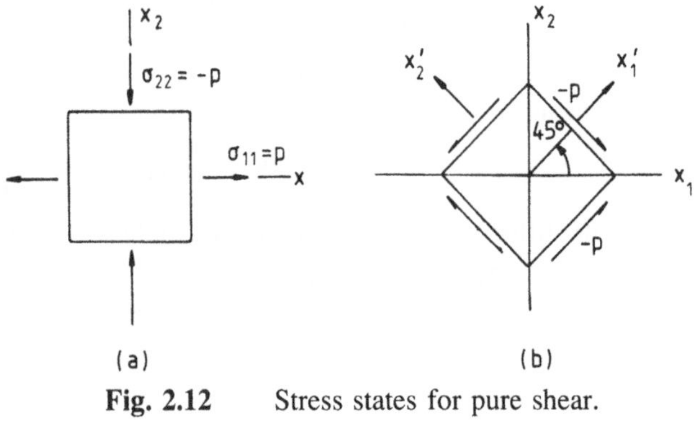

Fig. 2.12   Stress states for pure shear.

The stress components in the rotated axes are then given

$$[\sigma'] = \frac{p}{2} \begin{bmatrix} 1 & 1 \\ -1 & 1 \end{bmatrix} \begin{bmatrix} 1 & 0 \\ 0 & -1 \end{bmatrix} \begin{bmatrix} 1 & -1 \\ 1 & 1 \end{bmatrix} = p \begin{bmatrix} 0 & -1 \\ -1 & 0 \end{bmatrix}$$

This is the shear stress state shown in Fig. 2.12(b). Using eqn (2.125),

$$\varepsilon'_{ij} = \frac{1+\nu}{E}\sigma'_{ij} - \frac{\nu}{E}\sigma'_{kk}\delta_{ij}$$

In this case $\sigma'_{kk} \equiv 0$, so that

$$\varepsilon'_{ij} = -\frac{1+\nu}{E}p \quad \text{if} \quad i \neq j \; ; \quad = 0 \quad \text{if } i = j$$

That is,

$$\varepsilon'_{12} = \varepsilon'_{21} = -\frac{1+\nu}{E}p \tag{a}$$

However, the shear strain state in Fig. 2.12(b) is also described in terms of the shear stress and the shear modulus G; that is,

$$\varepsilon'_{12} = \varepsilon'_{21} = \frac{\sigma'_{12}}{2G} = \frac{-p}{2G} \tag{b}$$

Thus, comparing (a) and (b) gives the required relationship

$$G = \frac{E}{2(1+\nu)}$$

2.8.2 Matrix notation

The constitutive eqns (2.124), (2.125) etc. may be used in finite element formulations when the index notation is used. However, it is generally more convenient to use the vector notation, $\{\sigma\}$, $\{\varepsilon\}$, for the stresses and strains and eqns (2.105), (2.110), etc. In this case the constitutive relationships must be expressed in matrix form. These are easily deduced from eqns (2.125) and (2.128). However, it is usual to adopt the strength of materials definition of shear strain γ_{12} etc. such that $\gamma_{12} = 2\varepsilon_{12}$ etc. With this definition, γ_{12} is the change in the right angle X_1OX_2, when the body is deformed. Thus, the constitutive equations are written

Flexibility

$$\begin{Bmatrix} \varepsilon_{11} \\ \varepsilon_{22} \\ \varepsilon_{33} \\ 2\varepsilon_{12} \\ 2\varepsilon_{23} \\ 2\varepsilon_{31} \end{Bmatrix} = \frac{1}{E} \begin{bmatrix} 1 & \nu & \nu & \cdot & \cdot & \cdot \\ \nu & 1 & \nu & \cdot & \cdot & \cdot \\ \nu & \nu & 1 & \cdot & \cdot & \cdot \\ \cdot & \cdot & \cdot & 2(1+\nu) & \cdot & \cdot \\ \cdot & \cdot & \cdot & \cdot & 2(1+\nu) & \cdot \\ \cdot & \cdot & \cdot & \cdot & \cdot & 2(1+\nu) \end{bmatrix} \begin{Bmatrix} \sigma_{11} \\ \sigma_{22} \\ \sigma_{33} \\ \sigma_{12} \\ \sigma_{23} \\ \sigma_{31} \end{Bmatrix} \tag{2.131}$$

Stiffness

$$
\begin{Bmatrix} \sigma_{11} \\ \sigma_{22} \\ \sigma_{33} \\ \sigma_{12} \\ \sigma_{23} \\ \sigma_{31} \end{Bmatrix} = E^* \begin{bmatrix} (1-v) & v & v & \cdot & \cdot & \cdot \\ v & (1-v) & v & \cdot & \cdot & \cdot \\ v & v & (1-v) & \cdot & \cdot & \cdot \\ \cdot & \cdot & \cdot & \dfrac{1-2v}{2} & \cdot & \cdot \\ \cdot & \cdot & \cdot & \cdot & \dfrac{1-2v}{2} & \cdot \\ \cdot & \cdot & \cdot & \cdot & \cdot & \dfrac{1-2v}{2} \end{bmatrix} \begin{Bmatrix} \varepsilon_{11} \\ \varepsilon_{22} \\ \varepsilon_{33} \\ 2\varepsilon_{12} \\ 2\varepsilon_{23} \\ 2\varepsilon_{31} \end{Bmatrix} \qquad (2.132)
$$

Two special cases arise where the three–dimensional problem may be reduced to two dimensions.

2.8.3   Plane stress

This case is typified by a thin sheet of material loaded in the plane of the sheet. In this case,

$$\sigma_{33} = \sigma_{31} = \sigma_{32} = 0$$

The stress and strain tensors are thus of the form:

$$
[\sigma] = \begin{bmatrix} \sigma_{11} & \sigma_{12} & 0 \\ \sigma_{21} & \sigma_{22} & 0 \\ 0 & 0 & 0 \end{bmatrix} \quad \text{Stress} \qquad (2.133a)
$$

$$
[\varepsilon] = \begin{bmatrix} \varepsilon_{11} & \varepsilon_{12} & 0 \\ \varepsilon_{21} & \varepsilon_{22} & 0 \\ 0 & 0 & \varepsilon_{33} \end{bmatrix} \quad \text{Strain} \qquad (2.133b)
$$

2.8.4   Plane strain

This situation is typified by a hole in a continuum parallel with the x_3–axis extending for a distance many times greater than the hole diameter, with all sections being subjected to the same loading (e.g. internal pressure). In this case the approximation may be made,

$$\varepsilon_{33} = \varepsilon_{31} = \varepsilon_{32} = 0$$

$$
[\sigma] = \begin{bmatrix} \sigma_{11} & \sigma_{12} & 0 \\ \sigma_{21} & \sigma_{22} & 0 \\ 0 & 0 & \sigma_{33} \end{bmatrix} \quad \text{Stress} \qquad (2.134a)
$$

$$[\varepsilon] = \begin{bmatrix} \varepsilon_{11} & \varepsilon_{12} & 0 \\ \varepsilon_{21} & \varepsilon_{22} & 0 \\ 0 & 0 & 0 \end{bmatrix} \quad \text{Strain} \tag{2.134b}$$

In both cases the constitutive equation may be reduced to the two–dimensional form given in eqn (2.135); the constants E_r, τ, λ are given in eqn (2.136a–b).

$$\begin{Bmatrix} \sigma_{11} \\ \sigma_{22} \\ \sigma_{12} \end{Bmatrix} = E_r \begin{bmatrix} \tau & v & 0 \\ v & \tau & 0 \\ 0 & 0 & \lambda \end{bmatrix} \begin{Bmatrix} \varepsilon_{11} \\ \varepsilon_{22} \\ 2\varepsilon_{12} \end{Bmatrix} \tag{2.135}$$

For plane stress:

$$\tau = 1 \ , \ \lambda = \frac{1 - v}{2} \ , \ E_r = \frac{E}{1 - v^2} \tag{2.136a}$$

For plane strain:

$$\tau = 1 - v \ , \ \lambda = 0.5 - v \ , \ E_r = \frac{E}{(1 + v)(1 - 2v)} \tag{2.136b}$$

The other important stress–strain state for which it is worthwhile to set down a specific constitutive relationship occurs in continua possessing radial symmetry both in form and loading conditions. For this situation, three direct strains (ε_r, ε_{x_2}, ε_θ) occur together with the single shear strain $\varepsilon_{12} \equiv \varepsilon_{r2}$. The circumferential (or hoop) strain at any point is related to the radial displacement at that point by the equation,

$$\varepsilon_\theta = \frac{u_r}{r} \tag{2.137}$$

For the axisymmetric body the constitutive law is extracted from eqn (2.131) or (2.132). That is, for stiffness,

$$\begin{Bmatrix} \sigma_{rr} \\ \sigma_{22} \\ \sigma_{\theta\theta} \\ \sigma_{r_2} \end{Bmatrix} = E^* \begin{bmatrix} 1-v & v & v & \cdot \\ v & 1-v & v & \cdot \\ v & v & 1-v & \cdot \\ \cdot & \cdot & \cdot & \dfrac{1-2v}{2} \end{bmatrix} \begin{Bmatrix} \varepsilon_{rr} \\ \varepsilon_{22} \\ \varepsilon_{\theta\theta} \\ 2\varepsilon_{r_2} \end{Bmatrix} \tag{2.138}$$

2.9 THEORIES OF FAILURE

2.9.1 Deviatoric stress and strain tensors

In the development of the theories of failure of both ductile materials, for example, steel, and brittle materials, such as cast iron or rock, it is found beneficial to the understanding of the failure phenomena to divide the stress tensor into two components.

The first of these is the mean stress, which represents a hydrostatic stress state, and the second, the deviatoric stress tensor, which is a measure of the magnitude of the shear stress. Thus, let $s_{ij} = [s]$ be the deviatoric stress tensor, such that

$$s_{ij} = \sigma_{ij} - \frac{1}{3}\delta_{ij}\sigma_{kk} \tag{2.139}$$

In a similar way, the deviatoric strain tensor is defined

$$e_{ij} = \varepsilon_{ij} - \frac{1}{3}\delta_{ij}\varepsilon_{kk} \tag{2.140}$$

It is easily shown, by substituting eqns (2.125) and (2.127) in eqn (2.139), that

$$e_{ij} = \frac{s_{ij}}{2G} \tag{2.141}$$

Alternatively, from s_{ij} directly

$$e_{ij} = \frac{1+v}{E}s_{ij} - \frac{v}{E}s_{kk}\delta_{ij} = \frac{s_{ij}}{2G}$$

since $s_{kk} \equiv 0$ from eqn (2.139). Now from eqn (2.139) the stress tensor σ_{ij} is expressed as

$$\sigma_{ij} = s_{ij} + \frac{1}{3}\delta_{ij}\sigma_{kk} \tag{2.142}$$

The constitutive relationship, obtained by substituting eqns (2.141), (2.127) and (2.130), is

$$\sigma_{ij} = 2Ge_{ij} + \delta_{ij}K\varepsilon_{kk} \tag{2.143}$$

This alternative constitutive equation has divided the stress contribution into that from the deviatoric or shear strains e_{ij}, and that from the volume change ε_{kk}.

2.9.2   Invariants of the deviatoric stress tensor

In section 2.2.3 it has been shown that the principal stresses and directions arise from the solution of the equation,

$$\sigma_{ij}X_j = \lambda X_i \tag{2.144}$$

The solution exists if the determinant,

$$|\sigma_{ij} - \delta_{ij}\lambda| = 0$$

The resulting cubic equation in λ is given by

$$\lambda^3 - J_1\lambda^2 + J_2\lambda - J_3 = 0$$

in which,

$$J_1 = \sigma_{kk}$$

$$J_2 = \frac{1}{2}(\sigma_{ii}\sigma_{jj} - \sigma_{ij}\sigma_{ij})$$

$$J_3 = e_{ijk}\sigma_{1i}\sigma_{2j}\sigma_{3k} = |\sigma|$$

The coefficients J_1, J_2, J_3 must be invariant under coordinate transformation since the roots λ of eqn (2.145) are the principal stresses which are independent of the coordinate axes used to describe the stress tensor. Similarly, the invariants of the deviatoric stress tensor are

$$J_{e1} = s_{kk} = 0$$

$$J_{e2} = \tfrac{1}{2}s_{ij}s_{ij} \tag{2.145}$$

$$J_{e3} = e_{ijk}s_{1i}s_{2j}s_{3k} = |s|$$

It is found that J_1 and J_{e2} are expressions of the stress tensor which are used to describe:

J_1, the influence of the mean stress on failure of brittle materials

J_{e2}, the influence of shear stress on failure

2.9.3   The octahedral shear stress

Most modern theories of failure are based on relating the octahedral shear stress in the complex stress state to that failure shear stress obtained by testing the material in simple tension or compression. The octahedral shear is the shear stress on the plane (octahedral plane) whose normal is equally inclined to the principal stress directions (Fig. 2.13). The normal to the octahedral plane, $\hat{n}$ has direction cosines $(\frac{1}{\sqrt{3}}, \frac{1}{\sqrt{3}}, \frac{1}{\sqrt{3}})$.

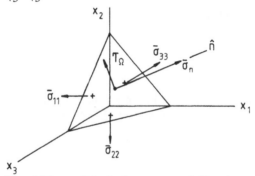

Fig. 2.13   Principal stresses and directions.

From eqn (2.42), the normal stress on the plane whose normal is equally inclined to the principal axes is given by

$$\sigma_{kn} = n_{ki}n_{jn}\bar{\sigma}_{ij} = \frac{1}{3}(\bar{\sigma}_{11} + \bar{\sigma}_{22} + \bar{\sigma}_{33}) \tag{2.146}$$

Furthermore, from eqn (2.40), the components of the force on unit area of the octahedral plane are

$$p_i = \sigma_{ji}n_j$$

and hence the square of the magnitude of this force is

$$F_n^2 = p_i p_i = \sigma_{ji}n_j\sigma_{ki}n_k = \frac{1}{3}(\bar{\sigma}_{11}^2 + \bar{\sigma}_{22}^2 + \bar{\sigma}_{33}^2) \tag{2.147}$$

If the shear stress on the octahedral plane is τ_Ω, then F_n^2 is also given by $\sigma_{kn}^2 + \tau_\Omega^2$, so that

$$\sigma_{kn}^2 + \tau_\Omega^2 = \frac{1}{3}(\bar{\sigma}_{11}^2 + \bar{\sigma}_{22}^2 + \bar{\sigma}_{33}^2) \qquad (2.148)$$

Substitution of eqn (2.145) in eqn (2.148) gives the expression for τ_Ω^2,

$$\tau_\Omega^2 = \frac{1}{3}(\bar{\sigma}_{11}^2 + \bar{\sigma}_{22}^2 + \bar{\sigma}_{33}^2) - \frac{1}{9}(\bar{\sigma}_{11} + \bar{\sigma}_{22} + \bar{\sigma}_{33})^2 \tau_\Omega^2$$

$$= \frac{1}{9}\{(\bar{\sigma}_{11} - \bar{\sigma}_{22})^2 + (\bar{\sigma}_{22} - \bar{\sigma}_{33})^2 + (\bar{\sigma}_{33} - \bar{\sigma}_{11})^2\} \qquad (2.149)$$

It is now shown that J_{e2} is a measure of τ_Ω^2. Consider the expression for J_{e2}, and write it in terms of the principal stress components. That is, from eqn (2.144),

$$J_{e2} = \frac{1}{2}s_{ij}s_{ij} = \frac{1}{2}(\sigma_{ij} - \frac{1}{3}\delta_{ij}\sigma_{kk})(\sigma_{ij} - \frac{1}{3}\delta_{ij}\sigma_{mm}) = \frac{1}{2}(\sigma_{ij}\sigma_{ij} - \frac{1}{3}\sigma_{ii}\sigma_{kk}) \qquad (2.150)$$

Substitute the principal stress tensor $\bar{\sigma}_{ij}$ in this equation, remembering that J_{e2} is invariant under coordinate transformation. Then,

$$J_{e2} = \frac{1}{6}\{3(\bar{\sigma}_{11}^2 + \bar{\sigma}_{22}^2 + \bar{\sigma}_{33}^2) - (\bar{\sigma}_{11} + \bar{\sigma}_{22} + \bar{\sigma}_{33})^2\} \qquad (2.151)$$

A comparison of eqns (2.149) and (2.151) gives

$$\tau_\Omega^2 = \frac{2}{3}J_{e2} = \frac{1}{3}s_{ij}s_{ij} \qquad (2.152)$$

2.9.4   Effective stress – effective strain

The effective stress concept arises when τ_Ω in the simple tension or compression test is related to the direct stress $\bar{\sigma}_{11}$ ($\bar{\sigma}_{22} = \bar{\sigma}_{33} = 0$). Then, from eqn (2.149), by definition of σ_e,

$$\tau_\Omega^2 = \frac{2}{9}\bar{\sigma}_{11}^2 = \frac{2}{9}\sigma_e^2 \qquad (2.153)$$

That is,

$$\sigma_e^2 = \frac{9}{2}\tau_\Omega^2$$

Substituting eqn (2.152) gives the effective stress,

$$\sigma_e^2 = \frac{3}{2}s_{ij}s_{ij} = 3J_{e2} \qquad (2.154)$$

Having defined the effective stress σ_e, in the complex stress state, the effective strain is given simply by eqn (2.155),

$$\varepsilon_e = \frac{1}{E}\sigma_e \qquad (2.155)$$

The effective strain is found also to be a function of the deviatoric strain tensor. Thus, the octahedral shear stress is given by

$$\tau_\Omega = \frac{1}{3}\{(\bar{\sigma}_{11} - \bar{\sigma}_{22})^2 + (\bar{\sigma}_{22} - \bar{\sigma}_{33})^2 + (\bar{\sigma}_{33} - \bar{\sigma}_{11})^2\}^{1/2}$$

Then the octahedral shear strain γ_Ω is given

$$\gamma_\Omega = \frac{\tau_\Omega}{G} = \frac{1}{3}\left\{\frac{(\bar\sigma_{11} - \bar\sigma_{22})^2}{G^2} + \frac{(\bar\sigma_{22} - \bar\sigma_{33})^2}{G^2} + \frac{(\bar\sigma_{33} - \bar\sigma_{11})^2}{G^2}\right\}^{1/2}$$ (2.156)

Now, $\tau_{12} = (\bar\sigma_{11} - \bar\sigma_{22})/2$ is the maximum shear stress in the $\bar x_1, \bar x_2$ plane, and similarly for τ_{23}, τ_{31}. However, $\tau_{12}/G = \gamma_{12}$ is the maximum shear strain in the same plane. With these substitutions eqn (2.156) becomes

$$\gamma_\Omega = \frac{2}{3}\{\gamma_{12}^2 + \gamma_{23}^2 + \gamma_{31}^2\}^{1/2}$$ (2.157)

The shear strain γ_{12} is related to the principal strains $\bar\varepsilon_{11}, \bar\varepsilon_{22}, \bar\varepsilon_{33}$, by $(\gamma_{12})/2 = (\bar\varepsilon_{11} - \bar\varepsilon_{22})/2$. Similarly, for γ_{23}, γ_{31}, so that, combining with eqn (2.157), it is seen that

$$\gamma_\Omega = \frac{2}{3}\{(\bar\varepsilon_{11} - \bar\varepsilon_{22})^2 + (\bar\varepsilon_{22} - \bar\varepsilon_{33})^2 + (\bar\varepsilon_{33} - \bar\varepsilon_{11})^2\}^{1/2}$$ (2.158)

Squaring both sides of this expression,

$$\gamma_\Omega^2 = \frac{4}{9}\{(\bar\varepsilon_{11} - \bar\varepsilon_{22})^2 + (\bar\varepsilon_{22} - \bar\varepsilon_{33})^2 + (\bar\varepsilon_{33} - \bar\varepsilon_{11})^2\}$$ (2.159)

By the same reasoning that led to eqn (2.152) for the octahedral shear stress, it is readily shown that eqn (2.159) leads to

$$\gamma_\Omega^2 = \frac{4}{3}e_{ij}e_{ij}$$ (2.160)

However, τ_Ω is related to γ_Ω simply by

$$\tau_\Omega = G\gamma_\Omega$$

or,

$$\tau_\Omega^2 = \frac{E^2}{[2(1+v)]^2}\gamma_\Omega^2$$

Using eqn (2.153), it is seen that

$$\frac{2}{9}\sigma_e^2 = \frac{E^2}{[2(1+v)]^2}\gamma_\Omega^2 = \frac{4E^2}{3[2(1+v)]^2}e_{ij}e_{ij}$$

Finally,

$$\sigma_e^2 = E^2\frac{3}{2(1+v)^2}e_{ij}e_{ij}$$ (2.161)

Comparing eqns (2.155) and (2.161), it is seen that

$$\varepsilon_e^2 = \frac{3}{2(1+v)^2}e_{ij}e_{ij}$$

Taking the square root of both sides of this equation, the expression for ε_e is obtained

$$\varepsilon_e = \frac{\sqrt{3}}{\sqrt{2}(1+v)}(e_{ij}e_{ij})^{1/2}$$ (2.162)

2.9.5   Plastic strain

In the simple theory of plasticity it is assumed that no volume change occurs as a result of plastic straining. That is, from eqn (2.130),

$$\frac{1}{K} = \frac{3(1 - 2\nu)}{E} = 0$$

This condition is satisfied for plastic strains if $\nu = 1/2$. With $\nu = 1/2$, from eqn (2.162),

$$\varepsilon_{ep}^2 = \frac{3}{2(3/2)^2} e_{ij} e_{ij} = \frac{2}{3} e_{ij} e_{ij} \tag{2.163}$$

The plastic strain increment is related to the effective stress by replacing $\dfrac{1}{E}$ by $d\lambda$, so that

$$d\varepsilon_{ep} = d\lambda \sigma_e \quad \text{or} \quad d\lambda = \frac{d\varepsilon_{ep}}{\sigma_e} \tag{2.164}$$

Now,

$$d\varepsilon_{ijp} = d\varepsilon_{ijp} - \frac{1}{3}\delta_{ij} d\varepsilon_{kkp} = d\varepsilon_{ijp}$$

From eqn (2.141), substituting $\gamma = 1/2$, $d\lambda = \dfrac{1}{E}$,

$$d\varepsilon_{ijp} = de_{ijp} = \frac{3d\lambda}{2} s_{ij} \tag{2.165}$$

Finally, from eqn (2.164),

$$d\varepsilon_{ijp} = \frac{3d\varepsilon_{ep}}{2\sigma_e} s_{ij} \tag{2.166}$$

This equation relates the Cartesian components of the plastic strain increment to the effective plastic strain increment $d\varepsilon_{ep}$. It is known as the Prandtl–Reuss strain relationship. Developing eqn (2.166) further, take derivatives of both sides of eqn (2.154) with respect to σ_{ij}, noting that

$$\frac{\partial s_{lm}}{\partial \sigma_{ij}} s_{lm} = s_{ij}$$

Thence

$$\frac{\partial \sigma_e}{\partial \sigma_{ij}} = \frac{3 s_{ij}}{2\sigma_e} \tag{2.167}$$

Substitution in eqn (2.166) gives the simple expression for $d\varepsilon_{ijp}$,

$$d\varepsilon_{ijp} = \frac{\partial \sigma_e}{\partial \sigma_{ij}} d\varepsilon_{ep} \tag{2.168}$$

2.9.6   Incremental stress – strain relationships

It is now possible to derive a relationship between the incremental stress and strain values $(d\sigma_{ij}, d\varepsilon_{ij})$. Firstly, assume that, as shown in Fig. 2.14, an experimental relationship has been established between σ_e and ε_{ep}. If the slope of this curve at any point is H', then,

$$d\sigma_e = H'd\varepsilon_{ep} \tag{2.169}$$

For a material such as mild steel exhibiting perfect plasticity with no strain hardening, $H' = 0$ over a considerable range of plastic strain. Any change in stress components is related to an increment in the elastic strain component, which, in turn, is equal to the total strain component less the plastic strain component. That is,

$$d\sigma_{ij} = C_{ijkl} d\varepsilon_{kl\ E} = C_{ijkl}(d\varepsilon_{kl} - d\varepsilon_{klp}) \tag{2.170}$$

From eqn (2.168),

$$d\sigma_{ij} = C_{ijkl}\left(d\varepsilon_{kl} - \frac{\partial\sigma_e}{\partial\sigma_{kl}}d\varepsilon_{ep}\right) \tag{2.171}$$

Using the chain rule for differentiation,

$$d\sigma_e = \frac{\partial\sigma_e}{\partial\sigma_{ij}}d\sigma_{ij} = H'd\varepsilon_{ep} \tag{2.172}$$

Then, from eqns (2.171) and (2.172),

$$\frac{\partial\sigma_e}{\partial\sigma_{ij}}d\sigma_{ij} = \frac{\partial\sigma_e}{\partial\sigma_{ij}}C_{ijkl}\left(d\varepsilon_{kl} - \frac{\partial\sigma_e}{\partial\sigma_{kl}}d\varepsilon_{ep}\right) = H'd\varepsilon_{ep} \tag{2.173}$$

Finally, the expression is obtained for $d\varepsilon_{ep}$,

$$d\varepsilon_{ep} = \frac{\dfrac{\partial\sigma_e}{\partial\sigma_{ij}}C_{ijkl}\,d\varepsilon_{kl}}{\left[H' + \dfrac{\partial\sigma_e}{\partial\sigma_{ij}}C_{ijmn}\dfrac{\partial\sigma_e}{\partial\sigma_{mn}}\right]} \tag{2.174}$$

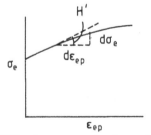

Fig. 2.14   Effective stress–plastic strain relationship.

Substitution for $d\varepsilon_{ep}$ in eqn (2.171) gives the required incremental stress–strain relationship in the elasto–plastic range:

$$d\sigma_{ij} = \left[C_{ijkl} - \frac{C_{ijrt}\dfrac{\partial\sigma_e}{\partial\sigma_{rt}}\dfrac{\partial\sigma_e}{\partial\sigma_{pq}}C_{pqkl}}{\left[H' + \dfrac{\partial\sigma_e}{\partial\sigma_{ij}}C_{ijmn}\dfrac{\partial\sigma_e}{\partial\sigma_{mn}}\right]}\right]d\varepsilon_{kl} \tag{2.175}$$

Although the expression in eqn (2.175) is quite usable, it is really more convenient to express the above theory in matrix notation, as in eqn (2.132) *et seq.* The matrix notation is more useful from the point of view of writing computer software for elasto–plastic analysis. For example, it is unrealistic to work in terms of C_{ijkl} when for a homogeneous, isotropic, material only the two elastic constants E, ν are required. Thus, reconstitute eqn (2.166) in matrix form, using the notation for shear strain $\gamma_{12} = (\varepsilon_{12} + \varepsilon_{21})$, etc. Then,

$$\begin{Bmatrix} d\varepsilon_{11} \\ d\varepsilon_{22} \\ d\varepsilon_{33} \\ d\gamma_{12} \\ d\gamma_{23} \\ d\gamma_{31} \end{Bmatrix}_P = \frac{3}{2} \frac{d\varepsilon_{ep}}{\sigma_e} \begin{Bmatrix} s_{11} \\ s_{22} \\ s_{33} \\ 2s_{12} \\ 2s_{23} \\ 2s_{31} \end{Bmatrix} \tag{2.176}$$

From eqn (2.170),

$$\{d\sigma\} = [k]\{d\varepsilon\}_E = [k][\{\varepsilon\} - \{\varepsilon\}_P] \tag{2.177}$$

Writing eqn (2.167) in matrix form,

$$\left\{ \frac{\partial \sigma_e}{\partial \sigma} \right\} = \frac{3}{2\sigma_e} \begin{Bmatrix} s_{11} \\ s_{22} \\ s_{33} \\ 2s_{12} \\ 2s_{23} \\ 2s_{31} \end{Bmatrix} \tag{2.178}$$

The flow rule, eqn (2.168), is now written

$$\{d\varepsilon\}_P = d\varepsilon_{ep} \left\{ \frac{\partial \sigma_e}{\partial \sigma} \right\} \tag{2.179}$$

and, from eqn (2.172),

$$d\sigma_e = \left\{ \frac{\partial \sigma_e}{\partial \sigma} \right\}^T \{d\sigma\} = H' \, d\varepsilon_{eP} \tag{2.180}$$

Then eqn (2.177) becomes

$$\{d\sigma\} = [k](\{d\varepsilon\} - d\varepsilon_{ep} \left\{ \frac{\partial \sigma_e}{\partial \sigma} \right\}) \tag{2.181}$$

Combining eqns (2.180) and (2.181),

$$\left\{ \frac{\partial \sigma_e}{\partial \sigma} \right\}^T \{d\sigma\} = \left\{ \frac{\partial \sigma_e}{\partial \sigma} \right\}^T [k](\{d\varepsilon\} - d\varepsilon_{ep} \left\{ \frac{\partial \sigma_e}{\partial \sigma} \right\}) = H' d\varepsilon_{ep} \tag{2.182}$$

Simplifying eqn (2.182) gives

$$d\varepsilon_{ep} = \dfrac{\left[\dfrac{\partial\sigma_e}{\partial\sigma}\right]^T [k]\{d\varepsilon\}}{\left[H' + \left[\dfrac{\partial\sigma_e}{\partial\sigma}\right]^T [k]\left[\dfrac{\partial\sigma_e}{\partial\sigma}\right]\right]} \qquad (2.183)$$

Finally, eqn (2.175) in matrix form becomes (with the substitution above)

$$\{d\sigma\} = \left[[k] - \dfrac{[k]\left[\dfrac{\partial\sigma_e}{\partial\sigma}\right]\left[\dfrac{\partial\sigma_e}{\partial\sigma}\right]^T [k]}{\left[H' + \left[\dfrac{\partial\sigma_e}{\partial\sigma}\right]^T [k]\left[\dfrac{\partial\sigma_e}{\partial\sigma}\right]\right]}\right]\{d\varepsilon\} \qquad (2.184)$$

In symbolic form,

$$\{d\sigma\} = [k_T]\{d\varepsilon\} \qquad (2.185)$$

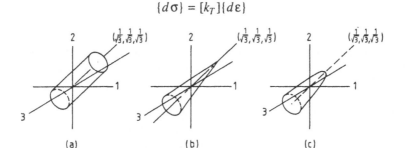

Fig. 2.15   Simple yield surfaces.

2.10 EXAMPLES OF YIELD SURFACES

2.10.1   Maxwell–Mohr–Von Mises criterion

This is perhaps the most widely known yield criterion because it applies to ductile materials, such as mild steel, brass and copper, which have well–defined yield points. The yield is assumed to be controlled by the shear stress on the octahedral plane, and the yield surface in the principal stress space is a cylinder equally inclined to the principal stress axes, as in Fig. 2.15(a). The stress in the material is assumed to be limited to lie within the surface defined by the yield function. For the Von Mises criterion,

$$F(\sigma) = \sqrt{3}J_{e2}^{1/2} - \sigma_{yp} = 0 \qquad (2.186)$$

From eqn (2.154), $\sqrt{3}J_{e2}^{1/2} = \sigma_e$, the effective stress, so eqn (2.186) is written

$$F = \sigma_e - \sigma_{yp} = 0 \qquad (2.187)$$

Thus, the effective stress is limited to be less than or equal to the yield point stress, assumed to be the same in tension and compression. The criterion is the most simple to apply because the only strength parameter to measure is σ_{yp}, and failure is independent of the mean stress present. Before the yield surfaces in Fig. 2.15(b), (c) are examined in detail, the concept of yield surface $F(\sigma) = 0$ in eqn (2.187) is developed further, and the ideas of associative and non–associative plasticity are introduced.

In the development of eqn (2.184), the flow rule was assumed to be given by eqn (2.179)

$$\{d\varepsilon\}_p = d\varepsilon_{ep}\left[\frac{\partial\sigma_e}{\partial\sigma}\right]$$

If the yield is based on the Von Mises criterion,

$$\left[\frac{\partial\sigma_e}{\partial\sigma}\right] = \left[\frac{\partial F}{\partial\sigma}\right]$$

and the components of the vector on the right–hand side are normal to the yield surface; that is, the normality rule of strain components applies. This flow rule can then be adapted for any yield surface by replacing σ_e by F the yield criterion. Then eqn (2.184) becomes

$$\{d\sigma\} = \left[[k] - \frac{[k]\left[\dfrac{\partial F}{\partial\sigma}\right]\left[\dfrac{\partial F}{\partial\sigma}\right]^T[k]}{\left[H' + \left[\dfrac{\partial F}{\partial\sigma}\right]^T[k]\left[\dfrac{\partial F}{\partial\sigma}\right]\right]}\right]\{d\varepsilon\} \tag{2.188}$$

This is the constitutive equation for stress strain increments in the plastic range for associative plasticity. Suppose, however, that there are two surfaces, F and $\bar{F}$, such that F is the usual stress space surface and $\bar{F}$ is a new surface to which the normality rule applies for the strain components. That is, it is non–associative plasticity because the strain components are no longer associated with the normal to the stress yield surface. The usual application of non–associative plasticity occurs when the yield surface is given as in Fig. 2.15(b) or (c), and the plastic strains are normal to the Von Mises cylinder in Fig. 2.15(a). That is, the flow rule is given

$$\{d\varepsilon\}_p = d\varepsilon_{ep}\left[\frac{\partial\bar{F}}{\partial\sigma}\right] \tag{2.189}$$

Proceeding with identical reasoning used to obtain eqn (2.183), it is found that the equivalent plastic strain increment is

$$d\varepsilon_{eP} = \left[\frac{\left[\dfrac{\partial F}{\partial\sigma}\right]^T[k]}{\left[H' + \left[\dfrac{\partial F}{\partial\sigma}\right]^T[k]\left[\dfrac{\partial \bar{F}}{\partial\sigma}\right]\right]}\right]\{d\varepsilon\} \tag{2.190}$$

Substitution in eqn (2.181) gives

$$\{d\sigma\} = \left[[k] - \frac{[k]\left[\dfrac{\partial \bar{F}}{\partial\sigma}\right]\left[\dfrac{\partial F}{\partial\sigma}\right]^T[k]}{\left\{H' + \left[\dfrac{\partial F}{\partial\sigma}\right]^T[k]\left[\dfrac{\partial \bar{F}}{\partial\sigma}\right]\right\}}\right]\{d\varepsilon\} \tag{2.191}$$

A problem now arises in that if eqn (2.191) is written

$$\{d\sigma\} = [k_T]\,\{d\varepsilon\} \tag{2.192}$$

the constitutive relationship $[k_T]$ is no longer a symmetric matrix, and the resulting tangent stiffness matrix for the whole structure in the elasto–plastic range will also be non–symmetric.

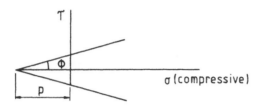

Fig. 2.16 Mohr–Coulomb c–ϕ strength relationship.

2.10.2 Linear Mohr–Coulomb or Drucker Prager yield criterion

This yield surface shown in Fig. 2.15(b) is a generalization of the Mohr–Coulomb law in two–dimensional stress space, relating the shear failure stress to the internal cohesion, c, and the angle of internal friction, ϕ, by the expression,

$$\tau = c + \sigma \tan \phi \tag{2.193}$$

where it is assumed that σ is compressive positive, c is the cohesion and ϕ is the angle of internal friction. The cone shown in Fig. 2.15(b) is defined by two of the three parameters, c, p, ϕ, given in the τ–σ space of plane strain analysis (Fig. 2.16). The yield surface in three–dimensional stress space is

$$F = \alpha J_1 + J_{e2}^{\frac{1}{2}} - \frac{\sigma_{yp}}{\sqrt{3}} = 0 \tag{2.194}$$

or,

$$\sqrt{3}\alpha J_1 + \sigma_e - \sigma_{yp} = 0 \tag{2.195}$$

The constants α and σ_{yp} are related to c and ϕ through the expressions,

$$\alpha = \frac{\sin \phi}{(9 + 3 \sin^2 \phi)^{\frac{1}{2}}} \tag{2.196}$$

and

$$\sigma_{yp} = c\,\{3(1 - 12\alpha^2)\}^{\frac{1}{2}} \tag{2.197}$$

2.10.3 Parabolic Mohr–Coulomb yield criterion

The linear Mohr–Coulomb criterion suffers from two obvious deficiencies. The first of these is the singular point at the apex of the cone where the normal to the yield surface is undefined. This problem is overcome by truncating the cone if $J_1 - p \le \delta$, where δ is an arbitrarily small value, and for this situation the normal is taken parallel with the cone axis.

The second problem is that the permissible octahedral shear stress increases indefinitely as the mean stress increases in the compressive direction. The parabolic yield surface, Fig. 2.15(c) overcomes these deficiencies to some extent. The plane strain condition is shown in Fig. 2.17.

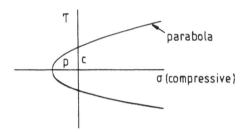

Fig. 2.17   Parabolic Mohr–Coulomb yield surface.

In this case the equation to the yield surface is

$$F = (3J_{e2} + \sqrt{3}\beta\sigma_{yp}J_1)^{\frac{1}{2}} - \sigma_{yp} = 0 \tag{2.198}$$

In this equation,

$$\beta = \frac{c^2/p}{\sqrt{3}\sigma_{yp}} \tag{2.199}$$

and

$$\sigma_{yp}^2 = 3c^2 - \left[\frac{c^2}{p}\right]^2 \tag{2.200}$$

Some properties of the above yield surfaces are given in the following sections.

2.10.4   Von Mises yield surface

$$\frac{\partial F}{\partial \sigma_{ij}} = \frac{\sqrt{3}}{2J_{e2}^{\frac{1}{2}}}s_{ij} = \frac{3}{2\sigma_e}s_{ij} \tag{2.201}$$

In matrix notation, this expression is written

$$\left\{\frac{\partial F}{\partial \sigma}\right\} = \frac{3}{2\sigma_e}\begin{bmatrix} \sigma_{11} - \sigma_m \\ \sigma_{22} - \sigma_m \\ \sigma_{33} - \sigma_m \\ 2\sigma_{12} \\ 2\sigma_{23} \\ 2\sigma_{31} \end{bmatrix} \tag{2.202}$$

In eqn (2.202), $\sigma_m = J_1/3 = $ mean stress.

2.10.5 Drucker Prager criterion

In this case,

$$\frac{\partial F}{\partial \sigma_{ij}} = \alpha \delta_{ij} + \frac{s_{ij}}{2J_{e2}^{\frac{1}{2}}} = \alpha \delta_{ij} + \frac{\sqrt{3}s_{ij}}{2\sigma_e} \tag{2.203}$$

In matrix notation,

$$\left\{\frac{\partial F}{\partial \sigma}\right\} = \alpha \begin{bmatrix} 1 \\ 1 \\ 1 \\ 0 \\ 0 \\ 0 \end{bmatrix} + \frac{\sqrt{3}}{2\sigma_e} \begin{bmatrix} \sigma_{11} - \sigma_m \\ \sigma_{22} - \sigma_m \\ \sigma_{33} - \sigma_m \\ 2\sigma_{12} \\ 2\sigma_{23} \\ 2\sigma_{31} \end{bmatrix} \tag{2.204}$$

2.10.6 Parabolic Mohr–Coulomb criterion

The function derivative is given from eqn (2.198):

$$\frac{\partial F}{\partial \sigma_{ij}} = \frac{1}{2f^*}[3s_{ij} + \sqrt{3}\beta\sigma_{yp}\delta_{ij}] \tag{2.205}$$

In matrix notation,

$$\left\{\frac{\partial F}{\partial \sigma}\right\} = \frac{3}{2f^*} \begin{bmatrix} \sigma_{11} - \sigma_m \\ \sigma_{22} - \sigma_m \\ \sigma_{33} - \sigma_m \\ 2\sigma_{12} \\ 2\sigma_{23} \\ 2\sigma_{31} \end{bmatrix} + \frac{\sqrt{3}\beta\sigma_{yp}}{2f^*} \begin{bmatrix} 1 \\ 1 \\ 1 \\ 0 \\ 0 \\ 0 \end{bmatrix} \tag{2.206}$$

In eqns (2.205) and (2.206),

$$f^* = (3J_{e2} + \sqrt{3}\beta\sigma_{yp}J_1)^{\frac{1}{2}} \tag{2.207}$$

2.11 TEMPERATURE AND CREEP EFFECTS

In this final discussion on constitutive relationships the effects of temperature on both the yield surface and the modulus of elasticity of the material are taken into consideration. For example, when a weld is deposited across a steel plate, not only are thermal strains introduced but also,

1. The yield surface is contracted at the elevated temperatures to expand again as the material cools.

2. The material softens with increase of temperature, so that $E_{t_1} < E_{t_0}$.

These effects when combined can produce both large residual stresses and unacceptable distortions. To examine the problem, in so far as the modification to the constitutive law of the material is concerned, suppose that ε_Δ is the total strain increment, given as the sum in eqn (2.208).

$$\{\varepsilon_\Delta\} = \{\varepsilon_{E\Delta}\} + \{\varepsilon_{T\Delta}\} + \{\varepsilon_{P\Delta}\} + \{\varepsilon_{C\Delta}\} \tag{2.208}$$

In this equation, $\{\varepsilon_{E\Delta}\}$ is the elastic strain increment, $\{\varepsilon_{T\Delta}\}$ is the thermal strain increment, $\{\varepsilon_{P\Delta}\}$ is the plastic strain increment, and $\{\varepsilon_{C\Delta}\}$ is the creep strain increment. The assumed yield surface will have the form,

$$F = \sigma_e(\sigma) - \sigma_{yp} \leq 0 \qquad (2.209)$$

It is now required to determine the plastic strain increment $\{\varepsilon_{P\Delta}\}$ and from the conditions of stress and strain on the yield surfaces deduce the relationship between the stress increment $\{\sigma_\Delta\}$ and the total strain increment ε_Δ. The relationship is found to be similar to that in eqn (2.191), complicated now by the additional effects. Thus, for the determination of the plastic strain increment, the yield criterion eqn (2.209) must be fulfilled not only at the beginning (−) but also at the end (+) of an incremental change of state. Notice that eqn (2.209) separates the stress terms from the strength terms. However, in practice, the yield surfaces such as that in eqn (2.198) do not satisfy this requirement. The yield conditions at the beginning (−) and end (+) of the interval are given

$$\phi_- = \sigma_{e-} - \sigma_{yp}(\varepsilon_{eP-},T_-) \leq 0 \qquad (2.210a)$$

$$\phi_+ = \sigma_{e+} - \sigma_{yp}(\varepsilon_{eP+},T_+) \leq 0 \qquad (2.210b)$$

These two conditions are shown in Fig. 2.18, for both strain hardening (a) and strain softening (b). These curves must be obtained by experimental observations.

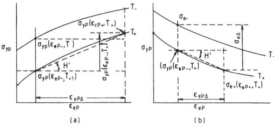

Fig. 2.18 Strain hardening and strain softening–effective stress–strain curves.

From Fig. 2.18, it is seen that

$$\sigma_{e+} = \sigma_{yp}(\varepsilon_{eP+},T_+) = \sigma_{yp}(\varepsilon_{eP-},T_+) + H'\varepsilon_{eP\Delta} \qquad (2.211)$$

Thence, the effective plastic strain increment is

$$\varepsilon_{eP\Delta} = \frac{1}{H'}[\sigma_{e+} - \sigma_{yp}(\varepsilon_{eP-},T_+)] , \quad \text{if } \phi_+ \text{ is positive; otherwise,} \quad \varepsilon_{ep\Delta} = 0. \qquad (2.212)$$

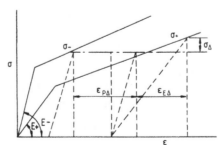

Fig. 2.19 Variation in modulus of elasticity.

In addition to the change in the strength parameter, σ_{yp}, the change in the elastic modulus from E_- to E_+ over the interval must be taken into account. This change may be a function of temperature and/or plastic strain. The situations are shown in Fig. 2.19, in which it has been assumed that there has been a decrease in the modulus of elasticity. Referring to Fig. 2.19,

$$\varepsilon_{E\Delta} = \frac{1}{E_+}\sigma_+ - \frac{1}{E_-}\sigma_- \qquad (2.213)$$

or,

$$\varepsilon_{E\Delta} = \frac{1}{E_+}(\sigma_+ - \sigma_-) + (\frac{1}{E_+} - \frac{1}{E_-})\sigma_- = \frac{1}{E_+}\sigma_\Delta + (\frac{1}{E_+} - \frac{1}{E_-})\sigma_-$$

This equation can be written

$$\varepsilon_{E\Delta} = f_+\sigma_\Delta + df_-\sigma_- \qquad (2.214)$$

In this equation, f_+ is the flexibility at the end of the increment and df, the change in flexibility. The inverse, stiffness relationship is

$$\sigma_\Delta = k_+(\varepsilon_{E\Delta} - f_-\sigma_-) \qquad (2.215)$$

In eqn (2.215), $k_+ = f_+^{-1}$. The eqn (2.215) has been developed for a single stress component from Fig. 2.19. For all stress components,

$$\{\sigma_\Delta\} = [k_+](\{\varepsilon_{E\Delta}\} - [df_-]\{\sigma_-\}) \qquad (2.216)$$

Now write,

$$\{\varepsilon_{r\Delta}\} = \{\varepsilon_\Delta\} - \{\varepsilon_{T\Delta}\} - \{\varepsilon_{C\Delta}\} - [dC_-]\{\sigma_-\} \qquad (2.217)$$

so that, using eqn (2.208),

$$\{\sigma_\Delta\} = [k_+](\{\varepsilon_{r\Delta}\} - \{\varepsilon_{P\Delta}\}) \qquad (2.218)$$

The normality rule, eqn (2.179), will be written

$$\{\varepsilon_{P\Delta}\} = \{\bar{a}_+\}\varepsilon_{eP\Delta} \qquad (2.219)$$

where $\{\bar{a}_+\}$ is the vector normal to the strain surface, and $\{a_+\}$ is the vector normal to the stress surface. With this definition, eqn (2.218) becomes

$$\{\sigma_\Delta\} = [k_+](\{\varepsilon_{r\Delta}\} - \{\bar{a}_+\}\varepsilon_{eP\Delta}) \qquad (2.220)$$

Now it is easily shown that since $\dfrac{\partial\sigma_e}{\partial\sigma_{ij}} = \dfrac{3\sigma_{ij}}{2\sigma_e}$ and $\{\sigma_+\} = s_{ij} + \dfrac{1}{3}\delta_{ij}\sigma_{kk}$,

$$\sigma_{e+} = \{a_+\}^T\{\sigma_+\} = \{a_+\}^T(\{\sigma_-\} + \{\sigma_\Delta\}) \qquad (2.221)$$

Thence, from eqn (2.211),

$$\{a_+\}^T\{\sigma_\Delta\} = \sigma_{yp}(\varepsilon_{eP-},T_+) + H'\varepsilon_{eP\Delta} - \{a_+\}^T\{\sigma_-\} \qquad (2.222)$$

Substitute in eqn (2.220) and collect terms in $\varepsilon_{eP\Delta}$ to give

$$\varepsilon_{eP\Delta} = \frac{1}{(H' + \{a_+\}^T [k_+]\{\bar{a}_+\})}[\{a_+\}^T(\{\sigma_-\} + [k_+]\{\varepsilon_{r\Delta}\}) - \sigma_{yp}(\varepsilon_{eP-}, T_+)] \qquad (2.223)$$

The similarity between eqn (2.190) and eqn (2.223) should be noted. Given $\varepsilon_{eP\Delta}$, it is now possible to develop the incremental tangent stiffness matrix including all effects. Remember, from eqns (2.217) and (2.218),

$$\{\sigma_\Delta\} = [k_+](\{\varepsilon_\Delta\} - \{\varepsilon_{T\Delta}\} - \{\varepsilon_{C\Delta}\} - \{\varepsilon_{P\Delta}\} - [dC_-]\{\sigma_-\})$$

Define

$$E^* = H' + \{a_+\}^T [k_+]\{\bar{a}_+\} \qquad (2.224)$$

Then, from eqn (2.220), substituting in eqn (2.223) for $\varepsilon_{eP\Delta}$,

$$\{\sigma_\Delta\} = [k_+](\{\varepsilon_\Delta\} - \{\varepsilon_{T\Delta}\} - \{\varepsilon_{C\Delta}\}) \qquad (2.225)$$

$$- \frac{\{\bar{a}_+\}}{E^*}[\{a_+\}^T(\{\sigma-\} + [k_+]\{\varepsilon_{r\Delta}\}) - \sigma_{yp}(\varepsilon_{eP-}, T_+)] - [dC_-]\{\sigma_-\}$$

Now define $[k_T]$ as

$$[k_T] = [k_+] - \frac{[k_+]\{\bar{a}_+\}\{a_+\}^T [k_+]}{E^*} \qquad (2.226)$$

Substitute for $\{\varepsilon_{r\Delta}\}$ from eqn (2.217) and group terms, using the definition for $[k_T]$, so that

$$\{\sigma_\Delta\} = [k_T][\{\varepsilon_\Delta\} - (\{\varepsilon_{T\Delta}\} + \{\varepsilon_{C\Delta}\} + [dC_-]\{\sigma_-\})]$$

$$+ \frac{[k_+]\{\bar{a}_+\}}{E^*}\sigma_{yp}(\varepsilon_{eP-}, T_+) - \frac{[k_+]\{\bar{a}_+\}\{a_+\}^T}{E^*}\{\sigma_-\} \qquad (2.227)$$

Again, the similarity between eqns (2.227) and (2.188) should be noted, with the added terms in eqn (2.227) arising from the more complex physical situation being modelled.

CHAPTER 3

The formulation of the finite element method

3.1 DISPLACEMENT FORMULATION

3.1.1 Introduction

Thus far, the text has been concerned with mathematical preliminaries and the fundamentals of solid mechanics. These fundamentals are common to all the various aspects of the finite element method, whether it be an application in planar or three–dimensional elasticity or in plates, shells, etc. In this chapter, however, it is the intention to describe briefly some of the strategies which may be used to apply Gauss' theorem to the approximate solution of the solid mechanics problem. Of course, some of these techniques have already been alluded to in Chapter 2, section 5, where the principle of virtual displacements and the principle of virtual forces are extended to the contragredient principle. The principle of virtual displacements leads naturally to the displacement formulation of the finite element method, but this is by no means the only viable approach. It is the intention to deal with the specific topics as they arise. However, in this chapter individual examples will be taken more or less at random to illustrate the principles involved.

3.1.2 Basic concepts

The displacement formulation depends on the application of the principle of virtual displacements. The contragredient principle eqns (2.104) or (2.105) may be used immediately to discretize the element surface and volume integrals. However, it is instructive to look first at the fundamental question of the application of Gauss' theorem, and also to develop certain element stiffness matrices from first principles in order to illustrate the difficulties which may be encountered in more advanced applications beyond simple linear elements and triangles in plane stress. Firstly, then, the principle of virtual displacements eqn (2.98) is reexamined; that is,

$$\int_V \{\sigma\}^T \{\bar{\varepsilon}\} dV - \int_V \{p_v\}^T \{\bar{u}\} dV - \int_{S_\sigma} \{p_s\}^T \{\bar{u}\} dS = 0 \tag{3.1}$$

Consider now the force residual $\{\delta\}$ obtained when $\{\sigma\}$, an approximate solution, is substituted in the equilibrium eqn (2.84). Then the integral of this residual over the volume is given by

$$\int_V \{\bar{u}\}^T \{\delta\} dV = \int_V \{\bar{u}\}^T ([D]\{\sigma\} + \{p_v\}) dV \tag{3.2}$$

Now the first integral on the right–hand side of eqn (3.2) is transformed via Gauss' theorem that is,

$$\int_V \{\bar{u}\}^T [D]\{\sigma\} dV = -\int_V \{\sigma\}^T [D]^T \{\bar{u}\} dV + \int_S \{\bar{u}\}^T (Dn)\{\sigma\} dS \tag{3.3}$$

Substituting the compatibility conditions in V for $\{\bar{u}\}$ (eqn (2.85)), and the boundary conditions for $\{\sigma\}$ on S, eqn (2.88),

$$[D]^T\{\bar{u}\} = \{\bar{\varepsilon}\}, \quad \text{in } V; \quad \text{and}$$

$$(Dn)\{\sigma\} = \{p_s\}, \quad \text{on } S,$$

the principle of virtual displacements has been obtained. That is, if, now, approximate functions are used for $\{\bar{u}\}$ and the integration carried out over subregions of the domain, the finite element approach can clearly be looked upon as simply an application of weighted residuals, a Galerkin approach. It must be noted that on interelement boundaries, $\{p_s\}$ is replaced by $(Dn)\{\sigma\}$, and thus, for contiguous elements, the element surface integrals combine as

$$\int_S \{\bar{u}\}^T [(Dn)\{\sigma_i\} - (Dn)\{\sigma_j\}] dS$$

Thus, if element displacements are compatible along the element boundaries and equilibrium is satisfied in the average, these integrals vanish, leaving only the integrals on the external surface, which must be set equal to the average applied tractions. The use of the method of weighted residuals technique will be explored in more detail later.

3.1.3   The direct formulation

Historically, the direct formulation was first used to develop the stiffness matrices of many simple elements, including the planar elasticity triangular element shown in Fig. 3.1. The element displacement field is assumed to be linear and is expressed as a function of the (x, y) coordinates; that is,

$$\tilde{u} = [1 \; x \; y] \begin{bmatrix} \alpha_1 \\ \alpha_2 \\ \alpha_3 \end{bmatrix} \quad \text{and} \quad \tilde{v} = [1 \; x \; y] \begin{bmatrix} \alpha_4 \\ \alpha_5 \\ \alpha_6 \end{bmatrix} \tag{3.4}$$

The generalized displacements $\alpha_1 \cdots \alpha_6$ may be expressed in terms of the nodal displacements u_i, v_i, etc. by substituting nodal coordinate values into eqn (3.4a), (b). This equation will be written

$$\{r\} = [B]\{\alpha\} \tag{3.5}$$

Inversion of this equation gives

$$\{\alpha\} = [B]^{-1}\{r\} \tag{3.6}$$

Using the notation in Fig. 3.1 for the triangle geometry, we find

$$[B]^{-1} = \frac{1}{2A} \begin{bmatrix} 2A & 0 & 0 & \cdot & \cdot & \cdot \\ f_j - f_k & f_k & -f_j & \cdot & \cdot & \cdot \\ e_k - e_j & -e_k & e_j & \cdot & \cdot & \cdot \\ \cdot & \cdot & \cdot & 2A & 0 & 0 \\ \cdot & \cdot & \cdot & f_j - f_k & f_k & -f_j \\ \cdot & \cdot & \cdot & e_k - e_j & -e_k & e_j \end{bmatrix} \tag{3.7}$$

Now the strain field in the triangle is given by

$$\{\varepsilon\}^T = \{\varepsilon_x, \varepsilon_y, \gamma_{xy}\}$$

Using eqn (3.4a), (b),

$$\{\varepsilon\} = [A]\{\alpha\} \tag{3.8}$$

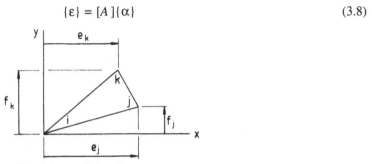

Fig. 3.1   Triangle geometry.

In this case, from eqn (3.4),

$$[A] = \begin{bmatrix} 0 & 1 & 0 & 0 & 0 & 0 \\ 0 & 0 & 0 & 0 & 0 & 1 \\ 0 & 0 & 1 & 0 & 1 & 0 \end{bmatrix} \tag{3.9}$$

Thence,

$$\{\varepsilon\} = [A][B]^{-1}\{r\} = [N]\{r\} \tag{3.10}$$

Combining eqns (3.7) and (3.9), the matrix $[N]$ is obtained

$$[N] = \frac{1}{2A} \begin{bmatrix} f_j - f_k & f_k & -f_j & 0 & 0 & 0 \\ 0 & 0 & 0 & e_k - e_j & -e_k & e_j \\ e_k - e_j & -e_k & e_j & f_j - f_k & f_k & -f_j \end{bmatrix}$$

The constitutive relationship for the material is given by

$$\{\sigma\} = [C]\{\varepsilon\} \tag{3.11}$$

where, for the isotropic material,

$$[C] = E^* \begin{bmatrix} \tau & \nu & 0 \\ \nu & \tau & 0 \\ 0 & 0 & \lambda \end{bmatrix}$$

See eqns (2.135) and (2.136) for details. Thus,

$$\{\sigma\} = [C][A][B]^{-1}\{r\} \tag{3.12}$$

Finally, statically equivalent node forces $\{R\}$ are used to replace the surface tractions arising from $\{\sigma\}$. The $\{\sigma\}$ stress fields and their nodal forces are shown in Fig. 3.2. From Fig. 3.2(b),

$$\{R\} = [D]\{\sigma\} \tag{3.13}$$

where the matrix $[D]$ is defined in eqn (3.14),

$$[D] = \frac{1}{2} \begin{bmatrix} f_j - f_k & 0 & e_k - e_j \\ f_k & 0 & -e_k \\ -f_j & 0 & e_j \\ 0 & e_k - e_j & f_j - f_k \\ 0 & -e_k & f_k \\ 0 & e_j & -f_j \end{bmatrix} \tag{3.14}$$

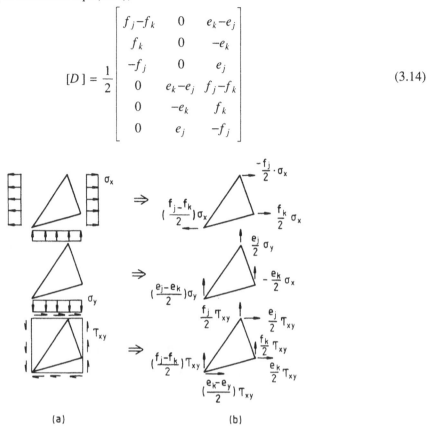

(a)   (b)

Fig. 3.2   Equivalent nodal forces for constant stress.

Thus, finally combining eqns (3.12) and (3.13),

$$\{R\} = [D][C][A][B]^{-1}\{r\} \tag{3.15}$$

The above procedure is cumbersome, to say the least, but nevertheless it retains some practical (as well as historical) interest, in certain special situations where the orthogonal interpolation functions introduced in section 1.4.2 are difficult to obtain. It will be the intention of the theory developed in later sections to avoid the above formulation wherever possible by using the mathematical tools developed in Chapters 1 and 2. If the simple triangular element is examined in the light of section 3.1.2, it will be seen that an assemblage of such elements will be able to approximate the Gauss' divergence integrals because inter-element displacement compatibility is satisfied; that is, the elements are C_0 compatible. This concept of element compatibility is fundamental to the theory of displacement finite elements, being a condition necessary to ensure the convergence of the solution to analytic results. An element with compatibility of the displacement functions and derivatives up to order n will be considered to possess C_n compatibility. Evidently, the higher the order of derivative compatibility required, the more will be the difficulties which arise in the determination of the orthogonal interpolation functions. Hermitian polynomials are a good example of such function construction [8], [11].

3.1.4   The contragredient principle

In this approach the discretization process applied to the principle of virtual displacements in eqns (2.98) and (2.99) is used directly in the integration process to obtain the relationship between nodal displacements and the lumped nodal (generalized) forces. That is, the virtual strain field $\{\bar{\varepsilon}\}$ is taken to be the actual strain field, so that, from eqn (2.100),

$$\{\varepsilon\} = [a]\{r\} \tag{3.16}$$

Then the equivalent nodal forces are given from eqn (2.105) as

$$\{R\} = \int_V [a]^T \{\sigma\} dV \tag{3.17}$$

As yet, no indication of how $[a]$ is formed is given; however, it is assumed that the element displacement field will be described in terms of orthogonal interpolation polynomials. For the simple triangle of Fig 3.1, these are

$$[\tilde{r}] = \left[\phi_1^T \ \phi_1^T \right] \begin{bmatrix} U \\ V \end{bmatrix} \tag{3.18}$$

where $\{\phi_1\}^T = \{\zeta_1 \ \zeta_2 \ \zeta_3\}$ is the linear interpolation function in area coordinates (see sections 1.4.5 and 1.4.6). Then

$$[a] = \frac{1}{2A} \begin{bmatrix} b_1 & b_2 & b_3 & 0 & 0 & 0 \\ 0 & 0 & 0 & a_1 & a_2 & a_3 \\ a_1 & a_2 & a_3 & b_1 & b_2 & b_3 \end{bmatrix} \tag{3.19}$$

Now the stresses $\{\sigma\}$ may arise from the strains $\{\varepsilon\}$, and in addition have an initial prestress value, $\{\sigma_0\}$, so that, from eqn (3.11),

$$\{\sigma\} = [C]\{\varepsilon\} + \{\sigma_0\} \tag{3.20}$$

Thus, finally, substitution in eqn (3.17) gives

$$\{R\} = \int_V [a]^T [C][a] \, dV \{r\} + \int_V [a]^T \{\sigma_0\} dV \tag{3.21}$$

In this case $[a]$ is constant over the triangle, so that, for thickness t,

$$\{R\} = \frac{t}{4A}[a]^T [C][a]\{r\} + At[a]^T \{\sigma_0\} \tag{3.22}$$

The eqn (3.21) forms the fundamental idea from which the major thrust of the book is developed.

3.2 THE NATURAL MODE TECHNIQUE

3.2.1   Theory

This technique may be looked upon as a variation of the displacement method with certain very special modifications. It has been developed largely by the research team at the University of Stutgaart headed by Prof. J.H. Argyris [1]. It has some important applications and is of particular value in non-linear, large displacement analysis. The appeal of the method is in its strong link with physical reasoning and intitution. Here a summary of the method is given, with further details later in the text.

The reasoning starts by considering the deflection of an element to be composed of

1. rigid–body motion;
2. natural modes describing the deformation (distortion) of the element.

For an arbitrary element with q degrees of freedom, and b rigid–body modes,

$$v = q - b \tag{3.23}$$

are the independent natural modes which describe pure deformation. A general state of displacements of an element may then be described in two distinct but nevertheless equivalent manners.

1. A displacement vector containing the individual degrees of freedom at the nodal points with respect to a coordinate system. The minimum condition is that the displacements be uniquely given at each node point.

$$\{\rho\}^T = \{\rho_1 \, \rho_2 \, \rho_3 \, \cdots \, \rho_i \, \cdots \, \rho_q\} \, (1 \times q) \tag{3.24}$$

The typical component ρ_i may be a displacement, a first–order derivative (rotation, strain), or a higher–order derivative, e.g. curvature.

2. In a form which recognizes the division of rigid–body and straining modes.

$$\{\rho'\}^T = \{\rho_0^T \, \rho_N^T\} \, (1 \times q) \tag{3.25}$$

In eqn (3.25), $\{\rho_0\}$ is the $(b \times 1)$ vector of the allowable rigid body modes and $\{\rho_N\}$ is the $(v \times 1)$ vector of the natural modes. Because the two representations are equivalent, it is possible to express the latter components in terms of the former. That is,

$$\{\rho'\} = [a_e]\{\rho\} = \begin{bmatrix} a_0 \\ a_N \end{bmatrix} \{\rho\} \tag{3.26}$$

and hence, of course,

$$\{\rho_0\} = [a_0]\{\rho\} \tag{3.27a}$$

and

$$\{\rho_N\} = [a_N]\{\rho\} \tag{3.27b}$$

Having selected $\{\rho_0\}$, $\{\rho_N\}$, it may be straightforward to set up the submatrices $[a_0]$, $[a_N]$. On the other hand there are many situations in which it is more convenient to find the inverse relationship, which may be written

$$\{\rho\} = [A_0 \, A_N] \begin{bmatrix} \rho_0 \\ \rho_N \end{bmatrix} = [A_e]\{\rho'\} = [a_e]^{-1} \{\rho'\} \tag{3.28}$$

Expanding this equation,

$$\{\rho\} = [A_0]\{\rho_0\} + [A_N]\{\rho_N\} \tag{3.29}$$

Inversion of eqn (3.28) is, as a rule, simple with many zero or integer terms in the matrix. Associated with the alternative definitions of the deformation vectors $\{\rho\}$ and $\{\rho'\}$ will be the generalized force vectors $\{P\}$ and $\{P'\}$. Their selection corresponding to $\{\rho\}$ and $\{\rho'\}$ may be obtained from the contragredient principle (eqns (2.114) and (2.116)).

The two vectors are again given

$$\{P\}^T = \{P_1\, P_2\, P_3\, \cdots\, P_i\, \cdots\, P_q\}\ (1 \times q) \tag{3.30}$$

and

$$\{P'\}^T = \{P_0^T\, P_N^T\}\quad (1 \times (b + v)) \tag{3.31}$$

Applying the contragredient principle, the displacement relationship gives the corresponding force relationship,

$$\{P\} = [a_e]^T\{P'\} = [a_0^T\ a_N^T]\begin{Bmatrix} P_0 \\ P_N \end{Bmatrix} = [a_0]^T\{P_0\} + [a_N]^T\{P_N\} \tag{3.32}$$

Now the forces $\{P_0\}$ correspond to the rigid modes $\{\rho_0\}$ and thus must be zero for equilibrium, i.e. $\{P_0 \equiv 0\}$, so that

$$\{P\} = [a_N]^T\{P_N\} \tag{3.33}$$

3.2.2 Example of natural modes for beam elements

To illustrate the application of the natural mode technique, the simple beam element given in Fig. 3.3 is used. The application to the triangular finite element is left to a more detailed discussion in Chapter 7. In Fig. 3.3(a) the element is referred to a set of local $\bar{x}$, $\bar{y}$ coordinates. The choice of rigid–body modes is shown in Fig. 3.3(c) and is taken with reference to a local global system, $O\ \bar{X}\ \bar{Y}$, set in the original, undeflected position.

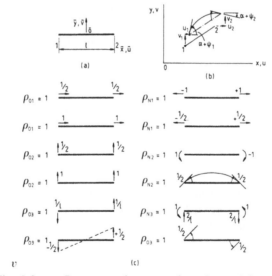

Fig. 3.3   Beam member natural modes and forces.

Then, the rigid–body modes are

$$\{\rho_0\}^T = \{\bar{u}_0\ \bar{v}_0\ \bar{\psi}_0\} \tag{3.34}$$

For the natural modes, the elongation and the symmetric and antisymmetric bending deformations are selected (Figs. 3.3(b) and 3.3(c)).

$$\{\rho_N\}^T = \{\rho_{N1}, \rho_{N2}, \rho_{N3}\} \tag{3.35}$$

From Fig. 3.3(c), the matrices $[\bar{A}_0]$, $[\bar{A}_N]$ can be written down by inspection;

$$[\bar{A}_0] = \begin{bmatrix} 1 & 0 & 0 \\ 0 & 1 & -\dfrac{l}{2} \\ 0 & 0 & 1 \\ 1 & 0 & 0 \\ 0 & 1 & \dfrac{l}{2} \\ 0 & 0 & 1 \end{bmatrix} \quad \text{and} \quad [\bar{A}_N] = \begin{bmatrix} -\dfrac{1}{2} & 0 & 0 \\ 0 & 0 & 0 \\ 0 & \dfrac{1}{2} & \dfrac{1}{2} \\ \dfrac{1}{2} & 0 & 0 \\ 0 & 0 & 0 \\ 0 & -\dfrac{1}{2} & \dfrac{1}{2} \end{bmatrix} \tag{3.36}$$

The order of the global displacements is

$$\{\bar{\rho}\}^T = \{\bar{u}_1\ \bar{v}_1\ \bar{\psi}_1\ \bar{u}_2\ \bar{v}_2\ \bar{\psi}_2\} \tag{3.37}$$

Inverting $[\bar{A}_c]$ gives the matrix $[\bar{a}_e]$,

$$[\bar{a}_e] = \begin{bmatrix} \bar{a}_0 \\ \bar{a}_N \end{bmatrix} = [\bar{A}_c]^{-1} = \begin{bmatrix} \dfrac{1}{2} & 0 & 0 & \dfrac{1}{2} & 0 & 0 \\ 0 & \dfrac{1}{2} & 0 & 0 & \dfrac{1}{2} & 0 \\ 0 & -\dfrac{1}{l} & 0 & 0 & \dfrac{1}{l} & 0 \\ -1 & 0 & 0 & 1 & 0 & 0 \\ 0 & 0 & 1 & 0 & 0 & -1 \\ 0 & \dfrac{2}{l} & 1 & 0 & -\dfrac{2}{l} & 1 \end{bmatrix} \tag{3.38}$$

It is seen in the above that all displacements have been referred to the local coordinate axes and must be obtained by transformation from the global coordinate system. Thus if $\{\rho\}$ is the vector of nodal displacements in the global axes,

$$\{\bar{\rho}\} = [T]\{\rho\} \tag{3.39}$$

where,

$$[T] = \begin{bmatrix} T_0 & 1 & T_0 & 1 \end{bmatrix} \quad \text{with} \quad [T_0] = \begin{bmatrix} \cos\alpha & \sin\alpha \\ -\sin\alpha & \cos\alpha \end{bmatrix} \tag{3.40}$$

Substituting from eqns (3.39) and (3.26),

$$\{\rho'\} = [\bar{a}_e][T]\{\rho\} = [a_e]\{\rho\} \tag{3.41}$$

Multiplying the various matrices in eqn (3.42) gives

$$[a_e] = [\bar{a}_e][T] = \begin{bmatrix} a_0 \\ a_N \end{bmatrix} = \begin{bmatrix} \dfrac{c_1}{2} & \dfrac{s_1}{2} & 0 & \dfrac{c_1}{2} & \dfrac{s_1}{2} & 0 \\[2mm] -\dfrac{s_1}{2} & \dfrac{c_1}{2} & 0 & -\dfrac{s_1}{2} & \dfrac{c_1}{2} & 0 \\[2mm] \dfrac{s_1}{l} & -\dfrac{c_1}{l} & 0 & -\dfrac{s_1}{l} & \dfrac{c_1}{l} & 0 \\[2mm] -c_1 & -s_1 & 0 & c_1 & s_1 & 0 \\[2mm] 0 & 0 & 1 & 0 & 0 & -1 \\[2mm] -2\dfrac{s_1}{l} & 2\dfrac{c_1}{l} & 1 & 2\dfrac{s_1}{l} & -2\dfrac{c_1}{l} & 1 \end{bmatrix} \qquad (3.42)$$

In eqn (3.42), $c_1 = \cos \alpha_1$ and $s_1 = \sin \alpha_1$. Applying the eqn (3.42) for the generalized forces $\{P_0 \, P_N\}$, it is easy to verify the expressions given in Fig. 3.3(c). It is seen now that having defined the natural modes as $\{\rho_N\}$, the integration of the virtual strains (or forces) over the element is reduced in dimension; in the present case, from six to three. This is further reduced here by the use of symmetric and antisymmetric bending modes. If the contragredient principle is now employed, the natural stiffness matrix $[k_N]$ is easily derived. Thus, let the strain, displacement transformation be given by

$$\{\varepsilon\} = [\alpha_N]\{\rho_N\} \qquad (3.43)$$

In this example $\{\varepsilon\}$ is the single term consisting of the axial strain $\{\bar{\varepsilon}_x\}$. The constitutive relationship is given by

$$\{\bar{\sigma}\} = [E]\{\bar{\varepsilon}\} \qquad (3.44)$$

Thus, using eqn (3.17),

$$[k_N] = \int_V [\alpha_N]^T [E][\alpha_N] dV \qquad (3.45)$$

Thence, the natural force displacement relationship is given by

$$\{P_N\} = [k_N]\{\rho_N\} \qquad (3.46)$$

To obtain the complete $(q \times q)$ elastic stiffness matrix $[k_E]$, which is associated with the load displacement expression,

$$\{P\} = [k_E]\{\rho\} \qquad (3.47)$$

It is seen that

$$[K_E] = [a_N]^T [k_N][a_N] \qquad (3.48)$$

For the beam in Fig. 3.3(a), the transverse deformation is given as the cubic equation,

$$v = l[\zeta_1^2(1-\zeta_1) \quad -\zeta_1(1-\zeta_1)^2]\begin{bmatrix} \phi_i \\ \phi_j \end{bmatrix} \qquad (3.49)$$

Double differentiation with respect to $\bar{x}$ gives

$$v_{,\bar{x}\bar{x}} = -\frac{2}{l}[-1+3\zeta_1 \quad -2+3\zeta_1]\begin{bmatrix}\phi_i \\ \phi_j\end{bmatrix} \tag{3.50}$$

Now, from Fig. 3.3(c),

$$\begin{Bmatrix}\phi_i \\ \phi_j\end{Bmatrix} = \frac{1}{2}\begin{bmatrix}1 & 1 \\ -1 & 1\end{bmatrix}\begin{Bmatrix}\rho_{N2} \\ \rho_{N3}\end{Bmatrix} \tag{3.51}$$

so that, finally,

$$v_{,\bar{x}\bar{x}} = \frac{1}{l}[-1 \quad 3\xi\,]\begin{Bmatrix}\rho_{N2} \\ \rho_{N3}\end{Bmatrix}; \quad \text{where} \quad \xi = 1 - 2\zeta_1 \tag{3.52}$$

Thence, using the assumption that plane sections remain plane, the longitudinal strain is given by

$$\varepsilon_{\bar{x}} = \frac{1}{l}[1 \quad y \quad -3y\,\xi]\begin{bmatrix}\rho_{N1} \\ \rho_{N2} \\ \rho_{N3}\end{bmatrix} \tag{3.53}$$

Finally, the expression on the right–hand side of eqn (3.46) is given

$$[\alpha_N]^T[E][\alpha_N] = \frac{E}{l^2}\begin{bmatrix}1 & y & -3y\,\xi \\ y & y^2 & -3y^2\xi \\ -3y\,\xi & -3y^2\xi & 9y^2\xi^2\end{bmatrix} \tag{3.54}$$

Noting that for axes through the centroid, $\int y\,dA = 0$ and $\int y^2 dA = I$, eqn (3.46) becomes

$$\int[\alpha_N]^T[E][\alpha_N]dV = \frac{1}{l}\lceil EA \quad EI \quad 3EI\rceil \tag{3.55}$$

The simplicity of this expression must be clearly noted. The natural mode technique will be examined in more detail for triangular elements in Chapter 7.

3.3 THE METHOD OF WEIGHTED RESIDUALS

As alluded to in section 3.1.2, the application of the principle of virtual displacements is, in fact, a method of weighted residuals if the trial functions are themselves the weighting functions. The formulation of the weighted residual technique is particularly useful in those applications where the force–displacement concept of work is not available. Thus, the method of weighted residuals presumes that a 'trial' function, e.g. a polynomial series which has been assumed to approximate the independent variable in a problem of mathematical physics, does not in general satisfy the relevant differential equation. However, substitution into the differential equation will result in a residual e. To obtain the best fit solution, the integral of the residual throughout the domain under consideration is minimized. That is,

$$\int_V e\, dV = \text{minimum} \tag{3.56}$$

In the Galerkin method, it is assumed that this error is orthogonal to certain functions ϕ. Orthogonality implies that

$$\int_V \phi_i e\, dV = 0 \tag{3.57}$$

In structural applications, e can be looked upon as being a force residual. Then the integral in eqn (3.57) represents the work done if ϕ_i is a generalized displacement. That is, although equilibrium may not be satisfied point by point throughout the continuum it is satisfied in the generalized modes. Hopefully, the more values taken for the shape functions ϕ_i, the better will be the point–by–point, convergence to equilibrium. Thus, let the governing differential equation be represented by

$$D(\tilde{r}) = 0 \tag{3.58}$$

Then the approximate function $\tilde{\tilde{r}}$ does not satisfy this differential equation, so that the criterion eqn (3.57) gives

$$\int_V \phi_i D(\tilde{\tilde{r}}) \, dV = 0 \tag{3.59}$$

The finite element approximation is now made, that $\tilde{\tilde{r}}$ can be expressed in terms of nodal values by the interpolation functions $\{N\}$. Thence,

$$\{\tilde{\tilde{r}}\} = [N]^T \{\bar{r}\} \tag{3.60}$$

Substitution in eqn (3.59) gives

$$\int_V N_i D([N]^T \{\bar{r}\}) dV = 0 \tag{3.61}$$

When this approximation is integrated over the element region and the appropriate boundary conditions substituted, a finite element formulation is obtained. As an example, consider Poisson's equation valid over a given domain,

$$\nabla^2 \phi = C \tag{3.62}$$

Choose the approximation $\bar{\phi}$ related to nodal values Φ,

$$\{\bar{\phi}\} = [N]^T \{\Phi\} \tag{3.63}$$

Then, from eqns (3.61) and (3.63),

$$\int_{area} (\nabla^2 \bar{\phi} - C) N_i \, dA = 0 \tag{3.64}$$

It is now possible to apply Gauss' theorem to this expression, see eqn (1.85). Thus,

$$\int_{area} \nabla^2 \bar{\phi} N_i \, dA = \int_C N_i (l_j \bar{\phi}_{,j}) \, dS - \int_{area} \bar{\phi}_{,j} N_{i,j} \, dA \tag{3.65}$$

In this equation substitute

$$\bar{\phi}_{,j} = [N]_{,j}^T \{\Phi\}$$

Then the second term on the right–hand side of eqn (3.65) is written

$$\int_{area} \bar{\phi}_{,j} N_{i,j} \, dA = \int_{area} N_{i,j} [N]_{,j}^T \, dA \, \{\Phi\}$$

For all N_i values, this expression becomes

$$\int_{area} \bar{\phi}_{,j} [N]_{,j} \, dA = \int_{area} [N]_{,j} [N]_{,j}^T \, dA \, \{\Phi\} \tag{3.66}$$

Similarly, for the contour integral,

$$\int_C N_i (l_i \bar{\phi}_{,j}) \, dS \to \int_C [N] l_j [N]_{,j}^T \, dS \, \{\Phi\}$$

Thus, eqn (3.64), written for all the shape functions N_i, becomes

$$[k_n]\{\Phi\} = \{F_n\} + \{F_c\} \tag{3.67}$$

Making the substitutions above, it is seen that

$$[k_n] = \int_{area} [N]_{,j} \, [N]_{,j}^T dA \tag{3.68}$$

$$\{F_n\} = \int_C [N] l_j [N]_{,j}^T dS \, \{\Phi\} \quad \text{and} \quad \{F_c\} = - \int_{area} [N] C dA \tag{3.69}$$

This formulation is precisely that used in the application to the heat–transfer problem given in Chapter 7.

3.4 HYBRID ELEMENT FORMULATION

3.4.1 Theory

It has been shown in section 3.1 that the compatible displacement models, while maintaining displacement compatibility across element boundaries, satisfy equilibrium only in the mean at discrete points. It should be evident that this is, by no means, the only approximation possible between elements. Since the displacement elements are in general too stiff, any relaxation in the inter–element connectivity conditions has at least the possibility of producing more accurate results for a given mesh subdivision. For example, in the hybrid element approach, the stress fields are assumed within the element, so that the differential equations of equilibrium are satisfied exactly. The resulting element deformations will no longer be simple functions along the boundaries. The choice is now made to define displacement patterns and corresponding generalized forces along the element boundaries. In this way compatibility is satisfied in the mean. For regular shapes, e.g. rectangles, these hybrid elements are easy to formulate and produce excellent results. They have also been used extensively in shell analysis (e.g. the semi–Loof element) with again high quality performance. However these particular semi–Loof elements are not easy to formulate, and practically the same result can be obtained by using a displacement model (discrete Kirchoff element with Loof nodes) which is detailed in Chapter 8. The theory for the hybrid models starts with the interpolation within the element,

$$\{\sigma\} = [P]\{\beta\} \tag{3.70}$$

The stresses $\{\sigma\}$ are chosen to satisfy the differential equations of equilibrium. The terms of $\{\beta\}$ are generalized forces. The strain field in the element is given from the constitutive relationship as

$$\{\varepsilon\} = [f][P]\{\beta\} \tag{3.71}$$

Applying the contragredient principle (principle of virtual forces in this case) to this strain field allows for the calculation of the generalized displacements $\{r_\beta\}$ corresponding to the forces $\{\beta\}$. That is,

$$\{r_\beta\} = \int_V [P]^T \{\varepsilon\} dV = \int_V [P]^T [f][P] dV \, \{\beta\} \tag{3.72}$$

This is the flexibility relationship,

$$\{r_\beta\} = [F]\{\beta\} \tag{3.73}$$

This equation is now inverted to give a stiffness relationship, and the method reverts to the direct stiffness method of analysis. Inverting eqn (3.73),

$$\{\beta\} = [F]^{-1}\{r_\beta\} = [k_\beta]\{r_\beta\} \tag{3.74}$$

The problem is now how to match tractions and displacements along element interfaces. Firstly, the surface tractions $\{G_s\}$ must be expressed in terms of the $\{\beta\}$ values by substitution of the boundary coordinates in eqn (3.70). That is,

$$\{G_s\} = [Q]\{\beta\} \tag{3.75}$$

The surface displacements are now interpolated in terms of the nodal parameters by an independent approximation, expressed as

$$\{\tilde{u}\} = [L]\{r\} \tag{3.76}$$

The nodal forces corresponding to $\{G_s\}$ are now given via the contragredient principle as

$$\{R\} = \int_S [L]^T [G_s]\, dS = \int_S [L]^T [Q]\, dS\ \{\beta\} \tag{3.77}$$

Write eqn (3.77) as

$$\{R\} = [T]^T \{\beta\} \tag{3.78}$$

in which

$$[T] = \int_S [Q]^T [L]\, dS \tag{3.79}$$

Again applying the contragredient principle to eqn (3.78), the generalized displacements corresponding to $\{R\}$ are given by

$$\{r_\beta\} = [T]\{r\} \tag{3.80}$$

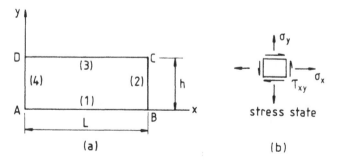

Fig. 3.4   Rectangular hybrid stress element.

Hence,

$$\{R\} = [T]^T[k_\beta][T]\{r\} \tag{3.81}$$

The requisite force displacement relationship has been obtained

$$\{R\} = [K]\{r\} \tag{3.82}$$

3.4.2 Example of the hybrid stress element formulation

The simple rectangular element dimensions (L, h) in plane stress are shown in Fig. 3.4. The $x,-y-$ coordinate axes are chosen to coincide with the sides of the rectangle, as shown in the figure: The stress field is chosen to satisfy the equations of equilibrium, eqn (2.84), giving eqn (3.70),

$$\begin{Bmatrix} \sigma_x \\ \sigma_y \\ \tau_{xy} \end{Bmatrix} = \begin{bmatrix} 1 & y & 0 & 0 & 0 & x & 0 & y^2 & 0 & x^2 \\ 0 & 0 & 1 & x & 0 & 0 & y & 0 & x^2 & y^2 \\ 0 & 0 & 0 & 0 & 1 & -y & -x & 0 & 0 & 2xy \end{bmatrix} \begin{Bmatrix} \beta_1 \\ \beta_2 \\ \cdot \\ \cdot \\ \cdot \\ \beta_{10} \end{Bmatrix} \tag{3.83}$$

It is easily shown that, with the material body force $\{p_v = 0\}$, the stress field satisfies the equation of equilibrium, eqn (2.84). If only linear terms are retained, then only the (3×5) submatrix of the coefficient matrix in eqn (3.83) is used, and there are five β terms. In this case, the flexibility matrix $[F]$ is given by

$$[F] = \int_0^{h} \int_0^{L} \begin{bmatrix} 1 & 0 & 0 \\ y & 0 & 0 \\ 0 & 1 & 0 \\ 0 & x & 0 \\ 0 & 0 & 1 \end{bmatrix} [f] \begin{bmatrix} 1 & y & 0 & 0 & 0 \\ 0 & 0 & 1 & x & 0 \\ 0 & 0 & 0 & 0 & 1 \end{bmatrix} dx\, dy \tag{3.84}$$

This integration is easily performed and the resulting (5×5) matrix inverted, giving $[k_\beta]$. Next the $[Q]$ matrix in eqn (3.75) is constructed. The surface tractions are obtained from Fig. 3.4(a), using the relevant faces of the infinitesimal cube in Fig. 3.4(b). The subscripts (1) to (4) indicate the four sides:

$$\begin{bmatrix} t_{x1} \\ t_{y1} \\ t_{x2} \\ t_{y2} \\ t_{x3} \\ t_{y3} \\ t_{x4} \\ t_{y4} \end{bmatrix} = \begin{bmatrix} -\tau_{xy1} \\ -\sigma_{y1} \\ \sigma_{x2} \\ \tau_{xy2} \\ \tau_{xy3} \\ \sigma_{y3} \\ -\sigma_{x4} \\ -\tau_{xy4} \end{bmatrix} = \begin{bmatrix} 0 & 0 & 0 & 0 & -1 \\ 0 & 0 & -1 & -x & 0 \\ 1 & y & 0 & 0 & 0 \\ 0 & 0 & 0 & 0 & 1 \\ 0 & 0 & 0 & 0 & 1 \\ 0 & 0 & 1 & x & 0 \\ -1 & -y & 0 & 0 & 0 \\ 0 & 0 & 0 & 0 & -1 \end{bmatrix} \{\beta\} \tag{3.85}$$

It remains now to define the nodal displacement parameters, generate the matrix $[L]$, of eqn (3.76), and perform the surface integration in eqn (3.77). In this simple model it is evidently satisfactory to use only the apex nodes A, B, C and D allowing a linear displacement variation along the sides (1) to (4). For example, let $(x/L) = \xi$ then along side (1),

$$
\begin{bmatrix} \bar{u} \\ \bar{v} \end{bmatrix} = \begin{bmatrix} 1-\xi & \xi & 0 & 0 \\ 0 & 0 & 1-\xi & \xi \end{bmatrix} \begin{bmatrix} u_A \\ u_B \\ v_A \\ v_B \end{bmatrix}
\tag{3.86}
$$

Similarly, with $\eta = (y/h)$, along side (2),

$$
\begin{bmatrix} \bar{u} \\ \bar{v} \end{bmatrix} = \begin{bmatrix} 1-\eta & \eta & 0 & 0 \\ 0 & 0 & 1-\eta & \eta \end{bmatrix} \begin{bmatrix} u_B \\ u_C \\ v_B \\ v_C \end{bmatrix}
\tag{3.87}
$$

With similar expressions for sides (3) and (4), the $[L]$ matrix is easily constructed. Integrating the product $[L]^T [Q]$, around the element perimeter $ABCDA$, gives the following value for $[T]$:

$$
[T] = \begin{bmatrix}
-\dfrac{h}{2} & \dfrac{h}{2} & \dfrac{h}{2} & -\dfrac{h}{2} & 0 & 0 & 0 & 0 \\
-\dfrac{h^2}{6} & \dfrac{h^2}{6} & \dfrac{h^2}{3} & -\dfrac{h^2}{3} & 0 & 0 & 0 & 0 \\
0 & 0 & 0 & 0 & -\dfrac{l}{2} & -\dfrac{l}{2} & \dfrac{l}{2} & \dfrac{l}{2} \\
0 & 0 & 0 & 0 & -\dfrac{l^2}{6} & -\dfrac{l^2}{3} & \dfrac{l^2}{3} & \dfrac{l^2}{6} \\
-\dfrac{l}{2} & -\dfrac{l}{2} & \dfrac{l}{2} & \dfrac{l}{2} & -\dfrac{h}{2} & \dfrac{h}{2} & \dfrac{h}{2} & -\dfrac{h}{2}
\end{bmatrix}
\tag{3.88}
$$

The calculation of $[k_\beta]$ and the pre– and post–multiplication with $[T]^T$ and $[T]$ in eqn (3.81) have been carried out, explicitly for this simple case, by Prezemieniecki [12]. Some of the computational aspects of the hybrid element can be observed from this simple example. Firstly, only in very simple cases can the flexibility matrix be inverted explicitly. As the order of this matrix increases in size (the number of β terms), the cost of forming the element stiffness matrix increases accordingly. The stress fields chosen in eqn (3.83) appear to be element oriented, and there is no apparent guarantee of the element efficiency if the shape is other than rectangular. Also, it is not obvious how to extend the element edge displacement to higher order functions (e.g. quadratic variation).

3.5 EQUILIBRIUM ELEMENTS

This class of elements extends the concept of element flexibility, solving for unknown stress components and generalized edge displacements, simultaneously. Thus, suppose that the element stresses are interpolated

$$
\{\bar{\sigma}\} = \{\xi\}^T \{\beta\}
\tag{3.89}
$$

and also, the element displacements in terms of nodal parameters are expressed

$$
\{\bar{u}\} = \{N\}^T \{r\}
\tag{3.90}
$$

From eqn (3.90), differentiation gives the expression for the relevent strain parameters:

$$\{\bar{\varepsilon}\} = [a]^T \{r\}$$

(3.91)

Thus, for this strain field the generalized displacements are given (using the contragredient principle)

$$\{r_\beta\} = \int_V [\xi][a]^T \, dV \{r\} = [F_{12}]\{r\}$$

(3.92)

However, given the constitutive relationship,

$$\{\bar{\varepsilon}\} = [f]\{\bar{\sigma}\}$$

(3.93)

the expression for $\{r_\beta\}$ is given by

$$\{r_\beta\} = \int_V [\xi][f][\xi]^T \, dV \{\beta\} = -[F_{11}]\{\beta\}$$

(3.94)

where

$$[F_{11}] = -\int_V [\xi][f][\xi]^T \, dV$$

(3.95)

Assuming that the values of $\{r_\beta\}$ are equal gives

$$[F_{11}]\{\beta\} + [F_{12}]\{r\} = 0$$

(3.96)

Furthermore, the generalized forces $\{R\}$ for the β stress fields, corresponding to the nodal displacements $\{r\}$, are given by

$$\{R\} = \int_V [a][\xi]^T \, dV \{\beta\} = [F_{21}]\{\beta\}$$

(3.97)

Combining eqns (3.96) and (3.97) for all elements gives the system of equations,

$$\begin{bmatrix} F_{11} & F_{12} \\ F_{21} & 0 \end{bmatrix} \begin{Bmatrix} \beta \\ r \end{Bmatrix} = \begin{Bmatrix} 0 \\ R \end{Bmatrix}$$

(3.98)

These equations are solved simultaneously for $\{\beta\}$ and $\{r\}$. If the variables are intermingled rather than partitioned as shown in eqn (3.98), care must be taken in the sequence in which these equations are solved. An example of the use of the equilibrium element is delayed until Chapter 8 where the Anderhaagen plate bending equilibrium element is studied in some detail.

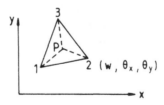

Fig. 3.5   Triangular plate element–nodal variables.

3.6 CONVERGENCE OF FINITE ELEMENT SOLUTIONS

3.6.1   Basic requirements

It is probably true to say that at the present time no completely general means is yet available by which a finite element formulation can be deemed satisfactory in terms of accuracy of performance for all situations likely to be encountered. For example, an element may give satisfactory prediction of a function (e.g. displacement), but be quite inaccurate for its first or second derivatives (strain or curvature).

A classic example of this is the Bazeley–Irons incompatible plate bending element which exhibits a large sawtooth effect in second derivatives (and hence bending moments). Alternatively, the element may appear to be excellent when used on regular meshes but fail hopelessly when the mesh is irregular. Some of the stress elements with incompatible modes are known to behave in this way, as do some of the triangular hybrid plate bending elements. Such elements must be treated with the greatest caution. It should be evident too that the shape and aspect ratio (length/breadth) of elements will have a significant bearing on their performance. In general, the less needle–like triangles are made and the more regular quadrilaterals are made, the better their performance. The reader should take an available finite element package and try various aspect ratios for triangles and isoparametric elements, extending the ratios of dimensions until results become unstable or singularity occurs in the solution. There are three simple criteria which can be applied before any comparative testing of an element formulation is undertaken. These are:

1. *The inclusion of constant strain terms*

These terms must be included, because without them it becomes impossible to model the constant stress (strain) field. The classic example of a failure to satisfy this criteria is in the formulation of a 9–node plate bending element using a quadratic displacement function in (x, y) coordinates, with (w, θ_x, θ_y) at the apices as the nodal values, as shown in Fig. 3.5. The complete cubic polynomial in two–dimensional space has ten terms

$$w = a_0 + a_1 x + a_2 y + a_3 x^2 + a_4 xy + a_6 y^2 + a_7 x^3 + a_8 x^2 y + a_9 xy^2 + a_{10} y^3$$

An early formulation of this element simply deleted the term $a_4 xy$, with disastrous consequences because the element was unable to model the constant twist stress state. When ζ coordinates are used (Chapter 1), it is seen that the cubic function cannot be interpolated correctly in this way because of the bubble function $\zeta_1 \zeta_2 \zeta_3$ which has zero boundary values and requires either an internal node or a side node derivative for its specification. All displacement models using complete polynomials, elements using the natural mode technique and hybrid models should easily accommodate this requirement.

2. *Rigid–body forces*

The second requirement is that the generalized forces corresponding to rigid– body motion of the element must be zero. For elasticity elements the requirement is simply that the sums of rows (columns) of the element stiffness matrix should be zero. Unless this requirement is met, rigid–body motion will obviously induce stress in the element. It should be easily satisfied, except that when elements whose nodal degrees of freedom involve first and second derivatives are used, the contributions to rigid–body motion (translation and rotation) may not be so easily identified.

3. *The patch test*

The element should pass the patch test. The patch test provides a means of testing element shape efficiency. It reasons that if an infinitely fine subdivision could be made, then the four elements shown in Fig. 3.6 can be considered to be in a state of uniform stress. Since the subdivision is infinitely small, it must reproduce this exactly. If the internal node position is altered, then the stress field should still be reproduced exactly. The two subdivisions in Fig. 3.6 should give σ_x in all four elements. Failure to do so indicates that results will be unreliable when irregularly shaped elements are used.

3.6.2 Testing of finite element solutions

In the previous section, conditions which help in the evaluation of an element's behaviour have been discussed. It remains now to examine practical implementation. Firstly, it must be remembered that the finite element method, in the final analysis, gives only an approximate solution. The resulting stress fields obtained are not exact and must be interpreted to give an approximate solution to the original continuum mechanics problem. The solution should satisfy in the mean the differential equations of equilibrium and compatibility.

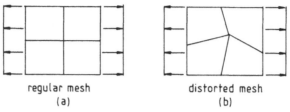

regular mesh   distorted mesh
(a)   (b)

Fig. 3.6   Patch test–four elements, regular and distorted.

Often, derivative smoothing techniques must be used to obtain satisfactory stress fields. For example, in isoparametric elements, stress fields should be sampled at a lower order of point density than that used for integration of the stiffness matrices. Other techniques of smoothing using weighted residuals may be employed. Practically, then, the finite element method can be shown to give satisfactory approximations to known analytic solutions which converge with the increase in the number of unknown degrees of freedom (i.e. mesh fineness) to these solutions. There has been a significant effort to develop more refined elements (and the process still continues), always with the intention of improving the approximation for a given degree of mesh fineness. The check should be for both displacements and stresses (derivatives), and in the latter case raw element data should be plotted to show within–element variation (the sawtooth effect). Convergence may be monotonic from above or below or oscillate as shown in Fig. 3.7. Various element groups usually fall in one or other of the catagories. The convergence characteristics shown in Fig. 3.7 are generally followed:
(A) by displacement compatible models,   (B) by equilibrium (or flexibility models),   (C) by hybrid element models. It will be expected from the nature of the finite element method that displacements will perform better than their derivatives, so that, as already mentioned, the convergence plots on stresses should be as rigorous as for displacements.

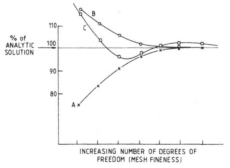

Fig. 3.7   Finite element convergence characteristics.

CHAPTER 4

The force method–line element structures

4.1 INTRODUCTION

4.1.1 Structural idealization

Thus far it has been the intention to derive the fundamental relationships necessary for the analysis of continua via the finite element discretization procedure. For the continuum it will be shown that it is not necessary to invoke the analogy between element discretization and a physical element of the same shape, and that the finite element method may be looked upon as simply a piecewise application of Gauss' divergence theorem to the region. However, in the case of line element structures such as frames and trusses the discretization process does coincide with the physical dimensions of the structure. The choice to begin the discussions with line element structures is thus made, not only because in the first instance of linear analysis the transformations are deceptively simple, leading to an easy understanding of the underlying principles, but, in addition, the theory presented herein embodies all the basic principles of matrix structural analysis. In this category of structures are found trusses, beams, grids, frames and cable nets. This last group needs special attention because of the necessity to undertake non–linear analysis of the node force–deflection relationships. The structures are interesting too, because low connectivity between nodes can lead to statical determinacy, for which member forces may be obtained without a knowledge of the deformation characteristics of the members. It is found, however, that even this simple class of structure is highly amenable to computer structural analysis. This inevitably leads to the possibilities of the introduction of computer methods as the very basis of structural mechanics. In this text, the emphasis will be on the stiffness analysis of structures based on the kinematic relationships between node displacements and member distortions. The reason for the choice of kinematic analysis in preference to static analysis is that the kinematic transformations of the form in eqn (2.122) are more readily determined than the force transformations in eqn (2.120). However, in this chapter a digression is made from the main theme of the book to examine statically determinate as well as statically indeterminate structures because of the usefulness of this class of structure in the study of elementary structural mechanics. It will be found that once the decision has been made to rely on the computer for the solution of the equilibrium equations it is expeditious to work always in terms of joints and joint equilibrium. This is in distinct contrast to hand methods of analysis, for example, in the analysis of determinate trusses where the equilibrium may be applied to either joints or sections depending on which is most advantageous for the analyst. Such arbitrary decisions are, of course, exceedingly difficult to incorporate into a general computer program.

4.1.2 Definition of structure

A structure is considered to be composed of rigid nodes connected by flexible members and supported from a rigid foundation so that the whole system is statically and geometrically stable. For a general discussion of the conditions of determinacy, indeterminacy and geometric instability, see [9].

The definition implies that the model to be analysed is only a mathematical one, and, in fact, as mentioned earlier in Chapter 3, for continuum problems, the reliance on the association with a physical interpretation is found to be unnecessary. In the development of the present chapter the mathematical model is easily identified with the line element (truss or frame member) joining the nodes. The important property of the element is that its force–displacement relationships are known. That is, if the stress field for the element can be expressed in terms of several generalized forces, then the force–displacement relationships at the points of application of the forces are calculable. Alternatively, the displacement fields in the element may be chosen. Then the member forces are calculable if the member displacements are known.

Example 4.1 *Force and displacement fields for linear elements*

The simplest example is the truss member shown in Fig. 4.1(a). For the axial force field,

$$F = \text{constant} \tag{a}$$

The corresponding displacement field is

$$u_1 = [\zeta_1 \ \zeta_2] \begin{bmatrix} u_i \\ u_j \end{bmatrix} \tag{b}$$

Then, calculating F from the strain field and Hooke's law,

$$F = EA\frac{\partial u_1}{\partial x_1} = \frac{EA}{l}\left[-\frac{\partial u_1}{\partial \zeta_1} + \frac{\partial u_1}{\partial \zeta_2}\right] = \frac{EA}{l}[-1 \ 1]\begin{Bmatrix} u_i \\ u_j \end{Bmatrix} \tag{c}$$

However, the extension of the member δ, is given by

$$\delta = (u_j - u_i) \tag{d}$$

Thence it follows that the definition in (a) for the displacement field leads to an identical force–field definition for linear elastic behaviour.

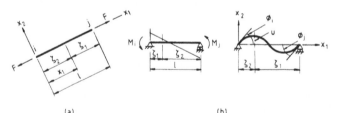

(a)   (b)

Fig. 4.1   Truss and beam members–axial force, moment and displacement fields.

Example 4.2 *Beam element with linear moment field*

For the beam element in Fig. 4.1(b), the moment and displacement field definitions are

Moment

$$M = [-\zeta_1 \ \zeta_2]\begin{Bmatrix} M_i \\ M_j \end{Bmatrix} \tag{e}$$

Note: The sign convention for M is tension on the bottom fibre positive.

Displacement

From Fig. 1.9, example 1.7, the cubic displacement interpolation is given by

$$u_2 = l \, [\zeta_1^2 \zeta_2 \quad -\zeta_1 \zeta_2^2] \begin{Bmatrix} \phi_i \\ \phi_j \end{Bmatrix} \tag{f}$$

If now the relationship between moment and deflection is given,

$$M = El u_{2,,} = \frac{EI}{l^2} \left\{ \frac{\partial^2 u_2}{\partial \zeta_1^2} - \frac{2\partial^2 u_2}{\partial \zeta_1 \partial \zeta_2} + \frac{\partial^2 u_2}{\partial \zeta_2^2} \right\} \tag{g}$$

Substitution in eqn (f) gives

$$u_{2,,} = \frac{2}{l} [(\zeta_2 - 2\zeta_1) \, (2\zeta_2 - \zeta_1)] \begin{Bmatrix} \phi_i \\ \phi_j \end{Bmatrix} \tag{h}$$

The values of M_i, M_j are given at ($\zeta_2 = 0$, $\zeta_1 = 1$) and ($\zeta_2 = 1$, $\zeta_1 = 0$).

$$M_i = -M = \frac{2EI}{l} [2 \quad 1] \begin{Bmatrix} \phi_i \\ \phi_j \end{Bmatrix}$$

$$M_j = M = \frac{2EI}{l} [1 \quad 2] \begin{Bmatrix} \phi_i \\ \phi_j \end{Bmatrix}$$

Combining these two equations and inverting gives

$$\begin{Bmatrix} \phi_i \\ \phi_j \end{Bmatrix} = \frac{l}{6EI} \begin{bmatrix} 2 & -1 \\ -1 & 2 \end{bmatrix} \begin{Bmatrix} M_i \\ M_j \end{Bmatrix} \tag{i}$$

Substitution of eqns (i) and (h) in (g) gives

$$M = \frac{1}{3} [(\zeta_2 - 2\zeta_1) \, (2\zeta_2 - \zeta_1)] \begin{bmatrix} 2 & -1 \\ -1 & 2 \end{bmatrix} \begin{Bmatrix} M_i \\ M_j \end{Bmatrix} = [-\zeta_1 \quad \zeta_2] \begin{Bmatrix} M_i \\ M_j \end{Bmatrix} \tag{j}$$

Thus, the cubic displacement field of eqn (f) produces the linear moment field (e) and vice–versa. This is true only for a prismatic member. If the member is tapered so that the flexural stiffness EI varies along the member, the results will be different, with the cubic assumption being in error. The nodes of the idealized structure may be of infinitely small dimensions, in which case they are used in the mathematical sense of a point to write the equations of force equilibrium (or of displacement compatibility) at that point. However, such a definition is an unnecessary restriction, and finite dimensions may be used; for example, in the analysis of a very stiff headstock supported by slender, flexible piles; (Fig. 4.2). The headstock may be considered to be a single rigid node with six equations of equilibrium (or six displacement degrees of freedom).

4.2 LINE ELEMENTS – MEMBER FORCES

The basis of the force method of analysis is the statically determinate structure in which the member forces (axial force, bending moments, shears and torque) can be calculated from the equations of statics alone. As mentioned in section 2.6, its basis is the establishment of eqn (2.120), written here as eqn (4.1), relating member forces $\{S\}$ to the nodal loads $\{R\}$.

In the form in which eqn (4.1) is used in computer structural analysis, the forces $\{R\}$ are usually stress resultants at the nodes. However, in applications in elementary mechanics, eqn (2.110) is extremely useful when the $[b]$ matrix may be a function of the variable along the length of the member; that is, it expresses the moment transformation for distributed loads.

$$\{S\} = [b]\{R\} \tag{4.1}$$

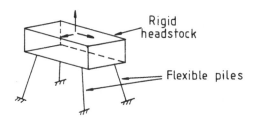

Fig. 4.2   Rigid headstock on flexible piles.

The question now arises as to the means by which the $[b]$ matrix may be calculated. Two methods come to mind (apart from non–automated versions). The first is the application of simple statics to the joint forces, and the second is the application of eqn (2.123), here eqn (4.2), to set up the equations of joint equilibrium.

$$\{R\} = [a]^T\{S\} \tag{4.2}$$

In this introductory treatment of the force method of analysis, only truss and plane frame structures are considered, although the extension to other forms is then merely a matter of detail. These member types, together with their fundamental forces, are shown in Fig. 4.3. Notice that whereas the truss member has the single force $\{F\}$, the beam member in two dimensions has three basic forces, $<M_i \; M_j \; F>$. For the member in Fig. 4.3(a), the transformation from $\{F\}$ to global components at the ends i–j of the member is

$$\begin{Bmatrix} F_{ix} \\ F_{iy} \\ F_{jx} \\ F_{jy} \end{Bmatrix} = \begin{bmatrix} \cos\alpha & -\sin\alpha & 0 & 0 \\ \sin\alpha & \cos\alpha & 0 & 0 \\ 0 & 0 & \cos\alpha & -\sin\alpha \\ 0 & 0 & \sin\alpha & \cos\alpha \end{bmatrix} \begin{bmatrix} -1 \\ 0 \\ 1 \\ 0 \end{bmatrix} F \tag{4.3}$$

Thus eqn (4.3) is written

$$\{R\} = [L_D]^T [T]^T \{S\} \tag{4.4}$$

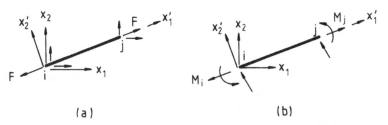

(a)   (b)

Fig. 4.3   Member forces–local and global axes components.

This transformation has been made, firstly, from $\{S\}$ to components $\{F'\}$ in the local coordinate axes and, thence, from $\{F'\}$ to $\{F\}$, the global components. It will be noticed that the premultiplication of $\{F'\}$ by $[L_D]^T$ is a simple application of the transformation of vector components and has been given before in eqns (1.39) and (1.54). The transposed notation $[L_D]^T$ is used because $[L_D]$ has rows which give direction cosines of the local coordinates of the member. When the frame element shown in Fig. 4.3(b) has its basic force components transformed to global components, the form of the transformation is identical with eqn (4.4). The dimensions of the matrices are different, however, reflecting the additional basic member forces and global components. In full, the transformation is given in eqn (4.5).

$$
\begin{bmatrix} F_{ix} \\ F_{iy} \\ M_i \\ F_{jx} \\ F_{jy} \\ M_j \end{bmatrix} = \begin{bmatrix} \cos\alpha & -\sin\alpha & 0 & 0 & 0 & 0 \\ \sin\alpha & \cos\alpha & 0 & 0 & 0 & 0 \\ 0 & 0 & 1 & 0 & 0 & 0 \\ 0 & 0 & 0 & \cos\alpha & -\sin\alpha & 0 \\ 0 & 0 & 0 & \sin\alpha & \cos\alpha & 0 \\ 0 & 0 & 0 & 0 & 0 & 1 \end{bmatrix} \begin{bmatrix} 0 & 0 & -1 \\ \dfrac{1}{l} & \dfrac{1}{l} & 0 \\ 1 & 0 & 0 \\ 0 & 0 & 1 \\ -\dfrac{1}{l} & -\dfrac{1}{l} & 0 \\ 0 & 1 & 0 \end{bmatrix} \begin{bmatrix} M_i \\ M_j \\ F \end{bmatrix} \qquad (4.5)
$$

When the multiplications in eqn (4.4) are carried out, the partitioned equation is written

$$
\begin{bmatrix} R_{iN} \\ R_{jN} \end{bmatrix} = \begin{bmatrix} A_{iN} \\ A_{jN} \end{bmatrix} \{S_N\} \qquad (4.6)
$$

The partitioning for member N is according to the force quantities at its ends i–j. The indices i, j may be merely symbolic. In the computer program, however, they take the actual node numbers at the ends of the members in the sequence, 1 to NJNTS (the number of joints of the structure). These indices are then used to add the contribution of the forces of member N to the total set of joint equilibrium equations for the whole structure. For the solution of these equilibrium equations by Gauss' elimination (or in small problems by matrix inversion), the number of unknowns must equal the number of joint equilibrium equations. Thus, if NMBS is the number of members, NRCT the number of reaction components, and NJNTS the number of joints, a necessary condition for statical determinacy is

for planar trusses: NMBS + NRCT $= 2 \times$ NJNTS
for plane frames : $3 \times$ NMBS + NCRT $= 3 \times$ NJNTS

The problem encountered when there are internal force releases (such as a pin) within a member, and, in fact, the general theory of statical determinacy, indeterminacy and geometric instability, are not developed herein, and the reader is referred to [9] for these details. In setting up the joint equilibrium equations, it is seen that the forces acting on the joint from the member are of opposite sign to those acting on the member in eqn (4.6). Thus, at a given joint,

$-\sum$ member force components + reaction + applied force $= 0$;   or, transposing,
$\sum$ member force components − reaction $=$ applied force

The addition of the coefficients of the member forces and reactions to the totality of equilibrium equations is shown diagrammatically in Fig. 4.4 for a structure with three reaction components.

The student is urged to write the necessary computer coding to generate the equilibrium equations for both truss and plane frame structures. Alternatively, the subroutine ADDELT achieves this result, and is given below.

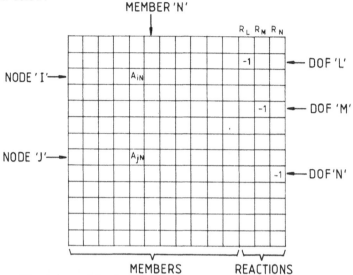

Fig. 4.4   Member contribution to nodal stiffness matrix.

```
SUBROUTINE ADDELT(NN,FORCE,N,NDF,NCN,NMF,NEQ,EQUIL)
*
* DEFINITION OF VARIABLES
* NN(NCN) NODE NUMBERS OF ELEMENT
* FORCE(NCN*NDFMNMF) COEFFICIENTS OF MEMBER FORCES
* EQUIL(NEQ,NEQ) MATRIX OF JOINT EQUILIBRIUM EQUATIONS
* N MEMBER NUMBER
* NCN DEGREES OF FREEDOM PER MEMBER
* NMF NUMBER OF BASIC FORCES PER MEMBER
* NEQ TOTAL NUMBER OF EQUATIONS
*
  DIMENSION EQUIL(NEQ,NEQ),FORCE(NCN*NDF,NMF),NN(NCN)
*
* LOOP ON NUMBER OF NODES PER ELEMENT
*
* COLUMN LOCATION IN EQUIL, 'JC'
*
JC = INP (N,NMF)
DO 10 IA = 1, NCN
* ROW LOCATIONS: JA IN FORCE, JB IN EQUIL
JA = INP (IA,NDF)
JB = INP (NN(IA),NDF)
```

```
DO 5 IB = 1, NDF
DO 5 IC = 1, NMF
* COLUMN LOCATION IC IN FORCE
EQUIL (JB+IB,JC+IC) = EQUIL (JB+IB,JC+IC) + FORCE (JA+IC,IC)
5 CONTINUE
10 CONTINUE
RETURN
END
* INTEGER FUNCTION TO GIVE SUBMATRIX START LOCATION
FUNCTION INP (I,J)
INP = (I-1) * J
RETURN
END
```

The above subroutine is not complete in that it contains no bounds check on the indices of either EQUIL or FORCE. The student should complete the program by including these checks with suitable warning print out and transfer of control if they are violated. When the complete matrix has been thus assembled the solution for member forces and reactions can be obtained by elimination or inversion. Symbolically, write the totality of equations,

$$[B]\begin{Bmatrix} S \\ S_R \end{Bmatrix} = \{R\} \tag{4.7}$$

In eqn (4.7), $\{S\}$ are member forces and $\{S_R\}$ the reactions. Thence, on inversion,

$$\begin{Bmatrix} S \\ S_R \end{Bmatrix} = [B]^{-1}\{R\} \tag{4.8}$$

Finally, then, extracting the appropriate rows from the inverse in eqn (4.9), the eqn (4.1) is obtained:

$$\{S\} = [b]\{R\} \tag{4.9}$$

The solution process shown symbolically in eqn (4.8) as matrix inversion must be carried out using pivotal search to locate the largest non–zero column member in the remaining equations to be reduced, because in the arrangement of the equations for member forces, zeros inevitably are placed on the diagonal. Failure to locate a non–zero term signals that the equations are singular and that the structure as input is geometrically unstable. It will be shown in section 4.3 that in obtaining eqn (4.9) a great deal more has been deduced than simply the member forces for a given loading pattern. It will be shown that a powerful method is now available for the calculation of deflections of the nodes of the structure. Before proceeding to the calculation of deflections and the related topic of statical indeterminacy, it is both important and useful to examine the transformation from joint displacements to member distortions. These member distortions correspond to the member forces $\{S\}$. Using the contragredient principle eqns (2.111) and (2.113) together with eqn (4.4), member distortions are given by

$$\{v\} = [T][L_D]\{r\} \tag{4.10}$$

Again, from eqn (4.6), for member N,

$$\{v_N\} = [A_{iN}^T \quad A_{jN}^T]\begin{Bmatrix} r_{iN} \\ r_{jN} \end{Bmatrix} \tag{4.11}$$

It is an interesting exercise to verify eqn (4.10) for both truss and frame structures from first principles, drawing the various displacement patterns corresponding to each $\{r_i\}$ displacement taken in turn to be unity. To this end, the basic joint and member displacement patterns in local (or member) coordinates are shown in Fig. 4.5. From Fig. 4.5(a), for the truss member:

$$\delta = [-1\ 0\ 1\ 0]\begin{Bmatrix} r'_{ix} \\ r'_{iy} \\ r'_{jx} \\ r'_{jy} \end{Bmatrix} \tag{4.12}$$

and from both Fig. 4.5(a) and (b), for the frame member,

$$\begin{bmatrix} \phi_i \\ \phi_j \\ \delta \end{bmatrix} = \begin{bmatrix} 0 & \dfrac{1}{l} & 1 & 0 & -\dfrac{1}{l} & 0 \\ 0 & \dfrac{1}{l} & 0 & 0 & -\dfrac{1}{l} & 1 \\ -1 & 0 & 0 & 1 & 0 & 0 \end{bmatrix} \begin{Bmatrix} r'_{ix} \\ r'_{iy} \\ \theta'_i \\ r'_{jx} \\ r'_{jy} \\ \theta'_j \end{Bmatrix} \tag{4.13}$$

The matrices in eqns (4.12) and (4.13) are obviously the variants of the $[T]$ matrix obtained from eqn (4.4). The transformation from global to local components follows directly from eqn (1.38) and is thus written

$$\{r'\} = [L_D]\{r\} \tag{4.14}$$

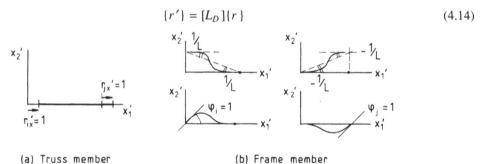

(a) Truss member   (b) Frame member

Fig. 4.5   Frame member distortions.

Eqn (4.10) has thus been verified from first principles. The power of the contragredient principle lies in the ability to reproduce eqn (4.10) from eqn (4.4), and vice versa, without having to resort to equilibrium equations on the one hand or displacement patterns on the other.

4.3 CALCULATION OF DEFLECTIONS

Once the transformation relating member forces to node forces has been obtained, the calculation of node deflections resulting from member distortions presents no further hazards and is a relatively simple task. Thus, the contragredient principle is invoked, and, using eqns (2.120) and (2.121), it follows that node deflections $\{r\}$ are given by

$$\{r\} = [b]^T\{v\} \tag{4.15}$$

It will be noticed that eqn (4.15) requires no connection to exist between $\{v\}$ the member distortions and $\{S\}$ the member forces, other than that $\{v\}$ terms correspond with $\{S\}$ directions. In general, it is possible to divide $\{v\}$ into two components, namely $\{v_s\}$, those member distortions produced by $\{S\}$, and $\{v^*\}$, those member distortions from any other source (temperature, volume change, distributed loads between nodes, etc.). Making this substitution, eqn (4.15) becomes

$$\{r\} = [b]^T \{v_s + v^*\} \tag{4.16}$$

The concept of the flexibility matrix is now introduced. The flexibility coefficient f_{ij} is the value of the distortion component v_i for a unit value of the member force S_j, so that the aggregate of all such relationships (using the summation convention on index j) is

$$v_i = f_{ij} S_j \tag{4.17}$$

Alternatively, in matrix notation,

$$\{v_N\} = [f_N]\{S_N\} \tag{4.18}$$

The subscript N indicates that the relationship is for member N. If all member forces and distortions are grouped sequentially in arrays $\{S\}$ and $\{V\}$, respectively, then the totality of all such relationships is,

$$\{v\} = [f]\{S\} \tag{4.19}$$

In eqn (4.19), $[f]$ is now a super–diagonal matrix with diagonal blocks given by the member flexibility matrices. Substitution of eqns (4.1) and (4.18) in (4.17) gives the nodal deflections $\{r\}$,

$$\{r\} = [b]^T [f][b]\{R\} + [b]^T \{v^*\} \tag{4.20}$$

The matrix $[F]$ will be defined as the flexibility matrix of the whole structure, and from eqn (4.20) it is calculated

$$[F] = [b]^T [f][b] \tag{4.21}$$

It should be noted that, although convenient to write the equations in the form of eqn (4.20), for practical purposes it is probably better to use the summation form of eqn (4.22).

$$\{r\} = \sum_{n=1}^{no\ members} [b_N]^T [f_N][b_N]\{R\} + \sum_{N=1}^{no\ members} [b_N]^T \{v_N\} \tag{4.22}$$

Using eqn (4.22) makes it unnecessary to form the matrices shown in eqn (4.20). Eqn (4.20) is generalized further by assuming that a second group of loads $\{\bar{R}\}$ has the force transformation

$$\{\bar{S}\} = [\bar{b}]\{\bar{R}\}$$

Then the deflections in the directions of $\{\bar{R}\}$ due to the $\{R\}$ forces are given by

$$\{\bar{r}\} = [\bar{b}]^T [f]\{b]\{R\} + [\bar{b}]^T \{v^*\} \tag{4.23}$$

Eqn (4.23) embodies the dummy unit load concept of classical structural analysis.

From eqn (2.131) it is seen that the constitutive matrix for an homogeneous isotrop~+~ material is symmetric; $f_{ij} = f_{ji}$. Since all flexibility matrices may be calculated in the form of eqn (4.12), it follows that $[F^T] = ([b^T][f][b])^T = b^T][f]^T[b] = [b^T][f][b]$, and $[F]$ is symmetric; that is, $F_{ij} = F_{ji}$. This relationship of symmetry is known as Maxwell's reciprocal theorem.

4.4 EFFECTS ON MEMBERS

4.4.1 Flexibility matrices

The use of eqns (4.20) and (4.23) presupposes that the flexibility matrices $[f]$ for members can be generated. For prismatic members with simple force fields given above, these flexibility matrices are easily generated. Thus, for member length l, area of cross section A, moment of inertia I, and Young's modulus of elasticity E, these flexibility matrices are

1. Truss member

$$[f] = \frac{l}{EA} \tag{4.24}$$

2. Frame member

$$[f] = \frac{l}{6EI} \begin{bmatrix} 2 & -1 & 0 \\ -1 & 2 & 0 \\ 0 & 0 & \dfrac{6I}{A} \end{bmatrix} \tag{4.25}$$

3. Grid member

$$[f] = \frac{l}{6EI} \begin{bmatrix} 2 & -1 & 0 \\ -1 & 2 & 0 \\ 0 & 0 & \dfrac{6EI}{GI_P} \end{bmatrix} \tag{4.26}$$

Example 4.3 *Calculation of beam deflections*

The cantilever beam shown in Fig. 4.6(a) has applied forces R_1, R_2. The $[b]$ matrix is generated not by computer solution of eqn (4.8), but by simple application of the unit values of R_1 and R_2, resulting in the bending moment diagrams shown in Fig. 4.6(b), (c). Thus, eqn (4.1) is given by

$$\begin{bmatrix} M_{1i} \\ M_{1j} \\ M_{2i} \\ M_{2j} \end{bmatrix} = \begin{bmatrix} -6 & -18 \\ 0 & 12 \\ 0 & -12 \\ 0 & 0 \end{bmatrix} \begin{Bmatrix} R_1 \\ R_2 \end{Bmatrix}$$

The matrix of member flexibilities is given b,

$$[f] = 10^{-5} \begin{bmatrix} 1.0 & -0.5 & 0 & 0 \\ -0.5 & 1.0 & 0 & 0 \\ 0 & 0 & 2.0 & -1.0 \\ 0 & 0 & -1.0 & 2.0 \end{bmatrix}$$

Performing the multiplications of eqns (4.21), the beam flexibility for the loads R_1 and R_2 is given by

$$[F] = 10^{-6} \begin{bmatrix} 1.92 & 7.67 \\ 7.67 & 51.84 \end{bmatrix}$$

In the above theory and example 4.1, the joint sign convention has been used for member moments and rotations relative to the chord. This results in alternating positive and negative moments along beams, and the negative off–diagonal flexibility term in eqn (4.25). In elementary strength of materials where non–computerized solutions are used, it is probably better to use the engineer's convention for signs of tension on the lower fibre positive. In this case the off–diagonal terms in the flexibility matrix of eqn (4.25) are positive.

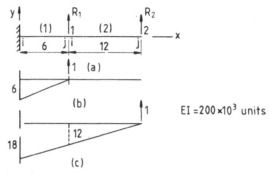

Fig. 4.6   Cantilever beam–dimensions and loads.

The use of eqn (4.20) without the $\{v^*\}$ term presupposes that loads are applied at the nodes of frame members so that the variation of member moment between nodes is linear. If such is not the case, as illustrated in Fig. 4.7(a), the term $\{v^*\}$ may be used to correct the member distortion terms for the variation from the linear diagram shown as the shaded area in Fig. 4.7(a).

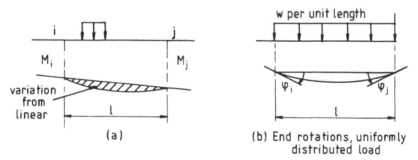

Fig. 4.7   Distributed load on member.

Because the effect of the bending moment diagrams is additive, in that to the end moments, the simply supported bending moment diagram for span l produces the variation from linearity, the values of $\{v^*\}$ for beam elements can be readily obtained from any text on the mechanics of materials. (for example, [1]). For the present illustration, for the uniformly distributed load w, shown in Fig. 4.7(b), the additional end rotations, $\{v^*\}$ (using engineer's sign convention), are

$$\{v^*\} = \begin{Bmatrix} \phi_i^* \\ \phi_j^* \end{Bmatrix} = \frac{l^2}{24EI} \begin{Bmatrix} 1 \\ 1 \end{Bmatrix} wl \qquad (4.27)$$

Example 4.4 *Calculation of beam deflections–distributed loads*

In this example, eqn (4.23) will be used in conjunction with eqn (4.26) to calculate the central deflection of the simply supported beam shown in Fig. 4.8, loaded with w per unit length over half the span.

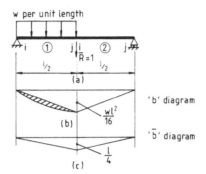

(a)

'b' diagram

(b) $\frac{wl^2}{16}$

'$\bar{b}$' diagram

(c) $\frac{l}{4}$

Fig. 4.8   Simply supported beam–distributed load deflections.

From Fig. 4.8(b), using designer's sign convention,

$$\{S\} = [b]\{R\} \rightarrow \begin{Bmatrix} M_{1i} \\ M_{1j} \\ M_{2i} \\ M_{2j} \end{Bmatrix} = \begin{bmatrix} 0 \\ l/8 \\ l/8 \\ 0 \end{bmatrix} \frac{wl}{2} \; ; \quad \text{and} \quad \{\bar{S}\} = [\bar{b}]\{\bar{R}\} \rightarrow \begin{Bmatrix} \bar{M}_{1i} \\ \bar{M}_{1j} \\ \bar{M}_{2i} \\ \bar{M}_{2j} \end{Bmatrix} = \begin{bmatrix} 0 \\ l/4 \\ l/4 \\ 0 \end{bmatrix} \{\bar{R}\} \tag{i}$$

The matrix of member flexibilities, using designer's sign convention, is, simply,

$$[f] = \frac{l}{12EI} \begin{bmatrix} 2 & 1 & 0 & 0 \\ 1 & 2 & 0 & 0 \\ 0 & 0 & 2 & 1 \\ 0 & 0 & 1 & 2 \end{bmatrix} \tag{ii}$$

From eqn (4.26), for both members,

$$\{v^*\} = \frac{l^2}{96EI} \begin{bmatrix} 1 \\ 1 \\ 0 \\ 0 \end{bmatrix} \frac{wl}{2} \tag{iii}$$

Substitution of eqns (i) to (iii) in eqn (4.23) gives

$$\{\bar{r}\} = [0 \; \frac{l}{4} \; \frac{l}{4} \; 0] \frac{l}{12EI} \begin{bmatrix} 2 & 1 & 0 & 0 \\ 1 & 2 & 0 & 0 \\ 0 & 0 & 2 & 1 \\ 0 & 0 & 1 & 2 \end{bmatrix} \begin{bmatrix} 0 \\ l/8 \\ l/8 \\ 0 \end{bmatrix} \frac{wl}{2} + \frac{l^2}{96EI} \begin{bmatrix} 1 \\ 1 \\ 0 \\ 0 \end{bmatrix} \frac{wl}{2} \tag{iv}$$

Multiplying the various matrices in eqn (iv) gives

$$\{\bar{r}\} = \frac{5wl^4}{768EI} \tag{v}$$

A check from [14] shows that this is the correct result, being one–half the value expected if the whole beam were to be loaded.

4.4.2   Member distortions due to temperature change

Two simple cases will be considered.

Axial distortion

A member length, l, with coefficient of thermal expansion, α, undergoes a temperature change, t, considered positive for an increase. Then the change in length of the member is, simply,

$$\Delta l = v^* = l\alpha t \tag{4.28}$$

Flexural distortion of a beam member

The beam member shown in Fig. 4.9(a) undergoes a temperature change, t_1, on its upper side and t_2 on its lower side, with an assumed linear variation through the section, as shown in Fig. 4.9(b). From Fig. 4.9(b), the temperature at distance y from the centroidal axis, with (c_1, c_2) both positive quantities, is

$$t = t_2 + \frac{(t_1 - t_2)}{(c_1 + c_2)}(y + c_2)$$

Substituting $y = 0$, gives the temperature at the centroid,

$$t_c = \frac{t_1 c_2 + t_2 c_1}{c_1 + c_2}$$

To calculate the end rotations $\{\phi_i^*, \phi_j^*\}$, use eqn (2.110); that is,

$$r = \int_V [b]^T \{\varepsilon\}\, dV \tag{2.110}$$

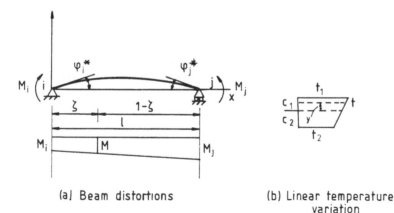

(a) Beam distortions   (b) Linear temperature variation

Fig. 4.9   Beam member–temperature distortions.

In the present instance,

$$\sigma_x = \frac{-My}{I} = \frac{-y}{I}[(1 - \zeta)\ \ \zeta]\begin{Bmatrix} M_i \\ M_j \end{Bmatrix}$$

Thus,

$$[b]^T = \frac{-y}{I}\left[1 - \zeta\atop \zeta\right], \quad \text{and} \quad \{\varepsilon\} = \varepsilon_x = \alpha t$$

Substituting in eqn (2.110) gives the rotations,

$$\left\{\phi_i^* \atop \phi_j^*\right\} = -\frac{\alpha}{I}\int_0^1\left[1 - \zeta\atop \zeta\right]\int_{-c_2}^{c_1} y\left\{t_2 + \frac{(t_1 - t_2)}{(c_1 + c_2)}(y + c_2)\right\} dA \; l d\zeta$$

Now, $\int_{-c_2}^{c_1} y \, dA = 0$, because y is measured from the centroid of the cross section, and

$\int_{-c_2}^{c_2} y^2 dA = I$ It follows that

$$\left\{\phi_i^* \atop \phi_j^*\right\} = \frac{-\alpha l}{c_1 + c_2}\int_0^1\left[(1 - \zeta)\atop \zeta\right]d\zeta(t_1 - t_2)$$

Finally, the rotations due to the temperature change are

$$\left\{\phi_i^* \atop \phi_j^*\right\} = \frac{\alpha l}{2(c_1 + c_2)}\begin{bmatrix} -1 & 1 \\ -1 & 1 \end{bmatrix}\begin{bmatrix} t_1 \\ t_2 \end{bmatrix} \tag{4.29}$$

4.4.3 Uncoupled member flexibility matrices

The member force–deflection relationship is given by eqn (4.17) as

$$v_i = f_{ij}S_j$$

Obviously, $f_{ii} \neq 0$, but also, in general, $f_{ij} \neq 0$ for $i \neq j$. That is, the off–diagonal terms are non–zero. The enquiry, then, is into the conditions for which these off–diagonal terms may be made to vanish. That is, $f_{ij} = 0$, for $i \neq j$. It will be noticed that this is a problem similar to that of finding the principal values of either the stress or strain tensors and the principal inertia axes of a beam cross section. Thus, let $[U]$ be the eigenvectors of $[f]$; that is, solutions of eqn (4.30)

$$[f]\{X\} = \lambda\{X\} \tag{4.30}$$

Then let

$$\{S\} = [U]\{s\}$$

be a force transformation which is obviously reversible if the eigenvectors are linearly independent. By using the contragredient principle, the generalized displacements corresponding to $\{s\}$ are given by

$$\{\zeta\} = [U]^T\{v\} \tag{4.31}$$

Solving eqn (4.31) for $\{v\}$,

$$\{v\} = [U^T]^{-1}\{\zeta\} \tag{4.32}$$

Substitute for $\{v\}$ and $\{S\}$ in eqn (4.19), so that

$$[U^T]^{-1}\{\zeta\} = [f][U]\{s\} \tag{4.33}$$

Premultiplying both sides of this equation by $[U]^T$ gives

$$\{\zeta\} = [U]^T[f][U]\{s\} = [f_\zeta]\{s\} \tag{4.34}$$

The flexibility matrix $[f_\zeta]$ is now a diagonal matrix because, from eqn (4.30),

$$[f][U] = [U]\lambda$$

Premultiplying both sides of this equation by $[U]^T$ gives

$$[U]^T[f][U] = [U]^T[U]\lambda = \lambda[I] \tag{4.35}$$

It is seen that $[f_{\zeta_i}] = [\lambda_i]$, with the eigenvalues as diagonal terms. The eigenvectors $[U]$ will be generalized displacement shapes composed of linear combinations of the nodal displacements. The forces $\{s\} = [U]^{-1}\{S\}$ will be the natural forces for the member or structure, and the response to the force s_i will be given only in mode ζ_i. The general principle of uncoupled flexibility matrices is used in the dynamic analysis of structures in the mode superposition method and is also the basis of the theory for principal axes of beam cross sections.

Example 4.5 *Natural modes for beam element*

Calculate the natural displacement modes and forces for the beam element shown in Fig. 4.10, given the end moments $<M_i\ M_j>$ and end rotations relative to the chord $<\phi_i\ \phi_j>$. For the beam member the flexibility matrix is

$$[f] = \frac{l}{6EI}\begin{bmatrix} 2 & -1 \\ -1 & 2 \end{bmatrix} \tag{i}$$

Eqn (4.26) becomes, in this instance,

$$\frac{l}{6EI}\begin{bmatrix} 2 & 1 \\ -1 & 2 \end{bmatrix}\{X\} = \lambda\{X\} \tag{ii}$$

In order to solve these equations, the determinant given in eqn (iii) must vanish.

$$\begin{vmatrix} 2-\lambda & -1 \\ -1 & 2-\lambda \end{vmatrix} = 0 \tag{iii}$$

On expansion of this equation, the resulting quadratic equation gives the values of λ; that is

$$\lambda^2 - 4\lambda + 3 = 0$$

Hence, $\lambda_1 = 3$ and $\lambda_2 = 1$. The matrix of eigenvectors is found to be

$$[U] = \begin{bmatrix} 1 & 1 \\ -1 & 1 \end{bmatrix} \tag{iv}$$

Substituting in eqn (4.29) gives

$$[f_\zeta] = \frac{l}{6EI}\begin{bmatrix} 1 & -1 \\ 1 & 1 \end{bmatrix}\begin{bmatrix} 2 & -1 \\ -1 & 2 \end{bmatrix}\begin{bmatrix} 1 & 1 \\ -1 & 1 \end{bmatrix} = \frac{l}{3EI}\begin{bmatrix} 3 & 0 \\ 0 & 1 \end{bmatrix} \tag{v}$$

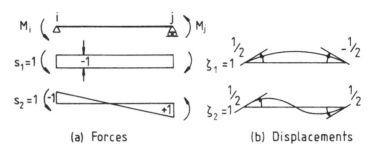

(a) Forces   (b) Displacements

Fig. 4.10   Natural modes of beam element.

It remains now to give a physical interpretation to $\{s\}$ and $\{\zeta\}$. Firstly, from eqn (4.27), the displacement modes are given

$$\{\zeta\} = \begin{Bmatrix} \zeta_1 \\ \zeta_2 \end{Bmatrix} = [U]^T\{v\} = \begin{bmatrix} 1 & -1 \\ 1 & 1 \end{bmatrix}\begin{Bmatrix} \phi_i \\ \phi_j \end{Bmatrix} = \begin{bmatrix} \phi_i - \phi_j \\ \phi_i + \phi_j \end{bmatrix}$$

and, of course,

$$\begin{Bmatrix} \phi_i \\ \phi_j \end{Bmatrix} = \frac{1}{2}\begin{bmatrix} 1 & 1 \\ -1 & 1 \end{bmatrix}\begin{bmatrix} \zeta_1 \\ \zeta_2 \end{bmatrix} = \begin{bmatrix} \frac{1}{2}(\zeta_1 + \zeta_2) \\ \frac{1}{2}(-\zeta_1 + \zeta_2) \end{bmatrix}$$

The generalized forces $\{s\}$ are given by

$$s = \begin{Bmatrix} s_1 \\ s_2 \end{Bmatrix} = [U]^{-1}\{S\} = \frac{1}{2}\begin{bmatrix} 1 & -1 \\ 1 & 1 \end{bmatrix}\begin{Bmatrix} M_i \\ M_j \end{Bmatrix} = \begin{Bmatrix} \frac{1}{2}(M_i - M_j) \\ \frac{1}{2}(M_i + M_j) \end{Bmatrix}$$

Alternatively,

$$\begin{Bmatrix} M_i \\ M_j \end{Bmatrix} = [U]\{s\} = \begin{bmatrix} 1 & 1 \\ -1 & 1 \end{bmatrix}\begin{Bmatrix} s_1 \\ s_2 \end{Bmatrix} = \begin{Bmatrix} s_1 + s_2 \\ -s_1 + s_2 \end{Bmatrix}$$

The usefulness of these natural modes will be demonstrated later with the calculation of non–linear effects produced by axial force on end rotations, and the effects of bowing on the axial shortening of members. In addition it forms the basis, in part, of the natural mode technique pioneered by Prof. J.H. Argyris [1], and given here in Chapter 7.

4.5 STATICALLY INDETERMINATE STRUCTURES

4.5.1   Introduction

Thus far, only stable, statically determinate structures have been discussed. For simple structures, stability may be assessed by inspection. However, if the analysis of an unstable structure is attempted, it will be found that in the generation of the force transformation matrix by inversion of the joint equilibrium coefficient matrix, singularity will be encountered.

That is, a zero pivot will be found during the elimination process. This, in fact, forms a convenient means of testing for stability. For the statically indeterminate structure there are more unknowns than equations of joint equilibrium. That is, the equilibrium matrix $[A]$ which possesses full row rank, is an $m \times n$ matrix with $m < n$ That is, rank $A = m$. The indeterminacy (or redundancy) of the structure may be classified as internal; that is, member forces are the unknowns (as in Fig. 4.11(a)): or as external (as in Fig. 4.11(b)), in which case the number of reaction components is in excess of that necessary for the equilibrium of the structure when considered in its entirety as a rigid body. If the redundant structure is to be analysed for member forces and node deflections due to applied loads, internal member strains, or settlement of supports, resort must be had to additional equations involving in some way the unknown force quantities independent of those already obtained from joint equilibrium. There will be, in fact, many independent ways in which these redundancies may be selected. The analysis of the structure may therefore be conceived in two stages. In the first stage, the structure is cut so that all the redundancies are removed; the structure remaining must, of course, be made stable as well as determinate. For the chosen determinate structure, deflections are calculated as functions of the external load system and the stress resultants at the cuts which have been selected as unknowns. From section 4.2, the $[b]$ force transformation matrices may now be generated for each of these force systems by using the joint equilibrium equations. In addition, from section 4.3, it is possible to calculate deflections throughout the structure for unit values of both force groups. Needless to say, in today's computer environment the whole of the above process must be automatic if it is to be competitive with other computer methods of analysis. In the second stage of the analysis the requirement is introduced that the continuity of deflections must be maintained between the various parts of the structure at the cut sections. For each condition of continuity, one equation must be formed involving the external (applied) loads and the unknown forces. These equations are known historically as the Maxwell–Mohr superposition equations.

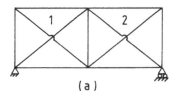

(a)   (b)

Fig. 4.11   Types of force redundancy: (a) member of redundancy, number of redundants = 2; (b) reaction redundancy, number of redundants = 3.

As an example, consider the continuous beam loaded as shown in Fig. 4.12. It has been made determinate by placing hinges at the two internal supports. The deflected shape for the determinate structure exhibits discontinuities of slopes at these two points (Fig. 4.12(b)), which may be removed by applying the unknown support moments X_1 and X_2. Notice that this type of slope discontinuity may be expressed in terms of the normals to member tangents and that when continuity is restored, these normals coincide or at least resume their original relative position. Before the general matrix theory for the calculation of the unknown forces is developed, a simple example is analysed with only one unknown.

Example 4.6 *Propped cantilever*

The beam shown in Fig. 4.13 is a propped cantilever. That is, it is built in (or fixed) at end i and has a roller at end j. Since the beam has only three equations of static equilibrium ($\sum F_x = 0$, $\sum F_y = 0$, and $\sum M = 0$ about any point) and there are four reactions, three at i and one at j, it is statically indeterminate with one redundancy. The redundant force may be taken as the reaction at end j. It is required to calculate the reaction at j for a distributed load w over the whole length l of the beam. In Fig. 4.13(b), it is seen that when the redundant support R_i is removed, the vertical deflection at j is, assuming w positive as shown,

$$\delta_{0j} = \frac{-wl^4}{8EI} \tag{i}$$

A unit value of R_j applied in its positive sense (Fig. 4.13(c)) produces the deflection at j,

$$\delta_{jj} = \frac{l^3}{3EI} \tag{ii}$$

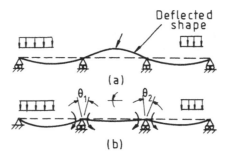

Fig. 4.12   Statically indeterminate structure: (a) continuous structure deflections; (b) discontinuity of normal slopes in cut structure.

The actual value of R_j must be such that the net deflection at j is zero, since the support is assumed to be unyielding and the deflected shape is as shown in Fig. 4.13(d). That is,

$$\delta_{0j} + R_j \delta_{jj} = 0 \tag{iii}$$

Substitute for the values of δ_{0j} and δ_{jj} from eqns (i) and (ii). Then

$$\frac{-wl^4}{8EI} + R_j \frac{l^3}{3EI} = 0$$

giving

$$R_j = \frac{3wl}{8}$$

4.5.2   Statical indeterminacy

The process given in example 4.6 may be generalized by using matrix notation for the calculation of deflections. In the first discussion of sign conventions, a force applied to a structure is considered positive in the direction given to it, and the deflection is positive when in the direction of the force. With this convention the forces applied on either side of a cut in a redundant structure are both considered positive.

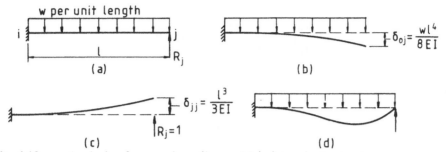

Fig. 4.13   Analysis of propped cantilever: (a) indeterminate structure;
(b) determinate structure; (c) deflection due to unit value of redundant;
(d) actual deflected shape.

This is in contrast with the sign convention in which forces and displacements in the directions of the positive coordinate axes are taken as positive. In programs which select the redundant force quantities automatically, this question of signs must be taken into account automatically. Consider a statically indeterminate structure with degree of redundancy m and loaded with the force system $\{R\}^T = <R_1\, R_2\, \cdots\, R_n>$. It is required to calculate member forces and node deflections, in the first instance, in the directions $1, 2, \ldots, n$ of the forces $\{R\}$. The structure is cut in such a way that deflections for each component part of the structure may be calculated from the $\{R\}$ forces acting on that part and also from the unknown forces $\{X\}^T = <X_1, X_2, \cdots, X_m>$ acting to restore the continuity of the structure at the cuts. In general, the structure obtained by introducing the $\{X\}$ forces will be statically determinate. This is, however, an unnecessary restriction since all that is required is that for each component part of the structure, b_0 and b_1, the force transformation matrices due to $\{R\}$ and $\{X\}$ should be known. Then,

$$\{S\} = [b_0\ b_1]\begin{bmatrix} R \\ X \end{bmatrix} \tag{4.36}$$

The deflections of the structure may be partitioned into $\{r_0\}$, those deflections at $\{R\}$, and $\{r_x\}$, the deflections at the cuts in the directions of $\{X\}$. The deflections are calculated from eqn (4.20),

$$\begin{bmatrix} r_0 \\ r_x \end{bmatrix} = \begin{bmatrix} b_0^T \\ b_1^T \end{bmatrix}[f\,][b_0\ b_1]\begin{bmatrix} R \\ X \end{bmatrix} + \begin{bmatrix} b_0^T \\ b_1^T \end{bmatrix}\{v^*\} \tag{4.37}$$

When the multiplication of the $[b]$ and $[f]$ matrices on the right-hand side of eqn (4.37) is carried out, it is seen that

$$\begin{bmatrix} r_0 \\ r_x \end{bmatrix} = \begin{bmatrix} b_0^T f b_0 & b_0^T f b_1 \\ b_1^T f b_0 & b_1^T f b_1 \end{bmatrix}\begin{bmatrix} R \\ X \end{bmatrix} + \begin{bmatrix} b_0^T \\ b_1^T \end{bmatrix}\{v^*\} \tag{4.38}$$

The submatrices in eqn (4.38) are simply the deflection influence coefficients of the structure for the forces $\{R\}$ and $\{X\}$. Since the $\{X\}$ values represent a pair of equal and opposite forces applied to the structure, the terms of $[b_1]^T[f][b_1]$ give the relative deflection coefficients at the cuts. The following shorthand notation is introduced for the four submatrices in eqn (4.38)

$$[D_{00}] = [b_0]^T[f][b_0]; \quad [D_{01}] = [b_0]^T[f][b_1]$$

$$[D_{10}] = [b_1]^T[f][b_0]; \quad [D_{11}] = [b_1]^T[f][b_1]$$

That is, with this notation,

$$\begin{bmatrix} r_0 \\ r_x \end{bmatrix} = \begin{bmatrix} D_{00} & D_{01} \\ D_{10} & D_{11} \end{bmatrix} \begin{bmatrix} R \\ X \end{bmatrix} + \begin{bmatrix} b_0^T \\ b_1^T \end{bmatrix} [v^*] \tag{4.39}$$

Thence,

$$\{r_0\} = [D_{00}]\{R\} + [D_{01}]\{X\} + [b_0]^T\{v^*\} \tag{4.40a}$$

$$\{r_x\} = [D_{10}]\{R\} + [D_{11}]\{X\} + [b_1]^T\{v^*\} \tag{4.40b}$$

However, it has been specified that continuity must be restored at the cuts, and for this to be accomplished, the necessary condition is

$$\{r_x\} \equiv 0 = [D_{10}]\{R\} + [D_{11}]\{X\} + [b_1]^T\{v^*\} \tag{4.41}$$

On solving eqn (4.41) for $\{X\}$,

$$\{X\} = -[D_{11}]^{-1}([D_{10}]\{R\} + [b_1]^T\{v^*\}) \tag{4.42}$$

When this value for $\{X\}$ is substituted in eqn (4.40a), the deflections $\{r_0\}$ are calculated as

$$\{r_0\} = ([D_{00}] - [D_{01}][D_{11}]^{-1}[D_{10}])\{R\} + ([b_0]^T - D_{01}][D_{11}]^{-1}[b_1]^T)\{v^*\} \tag{4.43}$$

Member forces are calculated from eqn (4.33),

$$\{S\} = [b_0]\{R\} + [b_1]\{X\} \tag{4.44}$$

Again, substitution for $\{X\}$ from eqn (4.42) gives

$$\{S\} = ([b_0] - [b_1][D_{11}]^{-1}[D_{10}])\{R\} - [b_1][D_{11}]^{-1}[b_1]^T\{v^*\} \tag{4.45}$$

The matrices in eqns (4.43) and (4.45) may now be interpreted as deflection and force influence coefficient matrices for the indeterminate structure; that is,

$$[F] = [D_{00}] - [D_{01}][D_{11}]^{-1}[D_{10}] \tag{4.46}$$

and

$$[b] = [b_0] - [b_1][D_{11}]^{-1}[D_{10}] \tag{4.47}$$

It should be emphasized that eqns (4.40a) and (4.44) are the usual equations used in deflection and stress calculation. Eqns (4.46) and (4.47) are used if the transformation matrices for the complete structure are required. For example, $[F]$ may be used in the calculation of natural frequencies of the structure. That is, the $[F]$ matrix is the flexibility matrix of the structure for the $\{R\}$ forces. In examples 4.7 and 4.8 which follow, designer's sign convention will be used for moments, and member flexibility matrices reflect this with positive off–diagonal terms.

Example 4.7 *Analysis of a portal frame*

The portal frame shown in Fig. 4.14(a) has fixed supports. In such a case there are six reaction components and only three equations of equilibrium. The portal frame is statically indeterminate with three unknown forces. These will be taken to be X_1, the moment at node *1*; X_2, the moment at node *4*; and X_3, the horizontal force at node *4*. The bending moment diagrams for unit values of each of these redundant forces are given in Fig. 4.14(b).

The portal frame is loaded with a distributed force of 1 kN/m run vertically downwards on the top member (2). It is required to calculate bending moments in the frame (at member ends) when bending distortions only are considered in deflection calculations. For this structure, eqn (4.42) will be used to calculate the redundant forces. From Fig. 4.14(b), the force transformation matrix $[b_0, b_1]$ is generated, connecting (M_i, M_j) for each member to (w, X_1, X_2, X_3). Notice that w produces no such member moments; that is,

$$[b_0\, b_1] = \begin{bmatrix} 0 & -1 & 0 & 0 \\ 0 & -1 & 0 & -1 \\ 0 & -1 & 0 & -1 \\ 0 & 0 & -1 & -1 \\ 0 & 0 & -1 & -1 \\ 0 & 0 & -1 & 0 \end{bmatrix}$$

The force $R = wl = 1 \times 20 = 20$ kN, so that

$$\{v^*\} = \{\delta_0\}\{R\} = \frac{50}{6EI}\begin{bmatrix} 0 \\ 0 \\ 1 \\ 1 \\ 0 \\ 0 \end{bmatrix} \qquad \text{Note that,} \quad \phi_i = \phi_j = \frac{wl^3}{24EI_2} = \frac{l^2}{24E(2I)}R$$

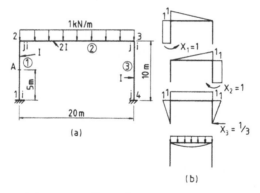

Fig. 4.14   Portal frame: (a) dimensions and loading; (b) moment diagrams for unit values of load and redundants.

The matrix of member flexibilities $[f]$ is given by

$$[f] = \frac{10}{6EI}\begin{bmatrix} 2 & 1 & 0 & 0 & 0 & 0 \\ 1 & 2 & 0 & 0 & 0 & 0 \\ 0 & 0 & 2 & 1 & 0 & 0 \\ 0 & 0 & 1 & 2 & 0 & 0 \\ 0 & 0 & 0 & 0 & 2 & 1 \\ 0 & 0 & 0 & 0 & 1 & 2 \end{bmatrix}$$

On calculation,

$$[D_{11}] = [b_1]^T[f][b_1] = \frac{10}{6EI}\begin{bmatrix} 8 & 1 & 6 \\ 1 & 8 & 6 \\ 6 & 6 & 10 \end{bmatrix}; \quad \text{and} \quad [D_{11}]^{-1} = \frac{6EI}{1260}\begin{bmatrix} 44 & 26 & -42 \\ 26 & 44 & -42 \\ -42 & -42 & 63 \end{bmatrix}$$

Also,

$$[b_1]^T\{\delta_0\} = \frac{50}{6EI}\begin{bmatrix} -1 \\ -1 \\ -2 \end{bmatrix}$$

Then, since $[b_1]^T[f][b_0] = [D_{10}] \equiv 0$, because $[b_0] = 0$

$$\{X\} = -[D_{11}]^{-1}([D_{10}] + [b_1]^T\{\delta_0\})\{R\}$$

giving,

$$\{X\} = \frac{-5}{126}\begin{bmatrix} 44 & 26 & -42 \\ 26 & 44 & -42 \\ -42 & -42 & 63 \end{bmatrix}\begin{bmatrix} -1 \\ -1 \\ -2 \end{bmatrix}20 = \begin{bmatrix} -11.1 \\ -11.1 \\ 33.3 \end{bmatrix}$$

Then,

$$\{S\} = [b_0]\{R\} + [b_1]\{X\} = \begin{bmatrix} -1 & 0 & 0 \\ -1 & 0 & -1 \\ -1 & 0 & -1 \\ 0 & -1 & -1 \\ 0 & -1 & -1 \\ 0 & -1 & 0 \end{bmatrix}\begin{bmatrix} -11.1 \\ -11.1 \\ 33.3 \end{bmatrix} = \begin{bmatrix} 11.1 \\ -22.2 \\ -22.2 \\ -22.2 \\ -22.2 \\ 11.1 \end{bmatrix}\text{kN/m}$$

Units in this equation are kN/m. These bending moments are plotted on the tension side of members in Fig. 4.15. The parabolic diagram with centre ordinate $wl^2/8$ has been added to the top member.

Example 4.8 *Portal frame–alternative method of analysis*

There exists an alternative method of calculating node deflections of structures to that employing the use of the $\{v^*\}$ matrix. Instead, the analysis proceeds rather as in the displacement method (Chapter 5), by first fixing all joints and calculating the fixed end forces for the loads between the joints. When these fixed end forces with reversed sign are applied to the nodes, the effect achieved is the release of the joints. Thus, the portal frame in Fig. 4.16 is reanalysed with the use of release forces. The condition of all joints fixed is shown in Fig. 4.16(a). For the frame shown in Fig. 4.16(b), the node forces producing bending distortions are the two node moments so that $\{R\}^T = \langle 33.33 \ 33.33 \rangle$. Then, by using the same release system as in Fig. 4.14, the $[b_0 \ b_1]$ matrices are given

$$[b_0 \ b_1] = \begin{bmatrix} 0 & 0 & -1 & 0 & 0 \\ 0 & 0 & -1 & 0 & -1 \\ 1 & 0 & -1 & 0 & -1 \\ 0 & 1 & 0 & -1 & -1 \\ 0 & 0 & 0 & -1 & -1 \\ 0 & 0 & 0 & -1 & 0 \end{bmatrix}$$

The member flexibility matrix $[f]$ and the $[b_1]$ matrix are the same as in example 4.7, so that $[D_{11}]^{-1}$ is as given before.

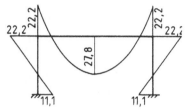

Fig. 4.15 Portal frame–bending moments.

The $[D_{10}]$ matrix is given

$$[D_{10}] = [b_1]^T [f][b_0] = \frac{10}{6EI} \begin{bmatrix} -2 & -1 \\ -1 & -2 \\ -3 & -3 \end{bmatrix}$$

Then, from eqn (4.29), $\{X\} = -[D_{11}]^{-1}[D_{10}]\{R\}$, so that

$$\{X\} = \frac{1}{126} \begin{bmatrix} 44 & 26 & -42 \\ 26 & 44 & -42 \\ -42 & -42 & 63 \end{bmatrix} \begin{bmatrix} 2 & 1 \\ 1 & 2 \\ 3 & 3 \end{bmatrix} \begin{bmatrix} 33.33 \\ 33.33 \end{bmatrix} = \begin{bmatrix} -11.11 \\ -11.11 \\ 33.33 \end{bmatrix}$$

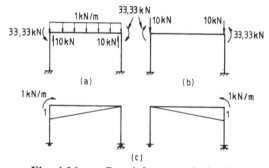

Fig. 4.16 Portal frame–node loads.

For the node–force solution,

$$\{S_N\} = [b_0]\{R\} + [b_1]\{X\}$$

That is,

$$\begin{bmatrix} M_{1i} \\ M_{1j} \\ M_{2i} \\ M_{2j} \\ M_{3i} \\ M_{3j} \end{bmatrix} = \begin{bmatrix} 0 & 0 \\ 0 & 0 \\ 1 & 0 \\ 0 & 1 \\ 0 & 0 \\ 0 & 0 \end{bmatrix} \begin{bmatrix} 33.33 \\ 33.33 \end{bmatrix} + \begin{bmatrix} -1 & 0 & 0 \\ -1 & 0 & -1 \\ -1 & 0 & -1 \\ 0 & -1 & -1 \\ 0 & -1 & -1 \\ 0 & -1 & 0 \end{bmatrix} \begin{bmatrix} -11.11 \\ -11.11 \\ 33.33 \end{bmatrix} = \begin{bmatrix} 11.11 \\ -22.22 \\ 11.11 \\ 11.11 \\ -22.22 \\ 11.11 \end{bmatrix}$$

To these moments must be added those fixed–end moments $\{S_F\}$, which exist before the nodes are released. Finally, the moments are identical to those calculated in example 4.7.

$$
\{S\} = \{S_N\} + \{S_F\} =
\begin{bmatrix}
11.11 \\
-22.22 \\
11.11 \\
11.11 \\
-22.22 \\
11.11
\end{bmatrix}
+
\begin{bmatrix}
0.00 \\
0.00 \\
-33.33 \\
-33.33 \\
0.00 \\
0.00
\end{bmatrix}
=
\begin{bmatrix}
11.11 \\
-22.22 \\
-22.22 \\
-22.22 \\
-22.22 \\
11.11
\end{bmatrix}
\text{kN/m}
$$

4.6 ALGORITHMS SUITABLE FOR COMPUTER ANALYSIS

4.6.1 Introduction

In section 4.5 it has been shown that the analysis of statically indeterminate structures may be considered to be separated into two basic problems. The first of these is the analysis of statically determinate structures as outlined in section 4.2. This process has been further complicated by the necessity of generating both the $[b_0]$ and $[b_1]$ matrices from the equilibrium equations, which now have more columns than rows. For the setting up of the equilibrium equations, it is required to know the structure geometry, the member–node connectivity (i.e. the structure's topology) and the support conditions. The computer program must be organized in some way to separate out the redundant forces from the totality of the unknowns. The second phase requires a deflection analysis of the structure for both the external and redundant forces, the actual values of these latter forces being such as to enforce compatibility of deformations. Evidently, this process requires a knowledge of the member flexibilities (i.e. their distortion–force relationship). As outlined herein, the automatic selection of redundants is quite feasible. Now a second difficulty arises in the force method in that unless the selection of redundants is undertaken in a very special way, the $[D_{11}]$ matrix of eqn (4.42) is fully populated and consequently involves not only large storage requirements but also significant CPU time for the solution of eqn (4.42). It must, of course, be realized that the formal inversion process of eqn (4.42) will not be carried out for large numbers of unknowns, but rather the equations will be solved by some elimination process. In this regard the force method has been rather neglected in its development because of the emphasis which has been placed on the direct stiffness method (Chapter 5, section 4). However, as shown herein, the possibility exists of arranging for the $[D_{11}]$ to be a banded matrix with skyline properties such that the solution of eqn (4.42) becomes a practical reality even for large systems. Not only that, but the flexibility matrix $[f]$ for the element is more readily formed than the corresponding direct stiffness matrix $[k]$, and, as shown in example 4.5, may in many instances be arranged to be a diagonal matrix. There are several methods for the automatic selection of redundants, of which the Gauss–Jordan (structure–cutter method by Denke [15]) is the most long–standing. More recently, a modification of the Gauss LU method called the turn–back LU procedure has been developed which enables the $[D_{11}]$ matrix to have the properties of bandedness mentioned above [16].

4.6.2 The structure–cutter method

Before dealing with the problem of the selection of redundants from a more mathematical point of view, the structure cutter–method is discussed because it introduces several interesting physical concepts which are useful to the general understanding of the problem. As already mentioned at the beginning of this section, the structure–cutter approach starts the analysis by setting up all the equations of joint equilibrium.

Both member forces and external reactions are included, and the resulting coefficient matrix has m rows and n columns, with $(n - m) = \rho$, expressing the degree of redundancy of the system. Each member will contribute forces to the joint equilibrium equations at its extremities, the type and number of independent member forces being dependent on the number of releases it contains. Since the arbitrary addition of releases in members may result in null joint equilibrium equations, a general program should check all joints for such conditions and delete all such equations. For example, in a plane frame, if all members meeting at a joint have a moment release there, the equation $\sum M = 0$ can never be satisfied for an arbitrary applied external moment. In this case the equation should be deleted and only the forces $\{R_x\}$ and $\{R_y\}$ applied at the joint. The equations of joint equilibrium, including reaction components $\{R_1\}$, are written

$$\{R\} + [A]\{R_1\} - [a^T]\{S\} = 0$$

That is, grouping $\{R_1\}$ and $\{S\}$, the unknown quantities,

$$\{R\} - [a^T \ -A]\begin{bmatrix} S \\ R_1 \end{bmatrix} = 0 \tag{4.48}$$

Suppose now that a determinate system is chosen so that its forces $\{S_0\}$ may be expressed in terms of $\{R\}$ and the unknown quanties $\{S_x\}$. These forces $\{S_0\}$ and $\{S_x\}$ are extracted from the vector $[S \ R_1]^T$, constituting this vector in rearranged form. With this notation, eqn (4.48) is now written

$$\{R\} - [a_0^T \ a_1^T]\begin{bmatrix} S_0 \\ S_x \end{bmatrix} = 0 \tag{4.49}$$

Thence,

$$\{S_0\} = ([a_0]^T)^{-1}\{R\} - ([a_0]^T)^{-1}[a_1]^T\{S_x\} \tag{4.50}$$

The inverse $([a_0]^T)^{-1}$ exists if the $\{S_0\}$ forces constitute a stable determinate system. It is seen that $([a_0]^T)^{-1} = [\bar{b}_0]$, the matrix which transforms the $\{R\}$ forces to member $\{S_0\}$ forces in the determinate structure. The product $-([a_0]^T)^{-1}[a_1]^T$ will be defined as $\bar{b}_1$ and it transforms the $\{S_x\}$ to forces in the determinate structure. However, if the $\{S_x\}$ values are member forces, they also produce forces in the indeterminate sections, and, hence,

$$[b_1] = \begin{bmatrix} \bar{b}_1 \\ I \end{bmatrix} \tag{4.51}$$

If these same sections are included in the $\{b_0\}$ transformation, then

$$[b_0] = \begin{bmatrix} \bar{b}_0 \\ 0 \end{bmatrix} \tag{4.52}$$

With $[b_0]$ and $[b_1]$ determined, the analysis proceeds via eqns (4.42) *et seq.* It is convenient to use Gaussian elimination in the solution of eqn (4.50), rather than to use the formal inversion of $[a_0]^T$. Then $[a_0]^T$ is, in effect, reduced to a form in which the first square submatrix has non–zero terms in its upper triangle with unit diagonal terms. At each stage of the reduction, the remaining columns of the matrix to be reduced are searched for the largest term and the rows and columns interchanged so that this value may be the next pivot term.

If instead of a load vector $\{R\}$, a series of unit loads is used such that $[R] = [I]$, the back substitution process produces $[\bar{b}_0]$. The back substitution process operating on $-[a_1]^T$ produces $[\bar{b}_1]$. If the elimination process is to produce a well–conditioned set of redundants, then the least flexible members should be chosen for their forces to be included in the determinate set. To achieve this, columns of $[a]^T$ may be weighted inversely to the member flexibility for the corresponding member force. In this way the largest terms will occur for members of greatest stiffness.

4.6.3 The Gauss–Jordan and LU procedures

The above structure cutter–procedure is now generalized and a method described in which the resulting $[b_1]$ matrix is such that the product $[b_1]^T[f][b_1]$ is both banded and of skyline profile. The joint equilibrium eqns (4.48) will be, for the present purposes, written as

$$[B][S] = \{R\} \qquad (4.53)$$

$[B]$ is an $m \times n$ matrix with $m < n$. In the Gauss–Jordan procedure an $m \times m$ identity matrix is produced in the first m columns of $[B]$. Because $[B]$ is a sparse matrix this will involve column changes as well as a sequence of m pivots. The process may be expressed as a series of matrix multiplications [16].

$$[G_m][G_{m-1}] \cdots [G_1][B]\{V\} = [I,M] \quad \text{or,} \quad [G][B][V] = [I,M] \qquad (4.54)$$

Where $[G_j]$ is the jth pivot matrix and $\{V\}$ is an $(n \times n)$ column permutation matrix such that $[V^T] = [V]$. Then,

$$[B][V] = [G^{-1}, G^{-1}M]$$

Thence,

$$[b_0] = [V]\begin{bmatrix} G \\ 0 \end{bmatrix}; \quad b_1 = [V]\begin{bmatrix} -M \\ I \end{bmatrix} \qquad (4.55)$$

since,

$$[B][b_0] = [B][V]\begin{bmatrix} G \\ 0 \end{bmatrix} = [G^{-1}, G^{-1}M]\begin{bmatrix} G \\ 0 \end{bmatrix}$$

and

$$[B][b_1] = [B][V]\begin{bmatrix} -M \\ I \end{bmatrix} = [G^{-1}, G^{-1}M]\begin{bmatrix} -M \\ I \end{bmatrix} = 0$$

Alternatively, by using an LU decomposition algorithm [16], the following factorization of $[B]$ is obtained.

$$[W][B] = [L][U], \quad [U][V] = [U_1 U_2] \qquad (4.56)$$

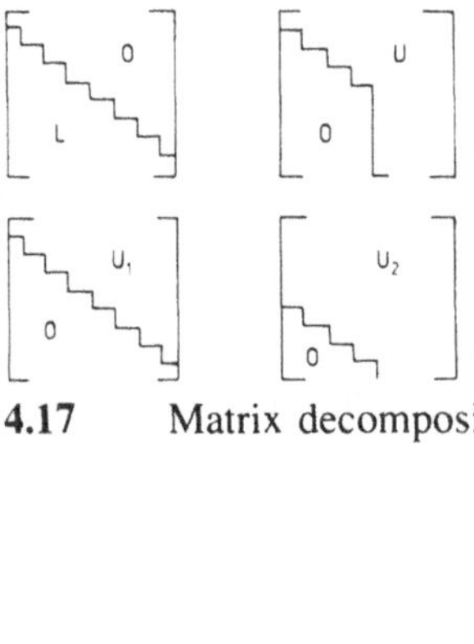

Fig. 4.17   Matrix decomposition.

In these equations $[W]$ is an $(m \times m)$ row permutation matrix, $[L]$ is a non-singular, $(n \times n)$ lower triangular matrix with unit diagonals, and $[U]$ is an $(m \times m)$ matrix whose columns can be rearranged, through an $(n \times n)$ column permutation matrix $[V]$, into $[U_1 \ U_2]$ where $[U_1]$ is a non-singular $(m \times m)$ upper triangular matrix and $[U_2]$ is an $(m \times p)$ matrix with zeros in its lower left corner. This is represented diagrammatically in Fig. 4.17 Then it follows that

$$[b_0] = [V] \begin{bmatrix} (U_1^{-1} L^{-1}) \ W \\ 0 \end{bmatrix} \quad \text{and,} \quad [b_1] = [V] \begin{bmatrix} -U_1^{-1} \ U_2 \\ I \end{bmatrix} \tag{4.57}$$

It is seen that this is so because

$$[B][b_0] = [W]^T [L][U][V] \begin{bmatrix} (U_1^{-1} L^{-1}) \ W \\ 0 \end{bmatrix} = [W]^T [L][U_1 U_2] \begin{bmatrix} (U_1^{-1} U_2) \ W \\ 0 \end{bmatrix} = [I] \tag{4.58}$$

and

$$[B][b_1] = [W]^T [L][U_1 U_2] \begin{bmatrix} -(U_1^{-1} U_2) \\ I \end{bmatrix} = 0 \tag{4.59}$$

It is now informative to examine the non–zero pattern of $[b_1]$ (and thus also of $[b_1]^T [f][b]_1$), obtained through the above procedures. In the LU decomposition of $[B]$, $[U_1]$ is upper triangular and $[U_2]$ has zeros in its lower left corner. Thus, $-[U_1^{-1}U_2]$ has the same non–zero pattern as $[U_2]$. The matrix $[b_1]$ is obtained by inserting $(n - m)$ rows of the identity matrix $[I]$ between the rows of $[-U_1^{-1}U_2]$, according to the row permutation $[V]$. It evolves that the permutation $[V]$ generated in the LU decomposition has the property that after the insertion of rows of $[I]$ into $[-U_1^{-1}U_2]$, the non–zero pattern is preserved, i.e. $[b_1]$ has zeros in its lower left corner. Unfortunately, when the product $[b_1]^T [f][b_1]$ is formed, the non–zero pattern tends to disappear. However, as shown later, a $[b_1]$ matrix can be formed during the decomposition process, which has the desired property of also having zeros in the upper right corner of the matrix. It should be noted in passing that the $[b_1]$ matrix produced in the structure–cutter method has no particular pattern, and thus $[b_1]^T [f][b_1]$ will be a dense matrix.

4.6.4 The turn back procedure

Assume that the LU factorization of the equilibrium matrix $[B]$ (eqn (4.57)) has been performed. Let $[b_1][L][U]$ be the redundant transformation matrix so obtained (eqn (4.57)). Thus, for $j + 1, \ \cdots \ , p$, let $[b^j]$ denote the jth column of $[b_1][L][U]$, such that

$$[b_1][L][U] = [b^1 b^2 \ \cdots \ b^p] \tag{4.60}$$

From the definition of $[b_1]$ (eqn (4.59)), each column $[b^j]$ is the solution of the homogeneous equations,

$$[B]\{y\} = 0 \tag{4.61}$$

and the solutions $[b^1, b^2, ..., b^p]$ are linearly independent. For each $[b^j]$, let k be the smallest index such that, $[b_i^j] = 0$ for all $i > k$. This implies that the first k columns of $[B]$ are linearly dependent; that is, $[B_1 \ B_2 \ \cdots \ B_k]$ since

$$[B][b^j] \rightarrow \sum_{i=1}^{k}[B_{l_i}][b_i^j] = \quad \text{for } l = 1 \text{ to } m \ , \ b_i^j \neq 0 \quad i \leq k \tag{4.62}$$

An attempt is now made to find a subset of eqn (4.61) which is still linearly dependent. The index k for each j in $(1, \cdots p)$ is determined by looking at the column permutation $[V]$ in eqn (4.56). In the turn–back procedure, the minimal linearly dependent set of columns of $[B]$ is obtained in the form,

$$[E_t] = [B_k, B_{k-1}, \cdots B_t], \ 1 \le t < k \tag{4.63}$$

such that B_k is linearly dependent on the others. That is, $[E_{t^*}]$ is the desired minimal set if $[B_k]$ is not linearly dependent on $[B_{k-1}]$ to $[B_{t^*}]$. It is crucial to reducing the band width to notice that the columns in $[E_t]$ are in the opposite order of the indices in $[B]$. Suppose that $[E_{t^*}]$ is the minimal dependent set. It follows, because $[B_k]$ is linearly dependent on columns $(k-1)$ to t^*, there exist scalars $(\beta_k \cdots \beta_{t^*})$ such that $\beta_k \ne 0$ and

$$\sum_{i=k}^{t^*} B_i \beta_i = 0$$

Let

$$\{y^j\}^T = [0, 0, 0, 0, \beta_{t^*}, \beta_{t^*+1}, \cdots, \beta_{k-1}, \beta_k, 0, 0, 0]^T$$

Then $\{y^j\}$ is a solution of eqn (4.61). This solution compares favourably with $\{b^j\}$, since the forms of $\{b_j\}$ and $\{y_j\}$ are

$$b^j = \begin{bmatrix} \cdot \\ \cdot \\ \cdot \\ \cdot \\ 0 \\ 0 \end{bmatrix}; \quad y^j = \begin{bmatrix} 0 \\ 0 \\ \cdot \\ \cdot \\ \cdot \\ 0 \end{bmatrix}$$

If this turn–back procedure is applied to each j contained in $(2, \cdots p)$ to obtain a solution $\{y^j\}$ of eqn (4.61) (for $j = 1$, use $\{y^1\} = \{b^1\}$), the resulting self–stress matrix $[b_{XTB}] = [y^1 \cdots y^p]$ tends to be sparse and banded. Diagrammatically the two matrices are represented thus:

Form of b_{1LU} and b_{1TB}

4.7 DEFLECTIONS OF CURVED MEMBERS

4.7.1 Introduction

The fundamentals of the force method of analysis have been developed in the preceding sections of this chapter. Members have been considered to be straight and prismatic and to exhibit only bending and axial distortions which are independent of the transverse deflection of the member (small displacements). In section 4.7, curved members are considered, and, in section 4.8, non–uniform members, shear distortions and axial force effects are analysed.

These topics extend the element library available to the analyst using the force method. A curved member may always be approximated by a series of straight members, the junction of two members becoming a node in the structure model, as shown in Fig. 4.18.

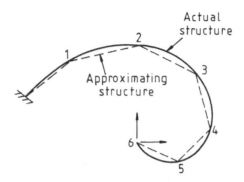

Fig. 4.18   Curved member.

Evidently, the more straight segments that are incorporated, the smaller their length and the closer the properties of the approximate structure approach those of the actual structure. The flexibility matrix of the structure for its terminal forces and moments can thus be approximated using eqn (4.21). The above procedure may be useful if the member curve is of non–mathematical form and defined at a few discrete points. In what follows, however, it can be assumed that the curved member can be approximated by a quadratic curve and the flexibility matrix calculated by numerical integration. Herein, two classes of curved members are considered which fit into the categories of frame and grid structures. These two classes are:

1.   Members curved in a plane and loaded in this plane. They possess a principal plane in the load plane and are usually referred to as arch structures.

2.   Members curved in a plane and loaded perpendicular to this plane. Again a principal axis will be assumed to be parallel to the load plane. These members are often referred to as beams curved in plan.

There is an analogy between the forces and displacements of the two systems, and once one is analysed, the other follows very simply. The detailed discussion here is restricted to arch members for which displacements are considered to be small only the essential differences of beams curved in plan are given.

4.7.2   Arch members

The curved arch member is shown in Fig. 4.19. In the first instance the effects throughout the arch will be calculated for the forces $<R_{x0}\ R_{y0}\ M_{z0}>$ acting at the free end. The calculation of the distribution of forces throughout the arch becomes an application of the transformation eqns (1.38) and (1.43). That is, transfer $<R_{x0}\ R_{y0}\ M_{z0}>$ first from O to P in axes parallel to XOY, and rotate into the set X', P, Y' at right angles to the member cross section at P. Thence,

$$\begin{bmatrix} R_x \\ R_y \\ M_z \end{bmatrix}_P = \begin{bmatrix} 1 & 0 & 0 \\ 0 & 1 & 0 \\ y & -x & 1 \end{bmatrix} \begin{bmatrix} R_x \\ R_y \\ M_z \end{bmatrix}_0 \tag{4.64}$$

or, in shorthand notation,

$$\{R_P\} = [b_1]\{R_0\} \tag{4.65}$$

Now transfer $\{R_P\}$ to components in the local axes X', Y'. From eqn (1.39),

$$\{R'_P\} = [L]\{R_P\} = [L][b_1]\{R_0\} \tag{4.66}$$

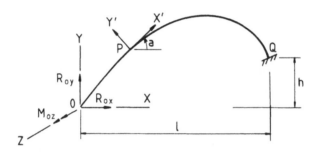

Fig. 4.19   Arch structure.

Here,

$$[L] = \begin{bmatrix} \cos\alpha & \sin\alpha & 0 \\ -\sin\alpha & \cos\alpha & 0 \\ 0 & 0 & 1 \end{bmatrix} \tag{4.67}$$

In eqn (4.66), the relationship in eqn (4.68) has been developed.

$$\{R'_P\} = [b]\{R_0\} \tag{4.68}$$

For the element length ds at P, the term $\{R'_P\}$ is the member force. The flexibility of the element is given by

$$[df] = \left| \frac{ds}{EA} \quad \frac{ds}{GA_{eff}} \quad \frac{ds}{EI} \right| = [f]ds \tag{4.69}$$

For the element ds, the distortion is given by

$$\{\delta v'_P\} = [f]\{R'_P\}ds \tag{4.70}$$

By the contragredient principle, the resulting deflection at O is given by

$$\{\delta r_0\} = [b^T]\{\delta v'_P\}ds = [b^T][f][b]\{R_0\}ds \tag{4.71}$$

For all such elements, integrate over the arch length from $x = 0$ to $x = l$. Therefore,

$$\{r_0\} = \int_0^l [b]^T [f][b]\ ds\ \{R_0\} \tag{4.72}$$

It remains to carry out the integration by some satisfactory numerical method. For example, if numerical quadrature such as Simpson's rule is used, coordinate values (x, y) for equal arc lengths ds must be calculated and thence $[b]^T[f][b]$ evaluated at each of these points. Then the approximation to eqn (4.72) will take the form,

$$\int_0^l = \frac{h}{3}[([b]^T[f][b])_1 + 4([b]^T[f][b])_2 + 2([b]^T[f][b])_3 + \cdots + ([b]^T[f][b])_n)] \tag{4.73}$$

When the deflection is required at O due to a load at another point on the arch, the eqn (4.73) is written,

$$\{r_{01}\} = \int_{x_1}^l [\bar{b}]^T [f][b]\ ds\ \{R_1\} \tag{4.74}$$

The integral takes place only over the region common to the diagrams $[b]$ and $[\bar{b}]$. It is convenient in arch analysis to subdivide the arch into an even number of intervals, and then limit the load $\{R_1\}$ to the even points, so that no discontinuity occurs at $\{R_1\}$ in the Simpson's integration.

4.7.2.1   Member flexibility matrices

In eqn (4.72) the integral gives the flexibility matrix in global coordinates at end Q; that is,

$$[F_0] = \int_0^l [b]^T [f][b]\ ds. \tag{4.75}$$

Transfer the force at the origin O to an equivalent force at Q. Then,

$$\begin{bmatrix} R_x \\ R_y \\ M_z \end{bmatrix}_Q = \begin{bmatrix} 1 & 0 & 0 \\ 0 & 1 & 0 \\ h & -l & 1 \end{bmatrix} \begin{bmatrix} R_x \\ R_y \\ M_z \end{bmatrix}_0 \tag{4.76}$$

This equation is written

$$\{R_Q\} = [B]\{R_0\} \tag{4.77}$$

The flexibility matrix for end Q is then given by

$$[F_Q] = [B]^T [F_0][B] \tag{4.78}$$

It is seen that if the flexibility matrix at one end is known, then that at the other end may be calculated from the arch geometry using eqn (4.78).

4.7.3   Beams curved in plan

Only beams whose centroid and shear centre coincide are considered. The beam is curved in one of its principal planes and the loads applied in the other principal plane. The axes chosen are shown in Fig. 4.20. It is seen that the X–Y plane is the plane in which the beam is curved, and the load is applied as a force F_z and moments M_x and M_y.

For the beam the origin of coordinates O has been taken at the fixed end. The coordinates of the free end will be (x_E, y_E), and those of the general point P, (x_P, y_P). Local and global coordinates at P are shown in Fig. 4.20(a) and (b). The distances α_x and α_y are defined by

$$\alpha_x = x_E - x_P ; \qquad \alpha_y = y_E - y_P \tag{4.79}$$

The distribution of forces throughout the beam for the forces $<F_z \; M_x \; M_y>$ on the free end is obtained by application of eqns (1.38) and (1.43). The transformation from E to P in global components is given by eqn (4.80)

$$\begin{bmatrix} F_z \\ M_x \\ M_y \end{bmatrix}_P = \begin{bmatrix} 1 & 0 & 0 \\ \alpha_y & 1 & 0 \\ -\alpha_x & 0 & 1 \end{bmatrix} \begin{bmatrix} F_z \\ M_x \\ M_y \end{bmatrix}_E \tag{4.80}$$

That is,

$$\{F_P\} = [b_1]\{F_E\} \tag{4.81}$$

The transformation from $\{F_P\}$ to local member components at P, that is, $\{F'_P\}$, is given by eqn (4.82),

$$\begin{bmatrix} F'_z \\ M'_x \\ M'_y \end{bmatrix} = \begin{bmatrix} 1 & 0 & 0 \\ 0 & \cos\theta & \sin\theta \\ 0 & -\sin\theta & \cos\theta \end{bmatrix} \begin{bmatrix} F_z \\ M_x \\ M_y \end{bmatrix}_P \tag{4.82}$$

That is,

$$\{F'_P\} = [L]\{F_P\} \tag{4.83}$$

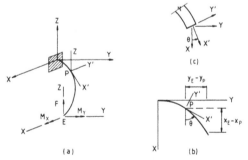

Fig. 4.20 Beam curved in plan.

Then, combined with eqn (4.81),

$$\{F'_P\} = [L][b_1]\{F_E\} \tag{4.84}$$

or,

$$\{F'_P\} = [b]\{F_E\} \tag{4.85}$$

For the principal axes shown in Fig. 4.20 the flexibility matrix is given by

$$[f]ds = \begin{bmatrix} \dfrac{ds}{GA_{eff}} & \dfrac{ds}{GI_P} & \dfrac{ds}{EI_x} \end{bmatrix} \tag{4.86}$$

The torsion stiffness term, I_P, may be calculated from St Venant's torsion theory (Chapter 7, section 4). For the element ds at P, the element distortions due to $\{F'_P\}$ are given by

$$\{\delta v'_P\} = [f][F'_P]ds = [f][b]\{F_E\}ds$$

By the contragredient principle, the resulting displacement at E is given by

$$\{\delta'_P\} = [b]^T[f][b]ds\{F_E\} \tag{4.87}$$

For all elements, the deflection at E is obtained by integrating $\{\delta_E\}$ from $x = 0$, to $x = x_E$. That is,

$$\{r_E\} = \int_0^{x_E} [b]^T[f][b]ds\{F_E\} \tag{4.88}$$

This equation may be integrated numerically in the same way as that for the arch problem.

4.8 MEMBER FLEXIBILITY MATRICES

4.8.1 Non–uniform members

For the general application of the equation $\{r\} = [b^T][f]\{R\}$ to the calculation of deflections of structures, means must be devised to generate $[f]$ for members other than those simple cases already discussed in section 4.7. Thus, in section 4.8, the flexibility matrix has been developed for arches and beams curved in plan. In this section, the theory is extended by considering further calculations of $[f]$ matrices. Perhaps the most common member form found in frame structures is that with non–uniform cross section and consequently with non–uniform moment of inertia. Three types of variable member inertia are shown in Fig. 4.21. Special note should be made of the type in Fig. 4.21, if it is to be included in three–dimensional structures, because in this case the curved centroidal axis will cause extra complications for forces in the Z' direction. However, in plane frame or grid structures, this difficulty generally does arise. In this section then, any variation in member properties will be assumed to be symmetrical about the member centroid when used in three–dimensional analyses.

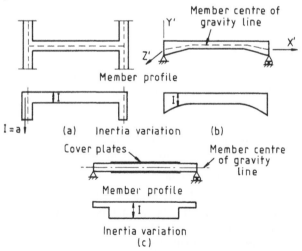

Fig. 4.21   Member types with variable inertia.

4.8.1.1 Member with infinite end inertia

This type of member is typical of rigid-frame structures in which column and beam thicknesses are taken into consideration. The member is shown in Fig. 4.22 and is considered to be composed of three members (1), (2) and (3) of lengths A, l_1, and B, respectively. Members (1) and (3) have infinite inertias; member (2) has inertia I.

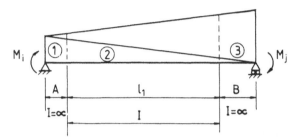

Fig. 4.22   Beam–member infinite end inertias.

The structure forces (M_i, M_j) applied at the ends i and j produce member forces in (1), (2) and (3), of which those of member (2) alone produce distortions. From Fig. 4.22 the $[b]$ matrix for member (2) is given by

$$\begin{bmatrix} M_{2i} \\ M_{2j} \end{bmatrix} = \begin{bmatrix} 1 - \dfrac{A}{l} & \dfrac{A}{l} \\ \dfrac{B}{l} & 1 - \dfrac{B}{l} \end{bmatrix} \begin{bmatrix} M_i \\ M_j \end{bmatrix}$$

The flexibility matrix for member (2) is

$$[f_2] = \frac{l_1}{6EI} \begin{bmatrix} 2 & 1 \\ 1 & 2 \end{bmatrix}$$

Hence, for the whole member,

$$[f] = \frac{l_1}{6EI} \begin{bmatrix} 1 - \dfrac{A}{l} & \dfrac{B}{l} \\ \dfrac{A}{l} & 1 - \dfrac{B}{l} \end{bmatrix} \begin{bmatrix} 2 & 1 \\ 1 & 2 \end{bmatrix} \begin{bmatrix} 1 - \dfrac{A}{l} & \dfrac{A}{l} \\ \dfrac{B}{l} & 1 - \dfrac{B}{l} \end{bmatrix} \qquad (4.89)$$

The multiplication of the matrices in eqn (4.89) may be undertaken explicitly; however, it is convenient in writing the software to retain the form of eqn (4.89).

4.8.1.2 Member with variable area and inertia

The member is shown in Fig. 4.23(a), and it will be assumed that the area and inertia values are continuous functions of x. Discontinuities may be accommodated either by considering each region separately or, alternatively, by subdividing the span into a sufficiently large number of intervals to isolate the discontinuity in only one small region. The member flexibility is calculated by $[F] = [b]^T[f][b]$.

Now, however, $[f]$ is expressed for the infinitesimal element dx, so that $[F]$ is calculated by matrix integration similar to that in eqn (4.74). Hence,

$$[F] = \int_0^l [b]^T[f][b]\,dx$$

Consider the axial force and moment terms independently. Simpson's rule is used in the integration, although any of the accepted methods such as Gauss' quadrature could be used. For purposes of illustration the member is divided into four equal intervals. In practice, the interval of subdivision may be finer for both the design process of the member and, of course, for the computer software. Then,

$$f_{axial} = \int_0^l 1 \frac{1}{EA} 1 \, dx = \frac{1}{EA_1} \int_0^l \alpha \, dx = \frac{h}{3EA_1}\{W\}^T\{\alpha\} ; \qquad \alpha = \frac{A_1}{A} \tag{4.90}$$

In eqn (4.90)

$$\{W^T\} = <1 \ 4 \ 2 \ 4 \ 1> \quad \text{and} \quad \{\alpha\}^T = <\alpha_1 \ \alpha_2 \ \alpha_3 \ \alpha_4 \ \alpha_5> \tag{4.91}$$

For moment flexibility,

$$M_x = M_i\left[1 - x/l\right] + M_j x/l = [1 - x/l \ \ x/l]\begin{bmatrix} M_i \\ M_j \end{bmatrix} = [b]\{R\}$$

Then,

$$[f]_{moment} = \begin{bmatrix} f_{11} & f_{12} \\ f_{21} & f_{22} \end{bmatrix} = \int_0^l \begin{bmatrix} 1 - x/l \\ x/l \end{bmatrix} 1/EI\left[1 - x/l \ \ x/l\right] dx$$

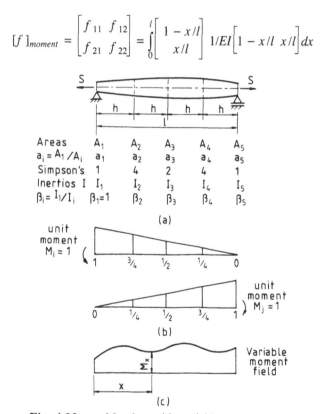

Fig. 4.23   Member with variable cross section.

Let $\beta = I_1/I$; then substituting for I in terms of β in the above equation and performing the matrix multiplications gives

$$[f]_{moment} = \frac{1}{EI_1} \int_0^l \beta \begin{bmatrix} (1-x/l)^2 & x/l(1-x/l) \\ x/l(1-x/l) & (x/l)^2 \end{bmatrix} dx$$

For sections 1 to 5,

$$\frac{x}{l} = (0, \frac{1}{4}, \frac{2}{4}, \frac{3}{4}, 1) = \{\zeta_1\}^T \; ; \; 1 - \frac{x}{l} = (1, \frac{3}{4}, \frac{2}{4}, \frac{1}{4}, 0) = \{\zeta_2\}^T$$

Define the matrix multiplication,

$$[A] * [B] = \begin{bmatrix} a_{11}b_{11} & a_{12}b_{12} \\ a_{21}b_{21} & a_{22}b_{22} \end{bmatrix} = [C]$$

That is, a multiplication such that the terms of $[C]$ are simply the products of corresponding terms of $[A]$ and $[B]$. Then, evaluated at the five points,

$$(x/l)^2 = \{\zeta_1\}*\{\zeta_1\} \; ; \; \text{and} \; (1-x/l)^2 = \{\zeta_2\}*\{\zeta_2\} \; ; \; x/l(1-x/l) = \{\zeta_1\}*\{\zeta_2\}$$

Then define $[\beta] = \begin{bmatrix} \beta_1 & \beta_2 & \beta_3 & \beta_4 & \beta_5 \end{bmatrix}$. Thence the flexibility matrix terms are

$$[f_{11}] = \{W\}^T [\beta]\{\zeta_2\}*\{\zeta_2\}$$

$$[f_{12}] = [f_{21}] = \{W\}^T [\beta]\{\zeta_2\}*\{\zeta_1\}$$

$$[f_{22}] = \{W^T\}[\beta]\{\zeta_1\}*\{\zeta_1\}$$

Collectively, these results are written

$$\begin{bmatrix} f_{11} & f_{12} \\ f_{21} & f_{22} \end{bmatrix} = \frac{1}{EI_1} \begin{bmatrix} W^T & \cdot \\ \cdot & W^T \end{bmatrix} \begin{bmatrix} \beta & \cdot \\ \cdot & \beta \end{bmatrix} \begin{bmatrix} \zeta_2*\zeta_2 & \zeta_1*\zeta_2 \\ \zeta_2*\zeta_1 & \zeta_1*\zeta_1 \end{bmatrix} \quad (4.92)$$

End rotations

If a variable moment field, m_x, shown in Fig. 4.23(c), is applied to the member, the end rotations $<\phi_i \; \phi_j>$ may be calculated by using eqn (4.75), in which case,

$$m_x = [b]\{R\}$$

Then the end rotations are calculated

$$\begin{bmatrix} \phi_i \\ \phi_j \end{bmatrix} = \frac{1}{EI_1} \int_0^l \begin{bmatrix} 1-x/l \\ x/l \end{bmatrix} \beta \, m_x \, dx$$

Let the moments along the member be written, $\{m\}^T = <m_1 \; m_2 \; m_3 \; m_4 \; m_5>$. Then, finally, the end rotations are given

$$\begin{bmatrix} \phi_i \\ \phi_j \end{bmatrix} = \frac{h}{3EI_1} \begin{bmatrix} W^T & \cdot \\ \cdot & W^T \end{bmatrix} \begin{bmatrix} \beta & \cdot \\ \cdot & \beta \end{bmatrix} \begin{bmatrix} \zeta_2*m \\ \zeta_1*m \end{bmatrix} \quad (4.93)$$

Example 4.9 *Beam end rotations – uniformly distributed load*

Let m_x be produced by the constant load w per unit length, so that

$$m_1 = m_5 = 0; \quad m_2 = m_4 = \frac{3wl^2}{32}; \quad m_3 = \frac{wl^2}{8}$$

In this case, for the four intervals, $h = l/4$, so that

$$\phi_i = \frac{l}{12EI}\left[4\frac{3}{4}\frac{3wl^2}{32} + 2\frac{2}{4}\frac{wl^2}{8} + 4\frac{1}{4}\frac{3wl^2}{32} \right] = \frac{wl^2}{24EI} = \phi_j$$

Equations (4.9) and (4.93) are readily incorporated in computer programs either for direct deflection calculation, as in section 4.3, eqn (4.16), or later in Chapter 5, for calculation of fixed–end moments.

Fig. 4.24   Beam end moments and shears.

4.8.2   Effect of shear distortions

Once again, for the calculation of the effects of shear distortion due to the end moments $<M_i \ M_j>$, eqn (2.110) is used. The beam is shown in Fig. 4.24, and the shear stress distribution due to the transverse stress resultant V is

$$\tau = \frac{VQ}{It}$$

However, the shear is expressed in terms of the end moments by

$$V = \left[-\frac{1}{l} \quad \frac{1}{l} \right]\left[\begin{matrix} M_i \\ M_j \end{matrix} \right]$$

Therefore, the shear stress is given by

$$\tau = \frac{Q}{It}\left[-\frac{1}{l} \quad \frac{1}{l} \right]\left[\begin{matrix} M_i \\ M_j \end{matrix} \right]$$

For the infinitesimal element, the shear distortion is given as $\gamma = \dfrac{\tau}{G}$. Thence,

$$\{r\} = \int_{vol} [b]^T[f][b]dV\{R\} = \frac{1}{G}\int_{vol}\left[\begin{matrix} -\dfrac{1}{l} \\ \dfrac{1}{l} \end{matrix} \right]\left(\frac{Q}{It} \right)^2\left[-\dfrac{1}{l} \quad \dfrac{1}{l} \right]dV\ \{R\}$$

$$= \frac{1}{G}\frac{1}{l^2}\frac{1}{l^2}\left[\begin{matrix} 1 & -1 \\ -1 & 1 \end{matrix} \right]\int_0^l\int_{c_1}^{c_2}\frac{Q^2}{t}dy\ dx\ \{R\}$$

Let A_{eff} be the effective shear area considered for shear deflections, such that

$$A_{eff} = \frac{l^2}{\int\limits_{c_1}^{c_2} \frac{Q^2}{t} \, dy}$$

With this notation,

$$\begin{bmatrix} \phi_i \\ \phi_j \end{bmatrix} = \frac{1}{GA_{eff}} \frac{1}{l} \begin{bmatrix} 1 & -1 \\ -1 & 1 \end{bmatrix} \begin{bmatrix} M_i \\ M_j \end{bmatrix}$$

That is, the shear flexibility contribution is

$$[f]_{shear} = \frac{1}{GA_{eff}} \frac{1}{l} \begin{bmatrix} 1 & -1 \\ -1 & 1 \end{bmatrix}$$

Notice that f_{shear} is a singular matrix. This implies that a beam with shear resistance to deformation alone cannot carry the moments (M_i, M_j). The complete member flexibility matrix, including shear and bending distortions, is written

$$[f] = \frac{l}{6EI} \begin{bmatrix} 2+\alpha & 1-\alpha \\ 1-\alpha & 2+\alpha \end{bmatrix}; \qquad \alpha = \frac{6EI}{GA_{eff}l^2}$$

4.8.3   Axial force effects

4.8.3.1   Axial compression

The effect of axial compression is clearly to increase bending flexibility, since the deflections produced by (M_i, M_j) cause P to induce bending moments which further increase the deflections and thus (ϕ_i, ϕ_j). The prismatic member of length l and with moments (M_i, M_j) and axial compression P is shown in   Fig. 4.25.

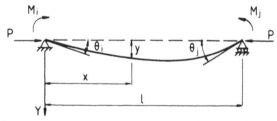

Fig. 4.25   Prismatic member axial compression and end moments.

The moment M_x along the member is given by

$$M_x = \left(1 - \frac{x}{l}\right) M_i + \frac{x}{l} M_j + Py$$

The differential equation to the deflected shape due to bending distortions is, thence,

$$EIy'' = -M_x = -(1 - \frac{x}{l}) M_i - \frac{x}{l} M_j - Py$$

Then,

$$y'' + k^2 y = \frac{-1}{EI}\left[(1 - \frac{x}{l})M_i + \frac{x}{l}M_j\right] \quad \text{with} \quad k^2 = \frac{P}{EI} \tag{4.94}$$

The complementary function for eqn (4.94) is

$$y_c = A\cos kx + B\sin kx$$

and the particular solution,

$$y_P = \frac{-1}{k^2 EI}\left[(1 - \frac{x}{l})M_i + \frac{x}{l}M_j\right]$$

The complete solution is the sum of these two solutions, so that

$$y = y_c + y_P = A\cos kx + B\sin kx - \frac{1}{k^2 EI}\left[(1 - \frac{x}{l})M_i + \frac{x}{l}M_j\right] \tag{4.95}$$

The constants A and B are determined from the boundary conditions, (1) when $x = 0$, $y = 0$, and, (2) when $x = l$, $y = 0$.

1. $0 = A - \dfrac{M_i}{k^2 EI}$ gives $A = \dfrac{M_i}{k^2 EI}$

2. $0 = \dfrac{M_i}{k^2 EI}\cos kl + B\sin kl - \dfrac{M_j}{k^2 EI}$ gives $B = \dfrac{1}{\sin kl\, k^2 EI}(M_j - M_i\cos kl)$

Substitute for A and B in eqn (4.95),

$$y = \frac{1}{k^2 EI}\left[M_i\cos kx + \frac{1}{\sin kl}(M_j - M_i\cos kl)\sin kx - (1 - \frac{x}{l})M_i - \frac{x}{l}M_j\right] \tag{4.96}$$

On differentiation of eqn (4.96) with respect to x, the equation to the slope of the deflected shape is obtained,

$$y' = \frac{1}{k^2 EI}\left[-M_i k\sin kx + \frac{k}{\sin kl}(M_j - M_i\cos kl)\cos kx + \frac{1}{l}M_i - \frac{1}{l}M_j\right] \tag{4.97}$$

The end slopes ϕ_i and ϕ_j are determined by substituting $x = 0$ and $x = l$, respectively, in eqn (4.97).

$$y'_{(0)} = \phi_i = \frac{1}{k^2 EI}\left[M_i(\frac{1}{l} - \frac{k\cos kl}{\sin kl}) - M_j(\frac{1}{l} - \frac{k}{\sin kl})\right] \tag{4.98a}$$

$$-y'_{(l)} = \phi_j = \frac{1}{k^2 EI}\left[-M_i(\frac{1}{l} - \frac{k}{\sin kl}) + M_j(\frac{1}{l} - \frac{k\cos kl}{\sin kl})\right] \tag{4.98b}$$

The notation is now changed slightly, so write

$$\beta = \frac{l}{2}\sqrt{\frac{P}{EI}}; \quad \text{that is,} \quad \beta = \frac{kl}{2}$$

and, thus,

$$\cos kl = \cos 2\beta = c; \quad \sin kl = \sin 2\beta = s$$

With this change in notation, eqn (4.98) is rewritten in matrix form

$$\begin{bmatrix} \phi_i \\ \phi_j \end{bmatrix} = \frac{1}{Pls} \begin{bmatrix} s - 2\beta c & -(s - 2\beta) \\ -(s - 2\beta) & s - 2\beta c \end{bmatrix} \begin{bmatrix} M_i \\ M_j \end{bmatrix} \qquad (4.99)$$

If now the ϕ functions used in [13] are introduced,

$$\phi_1 = \beta \cot \beta; \quad \phi_2 = \frac{1}{3} \frac{\beta^2}{1 - \beta \cot \beta}$$

$$\phi_3 = \frac{3\phi_2}{4} + \frac{\phi_1}{4}; \quad \phi_4 = -\frac{3\phi_2}{2} + \frac{\phi_1}{2}; \quad \phi_5 = \phi_1 \phi_2$$

After some algebra the end rotations are obtained

$$\begin{bmatrix} \phi_i \\ \phi_j \end{bmatrix} = \frac{l}{6EI} \begin{bmatrix} 2\left(\dfrac{3}{4\phi_1} + \dfrac{1}{4\phi_2}\right) & \dfrac{3}{2\phi_1} - \dfrac{1}{2\phi_2} \\ \dfrac{3}{2\phi_1} - \dfrac{1}{2\phi_2} & 2\left(\dfrac{3}{4\phi_1} + \dfrac{1}{4\phi_2}\right) \end{bmatrix} \begin{bmatrix} M_i \\ M_j \end{bmatrix} \qquad (4.100)$$

4.8.3.2   Axial tension

When the member is subjected to axial tension, the effect is the reverse of that for axial compression; that is, the flexibility of the member is decreased. In this case, the equation for the deflected shape is

$$y'' - k^2 y = \frac{-1}{EI}\left[(1 - \frac{x}{l})M_i + \frac{x}{l}M_j\right] \qquad (4.101)$$

Now the solution to this equation is

$$y = -\frac{M_i}{P}\cosh kx + \frac{\sinh kx}{P \sinh kl}(M_i \cosh kl - M_j)x + \frac{1}{Pl}(-M_i + M_j)x + \frac{M_i}{P} \qquad (4.102)$$

and hence the equation for member slope is,

$$y' = -\frac{kM_i}{P}\sinh kx + \frac{k \cosh kx}{P \sinh kl}(M_i \cosh kl - M_j) + \frac{1}{Pl}(-M_i + M_j) \qquad (4.103)$$

Finally, let $s = \sinh kl$, $c = \cosh kl$ and $\beta = kl/2$. The end rotations are obtained from y'_0 and y'_l,

$$\begin{bmatrix} \phi_i \\ \phi_j \end{bmatrix} = \frac{1}{P} \begin{bmatrix} \dfrac{kc}{s} - \dfrac{1}{l} & \dfrac{1}{l} - \dfrac{k}{s} \\ \dfrac{1}{l} - \dfrac{k}{s} & \dfrac{kc}{s} - \dfrac{1}{l} \end{bmatrix} \begin{bmatrix} M_i \\ M_j \end{bmatrix} \qquad (4.104)$$

Now, with the following definitions,

$$\phi_1 = \beta \coth \beta; \quad \phi_2 = \frac{1}{3} \frac{\beta^2}{\beta \coth \beta - 1}$$

$$\phi_3 = \frac{1}{4}(\phi_1 + 3\phi_2); \quad \phi_4 = \frac{1}{2}(\phi_1 - 3\phi_2)$$

After some algebra,

$$\begin{bmatrix} \phi_i \\ \phi_j \end{bmatrix} = \frac{l}{6EI} \begin{bmatrix} 2\left[\dfrac{3}{4\phi_1} + \dfrac{1}{4\phi_2}\right] & \dfrac{3}{2\phi_1} - \dfrac{1}{2\phi_2} \\[2mm] \dfrac{3}{2\phi_1} - \dfrac{1}{2\phi_2} & 2\left[\dfrac{3}{4\phi_1} + \dfrac{1}{4\phi_2}\right] \end{bmatrix} \begin{bmatrix} M_i \\ M_j \end{bmatrix} \tag{4.105}$$

It should be noted that when eqn (4.100) and (4.105) are used, the effect of P on bending flexibility has been taken into consideration, whereas the effects of $(M_i\ M_j)$ on axial flexibility have not. When the structure is statically determinate, the ϕ functions in eqns (4.100) and (4.105) may be calculated directly from the external loads, provided that gross changes in the nodal geometry do not occur. This is not possible in statically indeterminate structures in which the member flexibilities influence the values of the indeterminate forces, which, in turn, affect member forces and hence member flexibilities.

4.9 DEFLECTION ALONG MEMBER

Deflections at any point of a statically determinate structure may be calculated by using eqn (4.20) or (4.23) and simply including that point as a node of the structure. Thus, for the portal frame shown in Fig. 4.26(a) the deflections at point B are calculated by using nodes 1 to 5 and members (1) to (4) as shown.

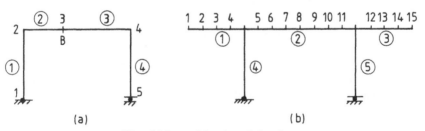

Fig. 4.26   Member deflections.

However, it will be realized that every point added as a node will necessarily increase the size of the statics matrix to be solved and also the sizes of the $[f]$ and $[b]$ matrices in eqn (4.20) etc. Since the size of the matrix increases by three in plane frames and six in three–dimensional frames for each node added and the storage required for a matrix is a function of the square of its size, it will be appreciated that this process of adding nodes soon becomes prohibitive in memory requirements. For example, in Fig. 4.26(b), the three members (1), (2) and (3) have 15 nodes for which deflections are required, and this necessitates a (45 × 45) matrix for these nodes alone. In addition, if member (2)–(3)–(4) in Fig. 4.26(a) is nonprismatic, it will be inconvenient to calculate inertias at equal intervals over (2)–(3) and over a different interval in member (3)–(4) for flexibility matrix generation. It is more practical to give inertias at equal intervals over the whole length (2) to (4). Members are then defined by their natural physical configuration. Thus, they may be terminated at changes in structure geometry as, for example, in Fig. 4.26(b), where the convenient number of members for the structure is five. When such a structure is analysed for node forces, node deflections and member bending moments are obtained.

From the terminal member moments and loads on the member, bending moments may be calculated at those sections for which the inertias are given. Deflections may be calculated at points along the member in the local member coordinates, given the node deflections, using numerical quadrature on the M/EI diagram to obtain deflections relative to the end values. If Simpson's rule is used as the quadrature formula, deflections will be calculated for every second point of the beam subdivision. Thus, for example, if M/EI is given at the tenth points, it is necessary to interpolate M/EI first to obtain values at the 20th points. In the development given here, quadratic interpolation is used to obtain these half–interval values. Furthermore, special provision can be made for members which have a hinge release along their length. It is assumed, then, that displacements (u_i, v_i, θ_i) and (u_j, v_j, θ_j) have been obtained for member i, j in local coordinates. This involves a transformation such as that in eqn (1.39) from the node displacements calculated in global coordinates.

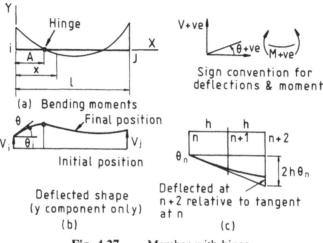

Fig. 4.27   Member with hinge.

Since the displacements (u_i, u_j) in the OX member coordinates (Fig. 4.27) will generally differ by only a small amount (the member distortions being small in its axial direction), the displacement u_x at distance x from i may be taken to be

$$u_x = u_i + (u_j - u_i)\frac{x}{l}$$

The v displacements will be of larger magnitude, so that v_x may differ significantly from v_i, v_j), and are shown in Fig. 4.27(b) for a member with a hinge at the point A. Notice that a slope discontinuity occurs at the hinge.

Interpolation for deflections

Suppose that $\alpha = M/EI$ is given at the intervals $n, n + 1, n + 4$, etc. To calculate deflections at these points, the values of α are interpolated at $n + 1, n + 3$, etc. If α is given at equally spaced intervals $(n - 1), n, (n + 1)$, it easily shown that the half interval values $(n - 1/2), (n + 1/2)$ are

$$\begin{bmatrix} \alpha_{(n-1/2)} \\ \alpha_{(n+1/2)} \end{bmatrix} = \frac{1}{8}\begin{bmatrix} 3 & 6 & -1 \\ -1 & 6 & 3 \end{bmatrix}\begin{bmatrix} \alpha_{(n-1)} \\ \alpha_{(n)} \\ \alpha_{(n+1)} \end{bmatrix}$$

A small segment of the beam is shown in Fig. 4.27(c). It is seen that the slope and deflection at station $n + 2$ may be written in terms of the various quantities at n, $n + 1$, $n + 2$. Assuming that α is parabolic in its variation over the interval n to $n + 2$, the slope at $n + 2$ is given by

$$\theta_{n+2} = \theta_n + \frac{h}{3}(\alpha_n + 4\alpha_{n+1} + \alpha_{n+2}) \qquad (4.106)$$

The deflection is given by

$$v_{n+2} = v_n + \theta_n 2h + \frac{h}{3}(\alpha_n 2h + 4\alpha_{n+1}h) \qquad (4.107)$$

By using these two equations it is possible to progress from the left–hand end, which has slope θ_i and deflection v_i, to the slope $\bar{\theta}_j$ and deflection $\bar{v}_j$ at the right–hand end of the beam. Of course, because numerical quadrature has been used, $\bar{v}_j$ will be, in general, slightly different from v_j. However, when a hinge is present, $\bar{v}_j$ will differ significantly from v_j because of the slope discontinuity there. Adjustment must then be made to the deflections for $x > A$, the quantity A being the distance from i to the hinge. The correction will be

$$x > A \qquad v_x = \bar{v}_x + (v_j - \bar{v}_j)\frac{(x - A)}{(l - A)} \qquad (4.108)$$

Finally, when all (u_x, v_x) have been calculated, they must be transformed back into global components (eqn (1.39)).

Example 4.10 *Interpolation of deflections*

Calculate deflections at the one–tenth points of member AC shown in Fig. 4.28. The member is prismatic with $I = 416 \times 10^3 mm^4$ and E = 206.8 GPa units. The deflections at A and C are calculated to be $\theta_A = -0.004586$ rad; $v_A = 0$; $v_C = 44.45$ mm.

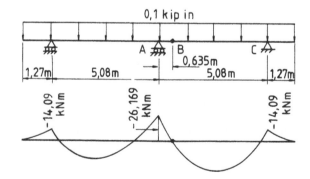

Fig. 4.28   Deflection of structure member with hinge.

Using the above theory, the deflections at the one–tenth points in member AC have been calculated and are given in Table 4.1.

Table 4.1

Distance x/l from A	Deflection mm
0.0	0.0
0.1	2.04
0.2	−3.86
0.3	−11.99
0.4	−19.39
0.5	−25.82
0.6	−31.17
0.7	−35.47
0.8	−38.89
0.9	− 41.74
1.0	− 44.45

4.10 NON–LINEAR STRUCTURE BEHAVIOUR

4.10.1 Causes of non–linearity in the load deflection relationship

The linear elastic theory developed so far assumes that material properties are ideally elastic and that geometry changes do not alter member flexibility or overall structure behaviour. In section 4.8 expressions have been developed for the non–linear effects of axial force in a frame member on its bending flexibility, (eqn (4.99)). The analytic expressions were obtained for this effect for both axial tension and compression. Since the axial forces to be determined enter into the deflection calculations of the analysis, the procedure will, in general, be an iterative one. When a structure, such as an arch or suspension cable, with a large dead load is given a load increment, it will be generally incorrect to base deflection calculations solely on the elastic properties of the structure, even though deflections are small and the behaviour over the load range is essentially linear. For example, the cable shown in Fig. 4.29(a) has axial tension, T, of such a magnitude that for practical purposes the cable may be considered to be straight.

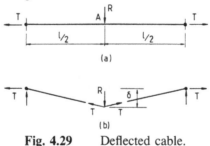

Fig. 4.29   Deflected cable.

In Fig. 4.29(b) the deflected shape required to carry the transverse load R is shown, and if the deflections are small, by considering the equilibrium in the deformed position,

$$\delta = \frac{l}{4T}R$$

Thus, a linear relationship exists between deflection and transverse force even though the geometric configuration gives the elastic flexibility at A to be infinite. When the cable is as shown in Fig. 4.30(a), both elastic and geometric flexibility of finite magnitude are available at A. For load increments at A both flexibilities should be considered in order to obtain the increment of the deflection.

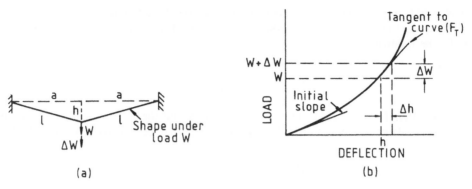

Fig. 4.30   Tangent flexibility.

From Fig. 4.30(b),

$$\Delta v = F_T \Delta W$$

where F_T is the tangent flexibility at load level W.

Changes in the geometry of the structure are manifested in ways other than modification of the member flexibility (stiffness) due to axial force. For example, suppose that the slender cantilever beam in Fig. 4.31 is composed of a material with a high yield point so that when the load R is applied its deflected shape is that shown in Fig. 4.31. Evidently, in this case, it would be a gross error to assume that from Fig. 4.31(a),

$$M_x = R(l - x)$$

Also from Fig. 4.31 it should be evident that to apply, in the case of large deflections v'_i, v'_j eqn (4.13) for calculating the chord rotation, ϕ, will cause considerable error. That is, for large displacements,

$$\phi \neq \frac{1}{l}(v'_j - v'_i)$$

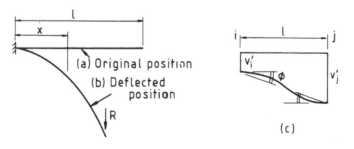

Fig. 4.31   Non–linear beam deflection.

4.10.2   Non–linear analysis by the force method

The method used here [17] adopts the general linear equations by introducing initial member strains and fictitious node forces to allow for non–linear effects. The general linear equations may be expressed for the following.

Member forces

$$\{S\} = [b]\{R\} + [\beta]\{H\} \tag{4.109}$$

In eqn (4.109), $\{R\}$ and $\{H\}$ are the applied forces and the initial member strains, respectively; see eqn (4.34) in which $\{\bar{v}\} \equiv \{H\}$. The matrix $[\beta]$ has been derived in eqn (4.39) and is given by

$$[\beta] = -[b_1][D_{11}]^{-1}[b_1]^T \tag{4.110}$$

Node deflections

$$\{r\} = [F]\{R\} + [b]^T\{H\} \tag{4.111}$$

The matrices $[\beta]$, $[F]$, and $[B]$ are all obtained as for the small deflection theory. It should be noted, however, that they may not be based on the original geometry, but rather on the geometry existing prior to the current load increment. To allow for non–linear effects, the fictitious, or ersatz, [17] node forces $\{R_i\}$ and initial member strains $\{H_i\}$ are introduced. That is, the structure is analysed with the original geometry and loads together with $\{R_i\}$ and $\{H_i\}$. However, $\{R_i\}$ and $\{H_i\}$ are functions of the member forces and the node displacements, so that the modified equations are non–linear. Thus, eqns (4.109) and (4.111) are written

$$\{S\} = [b](\{R\} + \{R_i\}) + [\beta](\{H\} + \{H_i\}) \tag{4.112}$$

$$\{r\} = [F](\{R\} + \{R_i\}) + [b^T](\{H\} + \{H_i\}) \tag{4.113}$$

The nature of $\{R_i\}$ and $\{H_i\}$ may be determined by considering the simple two–bar structure shown in Fig. 4.32. The first effect to be ascertained is that of the change in geometry on the equilibrium of the joint A. Suppose that S_0 is the member force in the original position (calculated from R on the assumption of no geometry change) and that it remains constant as the member rotates through the angle $\Delta\phi$, as shown in Fig. 4.32. The vector change ΔS_0 in S_0 is shown in Fig. 4.32(c), and, hence,

$$\Delta S_0 = S_0 \, \Delta\phi = \frac{S_0 \, r \cos \phi}{l} \tag{4.114}$$

Let

$$\cos \phi = c \, ; \quad \sin \phi = s \quad \text{and} \quad \frac{r}{y} = \rho$$

Then, with these substitutions,

$$S_0 = \frac{-1}{2s} R \, ; \quad \text{hence} \quad b = -\frac{1}{2s}$$

Substitution in eqn (4.114) gives

$$\Delta S_0 = -\frac{1}{2s} \frac{rc}{l} R = -\frac{1}{2} \rho c R$$

The vertical component of ΔS_0 is written ΔR_i

$$\Delta R_i = \Delta S_0 \cos \phi = -\frac{1}{2}\rho c^2 R$$

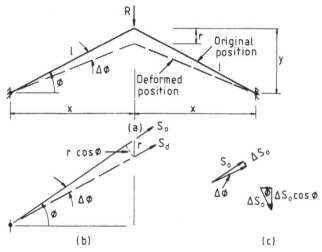

Fig. 4.32   Large deflections–forces.

For both members,

$$R_i = -\rho c^2 R \quad \text{or} \quad R_i = 2\rho s c^2 S_0 \tag{4.115}$$

In this derivation the same assumption has been made as in Fig. 4.32, that the deflection, r, is small, so that the above geometric approximations are valid. If large displacements occur for a given load, the structure must be analysed for a number of load increments when this type of approximation is used. Notice that in eqn (4.115) or (4.116), R_i is x a linear function of R or S_0 and also of the geometry change, r. Consider now the shortening of the members in Fig. 4.32 and the effect on r when the small deflection transformation, $\{r\} = [b]^T\{v\}$, is no longer valid. In Fig. 4.33 it is seen that when the member is shortened by the amount v, the vertical deflection r is produced by the rotation about end B. In order that the small deflection equation may be used, the additional strain, H_i, must be added, so that

$$\{r\} = [b]^T(\{v\} + \{H_i\}) \tag{4.116}$$

In Fig. 4.33, let the compressive strain $H_i = -X$, so that

$$2(l_d - X)X = (rc)^2$$

or, solving for X,

$$X \approx \frac{1}{2l_d}r^2c^2$$

Then the initial strain is given by

$$H_i \approx -\frac{r^2c^2}{2l_d} = \frac{y^2\rho^2c^2}{2l_d}$$

If the assumption $l_d \approx l$ is valid,

$$H_i \approx \frac{1}{2}ls^2\rho^2c^2 \qquad (4.117)$$

It is seen now, that if the displacements $\{r\}$ are relatively small when compared with the structure dimensions, the initial strains H_i are quadratic functions of the geometry change. In matrix form, eqn (4.116) will be written

$$\{R_i\} = [a_i]^T\{S\} \qquad (4.118)$$

Then eqn (4.112) becomes

$$\{S\} = [b](\{R\} + [a_i]^T\{S\}) + [\beta](\{H\} + \{H_i\}) \qquad (4.119)$$

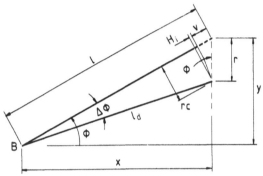

Fig. 4.33 Member distortions–large displacements.

Thence the member forces are

$$\{S\} = ([I] - [b][a_i]^T)^{-1}[b]\{R\} + ([I] - [b][a_i]^T)^{-1}[\beta](\{H\} + \{H_i\}) \qquad (4.120)$$

It is seen that

$$[b_d] = ([I] - [b][a_i]^T)^{-1}[b] \qquad (4.121)$$

is the force transformation matrix allowing for the $\{R_i\}$ forces.

$$\{S\} = [b_d]\{R\} + ([I] - [b][a_i]^T)^{-1}[\beta](\{H\} + \{H_i\}) \qquad (4.122)$$

Then, from eqn (4.118),

$$\{R_i\} = [a_i]^T([b_d]\{R\} + [I] - [b][a_i]^T)^{-1}[\beta](\{H\} + \{H_i\}) \qquad (4.123)$$

The introduction of eqn (4.123) into eqn (4.113) leads to a non–linear matrix equation in the unknown displacement vector $\{r\}$:

$$\{r\} = [F]\{R\} + [F][a_i]^T([I] - [b][a_i]^T)^{-1}[[b]\{R\}$$

$$+ [\beta](\{H\} + \{H_i\})] + [b^T](\{H\} + \{H_i\}) \qquad (4.124)$$

or, by using the expansion for the inverse of $([I] - [b][a_i])^{-1}$,

$$\{r\} = [F]\{R\} + [F][a_i][[I] + [b][a_i] + ([b][a_i])^2 + \cdots]$$

$$\times [[b]\{R\} + [\beta](\{H\} + \{H_i\})] + [b^T](\{H\} + \{H_i\}) \qquad (4.125)$$

In eqn (4.125), let

$$[a_i]^T[b] = [\phi_r] \quad \text{and} \quad [a_i]^T[\beta] = [\phi_H]$$

Substitution for ϕ_r and ϕ_H in eqn (4.125) give

$$\{r\} = [F]([I] + [\phi_r] + [\phi_r]^2 + \cdots)\{R\} +$$

$$[[b]^T + [F]([I] + [\phi_H] + [\phi_H]^2 + \cdots)]((\{H\} + \{H_i\}) \tag{4.126}$$

Now $\{R_i\} = [a_i]\{S\}$ and $\{S\} = [b]\{R\}$, so that $[a_i]^T[b]$ is a matrix of fictitious loads resulting from the linear element forces produced by unit external loads. Furthermore, $[a_i][\beta]$ is a matrix of fictitious loads resulting from the linear element forces produced by unit values of the element deformations. As an approximation, delete all powers of $[\phi_r]$ higher than the first and all products $[\phi_r]^n[\phi_H]$. Then, from eqn (4.126),

$$\{r\} = [F]([I] + [\phi_r])\{R\} + ([b^T] + [F][\phi_H])((\{H\} + \{H_i\}) \tag{4.127}$$

The matrices $[\phi_r]\{R\}$ and $[\phi_H]\{H\}$ involve the deflections $\{r\}$; therefore, for the first approximation as in eqn (4.115), assume that they are linear in $\{r\}$, so that

$$[\phi_r]\{R\} = [k_{ir}]\{r\} \quad \text{and} \quad [\phi_H]\{H\} = [k_{iH}]\{r\} \tag{4.128}$$

Then, by making these substitutions in eqn (4.115),

$$\{r\} = [F]\{R\} + [F][k_{ir}]\{r\} + [b]^T((\{H\}+\{H_i\}) + [F][k_{iH}]\{r\} + [F][\phi_H]\{H_i\}$$

Collecting terms involving $\{r\}$,

$$[[I] - F]([k_{ir}] + [k_{iH}])]\{r\} = ([F]\{R\} + [b^T]\{H\}) + ([b^T] + [F][\phi_H])\{H_i\} \tag{4.129}$$

For small deflections, $\{H_i\} = 0$, so that

$$\{r\} = [[I] - [F]([k_{ir}] + [k_{iH}])]^{-1}([F]\{R\} + [b^T]\{H\}) \tag{4.130}$$

When the matrix to be inverted tends to zero, the deflections tend to infinity. Hence, the small–deflection buckling conditions and modes are defined by the equation

$$[[I] - [F]([k_{ir}] + [k_{iH}])]\{r\} = 0 \tag{4.131}$$

The condition,

$$([I] - [F][k_{ir}]) = 0 \tag{4.132a}$$

governs the load buckling, and eqn (4.132b) the thermal buckling of the structure.

$$([I] - [F][k_{iH}])\{r\} = 0 \tag{4.132b}$$

Example 4.11 *Calculation of the critical load for a column*

The column of length l with EI constant is shown in Fig. 4.34. The load P on the free end causes the member force $S = P$ in segments 1 and 2 of the column. The matrix $[F]$ is the influence coefficient matrix for transverse loads at nodes 1 and 2, and it is easily shown that

$$[F] = \frac{l^3}{24EI}\begin{bmatrix} 8 & 5/2 \\ 5/2 & 1 \end{bmatrix}$$

Now, $[k_{ir}]$ is a matrix of fictitious loads resulting from unit deflections. From Fig. 4.34(b) it is seen that

$$[k_{ir}] = \frac{2P}{l}\begin{bmatrix} 1 & -1 \\ -1 & 2 \end{bmatrix}$$

This expression is derived as follows:

$$\begin{bmatrix} R_{1i} \\ R_{2i} \end{bmatrix} = \frac{2}{l}\begin{bmatrix} r_1 & -r_2 \\ -r_1 & 2r_2 \end{bmatrix}\begin{bmatrix} S_1 \\ S_2 \end{bmatrix} = \frac{2}{l}\begin{bmatrix} r_1 & -r_2 \\ -r_1 & 2r_2 \end{bmatrix}\begin{bmatrix} 1 \\ 1 \end{bmatrix}P$$

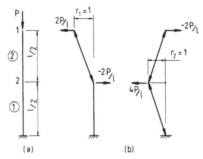

(a)   (b)

Fig. 4.34   Column buckling–ersatz forces.

Substitution for $[F]$ and $[k_{ir}]$ in eqn (4.132) gives

$$\left\{[I] - \frac{Pl^2}{24EI}\begin{bmatrix} 11 & -6 \\ 3 & -1 \end{bmatrix}\right\}r = 0$$

For $\{r\} \neq 0$, the determinant of the coefficient matrix must be zero, so that

$$\begin{vmatrix} \lambda - 11 & 6 \\ -3 & \lambda + 1 \end{vmatrix} = 0$$

where $\lambda = (24EI/Pl^2)$ and the values of λ are $\lambda = 5 \pm \sqrt{18}$. It is seen that the largest value of λ gives the smallest value of P, so that

$$P_{cr} = \frac{2.597EI}{l^2} \tag{4.133}$$

A disadvantage of the force method of analysis is apparent in that the lowest buckling load is associated with the largest root of eqn (4.132). It is more convenient to work with the stiffness of the structure, in which case the critical load is associated with the lowest root (Chapter 6 section 6) of the tangent stiffness matrix.

Example 4.12 *Fictitious node forces for a truss member*

In the truss example in Fig. 4.33, the force, R_i, to be applied at joint A was found to be

$$R_i = -2\rho sc^2 S_0$$

When the member forms part of a truss system X and Y, force components will be generated at both ends of the member, owing to the rotation of the member ($\delta\alpha$). Consider the member ij shown in Fig. 4.35 whose end displacements are

$$\{r^T\} = <r_{ix}\ r_{iy}\ r_{jx}\ r_{jy}> \tag{4.134}$$

When rigid–body motions are subtracted from these displacements, the displacement of end j relative to end i will be given by the components $<\delta x\ \delta y>$, such that

$$\begin{bmatrix} \delta x \\ \delta y \end{bmatrix} = \begin{bmatrix} -1 & 0 & 1 & 0 \\ 0 & -1 & 0 & 1 \end{bmatrix} \begin{bmatrix} r_{ix} \\ r_{iy} \\ r_{jx} \\ r_{jy} \end{bmatrix} \tag{4.135}$$

From Fig. 4.35, the rotation, $\delta\alpha$, measured in a anticlockwise direction, is found to be

$$\delta\alpha = \frac{1}{l}[-\sin\alpha\ \cos\alpha] \begin{bmatrix} \delta x \\ \delta y \end{bmatrix} \tag{4.136}$$

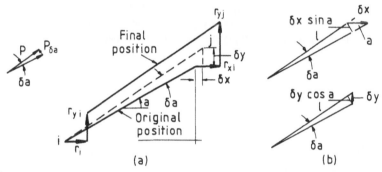

Fig. 4.35   Member displacement and rotation: (a) end displacements of member; (b) rotation for displacements δx, δy.

Combining eqns (4.135) and (4.136),

$$\delta\alpha = \frac{1}{l}[\sin\alpha\ -\cos\alpha\ -\sin\alpha\ \cos\alpha]\{r\} \tag{4.137}$$

Let P be the axial force in the member. The components of δP at end j for a rotation $\delta\alpha$ are

$$\begin{bmatrix} \Delta X \\ \Delta Y \end{bmatrix} = P \begin{bmatrix} -\sin\alpha \\ \cos\alpha \end{bmatrix} \delta\alpha \tag{4.138}$$

For ends i and j of the member,

$$\begin{bmatrix} \Delta X_i \\ \Delta Y_i \\ \Delta X_j \\ \Delta Y_j \end{bmatrix} = P \begin{bmatrix} \sin\alpha \\ -\cos\alpha \\ -\sin\alpha \\ \cos\alpha \end{bmatrix} \delta\alpha$$

When the value of $\delta\alpha$ is substituted from eqn (4.136), it is found that the the force increments are given by

$$
\begin{bmatrix} \Delta X_i \\ \Delta Y_i \\ \Delta X_j \\ \Delta Y_j \end{bmatrix} = \frac{P}{l} \begin{bmatrix} s^2 & -cs & -s^2 & cs \\ -cs & c^2 & cs & -c^2 \\ -s^2 & cs & s^2 & -cs \\ cs & -c^2 & -cs & c^2 \end{bmatrix} \begin{bmatrix} r_{ix} \\ r_{iy} \\ r_{jx} \\ r_{jy} \end{bmatrix}
\tag{4.139}
$$

In eqn (4.139),

$$ c = \cos\alpha ; \quad s = \sin\alpha $$

The general application of eqn (4.139) in eqns (4.127) etc. is not pursued further here. It should be noted that eqn (4.115) is obtained from eqn (4.138) by the extraction of the appropriate row and column value of the coefficient matrix. The above equations are used in the analysis of the free cable given in section 4.12.

4.11   Analysis of planar cable structure

The cable assumed, planar is shown in Fig. 4.36(a) under the influence of the loads R_1 and R_2. In the first instance the cable is assumed to be weightless. The self–weight may be accounted for by increasing the number of segments and applying segment dead loads as equivalent node forces.

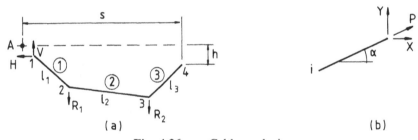

Fig. 4.36   Cable analysis.

If the unstrained lengths l_{10}, l_{20} etc. are known, and the end forces H and V are assumed, the position at end l may be calculated. If this differs from the required location of the support A, H and V must be modified in such a way as to bring l into coincidence with A. This is achieved in a manner which calculates flexibility coefficients at l including the effects of axial forces in the cable members. The first problem then is to calculate the tangent flexibility of the cable member with axial force P and components $<X\ Y>$ for small variations $<\Delta X\ \Delta Y>$. The member is shown in Fig. 4.36(b). That is, it is required to find $[F]_T$, the tangent–flexibility matrix, such that

$$
\begin{bmatrix} \delta x \\ \delta y \end{bmatrix} = [F]_T \begin{bmatrix} \Delta X \\ \Delta Y \end{bmatrix}
$$

Eqn (4.136) may be written

$$
[-\sin\alpha \ \cos\alpha] \begin{bmatrix} -\sin\alpha \\ \cos\alpha \end{bmatrix} \delta\alpha = \frac{1}{l}[-\sin\alpha \ \cos\alpha] \begin{bmatrix} \delta x \\ \delta y \end{bmatrix}
$$

Hence,

$$\left[\begin{array}{c} \delta x \\ \delta y \end{array} \right]_\alpha = l \left[\begin{array}{c} -\sin \alpha \\ \cos \alpha \end{array} \right] \delta\alpha \qquad (4.140)$$

If both sides of eqn (4.138) are multiplied by $[-\sin \alpha \cos \alpha]$, the expression is obtained for $\delta\alpha$ in terms of the changes in the components of P. That is,

$$\delta\alpha = \frac{1}{P}[-\sin \alpha \cos \alpha] \left[\begin{array}{c} \Delta X \\ \Delta Y \end{array} \right] \qquad (4.141)$$

Substitution for $\delta\alpha$ in eqn (4.140) gives

$$\left[\begin{array}{c} \delta x \\ \delta y \end{array} \right]_\alpha = \frac{l}{P} \left[\begin{array}{cc} \sin^2 \alpha & -\cos \alpha \sin \alpha \\ -\cos \alpha \sin \alpha & \cos^2 \alpha \end{array} \right] \left[\begin{array}{c} \Delta X \\ \Delta Y \end{array} \right] \qquad (4.142)$$

That is, if the cable rotates to cause $[\Delta X \ \Delta Y]$ variation in the components of P, eqn (4.132) gives the $[\delta x \ \delta y]$ changes in the member projections. The elastic flexibility may be calculated for the variation ΔP in P, caused by $[\Delta X \ \Delta Y]$ based on the original geometry. The variation in member force P causes the change in length v:

$$v = \frac{l \Delta P}{EA}$$

The X and Y components are

$$\left[\begin{array}{c} \delta x \\ \delta y \end{array} \right]_e = \left[\begin{array}{c} \cos \alpha \\ \sin \alpha \end{array} \right] v = \frac{l}{EA} \left[\begin{array}{c} \cos \alpha \\ \sin \alpha \end{array} \right] \Delta P \qquad (4.143)$$

Now, the components of ΔP are

$$\Delta X = \Delta P \cos \alpha; \quad \text{and} \quad \Delta Y = \Delta P \sin \alpha$$

Substitution in eqn (4.143) gives

$$\left[\begin{array}{c} \delta x \\ \delta y \end{array} \right]_e = \frac{l}{EA} \left[\begin{array}{c} \Delta X \\ \Delta Y \end{array} \right] = \left[\begin{array}{cc} \frac{l}{EA} & \frac{l}{EA} \end{array} \right] \left[\begin{array}{c} \Delta X \\ \Delta Y \end{array} \right] \qquad (4.144)$$

The expression in eqn (4.144) is the elastic flexibility relationship. The total effect on the changes in projections of l is obtained by adding the results of eqns (4.142) and (4.144). That is,

$$\left[\begin{array}{c} \delta x \\ \delta y \end{array} \right] = \left[\begin{array}{c} \delta x \\ \delta y \end{array} \right]_\alpha + \left[\begin{array}{c} \delta x \\ \delta y \end{array} \right]_e = \left[\begin{array}{cc} \dfrac{l}{P}\sin^2\alpha + \dfrac{l}{EA} & -\dfrac{l}{P}\cos \alpha \sin \alpha \\ -\dfrac{l}{P}\cos \alpha \sin \alpha & \dfrac{l}{P}\cos^2\alpha + \dfrac{l}{EA} \end{array} \right] \left[\begin{array}{c} \Delta X \\ \Delta Y \end{array} \right] \qquad (4.145)$$

If, then, H and V are varied at node l and produce variations of components $[\Delta X \ \Delta Y]$ in the cable members, the displacements at l will be, from eqn (4.145), with N = the number of members,

$$\left[\begin{array}{c} \delta u \\ \delta v \end{array} \right] = \sum_N \left[\begin{array}{c} \delta x \\ \delta y \end{array} \right]$$

Cable force

If the cable values for H and V are assumed, a recurrence relationship can be established to determine the cable tensions P. Note that H is constant for every member of the cable.

From Figs. 4.36 and 4.37 it is seen that the vertical components of force in the cable members are given by

$$V_1 = V$$

and for $i \neq 1$,

$$V_i = V_{i-1} - R_i = V - \sum_{j=1,i-1} R_j \qquad (4.146)$$

The cable tension for member i is

$$P_i = (H^2 + V_i^2)^{1/2} \qquad (4.147)$$

The member position is determined from

$$\sin \alpha_i = \frac{V_i}{P_i}; \quad \text{and} \quad \cos \alpha_i = \frac{H}{P_i} \qquad (4.148)$$

The strained length is calculated from

$$l'_i = l_i \left[1 + \frac{P_i}{EA} \right] \qquad (4.149)$$

The projections on the coordinate axes are

$$x_i = l'_i \cos \alpha_i \quad \text{and} \quad y_i = l'_i \sin \alpha_i \qquad (4.150)$$

If H and V are correctly chosen,

$$s = \sum x_i \quad \text{and} \quad h = \sum y_i$$

If these equalities are not satisfied, an iteration procedure must be used to find the correct values of H and V. That is, let

$$x = s - \sum x_i \quad \text{and} \quad y = h - \sum y_i$$

Apply increments ΔH and ΔV to produce displacements of the point 1, using the tangent flexibility at point 1.

$$[F_T] = \sum \begin{bmatrix} \dfrac{l}{P} s^2 + \dfrac{l}{EA} & -\dfrac{l}{P} cs \\[3mm] -\dfrac{l}{P} cs & \dfrac{l}{P} c^2 + \dfrac{l}{EA} \end{bmatrix} \qquad (4.151)$$

Then, for the first approximation,

$$\begin{bmatrix} x \\ y \end{bmatrix} = [F_T] \begin{bmatrix} \Delta H \\ \Delta V \end{bmatrix} \qquad (4.152)$$

Hence

$$\begin{bmatrix} \Delta H \\ \Delta V \end{bmatrix} = [F_T]^{-1} \begin{bmatrix} x \\ y \end{bmatrix} \qquad (4.153)$$

With the new H and V values, the process is repeated by starting with a recalculation of member tensions.

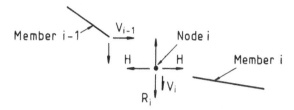

Fig. 4.37   Forces at a cable node.

Example 4.13 *Cable analysis*

A 15.88–mm–diameter cable of 304.8 m span and unstrained length of 335.28m has an area of 193.55 mm^2, and a modulus of elasticity of 206.8 GPa. The cable is analysed for:

1.   dead load of 292 kN/m.;

2.   load of 8.9 kN applied in 20 equal increments of 0.445 kN, each vertically upwards at the centre of the cable.

In the analysis, the cable has been divided into 20 equal lengths of 16.76 m each, as shown in Fig. 4.38, so that the node loads for the first part of the analysis are each equal to 0.489 kN. As an initial assumption, the trial values of H and V are

$$H = 7.42\,kN; \quad V = 4.45\,kN$$

The horizontal tension may be obtained by assuming the initial curve to be circular or parabolic, hence fixing the centre sag. When the load increment of $\Delta R_{11} = 0.445kN$ is applied, the trial values of H and V that are used are taken as those at the end of the previous analysis. The y cable coordinate of point *11* is tabulated in Table 4.2 for successive load cases, together with the y cable coordinates for the first and last load cases. The values in Table 4.2 have been plotted in Figs. 4.39 and 4.40.

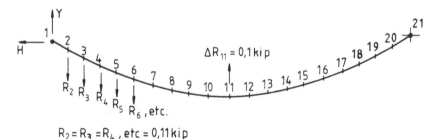

Fig. 4.38   Cable structure–node points and loads.

In Fig. 4.39 the central deflection is shown, and it should be noted that the curve is highly nonlinear; that is, the tangent flexibility at node 11 changes significantly over the range of load increments. The cable profiles are plotted in Fig. 4.40.

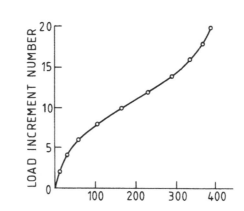

Fig. 4.39   Deflection, in metres, of node *11* from dead–load position.

The above procedure provides a practical means of calculation of the deflected shapes of single cables. The theory is readily developed for the case where the forces are non–coplanar. For cable nets, however, the use of the stiffness method in Chapter 5 is more attractive.

Table 4.2

(a) Cable deflections, metres		
Deflection at	Dead load	Load case 20
1	0.0	0.0
2	10.468	0.674
3	20.222	−0.299
4	29.170	−2.890
5	37.215	−7.030
6	44.257	−12.612
7	50.193	−19.510
8	54.930	−27.587
9	58.380	−36.706
10	60.480	−46.736
11	61.185	−57.555
(b) Deflections at point 3.353 m		
Load case increment No.	*y* ordinate at *11*	Deflection measured from dead–load position
Dead load	61.185	0.0
2	57.555	3.383
4	51.640	9.545
6	42.073	19.112
8	27.867	33.318
10	9.755	51.410
12	−9.775	70.655
14	−27.867	89.043
16	−42.073	103.258
18	−51.640	112.825
20	−57.555	118.740

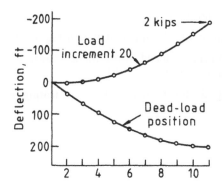

Fig. 4.40   Cable deflected shapes.

Problems

4.1   Use the force method to analyse the indeterminate structures shown in Fig. P4.1. Calculate member forces and deflections in terms of EI at the load points.

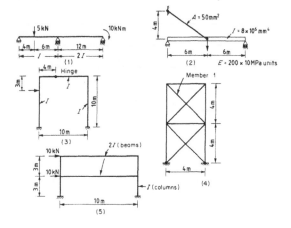

Fig. P4.1

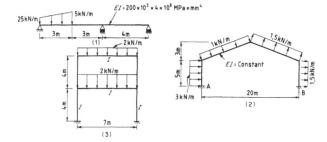

Fig. P4.2

4.2   The structures shown in Fig. P4.2 all have distributed loading. Calculate member forces and node deflections by the following methods:

a.   initial strains;

b.   analysis from a fixed–joint position.

4.3   Calculate the bending moments in the frame shown in Fig. P4.2 (3).

4.4   Calculate the bending moments induced in the frame shown in Fig. P4.2 (2) if support B settles 25 mm. Express the answer in terms of EI.

4.5   Analyse the frame and grid shown in Fig. P4.3 by tearing into suitable sub–structures as shown by the section $A-A$ in both figures.

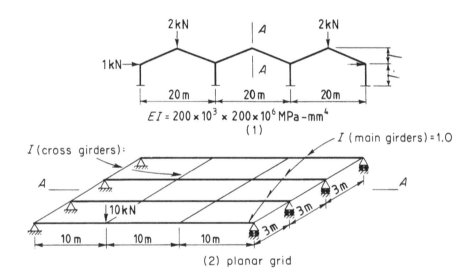

$$EI = 200 \times 10^3 \times 200 \times 10^6 \, \text{MPa} - \text{mm}^4$$
(1)

(2) planar grid

Fig. P4.3

4.6   Analyse structure (4) in Fig. P4.1, making allowance for the removal of member l.

CHAPTER 5

The displacement method–line element structures

5.1 STRUCTURE STIFFNESS

5.1.1   General concepts of structure stiffness

In eqns (4.21) and (4.43), the flexibility relationship expressed in eqn (5.1) has been obtained for a given elastic structure:

$$\{r\} = [F]\{R\} \tag{5.1}$$

The reverse process to that in eqn (5.1) is always possible. That is, given a set of node displacements, $\{r\}$, the node forces $\{R\}$ may be expressed as linear functions of these displacements. Let the inverse of the flexibility matrix be $[K]$, defined as a stiffness matrix, i.e. a displacement to force relationship. Then,

$$[K] = [F]^{-1} \tag{5.2}$$

This $[K]$ matrix must be specified as the stiffness matrix of the structure for the displacements $\{r\}$. It should be noted that whereas the flexibility coefficients in $[F]$ are independent of the number of loads chosen, those of $[K]$ will alter when the number of displacement points is changed. This will become apparent when the force–deflection relationship is examined in detail. The forces $\{R\}$ expressed as linear functions of the displacements $\{r\}$ are given by eqn (5.3):

$$\{R\} = [K]\{r\} \tag{5.3}$$

Expand the ith equation of this set, expressing the force R_i in terms of the displacements $<r_1\ r_2\ \cdots\ r_n>$.

$$R_i = K_{i1}r_1 + K_{i2}r_2 + \cdots + K_{ii}r_i + \cdots + K_{in}r_n \tag{5.4}$$

The coefficient K_{ij} is the value of the force R_i for a displacement $r_j = 1$ when all other displacements are held equal to zero. When the number of points n is changed, the coefficients K_{ij} will change, because of the changed constraint conditions.

Example 5.1 *The effect of constraints on stiffness terms*

The cantilever beam shown in Fig. 5.1 has uniform inertia EI and length l. The stiffness at end I may be calculated for a displacement $r_y = 1$, with the displacement θ_z either not constrained or constrained to be zero. These two displacement patterns are shown in Fig. 5.1(a) and (b), respectively. The stiffness terms for these two configurations are given by

1. $$K_Y = \frac{3EI}{l^3}$$

and

2. $$K_Y = \frac{12EI}{l^3}$$

The added constraint $\theta_z = 0$ in the second case obviously has a very marked effect on the stiffness, K_Y. It is noted that the structure flexibility matrix $[F]$ is symmetric (Maxwell's theorem). Then, so also, is the stiffness matrix $[K]$, since $[K]^T[F] = [K]^T[F]^T = ([F][K])^T = [I] = [K][F]$. Hence $[K]^T = [K]$.

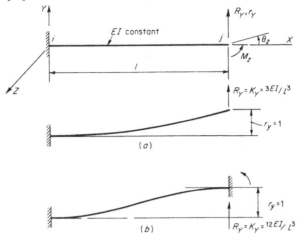

Fig. 5.1   Variation of stiffness with introduction of constraints.

In eqn (5.4) it has been shown that K_{ij}, the ijth term of the stiffness matrix $[K]$ of a structure, is the value of the force R_i for a displacement $r_j = 1$, all other specified displacements in $\{r\}$ being set to zero. This could provide a means of generation of the terms of $[K]$. Alternatively, they could be obtained by first calculating the flexibility matrix $[F]$ and, inverting to produce $[K]$. Such a method is then, in fact, used in the hybrid element formulation of the finite element method (Chapter 3). For the structure composed of many members, it is most convenient to build up $[K]$ from member stiffness matrices in a manner similar to that in which $[F]$ may be generated from the individual member flexibilities through the transformation of eqn (4.12).

5.1.2   Displacement transformation matrices

It has been shown in Chapter 2, eqn (2.122), that if the member distortions are expressed in terms of node displacements through the equation

$$\{v\} = [a]\{r\} \tag{5.5}$$

the equations of joint equilibrium may be expressed as

$$\{R\} - [a]^T\{S\} = 0 \tag{5.6}$$

In the particular case of static determinacy,

$$\{S\} = [a^T]^{-1}\{R\} = [b]\{R\} \tag{5.7}$$

In this way a method was devised for the generation of the $[b]$ matrix, and displacement transformations were calculated (section 4.2) for member distortions in terms of node displacements. In the displacement method of analysis, eqns (5.5) and (5.6) are applied to the general structure for which determinacy has not been assumed. Eqn (5.6) has been obtained by application of the contragredient principle, which states that when eqn (5.5) is true, then

$$\{R\} = [a]^T\{S\} \tag{5.8}$$

From eqns (5.5) and (5.6) it is evident that if a stiffness relationship relates member forces $\{S\}$ to member distortions $\{v\}$, the $\{R\}$ forces may be related to $\{r\}$. That is, let

$$\{S\} = [k]\{v\} \tag{5.9}$$

Then,

$$\{R\} = [a]^T[k]\{v\} = [a]^T[k][a]\{r\} \tag{5.10}$$

Before use is made of this equation in the generation of the structure stiffness matrix, further discussion is given to the displacement transformation matrix $[a]$ and the matrix of member stiffnesses $[k]$. Just as the force transformation matrix $[b]$ may be generated by successive applications of a unit load at various nodes in a structure, so the displacement transformation matrix $[a]$ may be produced by unit values being given, in turn, to each node displacement. This process was used in Chapter 4, in which it was shown that for a statically determinate structure

$$[b] = ([a]^T)^{-1} \tag{5.11}$$

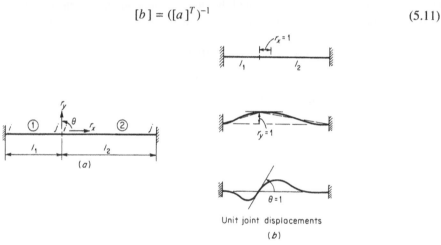

Fig. 5.2   Node and member displacements.

For example, in Fig. 5.2, the displacements of node _1_ will be $\{r\}^T = \langle r_x \ r_y \ \theta \rangle$. When unit values are given in turn to r_x, r_y and θ, the distortions of the members appear as in Fig. 5.2(b). If the member distortions are taken as $\{v\}^T = \langle \theta_i \ \theta_j \ \delta \rangle$, the $[a]$ matrix is given in eqn (5.12), being generated by inspection of Fig. 5.2(b).

$$
\begin{bmatrix} \theta_{1i} \\ \theta_{1j} \\ \delta_1 \\ \theta_{2i} \\ \theta_{2j} \\ \delta_2 \end{bmatrix}
=
\begin{bmatrix}
0 & -\dfrac{1}{l_1} & 0 \\
0 & -\dfrac{1}{l_1} & 1 \\
1 & 0 & 0 \\
0 & \dfrac{1}{l_2} & 1 \\
0 & \dfrac{1}{l_2} & 0 \\
-1 & 0 & 0
\end{bmatrix}
\begin{bmatrix} r_x \\ r_y \\ \theta \end{bmatrix}
\tag{5.12}
$$

In section 4.2 the relationships between member distortions and node displacements have been developed for the plane frame element. Therein it is shown how node displacements at the member ends in global coordinates are transformed to those in local coordinates for the member and thence to member distortions. The transformations are given in the form

$$\{v\} = [T][L_D]\{r\} \qquad (5.13)$$

It will be found later that these $[T]$ and $[L_D]$ matrices will be used to transform member stiffness matrices in their basic force deformation modes to those for the forces and displacements of the member ends in global coordinates. These stiffness matrices are used in the method of analysis generally designated the direct stiffness method. Transformation equations such as eqn (5.13) will be found useful also in certain structures such as the rigid headstock on flexible piles. In this chapter the main concern will be with the general formulation of the displacement method, and in the examples given, the $[a]$ matrices will be developed from elementary displacement patterns. These will sometimes be complicated by the assumption of no change in the axial length of members. Consider the general structure, and let $a_{i1}, a_{i2}, \cdots, a_{ik}$ be the values of the member distortion v_i due to $r_1 = 1, r_2 = 1, \cdots, s_k = 1$, respectively. That is, for all displacements $\{r\}^T = <r_1 \ r_2 \ \cdots \ r_k>$, the particular distortion v_i is given by the linear relationship,

$$v_i = a_{i1}r_1 + a_{i2}r_2 + \cdots + a_{ik}r_k \qquad (5.14)$$

For all such member distortions and joint displacements,

$$\begin{bmatrix} v_1 \\ v_2 \\ \cdot \\ \cdot \\ v_i \\ \cdot \\ \cdot \\ v_n \end{bmatrix} = \begin{bmatrix} a_{11} & \cdots & a_{1k} \\ a_{21} & \cdots & a_{2k} \\ \cdot & \cdots & \cdot \\ \cdot & \cdots & \cdot \\ a_{i1} & \cdots & a_{ik} \\ \cdot & \cdots & \cdot \\ \cdot & \cdots & \cdot \\ a_{n2} & \cdots & a_{nk} \end{bmatrix} \begin{bmatrix} r_1 \\ r_2 \\ \cdot \\ \cdot \\ r_k \end{bmatrix} \qquad (5.15)$$

Eqn (5.15) is, of course, simply,

$$\{v\} = [a]\{r\} \qquad (5.16)$$

For a typical member n of the structure with member forces S_n and member distortions v_n, a stiffness relationship will exist which relates S_n to v_n. This member stiffness matrix may be obtained from force or displacement patterns which have been assumed to exist in the member. It may be calculated from these directly or from inversion of the flexibility matrix $[f_n]$, if v_n includes only basic member distortions. It may be shown in such cases that

$$[k_n] = [f_n]^{-1} \quad \text{or} \quad [k_n] = [T]^T [f_n]^{-1} [T]$$

where rigid–body displacements are included in v_n. Notice that $[f_n]$ always refers to a basic set of forces since the equations of equilibrium must be satisfied when unit forces are applied to a member. In either case,

$$\{S_n\} = [k_n]\{v_n\} \qquad (5.17)$$

All such equations for members 1, 2, 3, etc. may be combined, with the member stiffness matrices forming diagonal submatrices.

$$\begin{bmatrix} S_1 \\ S_2 \\ S_3 \\ \cdot \end{bmatrix} = \begin{bmatrix} k_{11} & \cdot & \cdot & \cdot \\ \cdot & k_{22} & \cdot & \cdot \\ \cdot & \cdot & k_{33} & \cdot \\ \cdot & \cdot & \cdot & \cdot \end{bmatrix} \begin{bmatrix} v_1 \\ v_2 \\ v_3 \\ \cdot \end{bmatrix} \tag{5.18}$$

This is the equation,

$$\{S\} = [k]\{v\} \tag{5.19}$$

Then, from eqn (5.16),

$$\{S\} = [k][a]\{r\} \tag{5.20}$$

Eqn (5.20) provides a convenient method for determining member forces when the node displacements are known. The relationship,

$$\{R\} = [a]^T\{S\}$$

may be obtained, as in eqn (4.9), as a direct application of the contragredient principle or, alternatively, from the principle of virtual displacements. That is, the $\{R\}$ force system acting on the structure together with the member–node forces forms a system in equilibrium, and, hence, if the structure is given a set of compatible displacements,

$$\sum W_i = \sum W_e \tag{5.21}$$

As a suitable virtual displacement, choose $\bar{r}_i = \Delta$ (a small quantity) with all other displacements zero,

$$\sum W_e = R_i \times \bar{r}_i = R_i \Delta$$

To calculate $\sum W_i$, consider the member distortions $(\bar{v}_{1i}, \bar{v}_{2i}, \cdots, \bar{v}_{ni})$ associated with $\bar{r}_i = \Delta$. From eqn (5.15) it is seen that these distortions are simply Δ multiplied by the ith column of $[a]$. That is,

$$\begin{bmatrix} \bar{v}_{1i} \\ \bar{v}_{2i} \\ \cdot \\ \bar{v}_{ni} \end{bmatrix} = \begin{bmatrix} \bar{a}_{1i} \\ \bar{a}_{2i} \\ \cdot \\ \bar{a}_{ni} \end{bmatrix} \Delta$$

The work done by the member forces is

$$\sum W_i = (\bar{a}_{1i} S_1 + \bar{a}_{2i} S_2 + \cdots + \bar{a}_{ni} S_n)\Delta$$

That is, $\sum W_i = \{\bar{a}_i\}^T\{S\}\Delta$, where $\{\bar{a}_i\}$ is the ith column of $[a]$. Hence,

$$\{R_i\} = \{\bar{a}_i\}^T\{S\} \tag{5.22}$$

In the above case, the bar notation has been used to indicate that the formula $\bar{r}_i = \Delta$ need not be applied to the same constrained structure as the $\{r\}$ deflections. However, it is generally convenient to make $[\bar{a}] = [a]$. Finally,

$$\{R\} = [a]^T[k][a]\{r\} \tag{5.23}$$

The stiffness matrix for the displacements $\{r\}$ has been found to be

$$[K] = [a]^T [k][a] \tag{5.24}$$

The actual application of eqn (5.24) to the calculation of the structure stiffness matrix takes many forms, and it will be largely this topic which occupies the remainder of the present chapter. Generally, of course, the node forces $\{R\}$ are known, and the displacements $\{r\}$ are the required unknowns. The solution of eqn (5.23) may be written formally as

$$\{r\} = [K]^{-1}\{R\} \tag{5.25}$$

Some adjustment may be necessary to $[K]$ if it embodies the rigid–body displacements of the structure, before a solution may be obtained to eqn (5.25). Details of this will be given later.

5.1.3 Kinematically indeterminate structures

A structure is said to be kinematically determinate when all node displacements are known. Then node forces may be calculated directly from eqn (5.23). The condition of all joints fixed is a kinematically determinate state, since then $\{r\} \equiv 0$. As has been demonstrated in Chapter 4 (Example 4.7), this state is a convenient starting point in the analysis of structures for which loads appear on members between node points. In the more general case, suppose that only some of the nodes of an elastic structure are given specified displacements. Those at the remaining nodes will be unknown but may be calculated from the known displacements as follows. In the equation,

$$\{R\} = [a]^T [k][a]\ \{r\}$$

The vector $\{r\}$ will be partitioned into $\{r_0\}$, the known, and $\{r_x\}$, the unknown, joint displacements. Partition $\{R\}$ similarly into $\{R_0\}$ and $\{R_x\}$, and also $[a]$ into $[a_0]$ and $[a_1]$, so that

$$\begin{bmatrix} R_0 \\ R_x \end{bmatrix} = \begin{bmatrix} a_0^T \\ a_1^T \end{bmatrix} [k][a_0\ a_1] \begin{bmatrix} r_0 \\ r_x \end{bmatrix} \tag{5.26}$$

On multiplication of the partitioned matrices,

$$\begin{bmatrix} R_0 \\ R_x \end{bmatrix} = \begin{bmatrix} a_0^T k a_0 & a_0^T k a_1 \\ a_1^T k a_0 & a_1^T k a_1 \end{bmatrix} \begin{bmatrix} r_0 \\ r_x \end{bmatrix} \tag{5.27}$$

The notation is introduced:

$$[C_{00}] = [a_0]^T [k][a_0]$$

$$[C_{01}] = [C_{10}]^T = [a_0]^T [k][a_1]$$

$$[C_{11}] = [a_1]^T [k][a_1]$$

With this notation eqn (5.27) is written, omitting submatrix brackets in the combined matrices

$$\begin{bmatrix} R_0 \\ R_x \end{bmatrix} = \begin{bmatrix} C_{00} & C_{01} \\ C_{10} & C_{11} \end{bmatrix} \begin{bmatrix} r_0 \\ r_x \end{bmatrix} \tag{5.28}$$

In eqn (5.28), $\{r_0\}$ are the known displacements. The nodes corresponding to $\{r_x\}$ may be unloaded, in which case $\{R_x\} \equiv 0$ or have known forces applied. It is seen that the unknown quantities in eqn (5.28) are the forces $\{R_0\}$ and the displacements $\{r_x\}$.

On expansion of eqn (5.28) two equations are obtained, and from the second of these the $\{r_x\}$ displacements are given in terms of $\{R_0\}$ and $\{R_x\}$; that is,

$$\{r_x\} = [C_{11}]^{-1}(\{R_x\} - [C_{10}]\{r_0\}) \tag{5.29}$$

When $\{R_x\} \equiv 0$,

$$\{r_x\} = -[C_{11}]^{-1}[C_{10}]\{r_0\} \tag{5.30}$$

The $\{R_0\}$ forces are obtained from eqn (5.28),

$$\{R_0\} = [C_{00}]\{r_0\} + [C_{01}]\{r_x\}$$

Substitution for $\{r_x\}$ from eqn (5.28) gives

$$\{R_0\} = ([C_{00}] - [C_{01}][C_{11}]^{-1}[C_{10}])\{r_0\} + [C_{01}][C_{11}]^{-1}\{R_x\} \tag{5.31}$$

Again, when $\{R_x\} \equiv 0$,

$$\{R_0\} = ([C_{00}] - [C_{01}][C_{11}]^{-1}[C_{10}])\{r_0\} \tag{5.32}$$

The similarity between kinematic and static indeterminacy is shown when eqns (5.30) and (5.32) are compared with eqns (4.39) and (4.40). Member forces may be calculated from the expression,

$$\{S\} = [k][a]\{r\}$$

In the partitioned form of eqn (5.26),

$$\{S\} = [k][a_0 \ a_1]\begin{bmatrix} r_0 \\ r_x \end{bmatrix}$$

That is,

$$\{S\} = [k] [([a_0] - [a_1][C_{11}]^{-1}[C_{10}])\{r_0\} \tag{5.33}$$

An interesting class of problems arises in which both $\{r_0\}$ and $\{r_x\}$ are unknown, but $\{R_x\} \equiv 0$. That is, how are the displacements at the unloaded joints related to those of the load points? As in eqn (5.30),

$$\{r_x\} = -[C_{11}]^{-1}[C_{10}]\{r_0\}$$

and from eqn (5.32),

$$\{r_0\} = ([C_{00}] - [C_{01}][C_{11}]^{-1}[C_{10}])^{-1}\{R_0\} \tag{5.34}$$

Consider now eqn (5.28) in the form

$$\begin{bmatrix} R_0 \\ 0 \end{bmatrix} = [K]\begin{bmatrix} r_0 \\ r_x \end{bmatrix} \tag{5.35}$$

The flexibility matrix for the structure may be obtained from direct inversion of $[K]$, and in partitioned form,

$$[F] = \begin{bmatrix} F_{11} & F_{12} \\ F_{21} & F_{22} \end{bmatrix} \tag{5.36}$$

Then deflections are given

$$\begin{bmatrix} r_0 \\ r_x \end{bmatrix} = \begin{bmatrix} F_{11} & F_{12} \\ F_{21} & F_{22} \end{bmatrix} \begin{bmatrix} R_0 \\ 0 \end{bmatrix} \tag{5.37}$$

That is,

$$\{r_0\} = [F_{11}]\{R_0\} \quad \text{and} \quad \{r_x\} = [F_{21}]\{R_0\} \tag{5.38}$$

A comparison of eqn (5.34) and (5.38) shows that

$$[F_{11}] = ([C_{00}] - [C_{01}][C_{11}]^{-1}[C_{10}])^{-1} \tag{5.39}$$

Eqn (5.39) is a useful result since it means that the stiffness matrix for the displacements $\{r_0\}$ has been found with the displacements $\{r_x\}$ unconstrained; that is,

$$[K_{00}] = [F_{11}]^{-1} \tag{5.40}$$

This procedure may be used in the analysis of a plane frame structure in which $[K_{00}]$ is required in the analysis for dynamic loading. The frame shown in Fig. 5.3 is analysed for the unit loads shown at each floor level. Deflections, calculated for these unit loads shown when partitioned correctly, give $[F_{11}]$ and $[F_{21}]$.

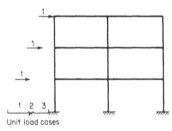

Fig. 5.3   Analysis to produce flexibility and stiffness matrices.

5.1.4   Member stiffness matrices

For a prismatic member of length l and area A with Young's modulus of elasticity E, the stiffness is the inverse of the flexibility. That is,

$$[k] = \frac{EA}{l} \tag{5.41}$$

For a beam element which is prismatic, of length l, and of bending stiffness EI, the sign convention for end moments on the member is anticlockwise moments positive when looking at the beam with end i on the left. This corresponds to M_z positive in a clockwise sense and is the convention used in the equilibrium equations in Chapter 4. The beam with end rotations ϕ_i and ϕ_j is shown in Fig. 5.4.

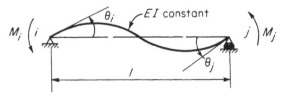

Fig. 5.4   Beam–element sign convention for end rotations and moments.

In this case the stiffness matrix is obtained as the inverse of the flexibility matrix given in eqn (4.25); that is,

$$[k] = \frac{2EI}{l} \begin{bmatrix} 2 & 1 & 0 \\ 1 & 2 & 0 \\ 0 & 0 & \dfrac{A}{2I} \end{bmatrix}$$

(5.42)

Example 5.2 *Analysis of a frame structure by the displacement method*

The portal frame is shown in Fig. 5.5 and is analysed for a 10–kN horizontal force on node *1*. Axial distortions are neglected, and the inertia values for members are shown in the figure. The analysis follows the sequence of operations:

1. Ascertain the relevant node displacements, and calculate the displacement transformation matrix.

$$\{v\} = [a]\{r\}$$

2. From member dimensions and properties, form the $[k]$ matrix of member stiffnesses.

$$\{S\} = [k]\{v\} = [k][a]\{r\}$$

3. Generate the structure stiffness matrix for the displacements $\{r\}$.

$$\{R\} = [a]^T[k][a]\{r\} = [K]\{r\}$$

4. Calculate displacements.

$$\{r\} = [K]^{-1}\{R\}$$

5. Calculate member forces.

$$\{S\} = [k][a]\{r\}$$

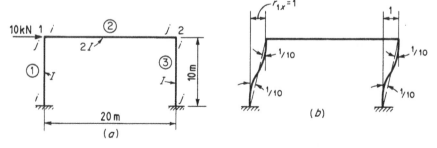

Fig. 5.5   Portal frame dimensions, loads and sway displacement.

In this example eqns (5.30) and (5.32) will be used to reduce the amount of hand calculation. There are three possible displacements of the two free nodes *1* and *2*. These are the two rotations θ_1 and θ_2 and the displacement r_{1x}. Notice that because the members are assumed inextensible,

$$r_{2x} = r_{1x} \quad \text{and} \quad r_{1y} = r_{2y} = 0$$

Thence,

$$\{v\} = \begin{bmatrix} \phi_{1i} \\ \phi_{1j} \\ \phi_{2i} \\ \phi_{2j} \\ \phi_{3i} \\ \phi_{3j} \end{bmatrix} = \begin{bmatrix} \cdot & \cdot & 0.1 \\ 1 & \cdot & 0.1 \\ 1 & \cdot & \cdot \\ \cdot & 1 & \cdot \\ \cdot & 1 & 0.1 \\ \cdot & \cdot & 0.1 \end{bmatrix} \begin{bmatrix} \theta_1 \\ \theta_2 \\ r_{1x} \end{bmatrix} \tag{5.43}$$

The matrix of member stiffnesses is

$$[k] = \frac{2EI}{10} \begin{bmatrix} 2 & 1 & \cdot & \cdot & \cdot & \cdot \\ 1 & 2 & \cdot & \cdot & \cdot & \cdot \\ \cdot & \cdot & 2 & 1 & \cdot & \cdot \\ \cdot & \cdot & 1 & 2 & \cdot & \cdot \\ \cdot & \cdot & \cdot & \cdot & 2 & 1 \\ \cdot & \cdot & \cdot & \cdot & 1 & 2 \end{bmatrix} \quad \text{and} \quad [k][a] = \frac{2EI}{10} \begin{bmatrix} 1 & 0 & 0.3 \\ 2 & 0 & 0.3 \\ 2 & 1 & 0 \\ 1 & 2 & 0 \\ 0 & 2 & 0.3 \\ 0 & 1 & 0.3 \end{bmatrix} \tag{5.44}$$

and the stiffness matrix is calculated

$$[K] = [a]^T[k][a] = \frac{2EI}{10} \begin{bmatrix} 4 & 1 & 0.3 \\ 1 & 4 & 0.3 \\ 0.3 & 0.3 & 0.12 \end{bmatrix}$$

Then, from step 3 above,

$$\begin{bmatrix} 0 \\ 0 \\ 10 \end{bmatrix} = \frac{2EI}{10} \begin{bmatrix} 4 & 1 & 0.3 \\ 1 & 4 & 0.3 \\ 0.3 & 0.3 & 0.12 \end{bmatrix} \begin{bmatrix} \theta_1 \\ \theta_2 \\ r_{1x} \end{bmatrix} \tag{5.45}$$

In this equation note that $[C_{11}]$ appears as the first submatrix. Then

$$\{0\} = \begin{bmatrix} 4 & 1 \\ 1 & 4 \end{bmatrix} \begin{bmatrix} \theta_1 \\ \theta_2 \end{bmatrix} + \begin{bmatrix} 0.3 \\ 0.3 \end{bmatrix} r_{1x}$$

Solve for θ_1 and θ_2 in terms of r_{1x}.

$$\begin{bmatrix} \theta_1 \\ \theta_2 \end{bmatrix} = -\frac{3}{50} \begin{bmatrix} 1 \\ 1 \end{bmatrix} r_{1x} \tag{5.46}$$

The second equation in (5.45) is written

$$10 = \frac{2EI}{10} \left\{ [0.3 \ 0.3] \begin{bmatrix} \theta_1 \\ \theta_2 \end{bmatrix} + 0.12 r_{1x} \right\}$$

Substitute for $\langle \theta_1 \ \theta_2 \rangle$ from eqn (5.46).

$$10 = \frac{2EI}{10}\left\{-[0.3 \ 0.3]\begin{bmatrix} 0.06 \\ 0.06 \end{bmatrix} + 0.12\right\}r_{1x} = \frac{2EI}{10}(0.084)r_{1x}$$

Thence,

$$r_{1x} = \frac{50\,000}{84EI}$$

From eqn (5.46),

$$\begin{bmatrix} \theta_1 \\ \theta_2 \end{bmatrix} = -\frac{3}{50}\begin{bmatrix} 1 \\ 1 \end{bmatrix}\frac{50\,000}{84EI} = \frac{-1000}{28EI}\begin{bmatrix} 1 \\ 1 \end{bmatrix}$$

Member moments are calculated from eqn (5.44):

$$\begin{bmatrix} M_{1i} \\ M_{1j} \\ M_{2i} \\ M_{2j} \\ M_{3i} \\ M_{3j} \end{bmatrix} = \frac{2EI}{10}\begin{bmatrix} 1 & 0 & 0.3 \\ 2 & 0 & 0.3 \\ 2 & 1 & 0 \\ 1 & 2 & 0 \\ 0 & 2 & 0.3 \\ 0 & 1 & 0.3 \end{bmatrix}\frac{1000}{EI}\begin{bmatrix} -1/28 \\ -1/28 \\ 50/84 \end{bmatrix} = \begin{bmatrix} 28.5 \\ 21.4 \\ -21.4 \\ -21.4 \\ 21.4 \\ 28.5 \end{bmatrix}$$

The bending moment diagram is drawn in Fig. 5.6.

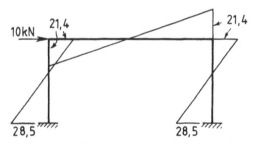

Fig. 5.6   Bending moment diagram in kN/m units.

5.1.5   Displacement analysis – loads on members between nodes

It has been shown in Chapter 4 that in the force method, loaded members may be treated by considering the structure to have all nodes fixed in space. For this condition the fixed–end forces are calculated for each member, and when all these are applied as release forces at the nodes, the total structure behaviour is the sum of the two analyses. The actual calculation of fixed–end forces has been left to the present discussion because fixed–end moments have traditionally been associated with the displacement analyses (for example, in slope deflection and moment distribution).

It is noted that the joint–fixed condition is the kinematically determinate solution, $\{r\} \equiv 0$. It is most convenient to transform member forces into local coordinate components for the calculation of fixed– end moments and shears. When these forces are used in fully automated computer programs, it will be found necessary to transform back to global coordinates, just as in general force method programs.

Example 5.3 *Cantilever beam–uniformly distributed load*

In this example the end deflection and rotation of the cantilever beam with uniformly distributed load are calculated by the displacement method. The loaded beam is shown in Fig. 5.7(a). The kinematically determinate situation $\{r\} \equiv 0$ is shown in Fig. 5.7(b), and, from this, release forces are determined for node *1* in Fig. 5.7(c). The displacement transformation for node *1* to member (1) is given by

$$\begin{bmatrix} \phi_i \\ \phi_j \end{bmatrix} = \begin{bmatrix} -1/4 & 0 \\ -1/4 & 1 \end{bmatrix} \begin{bmatrix} r_y \\ \theta_x \end{bmatrix}$$

The member stiffness is

$$[k] = \frac{2EI}{4} \begin{bmatrix} 2 & 1 \\ 1 & 2 \end{bmatrix}$$

Thence,

$$[k][a] = \frac{2EI}{4} \begin{bmatrix} -3/4 & 1 \\ -3/4 & 2 \end{bmatrix} \quad \text{and} \quad [a]^T[k][a] = \frac{EI}{16} \begin{bmatrix} 3 & -6 \\ -6 & 16 \end{bmatrix}$$

In the equation $\{R\} = [a]^T[k][a]\{r\}$, the forces $\{R\}$ are given by $\{R^T\} = <-2.00 \quad 4/3>$. Solving for displacements,

$$\begin{bmatrix} r_y \\ \theta_z \end{bmatrix} = \frac{16}{12EI} \begin{bmatrix} 16 & 6 \\ 6 & 3 \end{bmatrix} \begin{bmatrix} -2 \\ 4/3 \end{bmatrix} = \frac{32}{3EI} \begin{bmatrix} -3 \\ -1 \end{bmatrix}$$

Member forces are given as $\{S\} = [k][a]\{r\}$; that is,

$$\begin{bmatrix} M_i \\ M_j \end{bmatrix} = \frac{2EI}{4} \begin{bmatrix} -3/4 & 1 \\ -3/4 & 2 \end{bmatrix} \frac{32}{3EI} \begin{bmatrix} -3 \\ -1 \end{bmatrix} = \frac{4}{3} \begin{bmatrix} 5 \\ 1 \end{bmatrix}$$

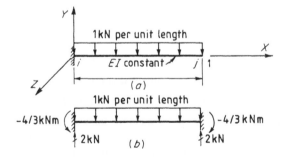

Fig. 5.7   Analysis for member loads: (a) cantilever beam;
(b) fixed–joint solution; (c) release forces.

The complete solution must include the fixed–end moments shown in Fig. 5.7(b), so that,

$$\begin{bmatrix} M_i \\ M_j \end{bmatrix}_{total} = \frac{4}{3} \begin{bmatrix} 5 \\ 1 \end{bmatrix} + \frac{4}{3} \begin{bmatrix} 1 \\ -1 \end{bmatrix} = \begin{bmatrix} 8 \\ 0 \end{bmatrix}$$

This result is to be expected since, for the statically determinate cantilever, it is seen by inspection that $M_i = wl^2/2$. It is noticed that in the displacement analysis, the equation $\{R\} = [K]\{r\}$ must be solved even when the structure is statically determinate. Of course, if the $[a]$ matrix had been checked, it would have been found to be square and non–singular, and the solution for member forces would follow directly from $\{R\} = [a]^T\{S\}$. It was found in Chapter 4 that in order to produce automated force–method programs (or to solve joint equilibrium equations), it is necessary that the axial forces be considered in all members. In an analogous way it is found that, generally, axial strains must be considered in frame members in the displacement method when automatic analysis procedures are developed. The complexity which arises in producing displacement patterns and equivalent node forces when bending distortions only are considered is illustrated in the analysis of the gable frame in example 5.4.

Example 5.4 *Gable frame – inextensible members*

The gable frame is shown in Fig. 5.8 and is loaded on member (2) only with a distributed load of 2 Kn of horizontal projection. The flexural properties EI are the same for all members. The fixed–end forces for member (2) are shown in Fig. 5.8(b). Independent node displacements must be chosen, and the inextensibility of the members taken into consideration. Evidently there are three rotation displacements $(\theta_{1z}, \theta_{2z}, \theta_{3z})$. There are, however, only two independent sidesway displacement modes since if node *1* is to have a horizontal displacement, it is necessary for node *2* also to be displaced, and similarly for node *3*. The two independent sidesway displacement modes $U_{1x} = 1$ and $U_{3x} = 1$ are shown in Fig. 5.9. It should be noted that any two independent patterns would suffice. For example, symmetric and antisymmetric patterns could be used to good advantage, and the reader should rework the problem, using these modes.

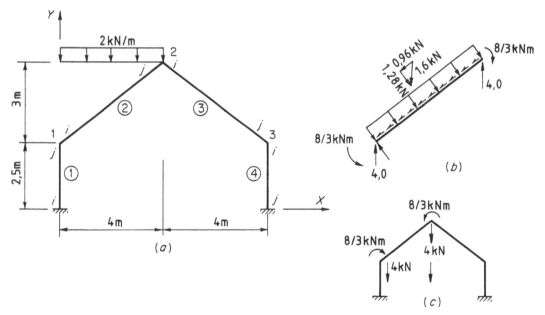

Fig. 5.8   Analysis of gable frame: (a) gable–frame dimensions and loading;
(b) fixed–end forces for member (2); (c) release forces.

From Fig. 5.9 and the rotation displacements, the displacement transformation $\{v\} = [a]\{r\}$ is found to be

$$
\begin{bmatrix} \phi_{1i} \\ \phi_{1j} \\ \phi_{2i} \\ \phi_{2j} \\ \phi_{3i} \\ \phi_{3j} \\ \phi_{4i} \\ \phi_{4j} \end{bmatrix}
=
\begin{bmatrix}
0 & 0 & 0 & 2/5 & 0 \\
1 & 0 & 0 & 2/5 & 0 \\
1 & 0 & 0 & -1/6 & 1/6 \\
0 & 1 & 0 & -1/6 & 1/6 \\
0 & 1 & 0 & 1/6 & -1/6 \\
0 & 0 & 1 & 1/6 & -1/6 \\
0 & 0 & 1 & 0 & 2/5 \\
0 & 0 & 0 & 0 & 2/5
\end{bmatrix}
\begin{bmatrix} \theta_{1z} \\ \theta_{2z} \\ \theta_{3z} \\ r_{1x} \\ r_{3x} \end{bmatrix}
\tag{5.47}
$$

The matrix of member stiffnesses, is the 8×8 matrix $[k]$.

$$
[k] = \frac{2EI}{5}
\begin{bmatrix}
4 & 2 & \cdot & \cdot & \cdot & \cdot & \cdot & \cdot \\
2 & 4 & \cdot & \cdot & \cdot & \cdot & \cdot & \cdot \\
\cdot & \cdot & 2 & 1 & \cdot & \cdot & \cdot & \cdot \\
\cdot & \cdot & 1 & 2 & \cdot & \cdot & \cdot & \cdot \\
\cdot & \cdot & \cdot & \cdot & 2 & 1 & \cdot & \cdot \\
\cdot & \cdot & \cdot & \cdot & 1 & 2 & \cdot & \cdot \\
\cdot & \cdot & \cdot & \cdot & \cdot & \cdot & 4 & 2 \\
\cdot & \cdot & \cdot & \cdot & \cdot & \cdot & 2 & 4
\end{bmatrix}
\tag{5.48}
$$

From eqns (5.47) and (5.48) the product $[k][a]$ is found to be

$$
[k][a] = \frac{2EI}{5}
\begin{bmatrix}
2 & \cdot & \cdot & 12/5 & \cdot \\
4 & \cdot & \cdot & 12/5 & \cdot \\
2 & 1 & \cdot & -1/2 & 1/2 \\
1 & 2 & \cdot & -1/2 & 1/2 \\
\cdot & 2 & 1 & 1/2 & -1/2 \\
\cdot & 1 & 2 & 1/2 & -1/2 \\
\cdot & \cdot & 4 & \cdot & 12/5 \\
\cdot & \cdot & 2 & \cdot & 12/5
\end{bmatrix}
\tag{5.49}
$$

Finally, the structure stiffness matrix (upper triangle only) is given:

$$
[a]^T[k][a] = \frac{2EI}{5}
\begin{bmatrix}
6 & 1 & 0 & 19/10 & 1/2 \\
\cdot & 4 & 1 & 0 & 0 \\
\cdot & \cdot & 6 & 1/2 & 19/10 \\
\cdot & \cdot & \cdot & 169/75 & -1/3 \\
\cdot & \text{symmetric} & \cdot & \cdot & 169/75
\end{bmatrix}
\tag{5.50}
$$

Before the equations $\{R\} = [a]^T[k][a]\{r\}$ may be solved, the consequence of using the displacement patterns in Fig. 5.9 must be evaluated in terms of the load vector $\{R\}$. The generalized force Q_1 corresponding to the displacement $r_{1x} = 1$ must be such that for a virtual displacement δr_{1x}, the work done, $\delta r_{1x} Q_1$ must be equal to the work done by node forces.

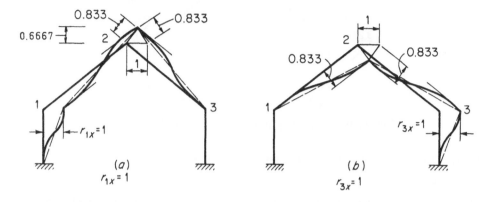

Fig. 5.9   Displacement patterns for horizontal node displacements.

It is seen that the displacement patterns are, in fact, producing side–sway type of equilibrium equations. That is, in the equations $\{R\} - [a]^T\{S\}$, the particular rows in $[a]^T\{S\}$ corresponding to r_{1x} are applicable to several node forces. In this case, from Figs 5.8(c) and 5.9,

$$Q_1 = (-4 \times 0.6667) = -8/3 \text{ kN}$$

$$Q_2 = (-4 \times -0.6667) = 8/3 \text{ kN}$$

Thence, using these values in the load vector,

$$\{R\}^T = <-8/3 \quad 8/3 \quad 0 -8/3 \quad 8/3>$$

Solving the equations $\{r\} = [K]^{-1}\{R\}$ gives deflection values, to be multiplied by $5/2EI$:

$$\theta_{1z} = -0.77904 \text{ rad}; \qquad \theta_{2x} = 1.06667 \text{ rad}; \quad \text{and} \quad \theta_{3x} = -0.82096 \text{ rad}$$

$$r_{1x} = -0.042293 \text{ m}; \quad \text{and} \quad r_{3x} = 2.042293 \ m$$

The bending moments in kN–m are (from $\{S\} = [k][a]\{r\}$)

$$\begin{bmatrix} M_{1i} \\ M_{1j} \\ M_{2i} \\ M_{2j} \\ M_{3i} \\ M_{3j} \\ M_{4i} \\ M_{4j} \end{bmatrix} = \begin{bmatrix} -1.6595 \\ -3.2175 \\ 0.5513 \\ 2.3893 \\ 0.2001 \\ -1.6173 \\ 1.6175 \\ 3.2595 \end{bmatrix} \quad \begin{matrix} M_{2i,total} = 0.5513 + 8/3 = 3.2177 \\ M_{2j,total} = 2.3973 - 8/3 = -0.2694 \end{matrix} \qquad (5.51)$$

The bending moment diagram is plotted in Fig. 5.10.

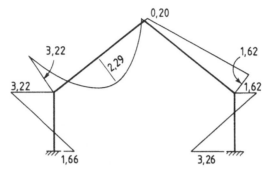

Fig. 5.10   Bending moment diagram in kN/m units.

5.2 INITIAL STRAINS – MODIFICATION OF MEMBER STIFFNESS

5.2.1   Initial strains

Initial strains may arise from a variety of causes, such as loads between nodes of a member, temperature change, lack of fit due to manufacture (either by chance or by intention), and by non–linear plastic strains. The last type will, of course, involve more detailed consideration at a latter stage, and it has been discussed in terms of yield surfaces etc. in Chapter 2. All these strains may be included in the analysis by starting from the kinematically determinate state $\{r\} \equiv 0$. Since $\{v\} = [a]\{r\}$, this implies that all member distortions must be eliminated by applying fixed–end forces. The structure analysis (as distinct from the member analysis) consists in determining the effects of the release forces in the structure. For any member, the total forces will be the sum of the fixed–end and the release forces; that is, the forces in the state $\{r\} \equiv 0$ together with those producing the deflections $\{r\}$. Suppose, then, that member i with stiffness $[k_i]$ has a set of initial strains $\{\theta_i\}$. For compatibility in the condition of all joints fixed, the displacements $-\{\theta_i\}$ must be applied to the member, and these will require forces,

$$\{J_i\} = -[k_i]\{\theta_i\} \tag{5.52}$$

To produce $\{J_i\}$, it will be necessary to apply joint forces,

$$\{R_i\}^F = [a_i]^T\{J_i\} \tag{5.53}$$

The release forces are

$$\{R_i\} = -\{R_i\}^F = [a_i]^T[k_i]\{\theta_i\} \tag{5.54}$$

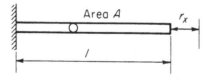

Fig. 5.11   Cantilever beam–unit displacement.

If $\{R\}$ is the vector of externally applied node loads, the structure is analysed for the loads,

$$\{R^*\} = \{R\} + \{R_i\} = \{R\} + [a_i]^T[k_i]\{\theta_i\} \tag{5.55}$$

The member forces $\{S^*\}$ calculated from $\{R^*\}$ will be

$$\{S^*\} = [k][a]\{r^*\} = [k][a][K]^{-1}\{R^*\} \tag{5.56}$$

That is,

$$\{S^*\} = [k][a][K]^{-1}(\{R\} + \{R_i\}) \tag{5.57}$$

For the member i,

$$\{S_i\} = \{S_i^*\} + \{J_i\} = \{S_i^*\} - [k_i]\{\theta_i\} \tag{5.58}$$

Hence,

$$\{S_i\} = [k_i][a_i][K]^{-1}(\{R\} + [a_i]^T[k_i]\{\theta_i\}) - [k_i]\{\theta_i\}$$

Finally,

$$\{S_i\} = [k_i][a_i][K]^{-1}\{R\} + [k_i]([a_i][K]^{-1}[a_i]^T[k_i] - [k_i])\{\theta_i\} \tag{5.59}$$

5.2.2   Removal of a member

The above theory may be used to calculate the effects of the removal of a member on the rest of the structure, since by suitable choice of initial strains, the forces in the member may be set to zero. Evidently, if the member has zero forces, it is not participating in the action of the structure and may be removed. For the whole structure with loads $\{R\}$, the force in member i will be

$$\{S_i\} = [k_i][a_i]\{r\}$$

Apply initial strains $\{\theta_i\}$ to the member i so that the new force in i, $\{\bar{S}_i\}$, is equal to zero. That is,

$$\{\bar{S}\}_i = 0 = \{S_i^*\} + \{J_i\}$$

Thence,

$$[k_i][a_i]\{r\} + (-[k_i] + [k_i][a_i][K]^{-1}[a_i]^T[k_i])\{\theta_i\} = 0$$

Therefore,

$$\{\theta_i\} = -(-[k_i] + [k_i][a_i][K]^{-1}[a_i]^T[k_i])^{-1}[k_i][a_i]\{r\} \tag{5.60}$$

The joint forces applied are

$$\{R^*\} = \{R\} + \{R_i\} = \{R\} + [a_i]^T[k_i]\{\theta_i\}$$

The total node deflections with member removed will be

$$\{r_{total}\} = [K]^{-1}(\{R\} + [a_i]^T[k_i]\{\theta_i\}) \tag{5.61}$$

Substitution for $\{\theta_i\}$ from eqn (5.60) gives

$$\{r_{total}\} = [K]^{-1}[\{R\} - [a_i]^T[k_i]((-k_i] + [k_i][a_i][K]^{-1}[a_i]^T[k_i])^{-1}[k_i][a_i]\{r\}] \tag{5.62}$$

Since $\{r\} = [K]^{-1}\{R\}$ gives the deflections before modification, it follows that

$$\{r_{total}\} = \{r\} + \{\bar{r}\}$$

$$= \{r\} - [K]^{-1}[a_i]^T[k_i](-[k_i] + [k_i][a_i][K]^{-1}[a_i]^T[k_i])^{-1}[k_i][a_i]\{r\}$$

or,

$$\{\bar{r}\} = -[K]^{-1}[a_i]^T[k_i](-[k_i] + [k_i][a_i][K]^{-1}[a_i]^T[k_i])^{-1}[k_i][a_i][K]^{-1}\{R\} \tag{5.63}$$

From this it is seen that

$$\{r_{total}\} = -[K]^{-1}[[I] - [a_i]^T[k_i](-[k_i] + [k_i][a_i][K]^{-1}[a_i]^T[k_i])^{-1}[k_i][a_i][K]^{-1}]\{R\} \tag{5.64}$$

That is, the modified flexibility matrix for the structure is

$$[F_{mod}] = [K]^{-1}[[I] - [a_i]^T(-[k_i]^{-1} + [a_i][K]^{-1}[a_i]^T)^{-1}[a_i][K]^{-1}] \tag{5.65}$$

Member forces are calculated

$$\{S_{total}\} = [k][a]\{r_{total}\} = [k][a](\{r\} + \{\bar{r}\}) \tag{5.66}$$

For member i,

$$\{S_{total,i}\} = \{S\} + \{\bar{S}\} - [k_i]\{\theta_i\} = 0$$

Example 5.5 *Truss member*

The pin–jointed truss member shown in Fig. 5.12 has an initial extension of a units. Calculate the fixed–end forces to restore the member to its original shape. The stiffness relationship between end displacements $<r'_{ix} \ r'_{jx}>$ in local coordinates and the forces $<F'_{ix} \ F'_{jx}>$ is easily shown to be

$$\begin{bmatrix} F'_{ix} \\ F'_{jx} \end{bmatrix} = \frac{EA}{l} \begin{bmatrix} 1 & -1 \\ -1 & 1 \end{bmatrix} \begin{bmatrix} r'_{ix} \\ r'_{jx} \end{bmatrix}$$

For initial strains, let $r'_{ix} = 0$, and then $r'_{jx} = a$. Applying eqn (5.53),

$$\begin{bmatrix} J'_{ix} \\ J'_{jx} \end{bmatrix} = \frac{-EA}{l} \begin{bmatrix} 1 & -1 \\ -1 & 1 \end{bmatrix} \begin{bmatrix} 0 \\ a \end{bmatrix} = \frac{EA}{l} \begin{bmatrix} a \\ -a \end{bmatrix}$$

Example 5.6 *Portal frame temperature strains*

The rigid frame shown in Fig. 5.13 has uniform load of 1/2 kN on the beam member, which also suffers a temperature variation of 122.4^0C on the outside and 61.2^0C on the inside. The frame is composed of a steel I section, 610×229 mm, with $I = 9.840 \times 10^8 mm^4$ and $A = 1.592 \times 10^4 mm^2$ The coefficient of thermal expansion is taken to be 0.2×10^{-6}mm/(mm)(C^o), and Young's modulus, $E = 2.1 \times 10^2$ kN/mm^2. The end rotations due to the uniform load will be a measure of the lack of fit introduced from this source. The temperature variation will also cause end rotations and, in addition, an axial extension.

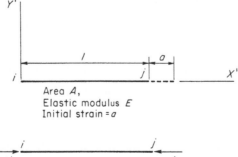

Fig. 5.12 Truss member with initial strain.

1. Rotations due to uniform load

$$\begin{bmatrix} \phi_{2i} \\ \phi_{2j} \end{bmatrix} = \frac{wl^3}{24EI} \begin{bmatrix} -1 \\ 1 \end{bmatrix} = \frac{6^3 \times (10^3)^3 \times 10^{-3}}{4.8 \times 21 \times 10^2 \times 9.84 \times 10^8} \begin{bmatrix} -1 \\ 1 \end{bmatrix} = 2.177 \times 10^{-5} \begin{bmatrix} -1 \\ 1 \end{bmatrix}$$

2. Rotations due to temperature strains

From eqn (4.29), noting the sign change in rotation ϕ_i due to the different sign convention for positive end moments,

$$\begin{bmatrix} \phi_i \\ \phi_j \end{bmatrix}_t = \frac{\alpha l}{2(c_1 - c_2)} \begin{bmatrix} 1 & -1 \\ -1 & 1 \end{bmatrix} \begin{bmatrix} t_1 \\ t_2 \end{bmatrix} = \frac{0.2 \times 10^{-6} \times 6000}{2 \times 610} \begin{bmatrix} 1 & -1 \\ -1 & 1 \end{bmatrix} \begin{bmatrix} 50 \\ 25 \end{bmatrix} = \frac{2.46 \times 10^{-5}}{2} \begin{bmatrix} 1 \\ -1 \end{bmatrix}$$

Axial extension is given by

$$\Delta l_{2t} = 0.2 \times 10^{-6} \times 6000 \times 75/2 = 0.045$$

The displacement transformation equation $\{v\} = [a]\{r\}$ is here given by

$$\begin{bmatrix} \phi_{1i} \\ \phi_{1j} \\ \phi_{2i} \\ \phi_{2j} \\ \delta_2 \\ \phi_{3i} \\ \phi_{3j} \end{bmatrix} = \begin{bmatrix} 0 & 0 & 1/2000 & 0 \\ 1 & 0 & 1/2000 & 0 \\ 1 & 0 & 0 & 0 \\ 0 & 1 & 0 & 0 \\ 0 & 0 & -1 & 1 \\ 0 & 1 & 0 & 1/2000 \\ 0 & 0 & 0 & 1/2000 \end{bmatrix} \begin{bmatrix} \theta_1 \\ \theta_2 \\ U_1 \\ U_2 \end{bmatrix}$$

The member stiffnesses are

$$k_1 = k_3 = \frac{2 \times 2.1 \times 10^2 \times 9.84 \times 10^8}{2 \times 10^3} \begin{bmatrix} 2 & 1 \\ 1 & 2 \end{bmatrix} = 2.0663 \times 10^8 \begin{bmatrix} 2 & 1 \\ 1 & 2 \end{bmatrix}$$

$$k_2 = \frac{2 \times 2.1 \times 10^2 \times 9.84 \times 10^8}{6000} \begin{bmatrix} 2 & 1 & 0 \\ 1 & 2 & 0 \\ 0 & 0 & \dfrac{1.594 \times 10^4}{2 \times 9.84 \times 10^8} \end{bmatrix} = 6.888 \times 10^7 \begin{bmatrix} 2 & 1 & 0 \\ 1 & 2 & 0 \\ 0 & 0 & 8.1 \times 10^{-6} \end{bmatrix}$$

Using the theory for initial strains eqn (5.52) *et seq.*,

$$\{R\,\delta t\} = a_i^T k_i\, \theta_i = \begin{bmatrix} 1 & 0 & 0 \\ 0 & 1 & 0 \\ 0 & 0 & 1 \end{bmatrix} \begin{bmatrix} 1.3762 \times 10^8 & 6.888 \times 10^7 & 0 \\ 6.888 \times 10^7 & 1.3762 \times 10^8 & 0 \\ 0 & 0 & 557.98 \end{bmatrix} \begin{Bmatrix} 2.46 \times 10^{-5} \\ -2.46 \times 10^{-5} \\ 4.5 \times 10^{-2} \end{Bmatrix} = \begin{Bmatrix} 1694.832 \\ -1694.832 \\ -25.108 \\ -25.108 \end{Bmatrix}$$

For the applied loads,

$$\{R_P\}^T = \{-1500 \quad 1500 \quad -1.5 \quad 1.5\}^T$$

so that the total release forces are given

$$\{R\}^T = \{194.832\ 194.832\ -26.6085\ 26.6085\}^T$$

The node displacements are calculated to be

$$\{r\} = \begin{Bmatrix} 1.4415 \times 10^{-5} \\ -1.4415 \times 10^{-5} \\ -2.179 \times 10^{-2} \\ 2.179 \times 10^{-2} \end{Bmatrix}$$

The final member forces are given

$$\begin{Bmatrix} M_{1i} \\ M_{1j} \\ M_{2i} \\ M_{2j} \\ F_2 \\ M_{3i} \\ M_{3j} \end{Bmatrix} = \begin{Bmatrix} -3775.570 \\ -796.418 \\ 990.887 \\ -990.887 \\ 24.314 \\ 796.418 \\ 3775.566 \end{Bmatrix} \qquad \begin{aligned} M_{2i\ total} &= 194.832 - 990.887 = -796.054 \\ M_{2j\ total} &= -194.832 + 990.887 = 796.054 \end{aligned}$$

The bending moment diagram is shown in Fig. 5.13.

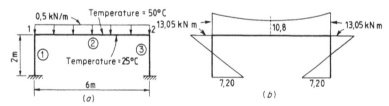

Fig. 5.13   Frame analysis for distributed load and temperature strain.

5.2.3   Modified member stiffness

The removal of a member is simply a special case in which the member stiffness is altered by an amount $[\Delta k_i]$. The variation of $[K]$, the structure stiffness, will be designated $[\Delta K]$, and

$$[K] + [\Delta K] = [K] + [a_i]^T [\Delta k_i][a_i] \tag{5.67}$$

The Householder algorithm for the modified inverse of a matrix is

$$([B] - [U][S][V]^T)^{-1} = [B]^{-1} + [B]^{-1}[U]([S]^{-1} - [V]^T[B]^{-1}[U])^{-1}[V]^T[B]^{-1} \tag{5.68}$$

In eqn (5.68), $[B]$ is an $n \times n$ matrix, $[S]$ is an $r \times r$ matrix, and $[U]$ and $[V]$ are $n \times r$ matrices. In the present case,

$$[S] = -[\Delta k_i]$$

$$[U] = [a_i]^T \quad \text{and} \quad [U]^T = [a_i]$$

$$[V]^T = [a_i] \quad \text{and} \quad [V] = [a_i]^T$$

Hence the modified structure stiffness is

$$([K] - [a_i]^T[[-\Delta k_i][a_i]])^{-1} = [K]^{-1} + [K]^{-1}[a_i]^T([-\Delta k_i])^{-1} - [a_i][K]^{-1}[a_i]^T]^{-1}[a_i][K]^{-1} \quad (5.69)$$

Notice that this expression is identical with eqn (5.65) if $[\Delta k_i] = -[k_i]$

Example 5.7 *Member modification*

The portal frame in Fig. 5.5 will be modified so that member (3) has zero bending stiffness. That is (omitting the multiplier EI),

$$[\Delta k_3] = -[k_3] = \frac{-2}{10}\begin{bmatrix} 2 & 1 \\ 1 & 2 \end{bmatrix}$$

Notice that member (3) is still assumed to have infinite axial stiffness so that the modified structure will be equivalent to that shown in Fig. 5.14. The matrices $[K]$, $[a_i]$, and $[k_i]$ are given in example 5.2 and are listed for reference below. The constant term EI is omitted from the calculations.

$$[K] = 2\begin{bmatrix} 0.4 & 0.1 & 0.03 \\ 0.1 & 0.4 & 0.03 \\ 0.03 & 0.03 & 0.012 \end{bmatrix} \quad \text{and} \quad [K]^{-1} = \frac{1}{2}\begin{bmatrix} 3.0953 & -0.2381 & -7.1435 \\ -0.2381 & 3.0953 & -7.1435 \\ -7.1435 & -7.1435 & 119.0581 \end{bmatrix}$$

$$[a_3] = \begin{bmatrix} 0 & 1 & 0.1 \\ 0 & 0 & 0.1 \end{bmatrix} \quad \text{and} \quad [\Delta k_3]^{-1} = [k_3]^{-1} = \frac{10}{6EI}\begin{bmatrix} 2 & -1 \\ -1 & 2 \end{bmatrix}$$

Then, following eqn (5.69),

$$[K][a_i]^T = \frac{1}{2}\begin{bmatrix} -0.95245 & -0.71435 \\ 2.38095 & -0.71435 \\ 4.76231 & 11.90580 \end{bmatrix}$$

Then let $[A]$ be calculated,

$$[A] = [\Delta k_i]^{-1} - [a_i][K]^{-1}[a_i]^T = \frac{10}{6}\begin{bmatrix} 2 & -1 \\ -1 & 2 \end{bmatrix} - \frac{1}{2}\begin{bmatrix} 2.85781 & 0.47623 \\ 0.47623 & 1.19058 \end{bmatrix} = \begin{bmatrix} 1.90443 & -1.90470 \\ -1.9047 & 2.73891 \end{bmatrix}$$

Then,

$$[A]^{-1} = \begin{bmatrix} 1.72085 & 1.20059 \\ 1.20059 & 1.20042 \end{bmatrix}$$

Then the modified flexibility matrix is calculated to be

$$[K_{mod}]^{-1} = \frac{1}{2}\begin{bmatrix} 3.0953 & -0.2381 & -7.1435 \\ -0.2381 & 3.0953 & -7.1435 \\ -7.1435 & -7.1435 & 119.058 \end{bmatrix} + \frac{1}{2}\begin{bmatrix} 1.9037 & -2.2563 & -17.8568 \\ -2.2563 & 3.1420 & 19.6260 \\ -17.8568 & 19.6760 & 172.665 \end{bmatrix}$$

$$= \frac{1}{2}\begin{bmatrix} 4.99898 & -2.4944 & -25.003 \\ -2.2944 & 6.2373 & 12.4825 \\ -25.003 & 12.4825 & 291.723 \end{bmatrix}$$

The modified structure is as shown in Fig. 5.14. This structure may be analysed by using the matrices in example 5.2, simply by omitting the member (3) from the $[a]$ and $[f]$ matrices.

In that case,

$$[K_{mod}] = [a]^T[k][a] = \begin{bmatrix} 0 & 1 & 1 & 0 \\ 0 & 0 & 0 & 1 \\ 0.1 & 0.1 & 0 & 0 \end{bmatrix} \begin{bmatrix} 2 & 1 & 0 & 0 \\ 1 & 2 & 0 & 0 \\ 0 & 0 & 2 & 1 \\ 0 & 0 & 1 & 2 \end{bmatrix} \begin{bmatrix} 0 & 0 & 0.1 \\ 1 & 0 & 0.1 \\ 1 & 0 & 0 \\ 0 & 1 & 0 \end{bmatrix}$$

$$= \frac{2}{10} \begin{bmatrix} 4 & 1 & 0.3 \\ 1 & 2 & 0 \\ 0.3 & 0 & 0.06 \end{bmatrix}$$

Finally, by inverting this matrix the modified flexibility matrix is obtained

$$[K_{mod}]^{-1} = \frac{10}{2} \begin{bmatrix} 12/100 & -6/100 & -6/100 \\ -6/100 & 15/100 & 30/100 \\ -60/100 & 30/100 & 700/100 \end{bmatrix}$$

These results agree within the precision of the calculations with those obtained via eqn (5.69).

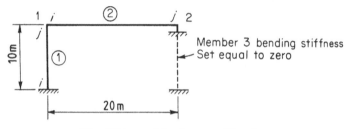

Fig. 5.14   Member modification.

5.3 MEMBER STIFFNESS MATRICES

So far, only two types of member stiffnesses have been used, namely, axial stiffness and bending stiffness of prismatic members. For the displacement method to be generally applicable, a variety of member stiffness matrices must be developed. In addition, a means of calculating fixed–end forces must now be introduced, since these exist in the kinematically determinate solution $\{r\} \equiv 0$.

5.3.1   Prismatic member in torsion

(a)   Circular cross section:

$$[k] = GI_p/l$$

(b)   Rectangular cross section:

$$[k] = k_1 \frac{G(2a)^3 2b}{l} \quad b > a$$

The value of k_1 taken from [18] is

$$k_1 = \frac{1}{3}\left(1 - \frac{192}{\pi^5}\frac{a}{b}\sum_{n=1,3,5}\frac{1}{n^5}\tanh\frac{n\pi b}{2a}\right)$$

Calculated values of k_1 are

$$\begin{bmatrix} b/a & 1 & 1.2 & 1.5 & 2.0 & 3.0 & 4.0 & 5.0 & 10.0 & \infty \\ k_1 & 0.1406 & 0.166 & 0.196 & 0.229 & 0.263 & 0.281 & 0.291 & 0.302 & 0.333 \end{bmatrix}$$

(c)   For the general prismatic member of closed cross section, the St Venant's torsion constant may be calculated by the finite element method (Chapter 7).

5.3.2   Inclusion of shear distortions in the bending stiffness

It has been shown in eqn (4.89) that the flexibility matrix for the prismatic member with end moments (M_i, M_j) and including shear distortions is

$$[f] = \frac{l}{6EI}\begin{bmatrix} 2+\alpha & 1-\alpha \\ 1-\alpha & 2+\alpha \end{bmatrix}; \quad \alpha = \frac{6EI}{GA_{eff}\,l^2}$$

In this expression the sign convention used for moments is that tension on one side is positive. When the convention that anticlockwise moment on the member ends is positive is used, the signs of the off–diagonal terms must be changed. To obtain the stiffness matrix, change the signs of these terms and invert. Thence the stiffness matrix with shear flexibility included is

$$[k] = \frac{2EI}{l(1+2\alpha)}\begin{bmatrix} 2+\alpha & 1-\alpha \\ 1-\alpha & 2+\alpha \end{bmatrix}$$

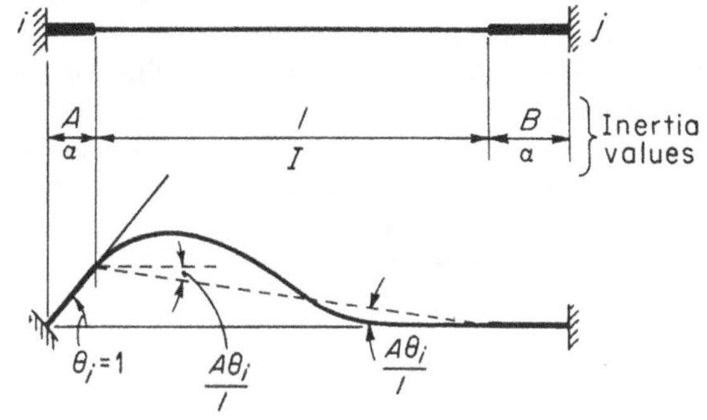

Fig. 5.15   Kinematic analysis of beam with infinite end inertias.

5.3.3   Bending stiffness of members with infinite end section inertias

The flexibility matrix for this member type was developed in eqn (4.96). The stiffness matrix could be obtained by inversion of this expression. However, it is an interesting and useful kinematic analysis to set up $[k]$ directly from eqn (5.24). Let $<\phi_i, \phi_j>$ be the rotations of the member tangents relative to the chord over the length l.

It should be noted that the end rotation $\phi_i = 1$ shown in Fig. 5.15 is possible by virtue of the flexibility of the centre section. From Fig. 5.15 it is seen that

$$\begin{bmatrix} \phi_i \\ \phi_j \end{bmatrix} = \begin{bmatrix} 1+\dfrac{A}{l} & \dfrac{B}{l} \\[2mm] \dfrac{A}{l} & 1+\dfrac{B}{l} \end{bmatrix}\begin{bmatrix} \theta_i \\ \theta_j \end{bmatrix}$$

The stiffness matrix for portion l of the member is of the form

$$[k] = \begin{bmatrix} k_{11} & k_{12} \\ k_{21} & k_{22} \end{bmatrix}$$

For the portion l, which is prismatic,

$$[k] = \frac{2EI}{l}\begin{bmatrix} 2 & 1 \\ 1 & 2 \end{bmatrix}$$

Then, for $<\theta_i \ \theta_j>$, the stiffness matrix is given by the general expression,

$$[K] = [a]^T[k][a] = \begin{bmatrix} 1 + \dfrac{A}{l} & \dfrac{A}{l} \\ \dfrac{B}{l} & 1 + \dfrac{B}{l} \end{bmatrix}\begin{bmatrix} k_{11} & k_{12} \\ k_{21} & k_{22} \end{bmatrix}\begin{bmatrix} 1 + \dfrac{A}{l} & \dfrac{B}{l} \\ \dfrac{A}{l} & 1 + \dfrac{B}{l} \end{bmatrix} \tag{5.70}$$

By multiplication of the matrices in eqn (5.60), explicit values can be obtained for the terms of $[K]$. However, in computer programming it is probably more convenient to use eqn (5.60).

5.3.4   Bending stiffness – member with intermediate hinge

In this case it is assumed that the stiffness matrix $[k]$ for end rotations $<\phi_i \ \phi_j>$ and moments $<M_i \ M_j>$ of a member is known. It is required to ascertain the effect on $[k]$ of a hinge placed in the span, as shown in Fig. 5.16(a). In addition the modification to fixed–end forces must be calculated. The usefulness of this type of transformation is in the fact that for non–prismatic members no special provisions need be made in the description of member properties. That is, the member with the hinge is treated as a normal member and is then modified. The member may be assumed to be non–prismatic and the stiffness matrix $[k]$ for the member calculated as in eqns (5.84) and (5.85).

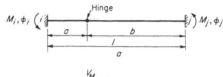

Fig. 5.16   Member with hinge.

The stiffness matrix will have the form:

$$[k] = \begin{bmatrix} k_{ii} & k_{ij} \\ k_{ji} & k_{jj} \end{bmatrix} \quad \text{and} \quad [k] = \frac{2EI}{l}\begin{bmatrix} 2 & 1 \\ 1 & 2 \end{bmatrix} \quad \text{for prismatic members}$$

The choice of displacements $<\phi_i \ \phi_j>$ and moments $<M_i \ M_j>$ to express the stiffness relationship of the member is an arbitrary one. Any two independent quantities may be used. In this case it will be convenient to use $<M_H \ V_H>$, the moment and shear at the release point. As yet, of course, the release has not been introduced. The transformation from end moments to the forces at the hinge is obtained by considering the equilibrium of forces on the portions A and B of the beam shown in Fig. 5.16(b). Then,

$$\begin{bmatrix} M_i \\ M_j \end{bmatrix} = \begin{bmatrix} -1 & -a \\ 1 & -b \end{bmatrix}\begin{bmatrix} M_H \\ V_H \end{bmatrix} \tag{5.71}$$

This equation will be written

$$\{S_{ij}\} = [T]\{S_H\} \tag{5.72}$$

Solving for $\{S_H\}$ gives

$$\{S_H\} = [T]^{-1}\{S_{ij}\} \tag{5.73}$$

The stiffness relationship for end forces and displacements will be

$$\{S_{ij}\} = [k]\{\phi_{ij}\} \tag{5.74}$$

From the contragredient principle, the expression for displacements at the releases in terms of $<\phi_i \ \phi_j>$ may be expressed

$$\{\delta_H\} = [T]^T\{\phi_{ij}\} \tag{5.75}$$

Combining eqns (5.73) and (5.74), and the inverse of (5.75), the stiffness expressed in terms of $\{\delta_H\}$ and $\{S_H\}$ is given by

$$[k_H] = [T]^{-1}[k][T^{-1}]^T \tag{5.76}$$

The reverse transformation also holds that

$$[k] = [T][k_H][T]^T$$

Let $[k_H]$ be of the form:

$$[k_H] = \begin{bmatrix} h_{ii} & h_{ij} \\ h_{ji} & h_{jj} \end{bmatrix}$$

At the hinge $M_H = 0$ hence,

$$\begin{bmatrix} 0 \\ V_H \end{bmatrix} = \begin{bmatrix} h_{ii} & h_{ij} \\ h_{ji} & h_{jj} \end{bmatrix} \begin{bmatrix} \theta_H \\ v_H \end{bmatrix} \tag{5.77}$$

From eqn (5.77),

$$\theta_H = -\frac{h_{ij}}{h_{ii}} v_H$$

and hence the shear at the hinge is expressed by

$$V_H = \left[h_{jj} - \frac{h_{ij} h_{ji}}{h_{ii}} \right] v_H = h'_{jj} v_H$$

The reduced stiffness matrix with the hinge introduced is

$$[k'_H] = \begin{bmatrix} 0 & 0 \\ 0 & h'_{jj} \end{bmatrix} \tag{5.78}$$

Then the stiffness modified for the hinge is

$$[k'] = [T][k'_H][T]^T \tag{5.79}$$

Example 5.8 *Hinge at centre of beam*

Suppose that for a prismatic member, the hinge is at the midpoint of the member, so that a = b = l/2. Then

$$[T] = \begin{bmatrix} -1 & -l/2 \\ 1 & -l/2 \end{bmatrix} \quad \text{and the inverse} \quad [T]^{-1} = \frac{1}{l}\begin{bmatrix} -l/2 & l/2 \\ -1 & -1 \end{bmatrix}$$

From eqn (5.76),

$$[k_H] = [T]^{-1}[k][T^{-1}]^T = \frac{2EI}{l^3}\begin{bmatrix} -l/2 & l/2 \\ -1 & -1 \end{bmatrix}\begin{bmatrix} 2 & 1 \\ 1 & 2 \end{bmatrix}\begin{bmatrix} -l/2 & -1 \\ l/2 & -1 \end{bmatrix} = \frac{2EI}{l^3}\begin{bmatrix} l^2/2 & 0 \\ 0 & 6 \end{bmatrix}$$

The reduced stiffness is simply

$$[k'_H] = \frac{2EI}{l^3}\begin{bmatrix} 0 & 0 \\ 0 & 6 \end{bmatrix}$$

For the end forces and rotations the stiffness is given

$$[k'] = [T][k'_H][T]^T = \frac{3EI}{l}\begin{bmatrix} 1 & 1 \\ 1 & 1 \end{bmatrix}$$

Notice that now $[k']$ is singular. This implies that a linear relationship exists between M_i and M_j. This must be the case since the bending moment diagram must always pass through the hinge.

5.3.5 The effect of the hinge on fixed–end moments

Suppose that the fixed–end moments $<M_{iF} \; M_{jF}>$ have been calculated for the member without the hinge. The moment and shear at the hinge point will be a combination of these together with the simply supported values of $<M_0 \; V_0>$. Then,

$$\begin{bmatrix} M_H \\ V_H \end{bmatrix} = [T]^{-1}\begin{bmatrix} M_{iF} \\ M_{jF} \end{bmatrix} + \begin{bmatrix} M_0 \\ V_0 \end{bmatrix}$$

To introduce the hinge, apply $-M_H$ to the member, allowing distortions θ_H, but maintaining $v_H = 0$. Then,

$$\begin{bmatrix} -M_H \\ V_H \end{bmatrix} = \begin{bmatrix} h_{ii} & h_{ij} \\ h_{ji} & h_{jj} \end{bmatrix}\begin{bmatrix} \theta_H \\ 0 \end{bmatrix}$$

Hence,

$$\theta_H = \frac{-M_H}{h_{ii}} \; ; \quad V_H = \frac{-h_{ji}M_H}{h_{ii}}$$

The effects of the release $-M_H$ on $<M_i \; M_j>$ is given in eqn (5.80).

$$\begin{bmatrix} M'_i \\ M'_j \end{bmatrix} = \begin{bmatrix} M_i \\ M_j \end{bmatrix} + [T]\begin{bmatrix} -1 \\ -\dfrac{h_{ji}}{h_{ii}} \end{bmatrix}M_H \tag{5.80}$$

Example 5.9 *Beam with hinge*

The beam shown in Fig. 5.17(a) has a hinge at the centre, a = b = 1/2 and is loaded there with a force P. The effect of the hinge on the original fixed–end moments $<M_i \; M_j> = (Pl/8)<1 \; -1>$ will be calculated by using eqn (5.80). The final condition is shown in Fig. 5.17(b).

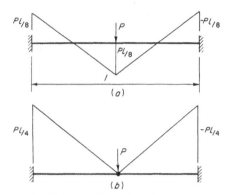

Fig. 5.17   Fixed–end moments

From example 5.8,

$$[k_H] = \frac{2EI}{l^3}\begin{bmatrix} l^2/2 & 0 \\ 0 & 6 \end{bmatrix}$$

and the transformation matrix $[T]$,

$$[T] = \begin{bmatrix} -1 & -l/2 \\ 1 & -l/2 \end{bmatrix} \quad \text{and the inverse} \quad [T]^{-1} = \frac{1}{l}\begin{bmatrix} -l/2 & l/2 \\ -1 & -1 \end{bmatrix}$$

Then,

$$\begin{bmatrix} M_H \\ V_H \end{bmatrix} = \frac{1}{l}\begin{bmatrix} -l/2 & l/2 \\ -1 & -1 \end{bmatrix}\begin{bmatrix} 1 \\ -1 \end{bmatrix}\frac{Pl}{8} + \begin{bmatrix} 2 \\ 0 \end{bmatrix}\frac{Pl}{8} = \begin{bmatrix} \dfrac{Pl}{8} \\ 0 \end{bmatrix}$$

In this case $h_{ji} = 0$, so that

$$\begin{bmatrix} M'_i \\ M'_j \end{bmatrix} = \begin{bmatrix} 1 \\ -1 \end{bmatrix}\frac{Pl}{8} + \begin{bmatrix} -1 & -l/2 \\ 1 & -l/2 \end{bmatrix}\begin{bmatrix} -1 \\ 0 \end{bmatrix}\frac{Pl}{8} = \begin{bmatrix} 1 \\ -1 \end{bmatrix}\frac{Pl}{8} + \begin{bmatrix} 1 \\ -1 \end{bmatrix}\frac{Pl}{8} = \begin{bmatrix} 1 \\ -1 \end{bmatrix}\frac{Pl}{4}$$

These moments are shown in Fig. 5.17(b).

5.3.6 Stiffness for non–uniform members

The same assumption is made here as was used in the calculation of flexibility matrices for non–uniform members. That is, the centroidal axis of the member is straight, and the material in the member is distributed symmetrically about the axis. The bending stiffness matrix could be obtained by inversion of the flexibility matrix. Herein eqn (2.110) will be used directly. The member with its area and inertia properties is shown in Fig. 5.18. These properties are given for four equal subdivisions of the member and Simpson's rule is used for the numerical quadrature.

5.3.6.1 Axial stiffness

From eqn (4.87), with axial stiffness $[k_{axial}] = [f_{axial}]^{-1}$, in which

$$f_{axial} = \frac{h}{3EA_1}\{W\}^T\{\alpha\}$$

In the above expression the terms are defined by

$$\{W\}^T = <1\ 4\ 2\ 4\ 1> \quad \text{and} \quad \{\alpha\}^T = <\alpha_1\ \alpha_2\ \alpha_3\ \alpha_4\ \alpha_5>\ ; \quad \alpha_i = \frac{A_1}{A_i}$$

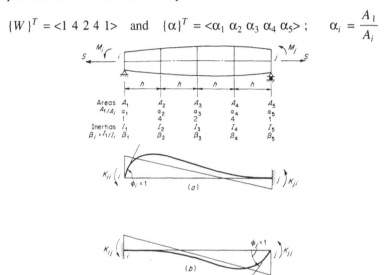

Fig. 5.18   Member properties with unit rotations of ends of member.

5.3.6.2 Bending stiffness

If a unit rotation is applied at end i while end j is held clamped, the force at end i will be k_{ii}, and that at end j will be k_{ji}. Similarly, when a unit rotation is given to end j, the force at end i will be k_{ij}, and that at j will be k_{jj}. Then, using the load system,

$$\left\{ \begin{array}{c} \bar{m}_i \\ \bar{R}_{yi} \end{array} \right\}$$

at the end i of the cantilever beam, fixed at end j, the moment $\bar{m}$ at x from i is given

$$\bar{m} = [-1\ x]\left\{ \begin{array}{c} \bar{m}_i \\ \bar{R}_{yi} \end{array} \right\}$$

and the bending strains $\{\varepsilon\}$ are

$$v = M\ \frac{dx}{EI} = \left\{ \frac{-k_{ii}(1 - \frac{x}{l}) + k_{ji}\frac{x}{l}}{EI} \right\}dx$$

From eqn (2.110),

$$\left\{\begin{array}{c}\phi_i\\\phi_j\end{array}\right\}=\left\{\begin{array}{c}1\\0\end{array}\right\}=\int_0^l\left\{\begin{array}{c}-1\\x\end{array}\right\}\left[\dfrac{-k_{ii}\left(1-\dfrac{x}{l}\right)+k_{ji}\dfrac{x}{l}}{EI}\right]dx \qquad (5.81)$$

Integrating numerically, using Simpson's rule for which the integration intervals are shown in Fig. 5.17(a), the first equation in (5.81) becomes

$$1=\dfrac{h}{3EI_1}\left\{k_{ii}\left[\beta_1+4\beta_2(3/4)+2\beta_3(2/4)+4\beta_4(1/4)\right]-k_{ji}\left[4\beta_2(1/4)+2\beta_3(2/4)+4\beta_4(3/4)+\beta_5\right]\right\}$$

Write this equation in the form,

$$1=\dfrac{h}{3EI_1}(k_{ii}A_1-k_{ji}A_2) \qquad (5.82)$$

As in section 4.9, let

$$\{\zeta_1\}^T=\dfrac{x}{l}=(0,\ 1/4,\ 2/4,\ 3/4,\ 1)\quad\text{and}\quad\{\zeta_2\}^T=1-\dfrac{x}{l}=(1,\ 3/4,\ 2/4,\ 1/4,\ 0)$$

Also

$$[\beta]=\left|\beta_1\ \beta_2\ \beta_3\ \beta_4\ \beta_5\right|$$

In this notation

$$A_1=[W]^T[\beta]\{\zeta_2\}\quad\text{and}\quad A_2=[W]^T[\beta]\{\zeta_1\}$$

From the second equation in (5.81),

$$0=k_{ii}B_1-k_{ji}B_2 \qquad (5.83)$$

where

$$B_1=4\beta_2(3/4)(1/4)+2\beta_3(2/4)(2/4)+4\beta_4(1/4)(3/4)$$

and

$$B_2=4\beta_2(1/4)(1/4)+2\beta_3(2/4)(2/4)+4\beta_4(3/4)(3/4)+\beta_5$$

That is, in a similar way to section 4.9,

$$B_1=[W]^T[\beta]\{\zeta_1\}*\{\zeta_2\}\quad\text{and}\quad B_2=[W]^T[\beta]\{\zeta_1\}*\{\zeta_1\}$$

Thence, from eqn (5.83),

$$k_{ji}=\dfrac{B_1k_{ii}}{B_2} \qquad (5.84)$$

Substitution for k_{ji} in eqn 5.66 gives k_{ii} as

$$k_{ii}=\dfrac{3EI_1B_2}{h(A_1B_2-A_2B_1)} \qquad (5.85)$$

The process is repeated to obtain k_{jj}. Remember that $k_{ij}=k_{ji}$ because of the symmetric property of the stiffness matrix.

5.3.6.3   Fixed–end moments and shears

Consider the case in which the load w_x is a continuous function of x producing moments m_x. End rotations are calculated in eqn (4.93). In the present case, the sign of ϕ_i is changed because of the different moment sign convention used herein, so that

$$\begin{Bmatrix} \phi_i \\ \phi_j \end{Bmatrix} = \frac{h}{3EI_1} \begin{bmatrix} -[W]^T & 0 \\ 0 & [W]^T \end{bmatrix} \lceil [\beta][\beta] \rfloor \begin{bmatrix} \{\zeta_2\}*\{m\} \\ \{\zeta_1\}*\{m\} \end{bmatrix} \tag{5.86}$$

It will be assumed that w_x has been calculated at the same intervals as m_x, as shown in Fig. 5.18. The reactions are given by

$$R_i = \frac{h}{3}[w_1 + 4w_2(3/4) + 2w_3(2/4) + 4w_4(1/4)]$$

That is,

$$R_i = \frac{h}{3}[W]^T[w]\{\zeta_2\} \quad \text{and} \quad R_j = \frac{h}{3}[W]^T[w]\{\zeta_1\} \tag{5.87}$$

In eqn (5.87) the matrix $[w]$ is defined

$$[w] = \begin{vmatrix} w_1 \ w_2 \ w_3 \ w_4 \ w_5 \end{vmatrix}$$

From eqn (5.86) the fixed–end moments will be such that the rotations $<\phi_i \ \phi_j>$ are constrained to be zero. Then

$$\begin{bmatrix} M_i \\ M_j \end{bmatrix}_F = -[k] \begin{bmatrix} \phi_i \\ \phi_j \end{bmatrix} \tag{5.88}$$

These moments will cause reactions

$$\begin{bmatrix} R_i \\ R_j \end{bmatrix}_F = \frac{1}{l} \begin{bmatrix} M_i + M_j \\ -(M_i + M_j) \end{bmatrix}_F = \frac{1}{l} \begin{bmatrix} 1 & 1 \\ -1 & -1 \end{bmatrix} \begin{bmatrix} M_i \\ M_j \end{bmatrix}_F \tag{5.89}$$

5.3.7   The effects of axial force on the bending stiffness

When the bending moments produced by the interaction between the axial force in the member and the deflected shape are taken into account, the behaviour is non–linear and requires some type of iteration procedure to arrive at a solution. That is, the bending stiffness depends on the axial load in the member, which, in turn, is calculated by using the bending stiffness. The classical expressions for flexibility matrices were derived for straight prismatic members in section 4.9. If the same theory is used to develop the stiffness matrix, it is found that

$$\begin{bmatrix} M_i \\ M_j \end{bmatrix} = \frac{2EI}{l} \begin{bmatrix} 2\phi_3 & \phi_4 \\ \phi_4 & 2\phi_3 \end{bmatrix} \begin{bmatrix} \phi_i \\ \phi_j \end{bmatrix} \tag{5.90}$$

For axial compression in the member, the ϕ functions are defined

$$\phi_1 = \beta\cot\beta \ ; \qquad \phi_2 = \frac{\beta^2}{3(1 - \beta\cot\beta)}$$

Then the remaining functions in eqn (5.90) are given:

$$\phi_3 = 1/4\,(\phi_1 + 3\phi_2) \ ; \qquad \phi_4 = 1/2(3\phi_2 - \phi_1)$$

Similarly, for axial tension in the member,

$$\phi_1 = \beta\coth\beta\ ; \qquad \phi_2 = \frac{\beta^2}{3(\beta\coth\beta - 1)}$$

and thence,

$$\phi_3 = 1/4(\phi_1 + 3\phi_2)\ ; \qquad \phi_4 = -1/2(\phi_1 - 3\phi_2)$$

The function β is defined

$$\beta = \frac{l}{2}\sqrt{\frac{P}{EI}}$$

Notice that in this approach no account is made for the effect of the bending moments on axial stiffness; that is, the axial stiffness is expressed as $k_{axial} = \dfrac{EA}{l}$.

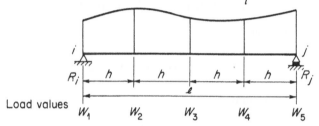

Fig. 5.19   Variable load intensity.

A useful alternative idea for the incorporation of the axial–force effects in the bending stiffness matrix assumes that the deflected shape is a given polynomial function. For example, if a cubic function is assumed, these effects are readily determined. The error in the method lies in the fact that the axial force will modify the assumed shape. This should not be a large source of error if the axial force is significantly less than the critical load. In Fig. 5.20 the member is shown with end rotations $<\phi_i\ \phi_j>$ and moments $<M_i\ M_j>$. Instead of using a single coordinate x to describe the position of a point on a beam, the two non–dimensional (natural) coordinates $<\zeta_1\ \zeta_2>$ will be used.

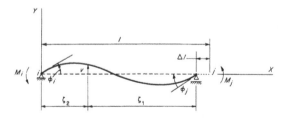

Fig. 5.20   Assumed deflected shape $v = f(\zeta_1, \zeta_2)$.

These are defined

$$\zeta_1 = 1 - \zeta_2 = 1 - \frac{x}{l} \quad \text{and} \quad \zeta_2 = \frac{x}{l}$$

The deflected shape v may be expressed as a cubic function of $<\zeta_1\ \zeta_2>$, so that

$$v = \{r\}^T \begin{bmatrix} l\zeta_1^2\zeta_2 \\ -l\zeta_1\zeta_2^2 \end{bmatrix}\ ; \quad \text{where,} \quad \{r\}^T = \{\phi_i\ \phi_j\}^T \qquad (5.91)$$

The vector containing the ζ functions in eqn (5.91) is an interpolation polynomial (Chapter 1), and will be designated $\{\Phi\}$. It is seen that

$$v = \{r\}^T\{\Phi\} = \{\Phi\}^T\{r\}$$

The slope of the curve v will be designated v_x, and is given

$$v_x = \{r\}^T \begin{bmatrix} \zeta_1(\zeta_1 - 2\zeta_2) \\ \zeta_2(\zeta_2 - 2\zeta_1) \end{bmatrix} = \{r\}^T\{\Phi_x\} = \{\Phi_x\}^T\{r\} \tag{5.92}$$

In eqn (5.92) write $\{\Phi_x\}$ in terms of ζ_2 only, using the relationship, $\zeta_1 = 1 - \zeta_2$, Then,

$$\{\Phi_x\}^T = <(1 - \zeta_2)(1 - 3\zeta_2) \quad \zeta_2(-2 + 3\zeta_2)>$$

Also the derivative of the deflection v is given by

$$v_x^2 = (\{\Phi_x\}^T\{r\})^2 = \{r\}^T\{\Phi_x\}\{\Phi_x\}^T\{r\}$$

The length l of the member is calculated by

$$l = \int_0^{l-\Delta l} (1 + v_x^2)^{1/2}dx \approx l - \Delta l + \frac{1}{2}\int_0^l v_x^2\, dx$$

The shortening due to bowing is, thus,

$$\Delta l = \frac{1}{2}\int_0^l v_x^2\, dx$$

However, the integral in this equation is expressed as

$$\int_0^l v_x^2\, dx = l\{r\}^T\int_0^1 \{\Phi_x\}\{\Phi_x\}^T d\zeta_2\{r\} = \frac{l\{r\}^T}{30}\begin{bmatrix} 4 & -1 \\ -1 & 4 \end{bmatrix}\{r\}$$

That is, performing the integration,

$$\int_0^l v_x^2\,dx = \frac{l}{30}(4\phi_i^2 - 2\phi_i\phi_j + 4\phi_j^2)$$

Then the change in length due to bowing is

$$\Delta l = \frac{1}{2}\int_0^l v_x^2\,dx = \frac{l}{30}(2\phi_i^2 - \phi_i\phi_j + 2\phi_j^2) \tag{5.93}$$

The small deflection flexibility matrix is given as

$$\begin{bmatrix} \phi_i \\ \phi_j \end{bmatrix} = \frac{l}{6EI}\begin{bmatrix} 2 & -1 \\ -1 & 2 \end{bmatrix}\begin{bmatrix} M_i \\ M_j \end{bmatrix}$$

With the deflected shape v given in terms of $<\phi_i\ \phi_j>$, the moments due to the axial force P (tension positive and hogging moment positive) are given (Fig. 5.21) by

$$M_P = -Pv = -P\{r\}^T \begin{bmatrix} l\zeta_1^2\zeta_2 \\ -l\zeta_1\zeta_2^2 \end{bmatrix}$$

For end moments M_i, M_j,

$$\overline{M} = [\zeta_1 \ -\zeta_2] \begin{bmatrix} M_i \\ M_j \end{bmatrix}$$

Thence, using eqn (2.110), the increment in end rotations $\{\Delta r\}$, due to M_P is

$$\{\Delta r\} = -\begin{bmatrix} \zeta_1 \\ -\zeta_2 \end{bmatrix} \frac{P\{r\}^T}{EI} \begin{bmatrix} l\zeta_1^2\zeta_2 \\ -l\zeta_1\zeta_2^2 \end{bmatrix} dx = \frac{-P}{EI} \begin{bmatrix} \zeta_1 \\ -\zeta_2 \end{bmatrix} [l\zeta_1^2\zeta_2 \ -l\zeta_1\zeta_2^2] \begin{Bmatrix} \phi_i \\ \phi_j \end{Bmatrix} dx$$

Integrating from 0 to l,

$$\begin{Bmatrix} \phi_{iP} \\ \phi_{jP} \end{Bmatrix} = -\frac{Pl^2}{60EI} \begin{bmatrix} 3 & -2 \\ -2 & 3 \end{bmatrix} \begin{Bmatrix} \phi_i \\ \phi_j \end{Bmatrix} \tag{5.94}$$

That is, given the deflected shape defined by $\{\phi_i \ \phi_j\}^T$, the axial force produces additional rotations $\{\phi_{iP} \ \phi_{jP}\}^T$ given by eqn (5.94). For the rotation to remain at $\{\phi_i \ \phi_j\}^T$, additional moments must be applied to produce rotations $-\{\phi_{iP} \ \phi_{jP}\}^T$. That is,

$$\begin{bmatrix} M_i \\ M_j \end{bmatrix} = \frac{2EI}{l} \begin{bmatrix} 2 & 1 \\ 1 & 2 \end{bmatrix} \left\{ \begin{Bmatrix} \phi_i \\ \phi_j \end{Bmatrix} + \frac{Pl^2}{60EI} \begin{bmatrix} 3 & -2 \\ -2 & 3 \end{bmatrix} \begin{bmatrix} \phi_i \\ \phi_j \end{bmatrix} \right\}$$

Finally, then, the total end moments are given

$$\begin{bmatrix} M_i \\ M_j \end{bmatrix} = \frac{2EI}{l} \begin{bmatrix} 2 & 1 \\ 1 & 2 \end{bmatrix} \begin{bmatrix} \phi_i \\ \phi_j \end{bmatrix} + \frac{Pl}{30} \begin{bmatrix} 4 & -1 \\ -1 & 4 \end{bmatrix} \begin{bmatrix} \phi_i \\ \phi_j \end{bmatrix} \tag{5.95}$$

It is seen that the first term on the right–hand side of eqn (5.95) is the normal small deflection stiffness matrix and that the second term gives the effect of the axial force. Eqn (5.95) will be used later in Chapter 6 in non–linear analysis of line element structures.

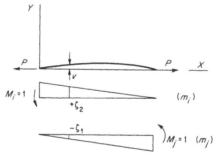

Fig. 5.21   Calculation of end rotations due to axial force.

5.4 THE DIRECT STIFFNESS METHOD

5.4.1   Transformation of stiffness matrices

In section 5.1 it has been shown how displacement transformation matrices may be developed so that the eqn (5.23),

$$[K] = [a]^T [k][a] = \sum_1^n [a_n]^T [k_n][a_n] \tag{5.23}$$

can be used to generate the structure stiffness matrix. A subtle variation to to the use of eqn (5.23) occurs if $[k_n]$, for member n is not the stiffness matrix for the fundamental member distortions but has been expanded to be the stiffness matrix for the member in the global coordinate system.

When the process in eqn (5.95) is carried out, rows of $[a_n]$ contain only zero or unit values. This implies that instead of the formal multiplication in eqn (5.23), the structure stiffness matrix $[K]$ may be generated by simply adding the element stiffness matrix terms to the appropriate row–column locations. Before the method can be used it is necessary to be able to obtain the member stiffness matrices in global coordinates. This requires a knowledge of how stiffness matrices transform. For example, consider the joint A of the plane structure shown in Fig. 5.22, and let the stiffness relationship between $<r_X \ r_Y>$ and $<R_X \ R_Y>$ be

$$\{R\} = [K]\{r\} \tag{5.96}$$

It is required to calculate the stiffness matrix $[K']$ when the forces and displacements are expressed in the $X' \ Y'$ coordinate axes obtained by a positive rotation α from XY. The relationships between (R, R') and (r, r') have been established to be (eqn (1.39))

$$\{R\} = [L]^T\{R'\} \ ; \quad \text{and, for displacement transformations,} \quad \{r\} = [L]^T\{r'\}$$

The transformation matrix $[L]^T$ is defined,

$$[L]^T = \begin{bmatrix} \cos \alpha & -\sin \alpha \\ \sin \alpha & \cos \alpha \end{bmatrix} \tag{5.97}$$

If these values are substituted in eqn (5.96), and it is noted that $[L]^{-1} = [L]^T$, then

$$\{R'\} = [L][K][L]^T\{r'\} \tag{5.98}$$

That is, the stiffness matrix in the rotated coordinates is

$$[K'] = [L][K][L]^T \tag{5.99}$$

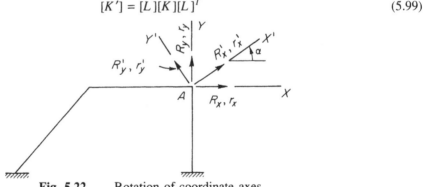

Fig. 5.22   Rotation of coordinate axes.

5.4.2   The stiffness matrix for plane frame members in global coordinates

The planar line element i, j is shown in Fig. 5.23 with components $<r_x \ r_y \ \theta_z>$ and force components $<R_x \ R_y \ M_z>$ at its ends. The fundamental member–force quantities $<M_i \ M_j \ F>$ and displacement quantities $<\phi_i \ \phi_j \ \delta>$ are defined in Fig. 5.23(b). The fundamental stiffness relationship is of the form,

$$\begin{bmatrix} M_i \\ M_j \\ F \end{bmatrix} = \begin{bmatrix} K_{11} & K_{12} & 0 \\ K_{21} & K_{22} & 0 \\ 0 & 0 & K_{33} \end{bmatrix} \begin{bmatrix} \phi_i \\ \phi_j \\ \delta \end{bmatrix} \tag{5.100}$$

Eqn (5.100) is written

$$\{S\} = [k']\{v\} \tag{5.101}$$

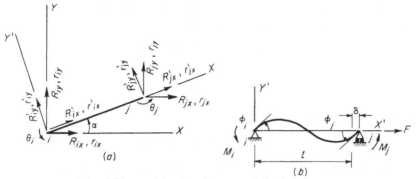

Fig. 5.23   Member forces and displacements.

For prismatic members without shear distortion,

$$[k'] = \frac{2EI}{l} \begin{bmatrix} 2 & 1 & 0 \\ 1 & 2 & 0 \\ 0 & 0 & \dfrac{A}{2I} \end{bmatrix}$$

The member distortions $\{v\}$ are obtained by the transformation eqn (3.63); that is,

$$\begin{bmatrix} \phi_i \\ \phi_j \\ \delta \end{bmatrix} = \begin{bmatrix} 0 & \dfrac{1}{l} & 1 & 0 & -\dfrac{1}{l} & 0 \\ 0 & \dfrac{1}{l} & 0 & 0 & -\dfrac{1}{l} & 1 \\ -1 & 0 & 0 & 1 & 0 & 0 \end{bmatrix} \begin{bmatrix} r'_{ix} \\ r'_{iy} \\ \theta_i \\ r'_{jx} \\ r'_{jy} \\ \theta_j \end{bmatrix} \qquad (5.102)$$

This equation is written,

$$\{v\} = [T]\{r'\} \qquad (5.103)$$

The transformation from $\{r\}$ to $\{r'\}$ has been developed in Chapter 1, eqn (1.139), such that for the two ends of the member

$$\{r'\} = [L_D]\{r\} \qquad (5.104)$$

The matrix $[L_D] = \lfloor L \rfloor_2$ is a super–diagonal matrix in which,

$$[L] = \begin{bmatrix} \cos \alpha & \sin \alpha & 0 \\ -\sin \alpha & \cos \alpha & 0 \\ 0 & 0 & 1 \end{bmatrix} \qquad (5.105)$$

Combining eqns (5.102) and (5.103),

$$\{v\} = [T][L_D]\{r\} \qquad (5.106)$$

The relationship between $\{R\}$ and $\{S\}$ is given by the contragredient principle as

$$\{R\} = [L_D]^T [T]^T \{S\} \qquad (5.107)$$

Hence, combining eqns (5.101), (5.105), and (5.106), the expression is obtained for the member stiffness matrix in global force components. That is,

$$\{R\} = [L_D]^T [T]^T [k'][T][L_D]\{r\} \tag{5.108}$$

or,

$$\{R\} = [k]\{r\}$$

where the member stiffness matrix is given by

$$[k] = [L_D]^T [T]^T [k'][T][L_D] \tag{5.109}$$

Since the member forces $\{S\}$ are of interest when the effects of stresses on the member material properties are being examined, they may be obtained from $\{r\}$ through the equation

$$\{S\} = [k'][T][L_D]\{r\} \tag{5.110}$$

5.4.3 The stiffness matrix for grid members in global components

In this particular case the transformation matrices will be developed for components in the coordinate axes shown in Fig. 5.24(a). The Z–axis will then coincide with the positive gravity force axis, and the grid structure will lie in the XY plane. For the grid member the basic member forces are $<M_{iy} \; M_{jy} \; M_T>$ and the corresponding displacements are $<\phi_{iy} \; \phi_{jy} \; \theta_T>$. The force M_T is the torque in the member, and θ_T is the relative rotation between ends i and j. The stiffness relationship is then given by

$$\begin{bmatrix} M_{iy} \\ M_{jy} \\ M_T \end{bmatrix} = \begin{bmatrix} K_{11} & K_{12} & 0 \\ K_{21} & K_{22} & 0 \\ 0 & 0 & K_{33} \end{bmatrix} \begin{bmatrix} \phi_{iy} \\ \phi_{jy} \\ \theta_T \end{bmatrix} \tag{5.111}$$

The term K_{33} is the torsion stiffness of the member. Eqn (5.111) will be written as for the plane frame member

$$\{S\} = [k']\{v\} \tag{5.112}$$

In this case the member distortions are related to the displacements $<\theta_x \; \theta_y \; r_z>$ of ends i and j of the member by eqn (5.113). That is,

$$\begin{bmatrix} \phi_{iy} \\ \phi_{jy} \\ \theta_T \end{bmatrix} = \begin{bmatrix} 0 & 1 & -\dfrac{1}{l} & 0 & 0 & \dfrac{1}{l} \\ 0 & 0 & -\dfrac{1}{l} & 0 & 1 & \dfrac{1}{l} \\ -1 & 0 & 0 & 1 & 0 & 0 \end{bmatrix} \begin{bmatrix} \theta_{ix} \\ \theta_{iy} \\ r_{iz} \\ \theta_{jx} \\ \theta_{jy} \\ r_{jz} \end{bmatrix}' \tag{5.113}$$

This is the equation

$$\{v\} = [T]\{r'\} \tag{5.114}$$

The transformation from global coordinate components to member coordinate components is identical with that in eqn (5.114) if the angle α measures the rotation of the $X'Y'$ set of axes from XY about the Z axis (Fig. 5.24(c)). Then,

$$\{v\} = [T][L_D]\{r\} \qquad (5.115)$$

The relationship between the forces $\{R\}$ for which

$$\{R\}^T = <M_{ix} \ M_{iy} \ R_{iz} \ M_{jx} \ M_{jy} \ R_{jz}>$$

on the member ends and the member forces $\{S\}$ is given by the contragredient principle to be

$$\{R\} = [L_D]^T [T]^T \{S\}$$

Hence, finally, the stiffness relation is

$$\{R\} = [L_D]^T [T]^T [k'][L_D][T]\{r\} \qquad (5.116)$$

A matrix expression for the stiffness matrix is obtained indentical to that in eqn (5.108). that is,

$$[K] = [L_D]^T [k'][L_D]$$

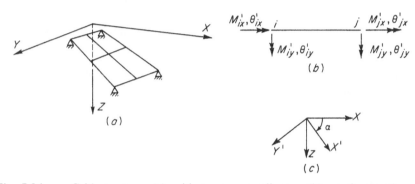

Fig. 5.24   Grid structure: (a) grid structure cordinates; (b) rotations and moment at member ends (local coordinates); (c) rotation of axes.

Because the transformations for the planar frame and grid members are identical in form, an analogy exists between the two, relating twisting moment and twist rotation, on the one hand, to axial force and axial extension, on the other. The analogy is easily extended to the forces and displacements on the member ends but is not pursued further here.

5.4.4 The stiffness matrix for – prismatic member in three–dimensional space

The procedures for the frame and grid structures may be extended to cover the general three–dimensional element. It is found that a certain amount of difficulty arises in defining the coordinate axes of the member relative to the global system. The basic member distortions will be taken as the following

δ = axial change in length

θ_T = relative twist of the ends of the member about the OX' axis

ϕ_{iz}

= relative rotations of tangents at end i relative to chord

ϕ_{iy}
ϕ_{jz}

= relative rotations of tangents at end j relative to chord

ϕ_{jy}

The displacement vectors at ends i and j in the local coordinates will be

$$\{r'_i\}^T = <r'_{ix}\ r'_{iy}\ r'_{iz}\ \theta'_{ix}\ \theta'_{iy}\ \theta'_{iz}> \quad \text{and} \quad \{r'_j\}^T = <r'_{jx}\ r'_{jy}\ r'_{jz}\ \theta'_{jx}\ \theta'_{jy}\ \theta'_{jz}>$$

Then the transformation to basic member distortions is given as

$$
\begin{bmatrix} \delta \\ \phi_{iz} \\ \phi_{jz} \\ \theta_T \\ \phi_{iy} \\ \phi_{jy} \end{bmatrix} =
\begin{bmatrix}
-1 & 0 & 0 & 0 & 0 & 0 & 1 & 0 & 0 & 0 & 0 & 0 \\
0 & \dfrac{1}{l} & 0 & 0 & 0 & 1 & 0 & -\dfrac{1}{l} & 0 & 0 & 0 & 0 \\
0 & \dfrac{1}{l} & 0 & 0 & 0 & 0 & 0 & -\dfrac{1}{l} & 0 & 0 & 0 & 1 \\
0 & 0 & 0 & -1 & 0 & 0 & 0 & 0 & 0 & 1 & 0 & 0 \\
0 & 0 & -\dfrac{1}{l} & 0 & 1 & 0 & 0 & 0 & \dfrac{1}{l} & 0 & 0 & 0 \\
0 & 0 & -\dfrac{1}{l} & 0 & 0 & 0 & 0 & 0 & \dfrac{1}{l} & 0 & 1 & 0
\end{bmatrix}
\begin{bmatrix} r'_{ix} \\ r'_{iy} \\ r'_{iz} \\ \theta'_{ix} \\ \theta'_{iy} \\ \theta'_{iz} \\ r'_{jx} \\ r'_{jy} \\ r'_{jz} \\ \theta'_{jx} \\ \theta'_{jy} \\ \theta'_{jz} \end{bmatrix}
\tag{5.117}
$$

This is the equation relating member distortions to local coordinate displacements eqn (5.114). The relationship between member forces $\{S\}$ for which,

$$\{S\}^T = <F\ M_{iz}\ M_{jz}\ M_T\ M_{iy}\ M_{jy}>$$

and the member distortions $\{v\}$ is given by

$$\{S\} = [k']\{v\} \tag{5.118}$$

where the basic member stiffness is written

$$
[k'] =
\begin{bmatrix}
\dfrac{EA}{l} & 0 & 0 & 0 & 0 & 0 \\
0 & \dfrac{4EI_z}{l} & \dfrac{2EI_z}{l} & 0 & 0 & 0 \\
0 & \dfrac{2EI_z}{l} & \dfrac{4EI_z}{l} & 0 & 0 & 0 \\
0 & 0 & 0 & \dfrac{GI_P}{l} & 0 & 0 \\
0 & 0 & 0 & 0 & \dfrac{4EI_y}{l} & \dfrac{2EI_y}{l} \\
0 & 0 & 0 & 0 & \dfrac{2EI_y}{l} & \dfrac{4EI_y}{l}
\end{bmatrix}
\tag{5.119}
$$

The member with its coordinate axes is shown in Fig. 5.25, and it is necessary (to transform the stiffness matrix to global coordinates) to construct the direction cosine matrix:

$$
[L] =
\begin{bmatrix}
l_{11} & l_{12} & l_{13} \\
l_{21} & l_{22} & l_{23} \\
l_{31} & l_{32} & l_{33}
\end{bmatrix}
$$

The member ij is, of course, simply the general directed line element. From the coordinates (x_i, y_i, z_i) and (x_j, y_j, z_j), the projections of ij, namely, $(\Delta x, \Delta y, \Delta z)$, on the coordinate axes are obtained. That is,

$$\Delta x = x_j - x_i ; \quad \Delta y = y_j - y_i ; \quad \text{and} \quad \Delta z = z_j - z_i$$

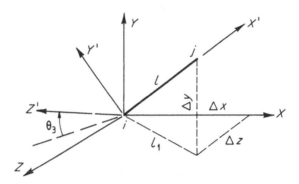

Fig. 5.25 Member orientation in three–dimensional space.

The member length is calculated by

$$l = \sqrt{(\Delta x)^2 + (\Delta y)^2 + (\Delta z)^2}$$

and the projected length on the XZ plane is

$$l_1 = \sqrt{(\Delta x)^2 + (\Delta z)^2}$$

Hence the length l is also given by

$$l = \sqrt{l_1^2 + (\Delta y)^2}$$

The $[L]$ matrix may be generated as the product of three transformation matrices which are defined by rotations about the axes OX', OY', and OZ'. These three rotations are given by the Euler angles of the X', Y', Z' axes. Only one of these three angles need be specified as data in addition to the coordinates of the ends i and j of the member. It is seen in Fig. 5.26 that the X', Y', Z' axes are completely specified when the X' direction is known, together with the angle θ_3 measured from the XZ plane to the axis Z'. Notice however, in Fig. 5.26 that a θ_3 angle can specify two positions for the $Z'Y'$ axes. In order to use a consistent transformation only one of these sets of coordinate positions will be permissible. Thus, θ_3 is defined as the angle for which the projection of Y' on the OY axis is positive when the $Z'Y'$ axes are rotated about iX' through $-\theta_3$ so that Z' lies in the XZ plane. The axes Z'_1, Y'_1 are, of course, quite acceptable but are defined by the rotation $(180^o + \theta_3)$. Notice also in this derivation of $[L]$ that θ_3 is measured positive in the positive sense about iX' from the XZ plane to iZ'. For the first transformation (Fig. 5.26), let

$$c_3 = \cos \theta_3 ; \quad s_3 = \sin \theta_3$$

$$[L_3] = \begin{bmatrix} 1 & 0 & 0 \\ 0 & c_3 & s_3 \\ 0 & -s_3 & c_3 \end{bmatrix} \qquad (5.120)$$

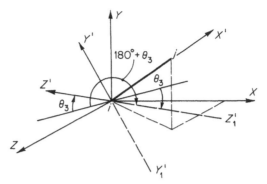

Fig. 5.26   Alternative positions of coordinate axes.

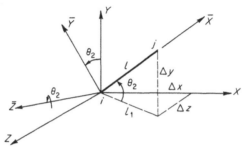

Fig. 5.27   Position after first rotation.

When Y', Z' have been rotated so that Z' lies in the XZ plane, let these new axes be designated $\bar{Z}$, $\bar{Y}$. The transformation of components of vectors in the X', $\bar{Y}$, $\bar{Z}$ system to those in X', Y', Z' is given by

$$\{U'\} = [L_3]\{\bar{u}\} \tag{5.121}$$

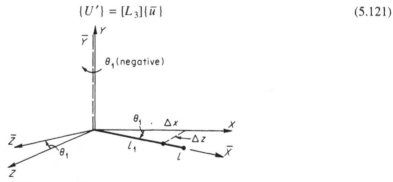

Fig. 5.28   Position of axes after second rotation.

The second rotation of the coordinate axes will be for $X'\bar{Y}$ about the $\bar{Z}$ axis so that $\bar{Y}$ coincides with the iY axis. These axes will be designated $\bar{Z}Y\bar{X}$ and are shown in Fig. 5.26. For this transformation, it is seen from Fig. 5.25 that

$$\cos \theta_2 = c_2 = \frac{l_1}{l} \quad \text{and} \quad \sin \theta_2 = s_2 = \frac{\Delta y}{l}$$

Define the matrix

$$[L_2] = \begin{bmatrix} c_2 & s_2 & 0 \\ -s_2 & c_2 & 0 \\ 0 & 0 & 1 \end{bmatrix} \qquad (5.122)$$

The transformation of components of vectors in the $\overline{X}\overline{Y}\overline{Z}$ system to those in the $X'\overline{Y}\overline{Z}$ system is given by

$$\{\overline{U}\} = [L_2]\{\overline{\overline{U}}\} \qquad (5.123)$$

The third rotation of the axes (Fig. 5.28) will be for $\overline{X}\overline{Z}$ about the $i\overline{Y}$, axis defined by the angle θ_1 such that

$$\cos\theta_1 = c_1 = \frac{\Delta x}{l_1} \quad \text{and} \quad \sin\theta_1 = s_1 = \frac{-\Delta z}{l_1}$$

Then,

$$[L_1] = \begin{bmatrix} c_1 & 0 & s_1 \\ 0 & 1 & 0 \\ -s_1 & 0 & c_1 \end{bmatrix} \qquad (5.124)$$

The transformation of components of the vector $\{U\}$ in the XYZ coordinate system to those of $\overline{\overline{U}}$ in the $\overline{X}\overline{Y}\overline{Z}$ axes is given by

$$\{\overline{\overline{U}}\} = [L_1]\{U\} \qquad (5.125)$$

It is seen from eqns (5.121), (5.123), and (5.125), the transformation from $\{U\}$ to $\{U'\}$ is given since

$$\{U'\} = [L_3][L_2][L_1]\{U\} \qquad (5.126)$$

However, it is known from eqn (1.38) that

$$\{U'\} = [L]\{U\} \qquad (5.127)$$

where $[L]$ is the matrix of direction cosines of the $X'Y'Z'$ axes with respect to XYZ. Hence,

$$[L] = [L_3][L_2][L_1] \qquad (5.128)$$

If the multiplication in eqn (5.128) is carried out, the explicit expression obtained for $[L]$ is

$$[L] = \begin{bmatrix} c_1 c_2 & s_2 & -s_1 c_2 \\ -s_1 c_1 c_2 + s_1 s_3 & c_2 c_3 & s_1 s_2 c_3 + c_1 s_3 \\ -c_1 s_2 s_3 + s_1 c_3 & -c_2 s_3 & -s_1 s_2 s_3 + c_1 c_3 \end{bmatrix} \qquad (5.129)$$

The terms of the first row of the $[L]$ matrix are simply the direction cosines of the member axis $i-j$; that is, $(\Delta x / l, \Delta y / l, \Delta z / l)$. Multiplication of the terms in eqn (5.128) produces these values. The above transformation matrices, $[L_3]$, $[L_2]$, $[L_1]$, are adequate except for the case when the member axis coincides with the global OY axis (Fig. 5.29).

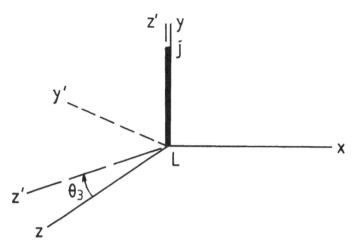

Fig. 5.29   Member axis parallel to global *OY* axis.

In this case θ_3 is taken to be the angle from the *OZ* axis to *OZ'*, as shown in Fig. 5.29. Then, $[L_3]$ and $[L_2]$ are as defined in eqns (5.120) and (5.122), and $[L_1]$ is the unit matrix. The matrix $[L_D]$ is now defined to be the (12×12) super–diagonal matrix with $[L]$ as submatrix diagonal blocks. That is,

$$[L_D] = \lceil [L] \rfloor_4$$

The development of the global stiffness matrix now follows the identical steps given for the plane frame in eqns (5.86 ff). That is, $[K]$ is now the (12×12) stiffness matrix,

$$[K] = [L_D]^T [T]^T [k][T][L_D]$$

The stiffness matrix generations for the plane frame, grid and space frame members are identical in form. The plane frame and grid member matrices are only condensed special cases of the general three–dimensional matrices and may be extracted from this. The stiffness matrices $[K]$ obtained above in all cases are singular since they have been expanded so that rigid–body displacements are contained in the vector $\{r\}$. Write the stiffness equation for the member

$$\{R\} = [K]\{r\}$$

In partitioned form so that the vectors $\{R\}$, $\{r\}$ are subdivided into the components at each end of the member. Then,

$$\begin{bmatrix} R_i \\ R_j \end{bmatrix} = \begin{bmatrix} k_{ii} & k_{ij} \\ k_{ji} & k_{jj} \end{bmatrix} \begin{bmatrix} r_i \\ r_j \end{bmatrix} \tag{5.130}$$

It is now shown that the submatrices k_{ii}, k_{ij}, k_{jj} are not independent, and, given any one of them, the other two may be determined. This gives a useful alternative method for calculating $[K]$ which may be used to good effect for curved members for which it is convenient to calculate the effects of forces at one end. The forces $\{R_i\}$ and $\{R_j\}$ acting on the member are in equilibrium. If, then, $\{R_i\}$ is transformed to end j, the components are equal and opposite to $\{R_j\}$. That is,

$$\{R_j\} = -[U]^{-1}\{R_i\} \tag{5.131}$$

The matrix $[U]^{-1}$ for the three–dimensional case is

$$[U]^{-1} = \begin{bmatrix} 1 & \cdot & \cdot & \cdot & \cdot & \cdot \\ \cdot & 1 & \cdot & \cdot & \cdot & \cdot \\ \cdot & \cdot & 1 & \cdot & \cdot & \cdot \\ \cdot & \Delta z & -\Delta y & 1 & \cdot & \cdot \\ -\Delta z & \cdot & \Delta x & \cdot & 1 & \cdot \\ \Delta y & -\Delta x & \cdot & \cdot & \cdot & 1 \end{bmatrix}$$ (5.132)

The $[U]^{-1}$ matrix for the plane frame and grid members is obtained by deletion of the appropriate rows and columns in eqn (5.132). Substitute for $\{R_j\}$ in eqn (5.132) so that

$$\begin{bmatrix} R_i \\ -U^{-1}R_i \end{bmatrix} = \begin{bmatrix} k_{ii} & k_{ij} \\ k_{ji} & k_{jj} \end{bmatrix} \begin{bmatrix} r_i \\ r_j \end{bmatrix}$$ (5.133)

On expansion of eqn (5.116),

$$\{R_i\} = [k_{ii}]\{r_i\} + [k_{ij}]\{r_j\}$$

and

$$\{R_i\} = -[U][k_{ji}]\{r_i\} - [U][k_{jj}]\{r_j\}$$

Comparing the coefficients of $\{r_i\}$, it is seen that

$$[k_{ji}] = -[U]^{-1}[k_{ii}]$$ (5.134)

Hence, from symmetry,

$$[k_{ij}] = -[k_{ii}][U^{-1}]^T$$ (5.135)

Comparing coefficients of $\{r_j\}$,

$$[k_{jj}] = -[U]^{-1}[k_{ij}] = [U]^{-1}[k_{ii}][U^{-1}]^T$$ (5.136)

It is seen that basic member force and displacement quantities could be chosen as $\{R_i\}$, $\{r_i\}$. It will be convenient to use the definitions in eqns (5.83), (5.94), etc. However, since these use the rotations ϕ_i, ϕ_j relative to the chord, these have been used extensively in the calculation of fixed–end forces when loads occur along the member or temperature strains are present. If, for example, the plane frame member shown in Fig. 5.28 has a distributed load w'_y over its whole length, the end rotations are

$$\phi_{iz} = -\phi_{jz} = \frac{w'_y l^3}{24EI}$$

The displacement vector for the member in the global coordinate system is thus given by

$$\{r\}^T = [r_{ix} \; r_{iy} \; \theta_{iz} \; r_{jx} \; r_{jy} \; \theta_{jz}] = \left\langle 0 \; 0 \; \frac{w'_y l^3}{24EI} \; 0 \; 0 \; -\frac{w'_y l^3}{24EI} \right\rangle$$

The multiplication

$$\{R\} = -[k]\{r\}$$

then gives the fixed–end forces acting to restrain the member, and the joint release forces are

$$\{R\} = [k]\{r\}$$

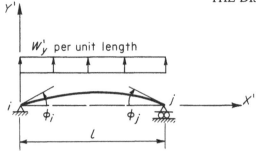

Fig. 5.30   Distributed load on plane frame member.

5.5 CONDENSATION OF STIFFNESS MATRICES

Consider first the stiffness matrix for the frame member given in eqn (5.130). It has been shown that R_i and R_j are dependent forces and that, consequently, $[k]$ can be generated from any one of its four submatrices. It is also found that $[k]$ is a singular matrix and that the submatrix of largest size that is non–singular is six for the three–dimensional member and three for the frame or grid member. This property can be demonstrated by calculating the stiffness at end j when end i has been released. This stiffness must necessarily be zero, since any displacement at end j, the member being unrestrained at i, will simply cause rigid–body motion. Thus, in eqn (5.130), to release end i, set $\{R_i\} = 0$, so that

$$0 = [k_{ii}]\{r_i\} + [k_{ij}]\{r_j\}$$

Hence,

$$\{r_i\} = -[k_{ii}]^{-1}[k_{ij}]\{r_j\}$$

Substitute now for $\{r_i\}$ in the second equation in eqn (5.130). Then,

$$\{R_j\} = ([k_{jj}] - [k_{ji}][k_{ii}]^{-1}[k_{ij}])\{r_j\} \tag{5.137}$$

The reduced matrix in this case is a null matrix. This is proved by substituting for $[k_{jj}]$, $[k_{ji}]$, and $[k_{ij}]$ in terms of $[k_{ii}]$, and the $[U]$ matrix from eqn (5.134). That is,

$$([k_{jj}] - [k_{ji}][k_{ii}]^{-1}[k_{ij}]) = [U]^{-1}[k_{ii}]([U]^{-1})^T - (-[U]^{-1}[k_{ii}])[k_{ii}]^{-1}[-[k_{ii}]([U^{-1}]^T)] = 0$$

Condensation of stiffness matrices can be of importance when a structure is composed of substructures which may be separated, so that internal subsets of nodes can be isolated in such a way that they do not affect the connection of the substructures. This is shown diagrammatically in Fig. 5.31. In Fig. 5.31(a) there are three substructures, *I, II* and *III*. The subsets A, B and C do not contribute stiffness at connection points between *I, II* and *III*, and, consequently their nodes may be released before the analysis proceeds from the kinematically determinate position $\{r\} \equiv 0$. In this case, $\{r\}$ will be a subset of the total number of nodes which does not contain the released nodes. As the nodes in A, B and C are released their forces will produce modified fixed–end forces on the remaining nodes, and in addition within subsets deflections will occur. Finally, when the deflections $\{r\}$ are found, the extra deflections in A, B and C must be calculated. Similarly, the finite element in Fig. 5.31(b) has four internal nodes, and the same procedure may be followed. The problem is first discussed from a partitioning viewpoint and then from the expediency of practical calculations.

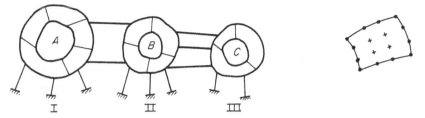

(a) Substructures   (b) Element with internal nodes

Fig. 5.31   Substructures and subsystems.

Consider one of the substructures, I with its subset A. The node forces and deflections may be partitioned into groups 1 and 2, according to whether they are in A or in the set connecting I to the rest of the system. Then,

$$\begin{bmatrix} R_1 \\ R_2 \end{bmatrix} = \begin{bmatrix} k_{11} & k_{12} \\ k_{21} & k_{22} \end{bmatrix} \begin{bmatrix} r_1 \\ r_2 \end{bmatrix} \tag{5.138}$$

1. First determine deflections $\{r_{10}\}$ which occur when the forces $\{R_1\}$ act on A with $\{r_2\}$ held at zero. Evidently, from eqn (5.138),

$$\{r_{10}\} = [k_{11}]^{-1}\{R_1\} \tag{5.139}$$

2. The forces $\{R_1\}$ now cause additional fixed–end forces at nodes $\{r_2\}$. With deflections $\{r_{10}\}$,

$$\{R_{2F}\} = [k_{21}]\{r_{10}\} = [k_{21}][k_{11}]^{-1}\{R_1\} \tag{5.140}$$

The release forces are the negative of $\{R_{2F}\}$. That is,

$$\{\bar{R}_2\} = \{R_2\} - [k_{21}][k_{11}]^{-1}\{R_1\} \tag{5.141}$$

3. The reduced stiffness matrix $[\bar{k}_{22}]$ is obtained by setting $\{R_1\} = 0$ in eqn (5.138). Then,

$$0 = [k_{11}]\{r_1\} + [k_{12}]\{r_2\}$$

$$\{r_1\} = -[k_{11}]^{-1}[k_{12}]\{r_2\} \tag{5.142}$$

This equation gives the deflections $\{r_1\}$ when $\{r_2\}$ occur, and the $\{r_1\}$ nodes are free to deflect. Substitution for $\{r_1\}$ in the eqn (5.138) gives

$$\{\bar{R}_2\} = ([k_{22}] - [k_{21}][k_{11}]^{-1}[k_{12}])\{r_2\} \tag{5.143}$$

or,

$$\{\bar{R}_2\} = [\bar{k}_{22}]\{r_2\} \tag{5.144}$$

The above process is conveniently carried out without resort to the formal inversion of the matrix $[k_{11}]$ by reducing the first submatrix $[k_{11}]$ to an upper triangular matrix by Gaussian reduction. That is, reduce eqn (5.138) to

$$\begin{bmatrix} \bar{R}_1 \\ \bar{R}_2 \end{bmatrix} = \begin{bmatrix} k_{11(u)} & \bar{k}_{12} \\ 0 & \bar{k}_{22} \end{bmatrix} \begin{bmatrix} r_1 \\ r_2 \end{bmatrix} \tag{5.145}$$

Then the deflections $\{r_{10}\}$ are obtained by back substitution in the equation

$$\{\bar{R}_1\} = [k_{11(u)}]\{r_{10}\} \tag{5.146}$$

The forces $\{\bar{R}_2\}$ are the required release forces; that is, the $\{\bar{R}_2\}$ in eqn (5.140) are the same as those in eqn (5.141). Finally, when $\{r_2\}$ are known, then $\{r_1\}$, the additional deflections of the released nodes, are calculated by back substitution in the equation,

$$[k_{11(u)}]\{r_1\} = -[\bar{k}_{12}]\{r_2\}$$

This is actually back substitution applied to the first $\{r_1\}$ terms in the system,

$$[k_{11(u)} \ \bar{k}_{12}]\begin{bmatrix} r_1 \\ r_2 \end{bmatrix} = [0] \tag{5.147}$$

To examine the Gaussian reduction process in more detail, write the stiffness equation in full instead of the partitioned form.

$$\begin{bmatrix} R_1 \\ R_2 \\ R_3 \\ \cdot \\ \cdot \\ R_n \end{bmatrix} = \begin{bmatrix} k_{11} & k_{12} & \cdots & k_{1n} \\ k_{21} & k_{22} & \cdots & k_{2n} \\ k_{31} & k_{32} & \cdots & k_{3n} \\ \cdot & \cdot & \cdots & \cdot \\ \cdot & \cdot & \cdots & \cdot \\ k_{n1} & k_{n2} & \cdots & k_{nn} \end{bmatrix}\begin{bmatrix} r_1 \\ r_2 \\ r_3 \\ \cdot \\ \cdot \\ r_n \end{bmatrix} \tag{5.148}$$

The first step in the reduction divides row *1* by $[k_{11}]$, that is, multiplies by $[k_{11}]^{-1} = 1/k_{11}$ for a single term. The first row becomes

$$\frac{R_1}{k_{11}} = \begin{bmatrix} 1 & \dfrac{k_{12}}{k_{11}} & \dfrac{k_{13}}{k_{11}} & \dfrac{k_{14}}{k_{11}} & \cdots & \dfrac{k_{1n}}{k_{11}} \end{bmatrix}\begin{bmatrix} r_1 \\ r_2 \\ r_3 \\ \cdot \\ \cdot \\ r_n \end{bmatrix} \tag{5.149}$$

It is seen that if all node deflections $r_2, \cdots, r_n$ are held equal to zero,

$$\frac{R_1}{k_{11}} = R_1 f_{11} = \text{ the deflection of } r_1 \equiv \bar{r}_{10} \tag{5.150}$$

Note that $\{r_{10}\}$ is modified to produce $\{r_{10}\}$ by the other deflections of nodes released. In addition if $r_2, r_3, \cdots, r_n$ are known, then, from eqn (5.149),

$$r_1 = -\begin{bmatrix} \dfrac{k_{12}}{k_{11}} & \dfrac{k_{13}}{k_{11}} & \cdots & \dfrac{k_{1n}}{k_{11}} \end{bmatrix}\begin{bmatrix} r_2 \\ r_3 \\ \cdot \\ \cdot \\ r_n \end{bmatrix} \tag{5.151}$$

This is the equation

$$\{r_1\} = -[k_{11}]^{-1}[k_{12}]\{r_2\}$$

and is simply the back substitution process referred to in eqn (5.142). Continuing the reduction process, for $i > 1$,

$$\bar{R}_i = R_i - R_1 \frac{k_{i1}}{k_{11}} \tag{5.152}$$

Once again, this is seen to be

$$\{\bar{R}_2\} = \{R_2\} - [k_{21}][k_{11}]^{-1}\{R_1\}$$

For the $(n-1) \times (n-1)$ submatrix with $[k_{22}]$ as its leading terms,

$$\bar{k}_{ij} = k_{ij} - \frac{k_{i1}k_{1i}}{k_{11}} \quad i = 2, n; \quad j = 2, n \tag{5.153}$$

This is the equation,

$$[\bar{k}_{22}] = [k_{22}] - [k_{21}][k_{11}]^{-1}[k_{12}]$$

When joint _1_ has been released, then the process can proceed to joint _2_, which is, in turn, released. Notice that the only effect that this and subsequent releases have on r_1 is to produce additional deflection that is accounted for in eqn (5.151).

5.6 THE DIRECT STIFFNESS METHOD

5.6.1 Application to line element structures

The direct stiffness method is one of the most useful tools of analysis to emerge from the current development of computer techniques. It has the advantage of being applicable to many areas of analysis, such as the finite element idealization of continua, as well as to the framed structures discussed in this chapter. It has been shown in Chapter 4 and at the beginning of the present chapter that the solution to the structures problem by the displacement method can be considered in the formal steps:

1.		$\{v\} = [a]\{r\}$	displacement transformation
2.		$\{S\} = [k]\{v\}$	member forces in terms of member displacements
3.		$\{R\} = [a]^T\{S\}$	joint equilibrium equations
4.		$\{r\} = ([a]^T[k][a])^{-1}\{r\}$	member forces
5.		$\{S\} = [k][a]\{r\}$	member forces

In section 5.4, expressions have been developed for the member stiffness matrix $[k]$, which relates forces $\{S\}$ and displacement components $\{v\}$ of the member expressed in their global components. In such cases the displacement transformation matrix $[a]$ in equation 1 above will contain only unit or zero values. Moreover, if the member displacements are given in the same order as the node displacements (for example, in the plane frame, r_x, r_y, θ_z), then the unit values form unit submatrices in $[a]$. Consider a particular member N connecting the nodes P and Q, and write out that part of $[a]^T[k][a]$ concerned with member N. Partition $\{r\}$ and $[a]$ in blocks so that the transformation $\{v_N\} = [a_n]\{r\}$ is as in eqn (5.154).

$$v_n = \begin{bmatrix} r_p \\ r_q \end{bmatrix} = \begin{bmatrix} 0 & 0 & 0 & 0 & I & 0 & 0 & 0 & 0 \\ 0 & 0 & 0 & 0 & 0 & 0 & 0 & I & 0 \end{bmatrix} \begin{bmatrix} r_1 \\ r_2 \\ \cdot \\ r_p \\ \cdot \\ r_q \\ \cdot \end{bmatrix} \qquad (5.154)$$

The member stiffness matrix is partitioned as in eqn (5.125) so that the contribution to the structure stiffness

$$[K] = [a]^T [k][a]$$

for member N is given in eqn (5.155) by $[K_N]$.

$$[K_N] = \begin{bmatrix} 0 & 0 \\ 0 & 0 \\ 0 & 0 \\ 0 & 0 \\ I & 0 \\ 0 & 0 \\ 0 & 0 \\ 0 & I \\ 0 & 0 \end{bmatrix} \begin{bmatrix} k_{ii} & k_{ij} \\ k_{ji} & k_{jj} \end{bmatrix} \begin{bmatrix} 0 & 0 & 0 & 0 & I & 0 & 0 & 0 & 0 \\ 0 & 0 & 0 & 0 & 0 & 0 & 0 & I & 0 \end{bmatrix} = \begin{bmatrix} 0 & 0 \\ 0 & 0 \\ 0 & 0 \\ 0 & 0 \\ I & 0 \\ 0 & 0 \\ 0 & 0 \\ 0 & I \\ 0 & 0 \end{bmatrix} \begin{bmatrix} 0 & 0 & 0 & 0 & k_{ii} & 0 & 0 & k_{ij} & 0 \\ 0 & 0 & 0 & 0 & k_{ji} & 0 & 0 & k_{jj} & 0 \end{bmatrix} \qquad (5.155)$$

Thence,

$$[K_N] = \begin{bmatrix} 0 & 0 & 0 & 0 & 0 & 0 & 0 & 0 & 0 \\ 0 & 0 & 0 & 0 & 0 & 0 & 0 & 0 & 0 \\ 0 & 0 & 0 & 0 & 0 & 0 & 0 & 0 & 0 \\ 0 & 0 & 0 & 0 & 0 & 0 & 0 & 0 & 0 \\ 0 & 0 & 0 & 0 & k_{ii} & 0 & 0 & k_{ij} & 0 \\ 0 & 0 & 0 & 0 & 0 & 0 & 0 & 0 & 0 \\ 0 & 0 & 0 & 0 & 0 & 0 & 0 & 0 & 0 \\ 0 & 0 & 0 & 0 & k_{ji} & 0 & 0 & k_{jj} & 0 \\ 0 & 0 & 0 & 0 & 0 & 0 & 0 & 0 & 0 \end{bmatrix} \qquad (5.156)$$

In eqn (5.156) each term is, of course, a submatrix whose size will depend on the type of structure being analysed. For example, for planar truss structures, the size is 2×2, whereas for plane frames and grids, it is 3×3, and for three–dimensional frames 6×6. From eqn (5.156) it is seen that it is not necessary to carry out the formation of the $[a]$ matrix in eqn (5.155). The blocks $[k_{ii}]$, $[k_{ij}]$, etc. can be added directly into $[K]$ by using the node numbers P and Q to locate the appropriate submatrix. If the order of $[k_{ii}]$ etc. is $n \times n$, then the row and column locations in $[K]$ for the first terms are as given in Table 5.3. Suitable computer program coding will then add in all the additional terms of the $[k]$ submatrices.

Table 5.1

		Row no.	Column no.
1	$(k_{ii})_1$	$n(p-1)+1$	$n(p-1)+1$
2	$(k_{ij})_1$	$n(p-1)+1$	$n(q-1)+1$
3	$(k_{ji})_1$	$n(q-1)+1$	$n(p-1)+1$
4	$(k_{jj})_1$	$n(q-1)+1$	$n(q-1)+1$

This property of being able to add the member stiffness matrix directly into the structure stiffness matrix rather than the formal matrix multiplication leads to the definition of the direct stiffness method of analysis.

5.6.2 Calculation of member forces

For the member N, node numbers p and q, the member displacements at ends i and j, are located in the n rows of $\{r\}$ following the locations $n(p-1)+1$ for i, and $n(q-1)+1$ for j. When these are loaded into the vector $\{r_N\}$, basic member forces are calculated with equations such as (5.93) and (5.93). That is,

$$\{S_N\} = [k'_N][T_N][L_{DN}]\{r_N\} \tag{5.157}$$

Example 5.10 *Structure stiffness matrix*

The frame shown in Fig. 5.32 has six members and four free joints (nodes). Since the three support joints are fixed, they may be omitted from the stiffness calculation $\{r\} = [K]\{r\}$. In Fig. 5.32 two different node numberings are shown, and the effect of this on the structure of $[K]$ will be examined. For the plane frame, the member stiffness $[k_N]$ is a 6×6 matrix composed of four 3×3 blocks. From Fig. 5.32(a) the table of member/node numbering is given, with support nodes shown as zero.

Table 5.2

Member	Node i	Node j
1	0	1
2	1	2
3	2	3
4	3	0
5	3	4
6	4	0

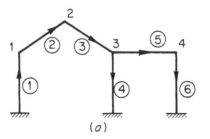 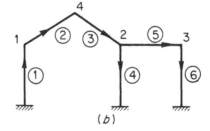

(a)   (b)

Fig. 5.32   Frame node numbering.

The structure stiffness matrix for the free joints *1, 2, 3,* and *4* is obtained by adding in the contributions from the member k_n. In this case [K] will be a 12×12 matrix composed of four 3×3 blocks. It is seen that all diagonal blocks will have stiffness terms, whereas off–diagonal blocks are non–zero only if the row–column number combination represents a physical connection of two nodes by a member. The row and column numbers then give the member connectivity. With the node numbering given in Fig. 5.32(a) the $[K]$ matrix will be as in eqn (5.158).

$$[K] = \begin{bmatrix} K_{11} & K_{12} & 0 & 0 \\ K_{21} & K_{22} & K_{23} & 0 \\ 0 & K_{32} & K_{33} & K_{34} \\ 0 & 0 & K_{43} & K_{44} \end{bmatrix} \tag{5.158}$$

In this case, non–zero terms are packed closely on either side of the diagonal, and the matrix is banded. The same stiffness matrix is shown in eqn (5.159) for the different node numbering given in Fig. 5.32. Note that with this numbering the non–zero off–diagonal terms are dispersed through the matrix.

$$[K] = \begin{bmatrix} K_{11} & 0 & 0 & K_{14} \\ 0 & K_{22} & K_{23} & K_{24} \\ 0 & K_{32} & K_{33} & 0 \\ K_{41} & K_{42} & 0 & K_{44} \end{bmatrix} \tag{5.159}$$

It will be a general feature of the direct stiffness method that the $[K]$ matrix can be banded as in eqn (5.158) by a suitable node–numbering sequence. Since the stiffness matrix is symmetrical, only the upper half of the band may be stored if the equation solution is carried out with an efficient band solver. If the equation solution is carried out by iteration methods, then the band property need not be exploited. For efficient storage, however, only non–zero blocks should be stored so that advantage can be taken of the general spacity of terms in the stiffness matrix.

Example 5.11 *Analysis of a riveted or bolted joint connection*

A typical joint detail is shown in Fig. 5.33. The load $\vec{P}$, applied at the point (x_P, y_P), produces shear stresses in the joint fasteners. The coordinates of fastener number *i* are, (x_{Ri}, y_{Ri}), $i = 1$ to N (the number of fasteners).

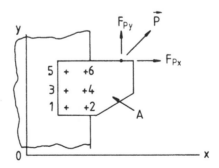

Fig. 5.33   Rivet group in under torque and shear.

The displacements of the rigid plate (A) to which the force $\vec{P}$ is applied has displacements and rotation $\vec{r}_0 = (u_0, v_0, \theta_0)$ referred to the origin of coordinates O. Then the displacements at the centre of a typical fastener (i) are given

$$\begin{bmatrix} u_{Ri} \\ v_{Ri} \end{bmatrix} = \begin{bmatrix} 1 & 0 & -y_{Ri} \\ 0 & 1 & x_{Ri} \end{bmatrix} \begin{bmatrix} u_0 \\ v_0 \\ \theta_0 \end{bmatrix} \tag{i}$$

This is eqn (5.7); that is,

$$\{v_i\} = [a_i]\{r_0\}$$

Suppose also that a stiffness relationship at each fastener exists between fastener shears and displacements of the form,

$$\begin{bmatrix} F_{xi} \\ F_{yi} \end{bmatrix} = \begin{bmatrix} k_{11} & k_{12} \\ k_{21} & k_{22} \end{bmatrix} \begin{bmatrix} u_{Ri} \\ v_{Ri} \end{bmatrix} \tag{ii}$$

It is usual to assume that $k_{12} = k_{21} = 0$, and that $k_{11} = k_{22} = k$. In all cases write eqn (ii) as

$$\{F_i\} = [k_i]\{v_i\} \tag{iii}$$

Thence, using eqn (5.22), the equivalent forces at O for all fasteners are given

$$\begin{bmatrix} R_{x0} \\ R_{y0} \\ M_{z0} \end{bmatrix} = \sum_{i=1,N} [a_i]^T [k_i][a_i] \begin{bmatrix} u_0 \\ v_0 \\ \theta_0 \end{bmatrix} \tag{iv}$$

That is the displacement/force relationship,

$$\{R_0\} = [K]\{r_0\} \tag{v}$$

Now the applied forces at P transformed to O become

$$\begin{bmatrix} R_{x0} \\ R_{y0} \\ M_{z0} \end{bmatrix} = \begin{bmatrix} 1 & 0 & 0 \\ 0 & 1 & 0 \\ -y_P & x_P & 1 \end{bmatrix} \begin{bmatrix} R_{xP} \\ R_{yP} \\ M_P \end{bmatrix} \tag{vi}$$

or,

$$\{R_0\} = [b_P]\{R_P\} \tag{vii}$$

Thence, finally, solving for $\{r_0\}$,

$$\{r_0\} = [K_0]^{-1}[b_P]\{R_P\} \tag{viii}$$

The shear force components in the ith fastener are given, using eqns. (iii) and (viii),

$$\{F_i\} = [k_i][a_i][K_0]^{-1}[b_P]\{R_P\} \tag{ix}$$

Suppose now that the fastener stiffnesses are $k_{11} = k_{22} = k$. Then, from (iv),

$$[K_0] = k \sum_{i=1,N} [a_i]^T [a_i] = k[\bar{K}_0] \tag{x}$$

Thence, substituting in eqn (viii),

$$\{r_0\} = \frac{1}{k}[\bar{K}_0]^{-1}[b_P]\{R_P\}$$

and the i th member force components are, from eqn (ix),

$$\{F_i\} = [a_i][\bar{K}_0]^{-1}[b_P]\{R_P\} \tag{xi}$$

It is seen from eqns (x) and (xi) that the usual expression for a rivet or bolted joint group in shear has been determined in matrix form. The process of analysis is automatic, and the maximum force is easily determined.

5.6.3 Support conditions

Fixed supports may be accounted for during the solution of the stiffness equations by simply deleting the corresponding rows and columns of the stiffness matrix. Since these rows and columns never appear in the assembled stiffness matrix, the storage requirements can be reduced considerably if there are many reaction points. Alternatively, a very simple method of handling degrees of freedom with prescribed displacements is simply to add 10^{20} to the diagonal term of the stiffness matrix and $10^{20} \times \delta$, where δ is the prescribed displacement, to the corresponding force term. This has the effect of having the structure supported by a set of very stiff springs, which are then elongated or compressed the required amounts.

Fig. 5.34   Applied displacements and forces.

Example 5.12 *Support displacements–line element*

The following very simple example illustrates the use of the spring supports. The cantilever member with two nodes *1, 2* has a force $R_{2X} = 100$ applied to node 2, as shown in Fig. 5.34. The stiffness equations are assumed to be

$$\begin{bmatrix} 100 & -100 \\ -100 & 100 \end{bmatrix} \begin{bmatrix} r_1 \\ r_2 \end{bmatrix} = \begin{bmatrix} R_1 \\ R_2 \end{bmatrix}$$

The boundary conditions ar, $r_1 = 1$, $R_2 = 100$. Since $r_1 = 1$ is a known displacement, add 10^{20} to the diagonal (1, 1) and 1.0×10^{20} to R_1, so that, now,

$$\begin{bmatrix} 100 + 10^{20} & -100 \\ -100 & 100 \end{bmatrix} \begin{bmatrix} r_1 \\ r_2 \end{bmatrix} = \begin{bmatrix} R_1 + 1.0 \times 10^{20} \\ 100 \end{bmatrix}$$

Now solve the equations by Gauss reduction.

$$\begin{bmatrix} 1 & 0 \\ 0 & 100 \end{bmatrix} \begin{bmatrix} r_1 \\ r_2 \end{bmatrix} = \begin{bmatrix} R_1 \times 10^{-20} + 1.0 \\ 100 + R_1 \times 100 \times 10^{-20} + 100 \end{bmatrix}$$

In this step, $100/10^{20} \to 0$; $R_1/10^{20} \to 0$. Back substitution now gives

$$r_2 = 2.0 ; \quad r_1 = 1.0$$

Substitution in the original equations gives the unknown node force,

$$R_1 = -100$$

A more elegant approach is to treat the solution of equations as that of a mixed system and arrange the computer algorithm accordingly, using the ideas given in eqns. (5.148) ff.

5.6.4 Support conditions in skew (rotated) axes

A support such as a roller on an inclined plane (Fig. 5.32) is not allowed for in the previous discussions, in which it has been assumed that the constraint condition $r_s = 0$ applies to displacements in the global coordinates. Suppose that for the planar support shown in Fig. 5.32, $R'_y = 0$. Consider the stiffness equations at joints T and U in detail. Joint U is a neighbouring joint to T, and has a connection to T. Then,

$$R_T = K_{T1}r_1 + K_{T2}r_2 + \cdots + K_{TT}r_T + K_{TU}r_U + \cdots \tag{5.160}$$

Transform R_T and r_T to the oblique coordinate axes, using eqns (5.160) and (5.161):

$$\{R'_T\} = [L]\{R_T\} \quad \text{and} \quad \{R_T\} = [L]^T\{R'_T\}$$

$$\{r'_T\} = [L]\{r_T\} \quad \text{and} \quad \{r_T\} = [L]^T\{r'_T\} \tag{5.161}$$

Substitute in eqn (5.159), so that

$$[L]^T\{R'_T\} = K_{T1}r_1 + K_{T2}r_2 + \cdots + K_{TT}[L]^T\{r'_T\} + \cdots + K_{TU}r_U$$

Finally, multiplying both sides of this equation by $[L]$ and substituting $[K'] = [L][K'_{TT}][L]^T$,

$$\{R'_T\} = [L][K_{T1}]\{r_1\} + \cdots + [K'_{TT}]\{r'_T\} + \cdots + [L][K_{TU}]\{r_U\} \tag{5.162}$$

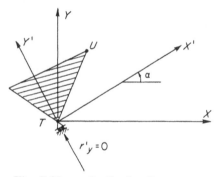

Fig. 5.35   Inclined roller support.

It is seen that all terms in the Tth row must be premultiplied by $[L]$, and, in addition, $[K'_{TT}]$ must be formed with the extra postmultiplication by $[L]^T$. Once the transformation has been made to eqn (5.162) for joint T, the support condition $r'_y = 0$ may be treated in any of the previously mentioned ways. Modification to all stiffness equations for joints such as U which are connected to T must be made. Write the equation for node U in full.

$$\{R_U\} = [K_{U1}]\{r_1\} + [K_{U2}]\{r_2\} + \cdots + [K_{UU}]\{r_U\} + \cdots + [K_{UT}]\{r_T\} + \cdots$$

Substitution for r_T gives

$$\{R_U\} = [K_{U1}]\{r_1\} + \cdots + [K_{UU}]\{r_U\} + \cdots + [K_{UT}][L]^T\{r'_T\} + \cdots \tag{5.163}$$

All terms in the Tth column must therefore be postmultiplied by the rotation transformation matrix $[L]$.

Example 5.13 *Analysis of a plane truss by the direct stiffness method*

The stiffness matrix for force and displacement components of the member $i-j$ (Fig. 5.36), with axial force as the only basic force (that is, the member has pins at each end), is readily deduced in the same way as that in eqn (5.107). From eqn (5.101) the transformation from displacements of i and j in member coordinates to member extension δ is given by

$$\delta = [-1\ 0\ 1\ 0]\begin{bmatrix} r'_{ix} \\ r'_{iy} \\ r'_{jx} \\ r'_{jy} \end{bmatrix} \tag{5.164}$$

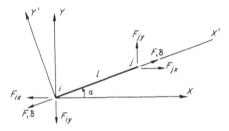

Fig. 5.36   Pin–jointed truss member.

The transformation from global to local components is

$$\begin{bmatrix} r'_{ix} \\ r'_{iy} \\ r'_{jx} \\ r'_{jy} \end{bmatrix} = \begin{bmatrix} \cos\alpha & \sin\alpha & 0 & 0 \\ -\sin\alpha & \cos\alpha & 0 & 0 \\ 0 & 0 & \cos\alpha & \sin\alpha \\ 0 & 0 & -\sin\alpha & \cos\alpha \end{bmatrix}\begin{bmatrix} r_{ix} \\ r_{iy} \\ r_{jx} \\ r_{jy} \end{bmatrix} \tag{5.165}$$

From (5.164) and (5.165),

$$\delta = [-\cos\alpha\ -\sin\alpha\ \cos\alpha\ \sin\alpha\ ]\begin{bmatrix} r_{ix} \\ r_{iy} \\ r_{jx} \\ r_{jy} \end{bmatrix} \tag{5.166}$$

Let $\tilde{c}^T = [\cos\alpha\ \sin\alpha\ ]$, the direction cosines of $i-j$; then, from (5.166),

$$\delta = [-\tilde{c}^T\ \tilde{c}^T]\begin{bmatrix} \tilde{r}_i \\ \tilde{r}_j \end{bmatrix} \tag{5.167}$$

The member stiffness is

$$[k'] = \frac{EA}{l}$$

Hence,

$$[k] = \begin{bmatrix} -\tilde{c} \\ \tilde{c} \end{bmatrix}\frac{EA}{l}[-\tilde{c}^T\ \tilde{c}^T] = \frac{EA}{l}\begin{bmatrix} \tilde{c}\tilde{c}^T & -\tilde{c}\tilde{c}^T \\ -\tilde{c}\tilde{c}^T & \tilde{c}\tilde{c}^T \end{bmatrix} \tag{5.168}$$

Example 5.14 *Truss analysis*

The truss to be analysed is shown in Fig. 5.34. Only node deflections and member forces will be calculated, so that equations related to the support conditions will be eliminated. Node and member numbering is given in Fig. 5.34. The area of all members will be taken to be 25 mm², and $E = 30 \times 10^3 kN/mm^2$ units. Hence $EA/l = 250$ for horizontal and vertical members and $EA/l = 176.77$ for diagonal members.

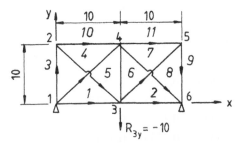

Fig. 5.37   Truss analysis–node and member numbers.

The positive member senses are shown in Fig. 5.37. Notice that they have been chosen so that node number j is always greater than node number i. This is a useful convention which allows $[k]$ to be formed in the arrangement which is added directly into the stiffness matrix $[K]$. In eqn (5.167) it is seen that the four submatrices are numerically the same. The off–diagonal blocks simply have their signs changed.

Table 5.3

Member number	Node i	Number j	Direction cosines cos α	sin α
1	1	3	1.0	0.0
2	3	6	1.0	0.0
3	1	2	0.0	1.0
4	2	3	0.707	−0.707
5	1	4	0.707	0.707
6	3	4	0.0	1.0
7	4	6	0.707	−0.707
8	3	5	0.707	0.707
9	5	6	0.0	−1.0
10	2	4	1.0	0.0
11	4	5	1.0	0.0

From eqn (5.168) and Table 5.3 of member properties, the stiffness matrices have been calculated for the members and listed below:

stiffness matrix   stiffness matrix
member number   member number

$$(1, 2, 10, 11) \qquad 250 \begin{bmatrix} 1 & 0 & -1 & 0 \\ 0 & 0 & 0 & 0 \\ -1 & 0 & 1 & 0 \\ 0 & 0 & 0 & 0 \end{bmatrix} \qquad\qquad (3, 6, 9) \qquad 250 \begin{bmatrix} 0 & 0 & 0 & 0 \\ 0 & 1 & 0 & -1 \\ 0 & 0 & 0 & 0 \\ 0 & -1 & 0 & 1 \end{bmatrix}$$

$$(5, 8) \qquad 250 \begin{bmatrix} 0.3535 & 0.3535 & -0.3535 & -0.3535 \\ 0.3535 & 0.3535 & -0.3535 & -0.3535 \\ -0.3535 & -0.3535 & 0.3535 & 0.3535 \\ -0.3535 & -0.3535 & 0.3535 & 0.3535 \end{bmatrix}$$

$$(4, 7) \qquad 250 \begin{bmatrix} 0.3535 & -0.3535 & -0.3535 & 0.3535 \\ -0.3535 & 0.3535 & 0.3535 & -0.3535 \\ -0.3535 & 0.3535 & 0.3535 & -0.3535 \\ 0.3535 & -0.3535 & -0.3535 & 0.3535 \end{bmatrix}$$

The total stiffness matrix, eqn (5.169), for the structure is constructed by the addition of the 2×2 submatrices composing the member–stiffness matrices above into the locations indicated by the node numbers given in Table 5.3. (That is, Table 5.3 gives the topology of the structure.) The stiffness matrix $[K]$ for the free nodes is obtained from $[K_{total}]$ by deleting those rows and columns for which displacements are zero; that is, for $r_{1x} = r_{1y} = r_{6y} = 0$. These give the row and column values 1, 2, and 12 to be deleted. The load vector $\{R\}$ is given by

$$\{R\}^T = <0\ 0\ 0\ -10\ 0\ 0\ 0\ 0\ 0>$$

The node displacements are calculated by

$$\{r\} = [K]^{-1}\{R\}$$

Then, given in metric units, we have

$$[K] = 250 \begin{bmatrix} 1.3535 & 0.3535 & 0 & 0 & -1 & 0 & -0.3535 & -0.3535 & 0 & 0 & 0 & 0 \\ 0.3535 & 1.3535 & 0 & -1 & 0 & 0 & -0.3535 & -0.3535 & 0 & 0 & 0 & 0 \\ 0 & 0 & 1.3535 & -0.3535 & -0.3535 & 0.3535 & -1 & 0 & 0 & 0 & 0 & 0 \\ 0 & -1 & -0.3535 & 1.3535 & 0.3535 & -0.3535 & 0 & 0 & 0 & 0 & 0 & 0 \\ -1 & 0 & -0.3535 & 0.3535 & 2.7070 & 0 & 0 & 0 & -0.3535 & -0.3535 & -1 & 0 \\ 0 & 0 & 0.3535 & -0.3535 & 0 & 1.7070 & 0 & -1 & -0.3535 & -0.3535 & 0 & 0 \\ -0.3535 & -0.3535 & -1 & 0 & 0 & 0 & 2.7070 & 0 & -1 & 0 & -0.3535 & 0.3535 \\ -0.3535 & -0.3535 & 0 & 0 & 0 & -1 & 0 & 1.707 & 0 & 0 & 0.3535 & -0.3535 \\ 0 & 0 & 0 & 0 & -0.3535 & -0.3535 & -1 & 0 & 1.3535 & 0.3535 & 0 & 0 \\ 0 & 0 & 0 & 0 & -0.3535 & -0.3535 & 0 & 0 & 0.3535 & 1.3535 & 0 & -1 \\ 0 & 0 & 0 & 0 & -1 & 0 & -0.3535 & 0.3535 & 0 & 0 & 1.3535 & -0.3535 \\ 0 & 0 & 0 & 0 & 0 & 0 & 0.3535 & -0.3535 & 0 & -1 & -0.3535 & 1.3535 \end{bmatrix}$$

$$\text{—(5.169)}$$

The node deflections are in millimetres, and the member forces are calculated, using eqn (5.166), by extracting the appropriate deflections, together with the zero values at the the supports.

$$
\text{deflections} \quad \{r\} = \begin{bmatrix} 0.02000 \\ -0.01094 \\ 0.00906 \\ -0.05282 \\ 0.00906 \\ -0.03469 \\ -0.00188 \\ -0.01094 \\ 0.01812 \end{bmatrix} ; \quad \text{member forces} \quad \{S\} = \begin{bmatrix} 2.265 \\ 2.265 \\ -2.734 \\ 3.867 \\ -3.203 \\ 4.530 \\ -3.203 \\ 3.867 \\ -2.734 \\ -2.734 \\ -2.734 \end{bmatrix} \qquad (5.170)
$$

Notice that some rounding–off error has occurred in these member forces, most probably due to truncation of the $\{r\}$ displacements. The above problem was worked using a matrix interpretative package. It is found that unless special assembly routines are incorporated in such interpretative packages, direct stiffness problems are not so conveniently worked in this way.

5.7 SPECIAL ELEMENT STIFFNESS MATRICES

In this section, two important topics relating to the calculation of member stiffness matrices are introduced. The first of these is the flexibile line element which has rigid links connecting it to the nodes of the structure. The second is that of the torsion stiffness of thin–walled, multi–cell closed tubes.

5.7.1   Line element with rigid end links

The flexible line element $I' - J'$ is shown in Fig. 5.33. The stiffness matrix for the member is calculated as in eqn (5.107) and will be denoted by $[K']$. The nodes I–J are the nodes of the structure to which the member I'–J' is attached through the rigid links II' and JJ'.

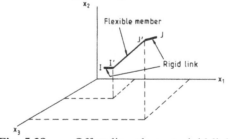

Fig. 5.38   Offset line element–rigid links.

The member joints I', J' have coordinates (x'_{im}, x'_{jm}), and the nodal joints I, J have (x_{im}, x_{jm}) $m = 1, 2, 3$. The nodal link coordinate components are, thus, $(\Delta x_{im}, \Delta x_{jm})$ where,

$$
\Delta x_{im} = x'_{im} - x_{im} \quad \text{and,} \quad \Delta x_{jm} = x'_{jm} - x_{jm} \qquad (5.171)
$$

Then using eqn (1.44), displacements at I' and J' can be related to those at I and J by considering the links to be rigid. For end I',

$$\{r'_i\} = \begin{bmatrix} I_3 & A_3 \\ 0_3 & I_3 \end{bmatrix} \{r_i\} = [T_{i-i'}]\{r_i\} \qquad (5.172)$$

where

$$[A_3] = \begin{bmatrix} 0 & \Delta x_{i3} & -\Delta x_{i2} \\ -\Delta x_{i3} & 0 & \Delta x_{i1} \\ \Delta x_{i2} & -\Delta x_{i1} & 0 \end{bmatrix} \qquad (5.173)$$

Similarly, for end J',

$$\{r'_j\} = [T_{j-j'}]\{r_j\} \qquad (5.174)$$

Combining eqns (5.172) and (5.174),

$$\begin{bmatrix} r'_i \\ r'_j \end{bmatrix} = \begin{bmatrix} T_{i-i'} & 0 \\ 0 & T_{j-j'} \end{bmatrix} \begin{bmatrix} r_i \\ r_j \end{bmatrix} \qquad (5.175)$$

That is,

$$\{r'\} = [T]\{r\} \qquad (5.176)$$

Applying the contragredient principle, the corresponding nodal force relationship is

$$\{R\} = [T]^T \{R'\} \qquad (5.177)$$

Thus, if from eqn (5.107), the member stiffness relationship is

$$\{R'\} = [K']\{r'\}$$

substitution of eqns (5.176), (5.177) gives

$$\{R\} = [T]^T [K'][T]\{r\}$$

That is, the required stiffness matrix at $I-J$ is

$$[K] = [T]^T [K'][T] \qquad (5.178)$$

The offsets may be provided by giving their actual values, $\Delta x_{im}, \Delta x_{jm}$, in which case x'_{im}, x'_{jm} are calculated, or, alternatively, both sets of coordinates for I, I', and J, J' are given and then $\Delta x_{im}, \Delta x_{jm}$ are calculated. The above procedure of catering for rigid links at the ends of members is better than the alternative of composing members with end sub–elements with large stiffnesses in the portions II' and JJ' and then removing the internal nodes by static condensation, because this process inevitably leads to numerical rounding–off errors.

5.7.2 Torsion stiffness of thin–walled closed tubes

The calculation of the St Venant's torsion stiffness for thin–walled tubes is important for the analysis of grid type structures frequently used in bridge construction as well as all types of aircraft assemblages. If the walls are moderately thick, then the finite element method is best used, both to locate the shear centre and to calculate the torsion stiffness (Chapter 7, section 4).

The thin walled closed tube is a special case of St Venant's torsion in which the shear stress τ due to a twisting moment can be considered to be constant across the wall thickness so that shear flow is equal to $\tau\delta$, where δ is the wall thickness. Herein the theory is first developed for a single cell (Fig. 5.34) and then extended to a tube with two cells. This process is readily generalized to multi–celled tubes. The concept of the stress function ϕ used in the St Venant's torsion is taken from [18]. For the closed tube the ϕ surface is constant across the inner void of the tube (Fig. 5.34(b)), and because the wall thickness δ is small compared with the other tube dimensions, the slope of the ϕ surface from the inner to outer perimeter of the tube can be considered to be constant. The area A is that enclosed by the mean perimeter of the tube, shown by the dotted line in Fig. 5.34(a). The longitudinal axis of the tube coincides with the x_3, axis.

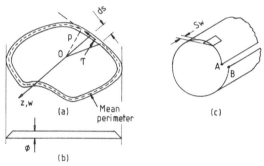

Fig. 5.39   Closed tube torsion.

Thus, from the theory of St Venant's torsion, M_T, the twisting moment about the x_3, axis, is given by

$$M_T = 2A\,\phi \tag{5.179}$$

The shear stress τ is calculated from

$$\tau = -\frac{\partial\phi}{\partial n} = \frac{\phi}{\delta} \tag{5.180}$$

Let G be the shear modulus of elasticity. Then shear strain is obtained from eqn (5.180)

$$\gamma_{zs} = \frac{\tau}{G} = \frac{\phi}{G\delta} \tag{5.181}$$

Let v be the displacement along the tangent due to the twist per unit length θ, and w the corresponding longitudinal warping. Then, since the cross section shape is considered to be rigid in the plane, the v displacement is given by

$$v = \theta\,z\,p$$

and w is an unknown function $w(s)$. The tangential shear strain, γ_{zs}, is expressed in terms of derivatives of the (v,w) displacements,

$$\gamma_{zs} = \frac{\partial v}{\partial z} + \frac{\partial w}{\partial s} = \theta p + \frac{\partial w}{\partial s}$$

From eqn (5.181),

$$\frac{\partial w}{\partial s} = \gamma_{zs} - \theta p = \frac{\phi}{G\delta} - \theta p \tag{5.182}$$

From Fig. 5.34(b) it is seen that the relative longitudinal deflection of two adjacent points on the perimeter, A and B must be zero. That is, around the perimeter,

$$\int \frac{\partial w}{\partial s} ds = 0$$

Integrating the right–hand side of eqn (5.182) gives

$$\phi \int \frac{ds}{\delta} = G \theta \int p \ ds = G \theta 2A$$

Thus, the magnitude ϕ of the stress function has been calculated to be

$$\phi = \frac{2AG\theta}{\int \frac{ds}{\delta}}$$

Substitution for ϕ in eqn (5.179) gives the required stiffness relationship between M_T and θ. That is,

$$M_T = \frac{4A^2 G \theta}{\int \frac{ds}{\delta}} \tag{5.183}$$

and, hence,

$$[k_T] = \frac{4A^2 G}{\int \frac{ds}{\delta}} \tag{5.184}$$

The shear stress is calculated from eqns (5.179) and (5.180)

$$\tau = \frac{M_T}{2A\delta}$$

5.7.3   The two – celled closed tube

The above theory is now extended to a two–celled box tube whence it may be easily generalized for multi–celled tubes. The two–cell tube with the ϕ surface, which now has two heights, ϕ_1 and ϕ_2, is shown in Fig. 5.35. The wall thickness δ is assumed to be constant.

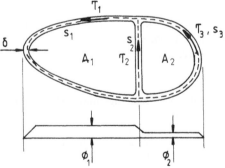

Fig. 5.40   Two–cell box tube.

Then, as in the previous example, the torque M_T is given by

$$M_T = 2 \times \text{ the volume under the } \phi \text{ surface}$$

That is,

$$M_T = [2A_1 \ 2A_2] \begin{bmatrix} \phi_1 \\ \phi_2 \end{bmatrix} \tag{5.185}$$

For the three section perimeters, lengths s_1, s_2, s_3, assuming constant wall thickness, the shear stresses are

$$\begin{Bmatrix} \tau_1 \\ \tau_2 \\ \tau_3 \end{Bmatrix} = \frac{1}{\delta} \begin{bmatrix} 1 & 0 \\ 1 & -1 \\ 0 & 1 \end{bmatrix} \begin{bmatrix} \phi_1 \\ \phi_2 \end{bmatrix} \tag{5.186}$$

Again, as in the previous case around each closed loop, the relative w deflection must be zero. That is,

1. $$(\phi_1 - \phi_2) \int_2 \frac{ds}{\delta} + \phi_1 \int_1 \frac{ds}{\delta} = G\theta \int_{1-2} p \ ds = 2A_1 G\theta$$

2. $$-(\phi_1 - \phi_2) \int_2 \frac{ds}{\delta} + \phi_2 \int_3 \frac{ds}{\delta} = G\theta \int_{2-3} p \ ds = 2A_2 G\theta$$

For δ constant,

$$\int_2 \frac{ds}{\delta} = \frac{s_2}{\delta}, \text{ etc.}$$

Thence, from 1 and 2,

3. $$\begin{bmatrix} (s_1 + s_2) & -s_2 \\ -s_2 & (s_2 + s_3) \end{bmatrix} \begin{bmatrix} \phi_1 \\ \phi_2 \end{bmatrix} = 2G\theta \begin{bmatrix} A_1 \delta \\ A_2 \delta \end{bmatrix}$$

Inversion of (3) gives the expression for ϕ_1, ϕ_2, in terms of θ as

$$\begin{bmatrix} \phi_1 \\ \phi_2 \end{bmatrix} = \frac{2G\theta}{s_1 s_2 + s_2 s_3 + s_3 s_1} \begin{bmatrix} (s_2 + s_3) & s_2 \\ s_2 & (s_1 + s_2) \end{bmatrix} \begin{bmatrix} A_1 \delta \\ A_2 \delta \end{bmatrix} \tag{5.187}$$

Substitution for these values of ϕ_1, ϕ_2 in eqn (5.185) gives the required stiffness relationship as

$$M_T = 2[A_1 \ A_2] \begin{bmatrix} (s_2 + s_3) & s_2 \\ s_2 & (s_1 + s_2) \end{bmatrix} \begin{bmatrix} A_1 \\ A_2 \end{bmatrix} \frac{\delta}{s_1 s_2 + s_2 s_3 + s_3 s_1} 2G\theta \tag{5.188}$$

That is,

$$k_T = \frac{2G \delta k}{s_1 s_2 + s_2 s_3 + s_3 s_1} \tag{5.189}$$

where

$$k = 2[A_1 \ A_2] \begin{bmatrix} (s_2 + s_3) & s_2 \\ s_2 & (s_1 + s_2) \end{bmatrix} \begin{bmatrix} A_1 \\ A_2 \end{bmatrix} \tag{5.190}$$

Using eqns (5.186), (5.187) etc., the shear stresses are found to be

$$
\begin{bmatrix} \tau_1 \\ \tau_2 \\ \tau_3 \end{bmatrix} = \frac{1}{k\delta} \begin{bmatrix} 1 & 0 \\ 1 & -1 \\ 0 & 1 \end{bmatrix} \begin{bmatrix} (s_2 + s_3) & s_2 \\ s_2 & (s_1 + s_2) \end{bmatrix} \begin{bmatrix} A_1 \\ A_2 \end{bmatrix} M_T
$$

(5.191)

The above expression for k_T is readily incorporated in the member stiffness matrices of eqns (5.109) (k_{33}) and (5.118) (for GI_P/l).

5.7.4   Thin–walled open sections – stiffness matrix formulation

The theory is developed herein for a special class of structure which includes I sections, channels and angles. These have particular significance because in many instances it is required to be able to calculate the lateral torsional buckling characteristics of continuous beams composed of such sections. The elastic stiffness is developed so that the element may be incorporated in an appropriate direct stiffness computer program. The analysis of the thin–walled open–section beam under transverse and torsional loads is complicated by the fact that a statically indeterminate relationship exists between the St Venant's and the warping torsion terms (see, for example the discussion in [19]). The problem has been studied very thoroughly in differential equation form by Vlasov [20] and from this work, for the beam referred to principal coordinates of its cross section, the fundamental differential equations are

$$
EA\, \zeta'' = 0
$$

(5.192)

$$
EI_{x_2} \xi^{iv} = q_{x_1}
$$

(5.193)

$$
EI_{x_1} \eta^{iv} = q_{x_2}
$$

(5.194)

$$
EJ_\omega \theta^{iv} - GJ_D \theta'' = m
$$

(5.195)

Eqn (5.192) gives the longitudinal displacement (ζ), and eqns (5.193) and (5.194) the transverse bending deflections (ξ, η) in the (x_1, x_2) directions, respectively, for a point A of the cross section. The superscripts in eqns (5.192–5.195) refer to differentiation with respect to the longitudinal axis variable x_3. The point A is such that for the cross section, the sectorial area $\omega(s)$ is orthogonal to the functions $x_1(s)$, $x_2(s)$. These orthogonality conditions are expressed by

$$
S_{x_2} = \int_A x_1\, dA = S_{x_1} = \int_A x_2\, dA = I_{x_1 x_2} = \int_A x_1 x_2 dA = 0
$$

(5.196)

$$
S_\omega = \int_A \omega dA = J_{\omega x_1} = \int_A x_1 \omega\, dA = J_{\omega x_2} = \int_A x_2 \omega dA = 0
$$

(5.197)

The functions $x_1(s)$, $x_2(s)$, $\omega(s)$ which satisfy the conditions of orthogonality are called the principal generalized coordinates of the cross section of a thin–walled open section. The fundamental generalized forces are

Axial force $F = \int_A \sigma_{x_3}\, dA$

(5.198)

Bending moments $M_{x_1} = \int_A \sigma x_2 dA$

(5.199)

$$
M_{x_2} = \int_A \sigma x_1\, dA
$$

(5.200)

Bimoment $B = \int_A \sigma_{x_3} \omega \, dA$ (5.201)

The bimoment on the section is statically equivalent to zero and is represented by a pair of equal and opposite moments. These are easily identified, in the case of the symmetric I section, as moments in the flanges. The torque T is equal to the St Venant's torque T_s plus the warping torque T_ω

$$T_\omega = \int_A \tau t d\omega$$ (5.202)

In the present application, in which a two–node beam element stiffness matrix is to be developed, only node loads and displacements are considered, so that $q_{x_1} = q_{x_2} = m = 0$. The general solution to eqn (5.195) is

$$\theta = C_1 + C_2 x_3 + C_3 \sinh \frac{k}{l} x_3 + C_4 \cosh \frac{k}{l} x_3 + \bar{\theta}(x_3)$$ (5.203)

where

$$\left[\frac{k}{l}\right]^2 = \frac{GJ_D}{EI_\omega}$$

and $\bar{\theta}(x_3)$ is the particular integral. For terminal loads only, $\bar{\theta}(x_3) = 0$. The torque T and bimoment B are expressed in terms of θ by the expressions,

$$T = T_\omega + T_S = -EJ_\omega \theta^{iii} + GJ_D \theta^i$$ (5.204)

and

$$B = -EJ_\omega \theta^{ii}$$ (5.205)

Thus, from eqn (5.203),

$$T(x_3) = GJ_D \{C_2 + \bar{\theta}'(x_3) - \frac{l^2}{k^2} \bar{\theta}^{iii}(x_3)\}$$ (5.206)

and

$$B(x_3) = -GJ_D \{C_3 \sinh \frac{k}{l} x_3 + C_4 \cosh \frac{k}{l} x_3 + \frac{l^2}{k^2} \bar{\theta}^{ii}(x_3)\}$$ (5.207)

These expressions may be solved exactly for known end conditions and lead to an 'exact' expression for the element stiffness matrix. The element forces are shown in Fig. 5.41. Notice that whereas, $B_L = B_0$, the bimoment at $x_3 = 0$; $B_R = -B_l$; the bimoment at $x_3 = l$.

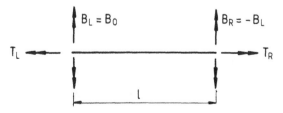

Fig. 5.41   Open section twist and bimoment forces.

The generalized forces and displacements at the ends of the member $i-j$ are $\{T_L \ B_L \ T_R \ B_R \ \}$ and $\{\theta_L \ \theta'_L \ \theta_R \ \theta'_R \ \}$. From eqns (5.206) and (5.207), it is found that

$$
\begin{bmatrix} T_L \\ B_L \\ T_R \\ B_R \end{bmatrix} = EJ_\omega \begin{bmatrix} 0 & \dfrac{k^2}{l^2} & 0 & 0 \\ 0 & 0 & 0 & -\dfrac{k^2}{l^2} \\ 0 & \dfrac{k^2}{l^2} & 0 & 0 \\ 0 & 0 & \dfrac{k^2}{l^2}\sinh k & \dfrac{k^2}{l^2}\cosh k \end{bmatrix} \begin{bmatrix} C_1 \\ C_2 \\ C_3 \\ C_4 \end{bmatrix}
\tag{5.208}
$$

In matrix notation write this equation as

$$
\{R\} = [B]\{C\}
\tag{5.209}
$$

Also, using eqns (5.203),

$$
\begin{bmatrix} \theta_L \\ \theta'_L \\ \theta_R \\ \theta'_R \end{bmatrix} = \begin{bmatrix} 1 & 0 & 0 & 1 \\ 0 & 1 & \dfrac{k}{l} & 0 \\ 1 & l & \sinh k & \dfrac{k}{l}\cosh k \\ 0 & 1 & \dfrac{k}{l}\cosh k & \dfrac{k}{l}\sinh k \end{bmatrix} \begin{bmatrix} C_1 \\ C_2 \\ C_3 \\ C_4 \end{bmatrix}
\tag{5.210}
$$

Write eqn (5.210) as

$$
\{\theta\} = [A]\{C\}
$$

so that

$$
\{C\} = [A]^{-1}\{\theta\}
\tag{5.211}
$$

Substitution for $\{C\}$ in eqn (5.209) gives

$$
\{R\} = [B][A]^{-1}\{\theta\}
\tag{5.212}
$$

That is, the required stiffness matrix $[K]$ is given

$$
[K] = [B][A]^{-1}
\tag{5.213}
$$

Performing the inversion of the matrix $[A]$ and premultiplication by $[B]$ gives the symmetric matrix,

$$
[K] = \frac{EJ_\omega}{D} \begin{bmatrix} K_{11} & K_{12} & K_{13} & K_{14} \\ K_{21} & K_{22} & K_{23} & K_{24} \\ K_{31} & K_{32} & K_{33} & K_{34} \\ K_{41} & K_{42} & K_{43} & K_{44} \end{bmatrix}
\tag{5.214}
$$

In eqn (5.214), $D = k \sinh k - 2 \cosh k + 2$, and

$$K_{11} = K_{33} = -K_{13} = (k/l)^3 \sinh k$$

$$K_{12} = K_{14} = -K_{23} = -K_{34} = (k/l)^2 (\cosh k - 1)$$

$$K_{22} = -K_{24} = (k/l)(k \cosh k - \sinh k)$$

$$K_{44} = -(k/l)(k - \sinh k)$$

5.7.5   Approximation to stiffness matrix using cubic interpolation functions

An examination of the twist terms $\{\theta_L \ \theta'_L \ \theta_R \ \theta'_R\}$ at the ends of the member shows that these terms will interpolate exactly a cubic function $\theta(x_3)$ between nodes $i-j$. The required interpolation functions are

$$\phi_\theta = \begin{bmatrix} 1 - 3\xi^2 + 2\xi^3 \\ l\xi(1 - 2\xi + \xi^2) \\ \xi^2(3 - 2\xi) \\ l\xi^2(-1 + \xi) \end{bmatrix} \tag{5.215}$$

Thence, derivatives of the cubic function are given

$$\begin{bmatrix} \theta^i \\ \theta^{ii} \end{bmatrix} = \begin{bmatrix} \dfrac{6\xi}{l}(-1 + \xi) & (1 - 4\xi + 3\xi^2) & \dfrac{6\xi}{l}(1 - \xi) & \xi(-2 + 3\xi) \\ \dfrac{6}{l^2}(-1 + 2\xi) & \dfrac{2}{l}(-2 + 3\xi) & \dfrac{6}{l^2}(1 - 2\xi) & \dfrac{2}{l}(-1 + 3\xi) \end{bmatrix} \begin{bmatrix} \theta_L \\ \theta'_L \\ \theta_R \\ \theta'_R \end{bmatrix}$$

This is eqn (5.15), $\{v\} = [a]\{r\}$. The corresponding internal forces are given

$$\begin{bmatrix} T_s \\ B \end{bmatrix} = \begin{bmatrix} GJ_D & 0 \\ 0 & -EJ_\omega \end{bmatrix} \begin{bmatrix} \theta^i \\ \theta^{ii} \end{bmatrix} \quad \text{or,} \quad \{S\} = [k]\{v\}$$

Application of eqn (2.105) gives

$$\{R\} = \int_0^l [a]^T [k][a] dx_3 \{r\}$$

On multiplication of the matrices and integrating over the length of the beam, it is found that

$$[K] = [K_\omega + K_S] \tag{5.216}$$

where,

$$[K_\omega] = \frac{-2EJ_\omega}{l} \begin{bmatrix} \frac{6}{l^2} & \frac{3}{l} & -\frac{6}{l^2} & \frac{3}{l} \\ \frac{3}{l} & 2 & -\frac{3}{l} & 1 \\ -\frac{6}{l^2} & -\frac{3}{l} & \frac{6}{l^2} & -\frac{3}{l} \\ \frac{3}{l} & 1 & -\frac{3}{l} & 2 \end{bmatrix} \qquad (5.217)$$

and

$$[K_S] = \frac{GJ_D}{l} \begin{bmatrix} 12 & l & -12 & l \\ l & \frac{4}{3}l^2 & l & -\frac{l^2}{3} \\ -12 & l & 12 & -l \\ l & -\frac{l^2}{3} & -l & \frac{4}{3}l^2 \end{bmatrix} \qquad (5.218)$$

A comparison of eqns (5.214) and (5.216) shows that using a cubic approximation to the twisted curve decouples the warping and St Venant's stiffness matrices. When the calculation of buckling loads is required, this cubic approximation is necessary for the calculation of the geometric stiffness matrix if a linear eigenvalue solution routine is used. There appears to be little reason to prefer eqn (5.216) to (5.214) for the calculation of the elastic stiffness.

Problems

5.1 Calculate node deflections and member forces for the truss shown in Fig. P5.1, using eqns (5.20), (5.23) and (5.25).

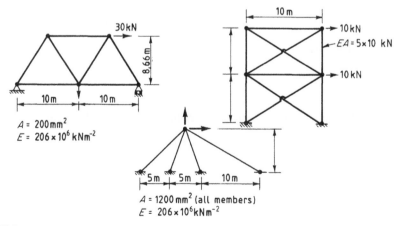

Fig. P5.1

5.2 Calculate the effect on node deflections and member forces if member (1) is removed from the structures b and c in Fig. P5.1.

5.3 Calculate the node deflections and member forces (bending moments only) for the frame structures shown in Figs P5.2 and P5.3.

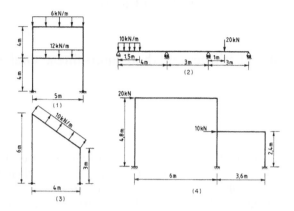

Fig P5.2

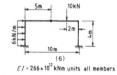

Fig P5.3

CHAPTER 6

The direct stiffness method–non–linear analysis

6.1 INTRODUCTION

In Chapter 4, section 11, some of the basic concepts of non–linear structural analysis were introduced. Therein the force method was used, and it was found that, although it is practical, some difficulties can arise. However, the single most compelling reason for using the direct stiffness method is that suitable computer codes are so well developed that it is a natural step to include non–linear analysis in their repertoire. In this chapter three distinct types of problem are discussed:

1. The analysis of tension structures. This leads to the shape finding and design of net and fabric structures.

2. The calculation of small displacement elastic critical loads (and/or bifurcation points) and mode shapes. This is a relatively small extension of the elastic analysis already developed in Chapter 5.

3. The tracing of the whole equilibrium path of a structure through snap–through phenomena. These analyses are inextricably linked with no. 2, as they invariably follow primary equilibrium paths. The load/deflection paths of various forms of behaviour are shown diagrammatically in Fig. 6.1.

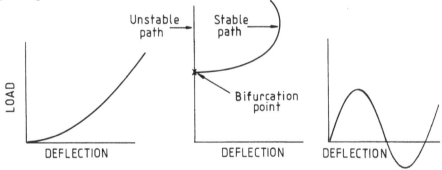

Fig. 6.1   (a) Tension net structure; (b) bifrucation phenomema; (c) snap–through phenomena.

All these classes of behaviour present special problems, but it is probably fair to comment that the level of difficulty increases from no. 1 to no. 3. The analyses will be concerned with two basically distinct problems. The first of these is the formulation of the structure equilibrium equations in their deflected configuration. The second is with the algorithms necessary for following the non–linear load/deflection path. It should be mentioned at the outset that all methods are incrementally linear and that increments of loads iterate on a series of linear displacement increments or vice versa. One of the obstacles to the understanding of the problem is that the situations depicted in Fig. 6.1 are one–dimensional and only partially represent the multi–dimensional problem. Each of the three topics mentioned above will be discussed in considerable detail in what follows, as will the iterative solution strategies.

6.2 TENSION STRUCTURES

6.2.1 Cable nets

Tension structures are particularly amenable to computer design in that the computer can be used in a creative sense in aiding the architect or engineer in the selection of a desired structural shape while starting with only a plan. Before giving details of analysis techniques, it is useful first to classify the basic types of tension structure. There are three categories:

1.   simply suspended cable net systems;

2.   double layer systems ;

3.   cable nets.

1.   *Simply suspended cable net systems*

Simply suspended cable systems are based on cables hanging in a catenary or similar shape, governed by the loads they support. The stiffness of such cables against upward forces and non–uniform loads depends on the applied dead loads. To provide sufficient stiffness of the structure, considerable dead weight, which is not provided by the cables alone, is required, and, as a consequence, roofs supported by cables in this way must be stiffened. This can be achieved by placing precast concrete panels (for example, 25 mm thick at 3–m intervals) during construction. Thus additional temporary dead load is applied, stretching the cables. The joints between the panels are concreted and the preload removed. The roof then, becomes, in effect, a precast concrete shell. Roofs of this type are really cable structures only until the insitu concrete is complete. After that they should be classified as precast concrete shells.

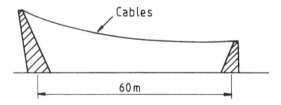

Fig. 6.2   Schematic section of Dulles Passenger Terminal.

2.   *Double layer systems*

An alternative method of providing stiffness to the cable system is to construct cable trusses such as those shown in Fig. 6.3. The ties can be tension or compression members, depending on whether the main cables are concave or convex outwards (Fig. 6.3(a) and (b)). The system shown in Fig. 6.3(b) has the advantage of a smaller depth of truss. Diagonal ties provide greater stiffness against unsymmetrical loading than do vertical ties. Cable tensions must be such that no cable becomes slack under any load condition. Identical trusses may be positioned parallel to one another to form a roof surface of single curvature. Cables can be arranged at right angles to the trusses and prestressed against them. Two sets of parallel trusses can be arranged at right angles to form a grid, or, alternatively, arranged radially between an inner tension and outer compression ring.

3.   *Cable nets*

Some outstanding structures have been built with pretensioned cable nets. Like cable trusses, cable nets have concave and convex cables stressed against each other, except that now these two sets of cables span in different directions.

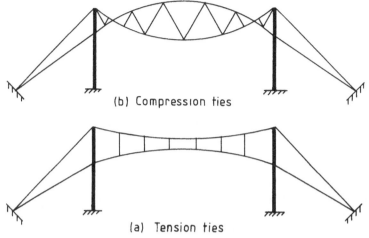

(b) Compression ties

(a) Tension ties

Fig. 6.3   Cable trusses.

Thus, prestressed cable nets always form anticlastic surfaces. Cables are joined where they intersect. They can be arranged to form a grid of quadrilaterals or triangles, as shown in Fig. 6.4. A net may be classified as doubly, triply or quadruply threaded, depending on whether two, three or four cables intersect at a joint. The doubly threaded net is the most popular. A grid consisting of quadrilaterals behaves differently from one consisting of triangles in that it has no stiffness against in–plane shear deformations. In practice, this enables an initially flat rectangular grid to take on a doubly curved shape without cable strains. The mesh size should be constant or nearly so, otherwise the manufacture is unnecessarily complicated.

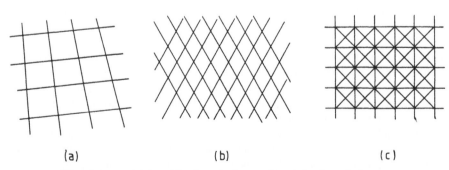

(a)   (b)   (c)

Fig. 6.4   (a) Doubly threaded net; (b) triply threaded net;
(c) quadruply threaded net.

Perhaps the most outstanding example of cable net structures is the Munich Olympic Stadium, in which 210 km of cable in a 3–m grid was used. Leonhardt and Schlaich [21] write, 'standardized prefabrication is a prior condition for the erection of a structure consisting of millions of individual members'. With this brief introduction we now turn to the practical analysis of such structures. The approach followed in the analysis is similar to that for the force method (section 4.11). That is, the non–linear effects are accounted for by the modification of the standard linear expressions. If the load is applied to the structure in small increments so that the displacements over any given load increment are small, the only effect to be considered will be the change in stiffness due to the axial forces in the members. In Fig. 6.5, for example, the tangent stiffness expression at the load level R may be written

$$\{\Delta R\} = [K_T]\{\Delta r\} \tag{6.1}$$

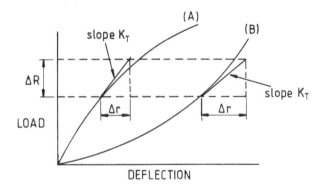

Fig. 6.5   Structure–load deflection curves.

In eqn (6.1), $[K_T]$ will be composed of two parts, an elastic stiffness $[K_E]$ and a geometric stiffness $[K_G]$ due to the presence of axial forces in the structure members at the load level at the beginning of the increment and the effects of the geometry changes on their components. Thus, the tangent stiffness $[K_T]$ may be written

$$[K_T] = [K_E] + [K_G] \tag{6.2}$$

and hence the relationship between displacement and load increments is expressed as

$$\{\Delta R\} = ([K_E] + [K_G])\{\Delta r\} \tag{6.3}$$

A simple illustration of elastic and geometric stiffness is given by the single element (Fig. 6.6) hinged at its base and with a transverse spring of stiffness k at its upper extremity. The member has an axial force S_0 and is loaded transversely with the force P as shown. When the force P is applied, the deflection at joint 2 is δ, so that the rotation of the member is δ/l.

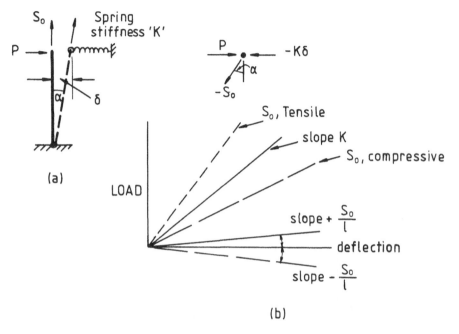

Fig. 6.6   Concept of geometric stiffness.

Writing the equation of equilibrium in the deformed position,

$$P - k\delta - S_0\alpha = 0$$

That is, since $\delta = \alpha l$,

$$P = (k + \frac{S_0}{l})\,\delta$$

Thus, an additional stiffness $k_G = S_0/l$ exists at joint 2. The various situations which occur, depending on whether S_0 is tensile or compressive, are shown in Fig. 6.6(b). It is seen that compressive forces decrease stiffness, leading to load deflection curves such as (A) in Fig. 6.5, whereas tensile forces increase stiffness, giving rise to curves such as (B) in Fig. 6.5. The critical load evidently occurs when

$$k + \frac{S_0}{l} = 0\;; \quad \text{that is,} \quad S_0 = -kl$$

In the present context of tension structures, the situation (B) of Fig. 6.5 occurs. The important feature of all large displacement and critical load problems is that the equilibrium equations are written in the displaced position. For example, if the equilibrium equation at node 2 in Fig. 6.6(a) is written in the undisplaced position the small deflection equations necessarily result.

6.2.2   Theory for the analysis of cable net structures

Expressions for the tangent stiffness

In the following theory, the expressions $[K_T]$ are developed in two distinct ways. In the first approach [22], the tangent stiffness matrix is calculated from the current deflected position, using the updated coordinates as the reference configuration. Member forces are calculated as the sum of the incremental contributions. In this updated approach, expressions for $[K_E]$ and $[K_G]$ are relatively simple. In the second approach [21], the displacements are summed and the member position defined with respect to a fixed set of local coordinates axes. Non–linear transformations result.

6.2.2.1   Method I

The theory for the analysis of cable structures in this method is based on small deflections from the current position.

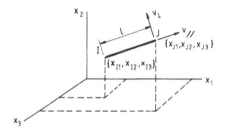

Fig. 6.7   Current position of tension net member.

The vector coordinates of the member ends (shown in Fig. 6.7) are given by

$$\{x_I\}^T = <x_{I1}, x_{I2}, x_{I3}> \quad \text{and} \quad \{x_J\}^T = <x_{J1}, x_{J2}, x_{J3}> \tag{6.4}$$

and, collectively, by

$$\{x\} = \begin{Bmatrix} x_I \\ x_J \end{Bmatrix} \tag{6.5}$$

Note in eqn (6.5), submatrices are left unbracketed. Then the member length is calculated (using the summation convention on repeated indices)

$$l = \{(x_{Ji} - x_{Ii})(x_{Ji} - x_{Ii})\}^{1/2} \tag{6.6}$$

The member direction cosines (current position) are

$$\{c\}^T = \frac{1}{l} <(x_{J1} - x_{I1})\ (x_{J2} - x_{I2})\ (x_{J3} - x_{I3})> \tag{6.7a}$$

or,

$$c_i = \frac{1}{l}(x_{Ji} - x_{Ii}) \tag{6.7b}$$

The vector of member displacements is written

$$\{v\} = \begin{bmatrix} v_I \\ v_J \end{bmatrix} \tag{6.8}$$

where $\{v_I\}$ and $\{v_J\}$ are defined in the same coordinate system as the coordinate vectors in eqn (6.4). Let $\{v_i\}$ be the increment in $\{v\}$ at the ith iteration so that the updated coordinates at the ends of the member are

$$\{x_{i+1}\} = \{x_i\} + \{v_i\} \tag{6.9}$$

Similarly, the force vectors at the member ends are given by

$$\{P\} = \begin{bmatrix} P_I \\ P_J \end{bmatrix}; \quad \text{with} \quad \{P_I\} = <P_{I1}, P_{I2}, P_{I3}>, \text{ etc.} \tag{6.10}$$

The member deformation, v_N, for the displacement $\{v\}$ is given by

$$v_N = [-c^T \ \ c^T]\begin{bmatrix} v_I \\ v_J \end{bmatrix} = [a_N]\{v\} \tag{6.11}$$

The member force is P_N and its global components are (using contragredience)

$$\begin{bmatrix} P_I \\ P_J \end{bmatrix} = \begin{bmatrix} -c \\ c \end{bmatrix} P_N = [a_N]^T P_N \tag{6.12}$$

The increment in member force is given in terms of v_{Ni}:

$$\Delta P_{Ni} = \frac{EA}{l_0} v_{Ni} = k_N v_{Ni} \tag{6.13}$$

Thence,

$$\{\Delta P_i\} = [a_{Ni}]^T k_N [a_{Ni}]\{v_i\} = \frac{EA}{l}\begin{bmatrix} cc^T & -cc^T \\ -cc^T & cc^T \end{bmatrix}\{v_i\} \tag{6.14}$$

That is,

$$[k_E] = \frac{EA}{l_0}\begin{bmatrix} cc^T & -cc^T \\ -cc^T & cc^T \end{bmatrix} \tag{6.15}$$

The stiffness matrix $[k_E]$ is the elastic stiffness matrix for the member, for increments of displacements from its position defined by the end coordinates $\{x\}$.

6.2.2.2   Geometric stiffness

The member displacement vectors v_I, v_J are resolved into components parallel and perpendicular to the member (Fig. 6.8). Let $\alpha = I$ or J, so that

$$\{v_\alpha\} = \{v_{\alpha\parallel}\} + \{v_{\alpha\perp}\} \tag{6.16}$$

This is a vector equation at each end of the member. From eqn (6.11), the magnitude of $v_{\alpha\parallel}$ is given,

$$\left|\{v_{\alpha\parallel}\}\right| = \{c\}^T\{v_\alpha\} \tag{6.17}$$

Thence, the components of $v_{\alpha\parallel}$ are given by

$$\{v_{\alpha\parallel}\} = \{c\}\{c\}^T\{v_\alpha\} \tag{6.18}$$

The expression for the component perpendicular to the member is thus given by

$$\{v_{\alpha\perp}\} = \{v_\alpha\} - \{c\}\{c\}^T\{v_\alpha\} \tag{6.19}$$

The vectors $v_{I\perp}$ and $v_{J\perp}$ at the two ends of the member are shown in Fig. 6.8(a). It is seen that a measure of the rotation of the member is given by $\{\bar{v}\}$, such that

$$\{\bar{v}\} = \frac{1}{l}[\{v_{J\perp}\} - \{v_{I\perp}\}]$$

Substituting from eqn (6.19) gives

$$\{\bar{v}\} = \frac{1}{l}\left\{-[I_3 - \{c\}\{c\}^T]\,[I_3 - \{c\}\{c\}^T]\right\}\begin{bmatrix} v_I \\ v_J \end{bmatrix} \tag{6.20}$$

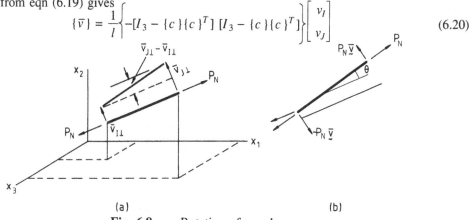

(a)   (b)

Fig. 6.8   Rotation of member.

The vectors $\{\Delta P_{IG}\}$, $\{\Delta P_{JG}\}$ are the change in the global components due to the rotation θ, (Fig. 6.8). Then

$$\Delta\begin{bmatrix} P_I \\ P_J \end{bmatrix} = P_N\begin{bmatrix} -\bar{v} \\ \bar{v} \end{bmatrix} = P_N\begin{bmatrix} -I_3 \\ I_3 \end{bmatrix}\{\bar{v}\} \tag{6.21}$$

Using eqns (6.20) and (6.21),

$$\Delta\begin{bmatrix} P_I \\ P_J \end{bmatrix} = \frac{P_N}{l}\begin{bmatrix} (I_3 - cc^T) & -(I_3 - cc^T) \\ -(I_3 - cc^T) & (I_3 - cc^T) \end{bmatrix}\begin{bmatrix} v_{Ii} \\ v_{Ji} \end{bmatrix} \tag{6.22}$$

This is the equation for the ith iteration:

$$\{\Delta P_G\} = [k_G]\{v_i\} \tag{6.23}$$

Combining eqns (6.14) and (6.23),

$$\{\Delta P\} = \{\Delta P_E\} + \{\Delta P_G\} = [k_E + k_G]\{v_i\} = [k_T]\{v_i\} \tag{6.24}$$

Finally, $[k_T]$ is written as the 6×6 partitioned matrix:

$$[k_T] = \begin{bmatrix} k_1 & -k_1 \\ -k_1 & k_1 \end{bmatrix} \tag{6.25}$$

where,

$$[k_1] = \left[\frac{EA}{l_0} - \frac{P_N}{l}\right]\{c\}\{c\}^T + \frac{P_N}{l}[I_3] \tag{6.26}$$

6.2.2.3 Iterative calculations

Simple Newton–Rhapson type iteration can be carried out as follows (Fig. 6.9). The total member force P_N is given by

$$P_N = \frac{EA}{l_0}(l - l_0) \tag{6.27}$$

For the load $\{\Delta R_0\}$ calculate

$$v_{N0} = [a_{N0}]\{\Delta r_0\} \tag{6.28}$$

In eqn (6.28) the increment of deflections is calculated from

$$\{\Delta r_0\} = [K_0]^{-1}\{\Delta R_0\} \tag{6.29}$$

Now update the nodal geometry

$$\{x_{i+1}\} = \{x_i\} + \{\Delta r_i\} \tag{6.30}$$

Calculate the out–of–balance node forces from

$$\{\Delta R_{i+1}\} = \{R\} - [a_N]_{i+1}^T P_N \tag{6.31}$$

Repeat the process, calculating $\{\Delta r_{i+1}\}$ etc. at each step.

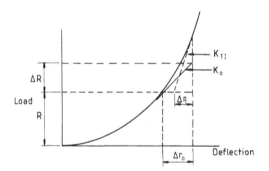

Fig. 6.9   Simple Newton–Rhapson iteration scheme.

6.2.2.4 Method II

The theory is now based on the initial geometry and the total displacements. Before giving examples of the above process, an alternative approach [21] is given in which the tangent stiffness matrix is developed using the initial geometry and total displacements. Because large displacements are now being used, the transformations are non–linear, in contrast to those in Method I. The development is particularly useful in that it leads to an understanding of the similar non–linear transformation matrices for framed structures (either planar or three–dimensional). A typical cable line element is shown in Fig. 6.10. There are now two sets of coordinate systems for each element, namely the global set (x_1, x_2, x_3) and a local set (x'_1, x'_2, x'_3). Both these coordinate sets are fixed in space. Note that the x'_1–axis coincides with the initial element location and that x'_2, x'_3 are arbitrary axes, so that (x'_1, x'_2, x'_3) form a right handed set. That is, the plane of x'_2, x'_3 is perpendicular to the initial direction $I–J$. The relative displacement of end J to end I in the local coordinate member axes is given by

$$\{v'\} = [-L \; L \;]\begin{bmatrix} v_I \\ v_J \end{bmatrix} \tag{6.32}$$

where $[L]$ is defined in eqn (1.37), and has as its rows the direction cosines of the primed axes. Write eqn (6.32) as

$$\{v'\} = [T]\{v\} \tag{6.33}$$

In eqn (6.32), $v'_1 = v_N$, as in eqn (6.11), and v'_2, v'_3 are relative sways of the member ends in the 2 and 3 axes directions, respectively. The member extension measured in the direction of the deformed member axes is given by

$$v_N = \sqrt{(l + v'_1)^2 + v'^2_2 + v'^2_3} - l \tag{6.34}$$

That is,

$$(l + v_N)^2 = (l + v'_1)^2 + v'^2_2 + v'^2_3 \tag{6.35}$$

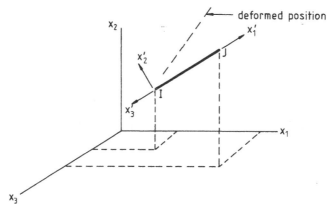

Fig. 6.10   Global and local coordinate axes.

From eqn (6.34), the variation in v_N, that is, Δv_N, for a variation $\{\Delta v'\}$ is given by

$$\Delta v_N = \frac{\partial v_N}{\partial v'_i} \Delta v'_i \tag{6.36}$$

In eqn (6.36), sum on $i = 1$ to 3. That is,

$$\Delta v_N = \left[\frac{l + v'_1}{l + v_N} , \frac{v'_2}{l + v_N} , \frac{v'_3}{l + v_N} \right] \begin{bmatrix} \Delta v'_1 \\ \Delta v'_2 \\ \Delta v'_3 \end{bmatrix} \tag{6.37}$$

Write this equation as

$$\Delta v_N = [A]\{\Delta v'\} \tag{6.38}$$

The total member force is

$$P_N = P_{N0} + \frac{EA}{l} v_N \tag{6.39}$$

So that

$$\Delta P_N = \frac{EA}{l} \Delta v_N = k \Delta v_N \tag{6.40}$$

In order to be able to calculate the tangent stiffness matrix for the element, it is necessary to establish the relationship between $\{\Delta \bar{R}\}$ and $\{\Delta v'\}$. The forces $\{\bar{R}\}$ are simply the generalized forces corresponding to the displacements $\{v'\}$. Using the contragredient principle eqn (2.16), it is seen immediately that from eqn (6.38),

$$\{\bar{R}\} = [A]^T P_N \tag{6.41}$$

Now $[A]^T P_N$ is a function of $(\{v'\}, P_N)$; hence, the variation $\{\Delta \bar{R}\}$ in $\{\bar{R}\}$ is given by partial differentiation as,

$$\{\Delta \bar{R}\} = P_N \frac{\partial [A]^T}{\partial v'_i} \Delta v'_i + [A]^T \Delta P \qquad \text{summing on the index } i \tag{6.42}$$

From the values of $[A]$ given in eqn (6.37),

$$P_N \frac{\partial [A]^T}{\partial v'_i} \Delta v'_i = \frac{P_N}{(l + v_N)^3} \begin{bmatrix} v'^2_2 + v'^2_3 & -v'_2(l + v'_1) & -v'_3(l + v'_1) \\ -v'_2(l + v'_1) & (l + v'_1)^2 + v'^2_3 & -v'_2 v'_3 \\ -v'_3(l + v'_1) & -v'_2 v'_3 & (l + v'_1)^2 + v'^2_2 \end{bmatrix} \{\Delta v'\} \tag{6.43}$$

Write the expression in eqn (6.43) as $[d] \Delta\{v'\}$, so that eqn (6.42) becomes

$$\{\Delta R'\} = [d]\{\Delta v'\} + [A]^T \Delta P_N \tag{6.44}$$

From eqns (6.38) and (6.40), substitution in eqn 6.44 gives

$$\{\Delta R'\} = [d]\{\Delta v'\} + [A]^T k [A]\{\Delta v'\} \tag{6.45}$$

However, from eqn (6.33), because $\{\Delta v'\}$ is referred to fixed member axes,

$$\{\Delta v'\} = [T]\{\Delta v\} \tag{6.46}$$

and, by contragredience,

$$\{\Delta R\} = [T]^T \{\Delta R'\} \tag{6.47}$$

Combining eqns (6.45) to (6.47),

$$\{\Delta R\} = [T]^T ([A]^T k [A] + [d])[T]\{\Delta v\} \tag{6.48}$$

The tangent stiffness matrix $[k_T]$ for the member calculated, using the original coordinates and the total displacements, is thus given by

$$[k_T] = [T]^T \left\{ [A]^T k [A] + [d] \right\} [T] \tag{6.49}$$

Assuming that the P_{N0} forces are in equilibrium under zero external node forces (that is, they are self–equilibrating), it follows that the node forces are given by

$$\{R\} = [T]^T [A]^T \frac{EA}{l} v_N \tag{6.50}$$

The iteration procedure is now similar to that outlined in Method I.

Example 6.1 *Analysis of a catenary*

Two examples are now given to demonstrate the validity of the above theory. The computer program referred to herein uses Method I, [24]. A catenary of 100 m span, with a uniformly distributed load of 100 kN, is analysed using 20 equal members, with a load of 5 kN per node. The equation for an inextensible catenary is

$$y = h \cosh(x/h) \qquad \qquad (i)$$

The constant h is calculated from

$$s = \{b^2 - h^2\}^{\frac{1}{2}} \qquad \qquad (ii)$$

where b is given in eqn (iii) and s is half the cable length.

$$b = h \cosh(l/h) \qquad \qquad (iii)$$

The analysis will not accept inextensible members (that is, infinite stiffness) however, using an elastic modulus of 10^8 kPa and a cable cross section area of 0.01 mm^2, gives a cable strain with an average value of 0.01% and increases the cable length from 104.219 to 104.230 m. One–half of this value is used as s in the equation (ii) above for the theoretical shape calculation. In Table 6.1 the theoretical sag and the computed sag are given at the node points and also at the element midpoints. It is seen that the computer model approximates the catenary very closely, being below the curve at nodes, and above at the midpoints, (Fig. 6.11(b)). The theoretical reaction is 108.1 kN, and the calculated value 106.9 kN.

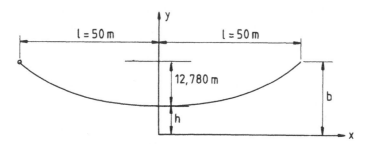

Fig. 6.11   Catenary dimensions.

Example 6.2 *Analysis of an experimental cable truss*

The experimental results are reported by Poskitt [25]. The comparison with the computed results is both interesting and informative because it shows the difficulties which arise in attempting such a correlation. experimental work on model cable structures is exacting because member lengths and node coordinates have to be measured with a high degree of precision. The experimental work reported in [25] is for the cable truss of dimensions shown in Fig. 6.12. Units in this investigation were originally in imperial units and have been converted to metric. The upper and lower cables are identical, each being 19–strand, 0.3048 mm diameter–cold–drawn cable, with $EA = 25.434 \times 10^4$ kN. The hangers are identical in area, being composed of 6.35×0.127 mm clock spring. The span is 3.0274 m and the depth at the supports, 0.810 m.

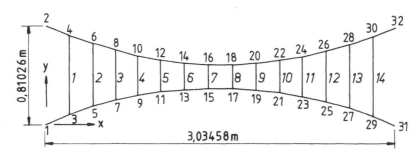

Fig. 6.12   Poskitt's experimental cable truss–
node and member numbering.

Table 6.1

Point	Distance from support	Computed sag	Theoretical sag (eqn(i))
11.0	15.240	3.900	3.895
11.5	14.446	3.879	3.885
12.0	13.652	3.858	3.854
12.5	12.860	3.796	3.802
13.0	12.069	13.734	3.730
13.5	11.281	3.631	3.637
14.0	10.493	3.529	3.524
14.5	9.712	3.386	3.392
15.0	8.931	3.243	3.239
15.5	8.158	3.061	3.068
16.0	7.385	2.879	2.876
16.5	6.621	2.660	2.667
17.0	5.858	2.441	2.438
17.5	5.106	2.185	2.193
18.0	4.354	1.930	1.928
18.5	3.614	1.640	1.648
19.0	2.875	1.351	1.349
19.5	2.149	1.028	1.037
20.0	1.423	0.706	0.705
20.5	0.711	0.353	0.362
21.0	0.000	0.000	0.000

   The span is 3.03458 m and the depth at the supports, 0.81026 m. The hangers were 'equally spaced along the span'. This has been interpreted to mean that the unstressed length of cable between hangers was 203.2 mm. The horizontal component of tension in both cables is 0.6895 kN, and data related to the hangers in their initial positions is summarized in Table 6.2.

Table 6.2

Hanger no.	EA/l_0	y coordinate bottom	y coordinate top	Force measured	Force computed
1	1390	76.962	733.04	44.171	15.480
2	1750	141.224	669.036	38.432	43.815
3	2180	194.056	616.204	34.296	37.498
4	2760	236.728	573.278	31.137	33.095
5	3280	270.510	539.750	45.372	46.973
6	3950	290.830	519.430	34.740	35.141
7	4370	300.990	508.270	28.068	28.335
8	4260	302.768	507.238	49.686	50.126
9	4030	290.068	519.938	23.887	24.198
10	3230	270.510	539.750	47.507	49.197
11	2750	236.728	573.278	32.027	34.029
12	2130	193.802	616.458	36.075	39.456
13	1770	140.208	670.052	38.788	44.482
14	1390	75.184	735.076	34.607	4.982

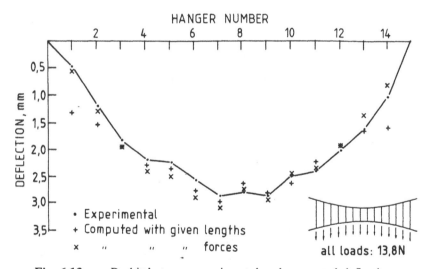

Fig. 6.13   Poskitt's truss–experimental and computed deflections.

For the purposes of analysis this data presented some problems. The unstressed lengths of the upper and lower cables were not given. Assuming that the unstressed lengths between the hangers are 203.2 mm leaves the lengths of the cables between the end hangers and the supports unknown. This problem was solved simply by analysing the structure with zero load, with the four end members being specified as variable initial length members, (section 6.4), with the required horizontal force components of 0.68947 kN. When this adjustment was made to member lengths to produce the required cable tension, the hanger forces were no longer equal to the values given by Poskitt [25]. The values are compared in Table 6.2 and are significantly different in the two outside hangers. The deflections of the lower chord for the uniform load of 13.8 N per hanger node are plotted in Fig. 6.13.

The discrepancy may well be due simply to the assumption that the unstressed cable lengths between hangers were 203.2 mm. Even if this assumption is correct, some discrepancy is to be expected because of the extreme accuracy required in measuring dimensions and properties of the structure. Measured and computed hanger unstressed lengths are given in Table 6.3.

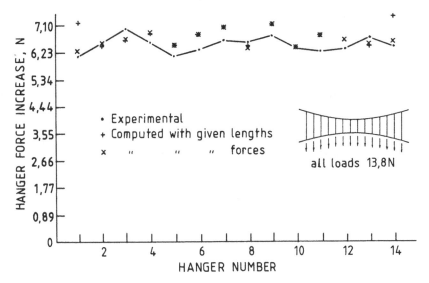

Fig. 6.14   Poskitt's truss–experimental and computed forces.

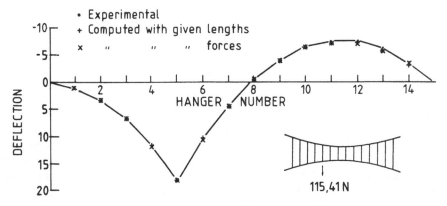

Fig. 6.15   Poskitt's truss–experimental and computed deflections.

Table 6.3

Hanger	Given hanger unstressed lengths	Computed hanger unstressed lengths
1	655.828	646.074
2	527.812	523.799
3	422.148	421.157
4	336.550	336.728
5	269.240	269.519
6	228.600	228.575
7	208.280	208.128
8	204.470	204.241
9	229.870	229.895
10	269.240	269.545
11	336.550	336.753
12	422.656	421.691
13	529.844	525.780
14	659.638	649.656

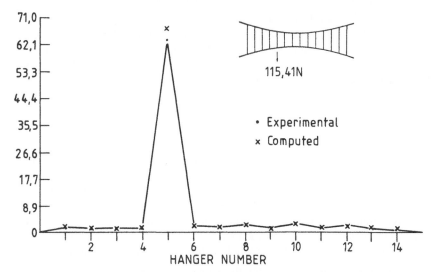

Fig. 6.16   Poskitt's truss–experimental and computed forces.

For the structure response to vertical loads, these hanger forces should not play a large part, however, because, being vertical initially, they will produce no geometric stiffness in the vertical direction. This observation is verified by the results. Two sets of analyses were carried out; the first using the computed initial member forces as explained above, the second simply using those reported by Poskitt. In Fig. 6.14, the increases in the hanger forces for equal loads of 13.76 N on all lower nodes are plotted. The results show some variation and are in best agreement when Poskitt's initial forces are used. In Figs. 6.15 and 6.16, the truss deflections and hanger forces increments for a vertical load of 115.41 N on node 5 are plotted. These results appear to be in excellent agreement.

6.3 PRACTICAL SHAPE FINDING OF CABLE NETS

Several problems which must be overcome arise in the analysis of tension net structures. These are:

1.   The initial member lengths are unknown.

2.   The structure stiffness on the initial analysis will be zero if there are no tensions in the members.

3.   Some portions of the net may become slack as the net is deformed to take up its required (or imagined) shape.

4.   The load/deformation curve for the cables may be non–linear.

These problems are overcome as follows:

1.   The initial position is considered to be a planar grid, so that the cable intersections and lengths are easily determined. This, however, leads directly to the second problem that for the initial run the transverse stiffness of the net is zero. This is overcome simply by supplying an initial, temporary prestress, which may be applied for the first few cycles until the net has taken up sufficient shape to possess both K_E and K_G stiffnesses. When the required shape is obtained the member lengths will, in general, be so strained from the initial position that the member forces are too high. At this stage, new initial lengths are specified and the iteration process reactivated to obtain a corrected equilibrium position of the joints under the applied loads and displacements.

2.   Variable initial length (VIL) members are very useful in adjusting cable forces to required values. The user specifies a member force, and the program adjusts the length of the member until it attains the required force. For the VIL member the calculated member force is replaced by the specified force, and the unstressed length of the member is changed to the value corresponding to its unstressed length and specified force. That is,

$$l_0 = \frac{l}{1 + \dfrac{P}{EA}} \tag{6.51}$$

3.   The third source of trouble arising from compressive strains in the cable members is potentially more serious, because it may indicate that the net shape and the prescribed support displacements are incompatible. On the other hand, compressive strains may appear in the first few cycles of iterations and then disappear. The present algorithm used in connection with this text allows compressive strains for a prescribed number of cycles, after which if still present the program defaults to error termination. The cure in such cases may simply be to supply shorter initial lengths for the offending members. Alternately the whole configuration may have to be redesigned.

4   Finally the load/deformation curve for the cable members is actually non–linear. This curve may be approximated by the expression,

$$\varepsilon = \frac{\sigma}{E} + \left[\frac{\sigma}{B}\right]^n \tag{6.52}$$

where ε is the strain, σ the corresponding stress, E the initial elastic constant and n and B constants which describe the non–linearity and are to be obtained experimentally.

Alternatively the curve, ((eqn (6.52)) may be approximated by a series of straight segments, as in Fig. 6.17, whose slopes E_0, E_1, E_2, etc. are supplied to the analysis as data. A further option is to approximate the curve with a B–spline function.

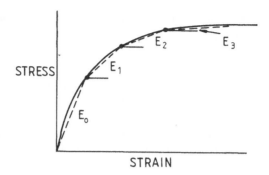

Fig. 6.17   Non–linear stress/strain curve.

Usually the analysis will be in the range E_0 of the curve. If the elasto–plastic analysis is used, the iteration process can be controlled by using a secant modulus of the form,

$$E_S = \alpha_1 E_T + \alpha_2 E_0$$

where E_T is the current tangent modulus and E_S an approximation to the secant modulus at the current stress level with $\alpha_1 + \alpha_2 = 1$. Values $\alpha_1 = 2/3$, and $\alpha_2 = 1/3$ have been found near optimum. Further details may be found in [24].

6.4   CALCULATION OF ELASTIC CRITICAL LOADS

6.4.1   Introduction

Critical or buckling loads (and bifurcation points) have already been discussed briefly in section 6.2, where it was found for the simple beam–spring structure given in Fig. 6.6, that the tangent stiffness $[K_T]$ vanishes when the axial force in the column S_0 is equal to $-kl$. In this case the vanishing of the $[K_T]$ matrix (in fact, it is the single diagonal term) is easily observed. In principle, the theory is similar for a complex structure with many degrees of freedom. However, in such cases the singularity in the stiffness matrix is not so readily detected. It is the purpose of this section first to develop the tangent stiffness matrices for beam type elements (planar, three–dimensional and torsional warping members), and then to give practical methods by which the critical load may be evaluated. It must be mentioned that in a multi–degree of freedom system there will be many load levels for which $[K_T]$ is singular. In general only the lowest load level is of interest. The exception is the situation (for example, in flat geodesic truss type structures) for which instability of both a local and general nature may be of importance. Again, the theory developed is only for the calculation of the bifurcation of the equilibrium path. If the nature of this path is required, then a complete non–linear analysis may be necessary. The tangent stiffness matrix for the truss element was developed in section 6.2, eqns (6.26) and (6.49). This matrix could be incorporated immediately in any critical load program, remembering, however, that eqn (6.49) is for large displacements and would have to be linearized, returning presumably to an identical expression as in eqn (6.26).

The bifurcation problem is then posed as follows:

> Given a load pattern on a structure for which member forces can be calculated, what is the magnitude of the scalar multiplier of the load pattern necessary such that the structure tangent stiffness is singular?

Mathematically this is expressed as: the matrix $[K_E + K_G]$ is singular. When this is the case, the determinant vanishes, leading to

$$|K_E + \lambda K_G| = 0 \tag{6.53}$$

There will be as many roots for λ as there are degrees of freedom of the structure, although λ values may occur in positive and negative pairs. Mathematically, eqn (6.53) is recognized as a linear eigenvalue problem. It is linear because λ appears as a scalar multiplier of the $[K_G]$ matrix. A situation different from this has already been encountered in section 5.3, eqn (5.74), in which the axial force in the member enters into the beam stiffness matrix through the transcendental functions, ϕ_3 and ϕ_4. That is, from section 5.3,

$$\begin{bmatrix} M_i \\ M_j \end{bmatrix} = \frac{2EI}{l} \begin{bmatrix} 2\phi_3 & \phi_4 \\ \phi_4 & 2\phi_3 \end{bmatrix} \begin{bmatrix} \phi_i \\ \phi_j \end{bmatrix} \tag{6.54}$$

where the definitions of ϕ_1, ϕ_2, ϕ_3 and ϕ_4 are given in section 5.3. The linearized expression equivalent to eqn (6.54) is (see eqn (5.94))

$$\begin{bmatrix} M_i \\ M_j \end{bmatrix} = \frac{2EI}{l} \begin{bmatrix} 2 & 1 \\ 1 & 2 \end{bmatrix} \left\{ [I_2] + \frac{Pl^2}{60EI} \begin{bmatrix} 3 & -2 \\ -2 & 3 \end{bmatrix} \right\} \begin{bmatrix} \phi_i \\ \phi_j \end{bmatrix} \tag{6.55}$$

Setting the determinant equal to zero, the transcendental eqn (6.54) has an infinite number of roots, the lowest of which is given by

$$P_{cr} = -\frac{\pi^2 EI}{l^2} \approx -\frac{9.86EI}{l^2} \tag{6.56}$$

Similarly, from eqn (6.55), the one root is

$$P_{cr} = -\frac{12EI}{l^2} \tag{6.57}$$

Obviously, this root is an inadequate approximation to π^2, and ways must be devised to improve the value if the linearized eigenvalue analysis is to be used.

It would appear that the use of the transcendental stiffness expressions is superior to the use of the linearized expressions of eqn (6.57). There are, however, compelling reasons for using the linearized stiffness matrix, not the least of which is that numerical techniques are well developed for solving this class of problem.

6.4.2   Plane frame stiffness

It was shown that for cable net structures it was necessary to add in the stiffness (or joint equilibrium correction) for the rotation of the chord $i-j$ of the member. The same expressions could be used here for frame structure members, but it is instructive to look again at the problem and redevelop these expressions from a slightly different point of view.

To this end, the plane frame member is shown in Fig. 6.18 in its local coordinate axes, with transverse displacements v'_i, v'_j of the ends i and j. From Fig. 6.18 the member rotation α is given by

$$\alpha = \frac{v'_j - v'_i}{l} = \frac{1}{l}[-1 \ 1] \begin{Bmatrix} v'_i \\ v'_j \end{Bmatrix} \tag{6.58}$$

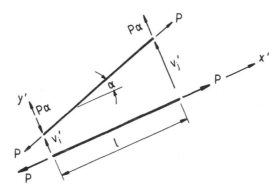

Fig. 6.18   Modification of node force components for rotation of member.

The generalized force for the displacement α, is Q_α, and, to calculate its magnitude, give the member a virtual displacement $\delta\alpha$. Then, the work expressions are

$$Q_\alpha \delta\alpha = (P\alpha)l \ \delta\alpha$$

Therefore,

$$Q_\alpha = Pl\alpha$$

The incremental relationship is found by taking differentials of this equation,

$$dQ_\alpha = Pl \ d\alpha$$

That is, K_{GQ}, the geometric stiffness,, is equal to Pl for small deformations $d\alpha$.

$$K_{GQ} = Pl \tag{6.59}$$

Now apply contragredience to eqn (6.58) to obtain the force components in the $X'Y'$ coordinates. That is,

$$\begin{bmatrix} R'_{iy} \\ R'_{jy} \end{bmatrix} = \frac{1}{l} \begin{bmatrix} -1 \\ 1 \end{bmatrix} Q_\alpha = \frac{1}{l} \begin{bmatrix} -1 \\ 1 \end{bmatrix} (Pl)\frac{1}{l}[-1 \ 1] \begin{bmatrix} v'_i \\ v'_j \end{bmatrix}$$

This simplifies to

$$\begin{bmatrix} R'_{iy} \\ R'_{jy} \end{bmatrix} = \frac{P}{l} \begin{bmatrix} 1 & -1 \\ -1 & 1 \end{bmatrix} \begin{bmatrix} v'_i \\ v'_j \end{bmatrix} \tag{6.60}$$

Thus, for all three force components at each of the ends of the frame member, R'_x, R'_y, M_z,

$$\begin{bmatrix} R'_i \\ R'_j \end{bmatrix} = \frac{P}{l} \begin{bmatrix} 0 & 0 & 0 & 0 & 0 & 0 \\ 0 & 1 & 0 & 0 & -1 & 0 \\ 0 & 0 & 0 & 0 & 0 & 0 \\ 0 & 0 & 0 & 0 & 0 & 0 \\ 0 & -1 & 0 & 0 & 1 & 0 \\ 0 & 0 & 0 & 0 & 0 & 0 \end{bmatrix} \begin{bmatrix} r'_i \\ r'_j \end{bmatrix} \tag{6.61}$$

where

$$\{r'_i\} = <r'_{ix}, r'_{iy}, \theta_{iz}>$$

and similarly for $\{R'_i\}$ etc.

In addition, the effect of bowing on the moment terms has already been considered in eqn (6.55), and this must be combined with eqn (6.61), so that

$$[k'_G] = \frac{Pl}{30} \begin{bmatrix} 0 & 0 & 0 & 0 & 0 & 0 \\ 0 & \dfrac{36}{l^2} & \dfrac{3}{l} & 0 & -\dfrac{36}{l^2} & \dfrac{3}{l} \\ 0 & \dfrac{3}{l} & 4 & 0 & -\dfrac{3}{l} & -1 \\ 0 & 0 & 0 & 0 & 0 & 0 \\ 0 & -\dfrac{36}{l^2} & -\dfrac{3}{l} & 0 & \dfrac{36}{l^2} & -\dfrac{3}{l} \\ 0 & \dfrac{3}{l} & -1 & 0 & -\dfrac{3}{l} & 4 \end{bmatrix} \tag{6.62}$$

The elastic stiffness in the member coordinates is given from eqns (5.104) and (5.107):

$$[k'_E] = [T]^T [k'][T] \tag{6.63}$$

and, hence,

$$[k'_T] = [k'_E] + [k'_G] \tag{6.64}$$

Alternatively, if eqns (6.54) and (6.61) are combined, the expression for $[k'_T]$ is given, with

$$q = \left(\frac{2\phi_3 + \phi_4}{l}\right)\frac{2EI}{l}; \quad r = 2\phi_3\frac{2EI}{l}; \quad t = \phi_4\frac{2EI}{l}$$

$$[k'_T] = \begin{bmatrix} EA/l & 0 & 0 & -EA/l & 0 & 0 \\ 0 & (2q+P)/l & q & 0 & -(2q+P)/l & q \\ 0 & q & r & 0 & -q & t \\ -EA/l & 0 & 0 & EA/l & 0 & 0 \\ 0 & -(2q+P)/l & -q & 0 & (2q+P)/l & -q \\ 0 & q & t & 0 & -q & r \end{bmatrix} \tag{6.65}$$

When the stability functions ϕ_3 and ϕ_4 are used, the chord rotation effect is completely described by eqn (6.65). There is no need to develop special sway function terms as given in many texts [13].

Finally, if the eqns (6.64) or (6.65) are to be used in a direct stiffness program, the member stiffness matrix must be transformed using eqn (5.107). That is,

$$[k_T] = [L_D]^T [k'_T][L_D] \tag{6.66}$$

Example 6.3 *Cantilever column*

This example calculates the critical load for the cantilever column loaded in uniform compression. Eqn (6.65) will be used to calculate P_{cr} for the column shown in Fig. 6.19. In this case the only free displacement variables of interest are r'_{2y}, θ_2 at the end 2 of the column.

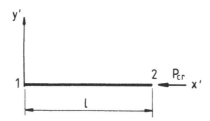

Fig. 6.19   Cantilever column.

From eqn (6.65) the stiffness relationship at joint 2 is written,

$$\begin{bmatrix} V \\ M \end{bmatrix}_2 = \begin{bmatrix} (2q + P)/l & -q \\ -q & r \end{bmatrix} \begin{bmatrix} v_2 \\ \theta_3 \end{bmatrix} \tag{i}$$

In eqn (6.65), P is taken as tension positive, and for the stability functions, the values of ϕ_1, ϕ_2 and hence ϕ_3, ϕ_4 are calculated using the cot or coth functions, using $|P|$ depending on whether the member is in compression or tension. At the critical load the determinant of the stiffness matrix in eqn (i) is equal to zero. That is,

$$\frac{r}{l}(2q + P) - q^2 = 0 \tag{ii}$$

With $\beta = \pi/4$ in eqn (6.56),

$$P = -\frac{\pi^2 EI}{l^2}$$

Thence,

$$\phi_1 = \frac{\pi}{4} \ ; \quad \phi_2 = \frac{\pi^2}{12(4 - \pi)} \ ; \quad \phi_3 = \frac{\pi}{2(4 - \pi)} \ ; \quad \phi_4 = \frac{\pi(\pi - 2)}{2(4 - \pi)}$$

Substitution for q, r and P using these values shows that eqn (ii) is indeed satisfied. The reader is urged to carry out the substitution, because it is found that even in this trivial example, the algebraic substitution is quite tedious.

6.4.3 Three–dimensional frame member geometric stiffness matrix

The theory is developed using eqns (6.54) or (6.57) applied independently to bending in each of the principal axes of the member cross section, just as was done for the simple elastic stiffness in eqn (5.118). Similarly, the effect of the chord rotation, as given for the plane frame member in eqn (6.61), is now applied independently to each of the principal axes chord rotations. Thus, $[k'_G]$ is given by

$$[k'_G] = \frac{Pl}{30}
\begin{bmatrix}
0 & 0 & 0 & 0 & 0 & 0 & 0 & 0 & 0 & 0 & 0 & 0 & 0 & 0 \\
0 & \frac{36}{l^2} & 0 & 0 & 0 & 0 & \frac{3}{l} & 0 & -\frac{36}{l^2} & 0 & 0 & 0 & 0 & \frac{3}{l} \\
0 & 0 & \frac{36}{l^2} & 0 & 0 & \frac{3}{l} & 0 & 0 & 0 & 0 & -\frac{36}{l^2} & 0 & \frac{3}{l} & 0 \\
0 & 0 & 0 & 0 & 0 & 0 & 0 & 0 & 0 & 0 & 0 & 0 & 0 & 0 \\
0 & 0 & \frac{3}{l} & 0 & \frac{3}{l} & 4 & 0 & 0 & 0 & \frac{3}{l} & \frac{3}{l} & 0 & -1 & 0 \\
0 & \frac{3}{l} & 0 & 0 & \frac{3}{l} & 0 & 4 & 0 & -\frac{3}{l} & \frac{3}{l} & 0 & 0 & 0 & -1 \\
0 & 0 & 0 & 0 & 0 & 0 & 0 & 0 & 0 & 0 & 0 & 0 & 0 & 0 \\
0 & -\frac{36}{l^2} & 0 & 0 & 0 & 0 & -\frac{3}{l} & 0 & \frac{36}{l^2} & 0 & 0 & 0 & 0 & -\frac{3}{l} \\
0 & 0 & -\frac{36}{l^2} & 0 & 0 & \frac{3}{l} & 0 & 0 & 0 & 0 & \frac{36}{l^2} & 0 & -\frac{3}{l} & 0 \\
0 & 0 & 0 & 0 & 0 & 0 & 0 & 0 & 0 & 0 & 0 & 0 & 0 & 0 \\
0 & 0 & \frac{3}{l} & 0 & \frac{3}{l} & -1 & 0 & 0 & 0 & \frac{3}{l} & -\frac{3}{l} & 0 & 4 & 0 \\
0 & \frac{3}{l} & 0 & 0 & 0 & 0 & -1 & 0 & -\frac{3}{l} & \frac{3}{l} & 0 & 0 & 0 & 4
\end{bmatrix} \qquad (6.67)$$

Again, using the same process as in eqn (6.63), with $[T]$ now defined by eqn (5.115),

$$[k'_E] = [T]^T [k'][T]$$

and

$$[k'_T] = [k'_E] + [k'_G]$$

To transform the stiffness matrix to global coordinates, $[L_D]$ is defined in eqn (5.128), and, hence,

$$[k_T] = [L_D]^T [k'_T][L_D] \qquad (6.68)$$

Example 6.4 *Effect of beam subdivision on accuracy to critical load approximation*

It was shown that use of the cubic approximation to the pin–ended column gives the eigenvalue,

$$P_{cr} \approx -\frac{12EI}{l^2}$$

When the member is further subdivided, and the effects of the chord rotations via eqn (6.61) included, the improvement in the approximation is quite marked. The values calculated are given in Table 6.4.

Table 6.4

Number of subdivisions	Approximation to π^2 for critical load
1	12.00
2	9.95
3	9.89
4	9.86

These values are plotted in Fig. 6.20.

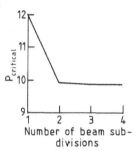

Fig. 6.20 Approximation to P_{cr} for beam subdivisions.

Example 6.5 *Cantilever column under self-weight*

The second example is for the cantilever column standing under its own weight, as shown in Fig. 6.21. The subdivisions of the column into 8 and 12 elements and the approximate axial forces in each subelement are shown in Fig. 6.21. The calculated critical loads are shown in Table 6.5.

Table 6.5

Number of subdivisions	(ql) critical load load
8	2920.0
12	2930.0
[27]	2939.0

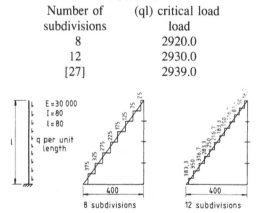

Fig. 6.21 Cantilever column distributed axial force.

Example 6.6 is of interest because, as shown later in lateral torsional beam buckling problems, (section 6.5), the flange compression forces vary linearly along the beam. Then example 6.5 gives some indication as to the number of subelements required in such problems.

6.4.4 Alternative derivation of geometric stiffness matrix

An alternative derivation of the geometric stiffness matrix of a framed member is given in which the effect of transverse forces is included. In Fig. 6.15 the influence of the chord rotation on the x', y' components of the axial force P in the member has been shown. However, the question arises as to the influence of the transverse shear S which is the result of node forces in the y' direction at the ends of the member. The components of these forces in the X', Y' coordinates will evidently change when the chord suffers a rotation $\Delta\alpha$. To study this effect, consider the forces on ends $i-j$ of the member in the local member axes. These will be

$$<F'_{ix} \ F'_{iy} \ F'_{jx} \ F'_{jy}>$$

For a rotation of axes from the X', Y' axes to the $\bar{X}$, $\bar{Y}$ axes through a small angle, $\Delta\alpha$, $\cos \Delta\alpha \approx 1$, $\sin \Delta\alpha \approx \Delta\alpha$, the transformation matrix for vector components is approximated by

$$[L] = \begin{bmatrix} \cos \Delta\alpha & -\sin \Delta\alpha \\ \sin \Delta\alpha & \cos \Delta\alpha \end{bmatrix} \approx \begin{bmatrix} 1 & -\Delta\alpha \\ \Delta\alpha & 1 \end{bmatrix} \tag{6.69}$$

In the local displacement components the increment in the angle α is given by

$$\Delta\alpha = \frac{1}{l}[0 \ -1 \ 0 \ 1 \,] \begin{bmatrix} r'_{ix} \\ r'_{iy} \\ r'_{jx} \\ r'_{jy} \end{bmatrix} = [B]\{r'\} \tag{6.70}$$

Assume now that under this transformation the axial force P_0 and the transverse shear S_0 remain constant. Write the force components in the $\bar{X}$, $\bar{Y}$ system, using the bar notation, so that

$$\begin{bmatrix} F'_{ix} \\ F'_{iy} \\ F'_{jx} \\ F'_{jy} \end{bmatrix} = \begin{bmatrix} \bar{F}_{ix} \\ \bar{F}_{iy} \\ \bar{F}_{jx} \\ \bar{F}_{jy} \end{bmatrix} = \begin{bmatrix} -1 & 0 \\ 0 & 1 \\ 1 & 0 \\ 0 & -1 \end{bmatrix} \begin{bmatrix} P_0 \\ S_0 \end{bmatrix} \tag{6.71}$$

The components of $\bar{F}$ in the $\bar{X}$, $\bar{Y}$ axes are given by the transformation eqn (1.38),

$$\{\bar{F}'\} = [L_D]\{\bar{F}\}$$

That is, the change in components is

$$\{\Delta F'\} = \{\bar{F}'\} - \{F'\} = \begin{bmatrix} 1 & -Br' & 0 & 0 \\ Br' & 1 & 0 & 0 \\ 0 & 0 & 1 & -Br' \\ 0 & 0 & Br' & 1 \end{bmatrix} \{F'\} - \{F'\} \tag{6.72}$$

Simplifying eqn (6.72),

$$\{\Delta F'\} = \begin{bmatrix} 0 & -1 & 0 & 0 \\ 1 & 0 & 0 & 0 \\ 0 & 0 & 0 & -1 \\ 0 & 0 & 1 & 0 \end{bmatrix} \{F'\}[B]\{r'\}$$

Substitution for $\{F'\}$ from eqn (6.71) and $[B]$ from eqn (6.70) gives, separating terms in P_0 and S_0,

$$\{\Delta F'\} = \left\{ \frac{P_0}{l} \begin{bmatrix} 0 & 0 & 0 & 0 \\ 0 & 1 & 0 & -1 \\ 0 & 0 & 0 & 0 \\ 0 & -1 & 0 & 1 \end{bmatrix} + \frac{S_0}{l} \begin{bmatrix} 0 & 1 & 0 & -1 \\ 0 & 0 & 0 & 0 \\ 0 & -1 & 0 & 1 \\ 0 & 0 & 0 & 0 \end{bmatrix} \right\} \{r'\} \tag{6.73}$$

The first term is the expression already given in eqn (6.61) for the axial force effect. The second expression is seen to be unsymmetric. Symmetry can be restored if account is taken of the change in the length, of the member and of the fact that the moments M_i and M_j producing the shear S_0 are constant, rather than S_0 itself. That is, after the change in length δ the product of shear times length is constant. If S_0 is the initial shear force and ΔS_0 its variation, then

$$(S_0 + \Delta S_0)(l + \delta) = S_0\, l$$

Neglecting the second–order term, $\Delta S_0\, \delta$,

$$\Delta S_0 = -\frac{S_0\, \delta}{l}$$

Now,

$$\delta = [-1\ 0\ 1\ 0\,]\{r'\}$$

The variation in F' force components due to ΔS_0 will be $\Delta \bar{F}'_s$, and, hence,

$$\{\Delta \bar{F}'_s\} = \begin{bmatrix} 1 & -Br' & 0 & 0 \\ Br' & 1 & 0 & 0 \\ 0 & 0 & 1 & -Br' \\ 0 & 0 & Br' & 1 \end{bmatrix} \begin{bmatrix} 0 \\ \Delta S_0 \\ 0 \\ -\Delta S_0 \end{bmatrix} \tag{6.74}$$

Neglecting $[B]\{r'\}\Delta S_0$ as a second–order term gives

$$\{\Delta \bar{F}'_s\} = -\frac{S_0}{l} \begin{bmatrix} 0 \\ 1 \\ 0 \\ -1 \end{bmatrix} [-1\ 0\ 1\ 0]\{r'\} = \frac{S_0}{l} \begin{bmatrix} 0 & 0 & 0 & 0 \\ 1 & 0 & -1 & 0 \\ 0 & 0 & 0 & 0 \\ -1 & 0 & 1 & 0 \end{bmatrix} \{r'\} \tag{6.75}$$

Adding the effect in eqn (6.75) to eqn (6.73) gives the now symmetric expression,

$$[k'_G] = \frac{P_0}{l} \begin{bmatrix} 0 & 0 & 0 & 0 \\ 0 & 1 & 0 & -1 \\ 0 & 0 & 0 & 0 \\ 0 & -1 & 0 & 1 \end{bmatrix} + \frac{S_0}{l} \begin{bmatrix} 0 & 1 & 0 & -1 \\ 1 & 0 & -1 & 0 \\ 0 & -1 & 0 & 1 \\ -1 & 0 & 1 & 0 \end{bmatrix} \tag{6.76}$$

Notice that the second term in eqn (6.76) with $\dfrac{S_0}{l} = \dfrac{M_i + M_j}{l^2}$ will generally have only a minor effect because $S_0 \ll P_0$.

6.5 GEOMETRIC STIFFNESS FOR THIN WALLED OPEN SECTIONS

In section 5.8 the theory was developed by which the elastic stiffness of the thin–walled open cross section line element $i-j$ can be calculated. This was achieved either by solving the indeterminate relationship between the St. Venant's and warping torsion effects, (eqn (5.214)), or, alternatively, by using a cubic interpolation for the member twist, in which case the approximate elastic stiffness is expressed as the sum of the warping and St Venant's stiffnesses (eqn (5.214)). The calculation of the geometric stiffness is best approximated by using the cubic interpolation functions. In section 5.8 the theory was developed only for the twist variables (θ_L, θ'_L, θ_R, θ'_R) at the ends of the member(θ being the rotation about the member Z–axis and θ' its first derivative). In order to derive the geometric stiffness, the principle of virtual displacements is used to examine the equilibrium of the stresses in the beam in the deformed position. Continuum theory is adapted for the thin–walled beam, considering both the twist and the bending about the principal axes. When this general case has been developed, the appropriate terms can be extracted, if so desired, for the limited analysis associated only with the bending in one principal plane.

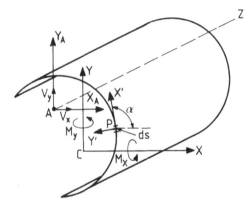

Legend

A	shear centre
C	centroid of cross section
(X, Y)	principal axes through centroid
(X', Y')	tangential and normal axes to element ds at P
α	direction of X' to X

Fig. 6.22   Thin–walled beam cross section.

It can be shown [26] that the expression for the virtual work written in the displaced position is

$$\int_{A_0} (t_n + \Delta t_n)\delta(\Delta u_n)dA_0 - \int_{vol} S_{ij}\,\delta(\Delta\varepsilon_{ij})dV_0 = \int_{vol} S_{ij}\,\delta(\Delta\eta_{ij})dV_0 + \int_{vol} \Delta S_{ij}\,\delta(\Delta E_{ij})dV_0 \qquad (6.77)$$

In this equation the second term on the right–hand side leads to the elastic stiffness matrix calculation in the deformed position, if so required. The first term, however, is the correction for the displaced position and is used in the calculation of the initial tangent stiffness for the stability problem. The strain tensor is written

$$E_{ij} = \frac{1}{2}(u_{i,j} + u_{j,i}) + \frac{1}{2}u_{k,i}u_{k,j} \qquad (6.78)$$

where the subscript (,) denotes differentiation with respect to the following index variable. That is,

$$E_{ij} = \varepsilon_{ij} + \eta_{ij}$$

Taylor series expansion, including only first–order differentials and up to quadratic terms, gives

$$\Delta\varepsilon_{ij} = \frac{1}{2}(\Delta u_{i,j} + \Delta u_{j,i}) + \frac{1}{2}u_{k,i}\Delta u_{k,j} + \frac{1}{2}\Delta u_{k,i}u_{k,j} \tag{6.79}$$

and

$$\Delta\eta_{ij} = \frac{1}{2}\Delta u_{k,i}\Delta u_{k,j} \tag{6.80}$$

For the continuum it is relatively easy to proceed from these expressions to the calculation of the tangent stiffness once simple interpolation functions have been chosen. The beam problem is more difficult because of the assumptions already made concerning the state of strain. In the thin–walled section the complication also arises that the shear stress is given in the local rather than the global coordinate axes. Consider, then, the open section shown in Fig. 6.22.

The displacement of a point P on the cross section is given as $(\tilde{u}, \tilde{v}, \tilde{w})$, the coordinates of P being (x, y, z). The displacements of the centroid are (u, v, w). Thence,

$$\tilde{u} = u - y\theta \tag{6.81}$$

$$\tilde{v} = v + x\theta \tag{6.82}$$

$$\tilde{w} = w - xu_{,z} - yv_{,z} + \theta_{,z}\omega_n \tag{6.83}$$

where the subscript again denotes differentiation with respect to the variable following and ω_n is the normalized warping function. The stresses are given by

$$\sigma_z = \frac{P}{A} - \frac{M_y x}{I_y} + \frac{M_x y}{I_x} \tag{6.84}$$

$$\tau = -\frac{1}{t}\left\{\frac{Q_x V_y}{I_y} + \frac{Q_y V_x}{I_x}\right\} \tag{6.85}$$

The displacements at the ends of the member are now defined and their interpolation functions with respect to Z established. In what follows, the subscripts L and R refer to the left– and right–hand ends of the element, respectively.

1. Rotation about the Z, axis, θ

$$\{r_\theta\} = \{\theta_L \ \theta'_L \ \theta_R \ \theta'_R\} \tag{6.86}$$

The prime here denotes the derivative of the function with respect to Z.

2. Displacement in the X direction, u.

Assume a cubic variation as in no. 1, and the relevant end parameters are

$$\{r_u\} = \{u_L \ \theta_{YL} \ u_R \ \theta_{YR}\} \tag{6.87}$$

Note that θ_{YL}, θ_{YR} are the end rotations in the (X, Z) plane defined to correspond with M_{YL} and M_{YR} and thus $\theta_{YL} = -(\frac{du}{dz})_L$ and $\theta_{YR} = -(\frac{du}{dz})_R$.

3. Displacement in the Y direction, v.

Assume a cubic variation again as in no. 1 and the relevant end parameters are

$$\{r_v\} = \{v_L \ \theta_{XL} \ v_R \theta_{XR}\} \tag{6.88}$$

Note, again, that θ_{XL} and θ_{XR} are the end rotations in the (Y, Z) plane defined to correspond with M_{XL} and M_{YL}, and thus $\theta_{XL} = (\dfrac{dv}{dz})_L$ and $\theta_{XR} = \dfrac{(dv)}{dz})_R$.

4. Displacement in the Z direction, w.

Assume a linear variation with relevant parameters,

$$\{r_w\} = \{w_L, \ w_R\} \tag{6.89}$$

The cubic interpolation function ϕ may now be used for the terms r_θ, r_u and r_v, and, noting the appropriate signs referred to above, it follows that

$$\{\phi_1\} = \{\phi_\theta\} = \{\phi_v\} = \begin{bmatrix} 1 - 3\xi^2 + 2\xi^3 \\ l\xi(1 - 2\xi + \xi^2) \\ \xi^2(3 - 2\xi) \\ l\xi^2(-1 + \xi) \end{bmatrix} \tag{6.90}$$

$$\{\phi_2\} = \{\phi_u\} = \begin{Bmatrix} 1 - 3\xi^2 + 2\xi^3 \\ -l\xi(1 - 2\xi + \xi^2) \\ \xi^2(3 - 2\xi) \\ -l\xi^2(-1 + \xi) \end{Bmatrix} \tag{6.91}$$

where the displacement and rotation components are interpolated

$$u = \{\phi_2\}^T \{r_u\} ; \qquad v = \{\phi_1\}^T \{r_v\} ; \qquad \theta = \{\phi_1\}^T \{r_\theta\} \tag{6.92}$$

and, in eqn (6.90), ξ is the non–dimensional coordinate $\xi = z/l$. Finally, for interpolation of w

$$\{\phi_w\} = \begin{Bmatrix} 1 - \xi \\ \xi \end{Bmatrix} \tag{6.93}$$

Substitution of these interpolation functions into eqns (6.83) for $\tilde{u}$, $\tilde{v}$ and $\tilde{w}$ gives

$$\begin{bmatrix} \tilde{u} \\ \tilde{v} \\ \tilde{w} \end{bmatrix} = \begin{bmatrix} \phi_2^T & 0 & 0 & -y\phi_1^T \\ 0 & \phi_1^T & 0 & x\phi_1^T \\ -x\phi_{2,z}^T & -y\phi_{1,z}^T & \phi_w^T & \omega_n\phi_{1,z}^T \end{bmatrix} \begin{bmatrix} r_u \\ r_v \\ r_w \\ r_\theta \end{bmatrix} \tag{6.94}$$

The transformations into local axes (X', Y') give

$$
\begin{bmatrix} \tilde{u}' \\ \tilde{v}' \end{bmatrix} = \begin{bmatrix} \cos\alpha & \sin\alpha \\ -\sin\alpha & \cos\alpha \end{bmatrix} \begin{bmatrix} \tilde{u} \\ \tilde{v} \end{bmatrix}
\tag{6.95}
$$

and

$$
\begin{bmatrix} x' \\ y' \end{bmatrix} = \begin{bmatrix} \cos\alpha & \sin\alpha \\ -\sin\alpha & \cos\alpha \end{bmatrix} \begin{bmatrix} x \\ y \end{bmatrix}
\tag{6.96}
$$

Use of these transformations in eqn (6.94) gives

$$
\begin{bmatrix} \tilde{u}' \\ \tilde{v}' \\ \tilde{w}' \end{bmatrix} = \begin{bmatrix} \cos\alpha\,\phi_2^T & \sin\alpha\,\phi_1^T & 0 & -y'\phi_1^T \\ -\sin\alpha\,\phi_2^T & \cos\alpha\,\phi_1^T & 0 & x'\phi_1^T \\ -x\phi_{2,z}^T & -y\phi_{1,z}^T & \phi_w^T & \omega_n\phi_{1,z}^T \end{bmatrix} \begin{bmatrix} r_u \\ r_v \\ r_w \\ r_\theta \end{bmatrix}
\tag{6.97}
$$

In the calculation of the first term on the right–hand side of eqn (6.79), eqns (6.94) and (6.97) will be used, as appropriate to the axes used for the stress tensor. The stress tensor is written as below, with the prime indicating local axes,

$$
S_{ij} = \begin{bmatrix} 0 & 0 & \tau'_{13} \\ 0 & 0 & 0 \\ \tau'_{31} & 0 & \sigma_{33} \end{bmatrix}
\tag{6.98}
$$

In eqn (6.80) it is necessary to examine the η_{ij} terms and evaluate them in terms of the nodal parameters $\{r_u, r_v, r_w, r_\theta\}$. The stresses in eqn (6.98) show that only η'_{13}, η'_{31} and η_{33} need be considered. Now,

$$
\eta'_{13} = \eta'_{xz} = \frac{1}{2}(\tilde{u}_{,x}\tilde{u}_{,z} + \tilde{v}_{,x}\tilde{v}_{,z} + \tilde{w}_{,x}\tilde{w}_{,z})
\tag{6.99}
$$

In this expression, $\tilde{u}_{,x'} = 0$, from eqn (6.97), and the term in $\tilde{w}$ derivatives is neglected in what follows. Further,

$$
\eta_{33} = \eta_{zz} = \frac{1}{2}(\tilde{u}_{,z}^2 + \tilde{v}_{,z}^2 + \tilde{w}_{,z}^2)
\tag{6.100}
$$

The procedure to be adopted is straightforward. Derivatives of η'_{13}, η'_{31} and η_{33} are obtained with respect to each of the components of $\{r_u\}$, $\{r_v\}$ and $\{r_\theta\}$, and the first term of the right–hand side of eqn (6.79) is used to yield the terms of the geometric stiffness matrix. Consider first η'_{13}, noting that η'_{31} is identical. From eqn (6.97),

$$
\tilde{v}'_{,x} = [0\ 0\ 0\ \phi_1^T]\{r\}
\tag{6.101}
$$

where

$$
\{r\}^T = \{r_u\ r_v\ r_w\ r_\theta\}
$$

This leads to

$$
\tilde{v}'_{,x} = \{\phi_1\}^T\{r_\theta\}
\tag{6.102}
$$

Also

$$\tilde{v}'_{,z} = [-\sin \alpha \; \phi_{2,z}^T \; \cos \alpha \; \phi_{1,z}^T \; 0 \; x' \phi_{1,z}^T \;]\{r\} \tag{6.103}$$

Thence,

$$\eta'_{13} = \frac{1}{2} \tilde{v}'_{,x} \tilde{v}'_{,z} = \frac{1}{2} \{r_\theta\}^T [\phi_1][-\sin \alpha \; \phi_{2,z}^T \; \cos \alpha \; \phi_{1,z}^T \; 0 \; x' \phi_{1,z}^T]\{r\}$$

After multiplication,

$$\eta'_{13} = \eta'_{31} = -\frac{1}{2} \sin \alpha \{r_\theta\}^T \{\phi_1\} \{\phi_{2,z}^T\} \{r_u\}$$

$$+ \frac{1}{2} \cos \alpha \{r_\theta^T\} \{\phi_1\} \{\phi_{1,z}^T\} \{r_v\} + \frac{x'}{2} \{r_\theta\}^T \{\phi_1\} \{\phi_{1,z}\}^T \{r_\theta\} \tag{6.104}$$

After derivatives are taken with respect to the components of $\{r\}$, substitution in the first term of eqn (6.79) gives the contributions to the member geometric stiffness as below, from η'_{13} and η'_{31}.

$$\begin{bmatrix} R_u \\ R_v \\ R_w \\ R_\theta \end{bmatrix} = \int\limits_{vol} \tau'_{13} \begin{bmatrix} 0 & 0 & 0 & -\phi_{2,z}\phi_1^T \sin \alpha \\ 0 & 0 & 0 & \phi_{1,z}\phi_1^T \cos \alpha \\ 0 & 0 & 0 & 0 \\ -\phi_{2,z}\phi_1^T \sin \alpha & \phi_{1,z}\phi_1^T \cos \alpha & 0 & 2x'\phi_{1,z}\phi_1^T \end{bmatrix} \begin{bmatrix} r_u \\ r_v \\ r_w \\ r_\theta \end{bmatrix} dv \tag{6.105}$$

The eqn (6.105) is then rewritten in terms of V_x and V_y, the shear forces at the shear centre. That is,

$$\begin{bmatrix} R_u \\ R_v \\ R_w \\ R_\theta \end{bmatrix} = \begin{bmatrix} 0 & 0 & 0 & -V_y \Phi_A \\ 0 & 0 & 0 & V_x \Phi_B \\ 0 & 0 & 0 & 0 \\ -V_y \Phi_A & V_x \Phi_B & 0 & \Omega \Phi_B \end{bmatrix} \begin{bmatrix} r_u \\ r_v \\ r_w \\ r_\theta \end{bmatrix} \tag{6.106}$$

In eqn (6.106),

$$\Phi_A = \int\limits_0^l \{\phi_{2,z}\} \{\phi_1\}^T \; dz \; ; \quad \Phi_B = \int\limits_0^l \{\phi_{1,z}\} \{\phi_1\}^T \; dz \; ; \quad \text{and} \quad \Omega = 2 \int\limits_{area} \tau'_{13} x' \; dA \tag{6.107}$$

Explicit results for eqns (6.107) are given later. Turning now to η_{33}, given by eqn (6.100),

$$\tilde{u}_{,z} = [\phi_{2,z}^T \; 0 \; 0 \; -y \phi_{1,z}^T]\{r\} \tag{6.108}$$

$$\tilde{v}_{,z} = [0 \; \phi_{1,z}^T \; 0 \; x \phi_{1,z}^T]\{r\} \tag{6.109}$$

These results are taken directly from eqn (6.94). After rearrangement, the following results emerge:

$$\tilde{u}_{,z}^2 = \{r_u\}^T \{\phi_{2,z}\} \{\phi_{2,z}\}^T \{r_u\}$$

$$- 2y \{r_u\}^T \{\phi_{2,z}\} \{\phi_{1,z}\}^T \{r_\theta\} + y^2 \{r_\theta\}^T \{\phi_{1,z}\} \{\phi_{1,z}\}^T \{r_\theta\} \tag{6.110}$$

$$\tilde{v}_{,z}^2 = \{r_v\}^T \{\phi_{1,z}\}\{\phi_{1,z}\}^T \{r_v\}$$

$$+ 2x\{r_v^T\}\{\phi_{1,z}\}\{\phi_{1,z}^T\}\{r_\theta\} + x^2\{r_\theta\}^T \{\phi_{1,z}\}\{\phi_{1,z}\}^T \{r_\theta\} \tag{6.111}$$

Proceeding, as before, by taking derivatives of η_{33} with respect to each of the components of $\{r\}$ and integrating the first term of eqn (6.77) over the volume leads to

$$\begin{bmatrix} R_u \\ R_v \\ R_w \\ R_\theta \end{bmatrix} = \begin{bmatrix} P\Phi_C & 0 & 0 & -M_x\Phi_D \\ 0 & P\Phi_E & 0 & -M_y\Phi_E \\ 0 & 0 & 0 & 0 \\ -M_x\Phi_D & -M_y\Phi_E & 0 & \gamma\Phi_E \end{bmatrix} \begin{bmatrix} r_u \\ r_v \\ r_w \\ r_\theta \end{bmatrix} \tag{6.112}$$

In eqn (6.112),

$$\Phi_c = \int_0^l \{\phi_{2,z}\}\{\phi_{2,z}\}^T \ dz \ ; \qquad \Phi_D = \int_0^l \{\phi_{2,z}\}\{\phi_{1,z}\}^T \ dz \tag{6.113a}$$

$$\Phi_E = \int_0^l \{\phi_{1,z}\}\{\phi_{1,z}\}^T \ dz \ ; \qquad \gamma = \int_{area} (x^2 + y^2)\sigma_{33} \ dA \tag{6.113b}$$

In eqn (6.112), M_x and M_y are the average moments along the member. Eqs (6.106) and (6.112), when added, give the final statement of the geometric stiffness matrix for the member. An examination of the terms of eqns (6.107) and (6.113) is revealing. The following values of Φ_A, Φ_B, Φ_C, Φ_D and Φ_E are easily obtained from eqns (6.90) and (6.91):

$$[\Phi_A] = \begin{bmatrix} -\dfrac{1}{2} & -\dfrac{l}{10} & -\dfrac{1}{2} & \dfrac{l}{10} \\ -\dfrac{l}{10} & 0 & \dfrac{l}{10} & -\dfrac{l^2}{60} \\ \dfrac{1}{2} & \dfrac{l}{10} & \dfrac{1}{2} & -\dfrac{1}{10} \\ \dfrac{l}{10} & \dfrac{l^2}{60} & -\dfrac{l}{10} & 0 \end{bmatrix} \tag{6.114}$$

$$[\Phi_B] = \begin{bmatrix} -\dfrac{1}{2} & -\dfrac{l}{10} & \dfrac{1}{2} & \dfrac{l}{10} \\ \dfrac{l}{10} & 0 & -\dfrac{l}{10} & \dfrac{l^2}{60} \\ \dfrac{1}{2} & \dfrac{l}{10} & \dfrac{1}{2} & -\dfrac{l}{10} \\ -\dfrac{l}{10} & -\dfrac{l^2}{60} & \dfrac{l}{10} & 0 \end{bmatrix} \tag{6.115}$$

$$[\Phi_C] = \begin{bmatrix} \dfrac{6}{5l} & -\dfrac{1}{10} & -\dfrac{6}{5l} & -\dfrac{1}{10} \\[2ex] -\dfrac{1}{10} & \dfrac{2l}{15} & \dfrac{1}{10} & -\dfrac{l}{30} \\[2ex] -\dfrac{6}{5l} & \dfrac{1}{10} & \dfrac{6}{5l} & \dfrac{1}{10} \\[2ex] -\dfrac{1}{10} & -\dfrac{l}{30} & \dfrac{1}{10} & \dfrac{2l}{15} \end{bmatrix} \tag{6.116}$$

$$[\Phi_D] = \begin{bmatrix} \dfrac{6}{5l} & \dfrac{1}{10} & -\dfrac{6}{5l} & \dfrac{1}{10} \\[2ex] \dfrac{1}{10} & -\dfrac{2l}{15} & \dfrac{1}{10} & \dfrac{l}{30} \\[2ex] -\dfrac{6}{5l} & \dfrac{1}{10} & \dfrac{6}{5l} & -\dfrac{1}{10} \\[2ex] \dfrac{1}{10} & \dfrac{l}{30} & -\dfrac{1}{10} & \dfrac{2l}{15} \end{bmatrix} \tag{6.117}$$

$$[\Phi_E] = \begin{bmatrix} \dfrac{6}{5l} & \dfrac{1}{10} & -\dfrac{6}{5l} & \dfrac{1}{10} \\[2ex] \dfrac{1}{10} & \dfrac{2l}{15} & -\dfrac{1}{10} & -\dfrac{l}{30} \\[2ex] -\dfrac{6}{5l} & -\dfrac{1}{10} & \dfrac{6}{5l} & \dfrac{1}{10} \\[2ex] \dfrac{1}{10} & -\dfrac{l}{30} & -\dfrac{1}{10} & \dfrac{2l}{15} \end{bmatrix} \tag{6.118}$$

Finally, eqn (6.113d) is worthy of further investigation. If warping direct stresses are ignored, eqn (6.84) gives σ_{33} (σ_{zz}):

$$\sigma_{zz} = \frac{P}{A} - \frac{M_y x}{I_y} + \frac{M_x y}{I_x} \tag{6.119}$$

where P is the tensile load in the section. Thus, in eqn 6.113(d) γ is given by

$$\gamma = \int_{area} (x^2 + y^2)\left\{ \frac{P}{A} - M_y \frac{x}{I_y} + M_x \frac{y}{I_x} \right\} dA$$

So that,

$$\gamma = \frac{P}{A}(I_x + I_y) - \frac{M_y}{I_x}\beta_1 + \frac{M_x}{I_y}\beta_2 \tag{6.120}$$

The terms β_1 and β_2 are defined as follows:

$$\beta_1 = \int\limits_{area} (x^3 + xy^2)\, dA \;\; ; \qquad \beta_2 = \int\limits_{area} (y^3 + yx^2)\, dA \tag{6.121}$$

It follows that β_1, β_2 are zero for bisymmetric sections; eqns (6.106) and (6.112) show clearly the structure of the geometric stiffness matrices. The limited problem for lateral torsional buckling analysis is shown in Fig. 6.23, in which it is assumed that the YZ plane is a plane of symmetry. In this case the variables at the ends $i-j$ of the member entering into the problem are $(u, \theta_y, \theta_z, \theta_z')$. The resulting geometric stiffness matrix is an 8×8 matrix, and its terms may be extracted from eqns (6.106) and (6.112) with the aid of eqns (6.114) to (6.118) included. Note: in this case, $\gamma = PI_x/A$.

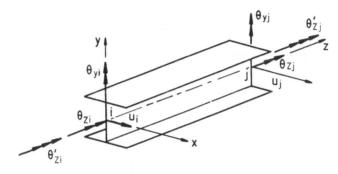

Fig. 6.23 Torsional buckling–nodal variables.

Now define the following constants.

$$A = \frac{6P}{5l} \; ; \quad B = \frac{P}{10} \; ; \quad C = \frac{V_y}{2} \; ; \quad D = \frac{6M_x}{5l} \; ; \quad E = \frac{V_y l}{10}$$

$$F = \frac{M_x}{10} \; ; \quad G = \frac{2PL}{15} \; ; \quad H = \frac{2M_x l}{15} \; ; \quad I = \frac{PL}{30} \; ; \quad J = \frac{V_y l^2}{60} \tag{6.122}$$

$$K = \frac{6PI_x}{5lA} ; \quad L = \frac{PI_x}{10A} ; \quad M = \frac{2PI_x l}{15A}$$

Then the $[K_G]$ matrix is written

$$[K_G] = \begin{bmatrix}
A & -B & C-D & E+F & -A & -B & C+D & -(E+F) \\
-B & G & E+F & H & B & -I & -(E+F) & J \\
C-D & E+F & K & L & D-C & -(E+F) & -K & L \\
E+F & H & L & M & -(E+F) & -J & -B & -I \\
-A & B & D-C & -(E+F) & A & B & C-D & E+F \\
-B & -I & -(E+F) & -J & B & G & E+F & H \\
C+D & -(E+F) & -K & -B & C-D & E+F & K & -L \\
-(E+F) & J & L & -I & E+F & H & -L & M
\end{bmatrix} \tag{6.123}$$

The effect of load offset on the buckling load

If, in Fig. 6.24, the load in the YZ plane is applied at a distance y_R from the centroidal axis, it has the effect of introducing a rigid truss member at the joint. However, the geometric stiffness for the rotation of such a member is given by eqn (6.59). In the present case, the term to be added to the θ term of $[K_G]$ is $\pm R_y y_R$, the $+$ or $-$ sign depending on whether R_Y acts away or towards the centroid. The beam shown in Fig. 6.24 was analysed for the various load conditions shown.

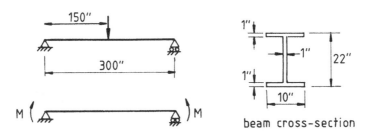

Fig. 6.24   Simply supported beam–lateral torsional buckling.

The values of the critical load values for the beam subdivided into 12 segments are given in Table 6.6 and are compared with those from [27] and [28].

Table 6.6

	Timoshenko [27]	Powell [28]	Present case	Load bottom flange	Load top flange
Concentrated load	198	200	201.7	172.3	232.2
uniform moment	11 200	11 200	11 087	–	–

6.6 PRACTICAL EVALUATION OF EIGENVALUES AND EIGENVECTORS

6.6.1 Determinant search

It has been shown that the formulation of the bifurcation problem for small displacements from the initial position of the structure leads to an eigenvalue problem of the type, in the non–linear formulation,

$$[K_T]\{r\} = 0 \tag{6.124a}$$

or, if the problem has been linearized,

$$[K_E]\{r\} = -\lambda[K_G]\{r\} \tag{6.124b}$$

If the problem size is small, then there are many standard computer subroutines available which can be used to obtain one or several of the eigenvalues and vectors of eqn (6.124b). It should be noted that the lowest value of λ is required, and, hence, in any iteration scheme, convergence should be to this value. It will be seen that in eqn (6.124a), in which transcendental functions occur, an infinite number of eigenvalues are available. The general solution of eqn (6.124b) may be reduced to the standard form as follows. Thus, consider the equation,

$$[A]\{x\} = \lambda[B]\{x\} \tag{6.125}$$

For stability problems $[A] \equiv [K_E]$ and $[B] \equiv -[K_G]$. The matrix $[A]$ is positive definite, and $[B]$ is symmetric but not necessarily positive definite. The problem is then reversed to solve

$$[B]\{x\} = \Omega[A]\{x\} ; \quad \text{with} \quad \Omega = \frac{1}{\lambda} \tag{6.126}$$

It is well known that if $[A]$ is positive definite it may be decomposed into the product of an upper triangular matrix and its transpose. Thus,

$$[A] = [L][D][L]^T = [U]^T[U] \tag{6.127}$$

Then,
$$[U] = [D]^{\frac{1}{2}}[L]^T$$

Now the coordinate substitution is made

$$\{y\} = [U]\{x\} \tag{6.128a}$$

or,
$$\{x\} = [U]^{-1}\{y\} \tag{6.128b}$$

Substitution of eqns (6.127) and (6.128) in eqn (6.126) gives

$$[B][U]^{-1}\{y\} = \Omega[U]^T[U]\{x\}$$

Hence,

$$[U^{-1}]^T[B][U^{-1}]\{y\} = \Omega\{y\}$$

That is,
$$[B^*]\{y\} = \Omega\{y\} \tag{6.129}$$

where
$$[B^*] = [U^{-1}]^T[B][U^{-1}] \tag{6.130}$$

In eqn (6.129), it is seen that $[B^*]$ is a generalized flexibility matrix corresponding to the variable $\{y\}$. Eqn (6.129) is now the standard linear eigenvalue problem. It is shown that the normalized eigenvectors $\{Y\}$ of eqn (6.129) produce eigenvectors $\{X\}$ of eqn (6.125). That is, let the $\{Y\}$ vectors be normalized such that
$$\{Y\}^T\{Y\} = [I]$$

Then, from eqn (6.128a)
$$\{Y\} = [U]\{X\}$$

so that substitution for $\{Y\}$ gives
$$\{X\}^T[U]^T[U]\{X\} = [I]$$

Thence, from eqn (6.127),
$$\{X\}^T[A]\{X\} = [I] \tag{6.131}$$

It is seen also from eqns (6.125) and (6.131) that
$$\{X\}^T[A]\{X\} = \{X\}^T[B]\{X\}\lfloor\lambda\rfloor = [I]$$

Hence,

$$\{X\}^T [B]\{X\} = \left| \frac{1}{\lambda} \right| \qquad (6.132)$$

In real applications, however, the size of the problem may be large, and $\{r\}$ in eqn (6.124b) may be a vector of hundreds or thousands of variables. If only the lowest critical load value is required, the solution of either eqns (6.124a) or (b) is relatively straightforward. First an elastic analysis is carried out, and member forces for a given level of the applied load determined. A simple interval–halving routine is used in which the stiffness matrix,

$$[K_E + \lambda K_G]$$

is reduced by Gaussian, (or other) elimination for successive choices of λ. To start the sequence, choose $\lambda = 1$. If the matrix proves to be positive definite, that is, no zeros or negatives appear on the diagonal during the reduction, the value of λ is doubled and the process is repeated, until such a value of λ, say, λ_{n+1}, is reached for which a negative does occur. The interval between the new λ_{n+1} value and λ_n is now halved, so that

$$\lambda_{n+2} = \frac{1}{2}(\lambda_n + \lambda_{n+1})$$

This process of interval halving is now repeated. The interval to be halved is always chosen to be that for which the values λ_n, λ_{n+1} straddle the zero pivot point. The process may simply be continued until a tolerance limit on the accuracy of λ is reached. The mode shape is now obtained with this load level (λ_m) by recording which pivot term is approaching zero, and then finally applying a unit displacement to this degree of freedom with the tangent stiffness $[K_E + \lambda_m K_G]$. The resulting mode shape is then normalized so that the maximum displacement is unity. This method was used in all the given examples, for example, the frame shown in Fig. 6.25. For the arch bridge shown in Fig. 6.26, this process took 14 iterations to reach $\lambda_m = 7.9 \times$ the base load.

Example 6.6 *Critical load of frame*

In this example the results are given for the calculation of the critical load of four–storey framed structure shown in Fig. 6.25(a). The axial compressive forces in the columns are 20, 40, 60 and 80 $\times 10^3$lb, from top to bottom. The members were subdivided, two subelements per member. The value of P_{cr}, as calculated, equalled 3.96 $\times$ base load, and compares favourably with the value 3.97 quoted in [29]. The mode shape is shown in Fig. 6.25(b). This example shows that for framed structures with sway the two subdivisions per member give adequate approximation to the critical load when the eigenvalue problem has been linearized. Of course, if stability functions are used, no subdivision of members is required, but the penalty is incurred of having to solve a non–linear eigenvalue problem.

Example 6.7 *Buckling analysis of arch bridge*

In this last example the critical load and mode shape for a tied arch bridge are given. The structure and the mode shape are shown in Fig. 6.26. The member forces, compression in the arch, tension in the hangers and tie have been computed for the dead load of the bridge and the superimposed live load. The axial forces in the members are taken to be $F_0 + \lambda_{cr} F_l$, where F_0 is the force due to dead load, F_l is the force due to base live load, and λ_{cr}, is the live load scalar multiplier.

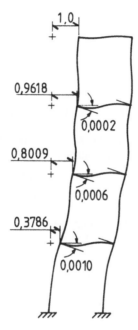

Fig. 6.25   Four–storey frame–mode shape.

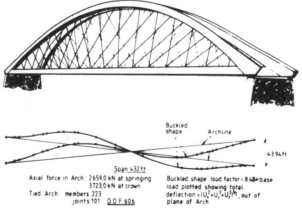

Fig. 6.26   Tied arch bridge–dimensions and mode shape.

The process may be accelerated by interpolation. During the reduction of the stiffness equations, the determinant D is evaluated by multiplying together the diagonal terms. For various λ values, this determinant D will be as plotted in Fig. 6.27.

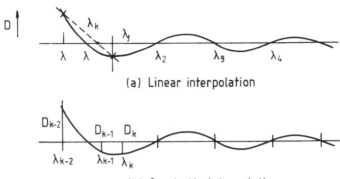

(a) Linear interpolation

(b) Quadratic interpolation

Fig. 6.27   Quadratic interpolation.

Thus, in Fig. 6.27(a), if two successive λ values are known, λ_{k-1}, λ_k, where λ_{k-1} gives D_{k-1} positive and D_k, negative, and with only one negative pivot value, then the approximation to λ_{k+1} is

$$\lambda_{k+1} = \frac{D_{k-1}\lambda_k + |D_k|\lambda_{k-1}}{D_{k-1} + |D_k|} \tag{6.133}$$

Alternatively, three consecutive points may be used with quadratic interpolation, as shown in Fig. 6.27(b). In this case, at least one, the D_k values, should be negative with only one negative pivot. Then, assuming the quadratic function,

$$D = A + B\lambda + C\lambda^2$$

the coefficients A, B, and C may be obtained,

$$\begin{bmatrix} A \\ B \\ C \end{bmatrix} = \begin{bmatrix} 1 & \lambda_{k-2} & \lambda_{k-2}^2 \\ 1 & \lambda_{k-1} & \lambda_{k-1}^2 \\ 1 & \lambda_k & \lambda_k^2 \end{bmatrix}^{-1} \begin{bmatrix} D_{k-2} \\ D_{k-1} \\ D_k \end{bmatrix}$$

Then the approximation to D equal to zero is given by,

$$\lambda_{k+1} = \frac{-B + \sqrt{(B^2 - 4AC)}}{2A}$$

These methods can reduce the number of trial λ values dramatically if the criteria mentioned above are satisfied. In theory, the process can be extended to calculate several eigenvalues and eigenvectors and is really only the Sturm sequence process. It becomes inefficient if many eigenvalues are required because of the computational time taken to reduce the stiffness matrix.

6.6.2   Matrix iteration

Simple matrix iteration is also useful if only one (the dominant) eigenvalue is required. Then, if P_{cr}, the smallest critical load, is required, from the equations,

$$[K_E + \lambda K_G]\{r\} = \{R\}$$

Iteration must take place on $1/P_{cr}$. Thus, for $\{R\} = 0$,

$$[K_E]\{r\} = -[K_G]\{r\}$$

Premultiplying by $[K_E]^{-1} = [F]$, the structure flexibility matrix gives

$$\frac{1}{P_{cr}}\{r\} = -[K_E]^{-1}[K_G]\{r\} = -[F][K_G]\{r\} \tag{6.134}$$

This equation should be compared with eqn (4.141), and it will be seen that the method has been reduced to the flexibility method of calculation of critical loads where $(-[K_G]\{r\}$ is now the matrix of fictitious joint forces. Thus, eqn (6.134) leads to the solution of the equations,

$$[A]\{r\} = \lambda\{r\} \tag{6.135}$$

where $[A]$ is now an unsymmetric matrix. The iteration process starts with a choice of a starting vector r_0, which approximates the first mode shape. Consider now the sequence of iterates,

$$\{r_0\}, \{r_1\} = [A]\{r_0\}, \cdots, \{r_k\} = [A]^k\{r_0\} = [A]\{r_{k-1}\}$$

and also the transposed sequence,

$$\{r_0\}, \{r'_1\} = [A]^T\{r_0\}, \cdots, \{r'_k\} = ([A]^T)^k\{r_0\} = [A]^T\{r_{k-1}\}$$

Let $a_1, a_2, a_3, \cdots, a_n$ be the coordinates of the vector r_0 with respect to the basis $X_1, X_2, X_3, \cdots, X_n$, and $b_1, b_2, b_3, \cdots, b_n$ the coordinates of r_0 with respect to $X'_1, X'_2, X'_3, \cdots, X'_n$. Form the scalar product,

$$(r'_k, r_k) = ([A^T]^k\{r_0\}, [A]^k\{r_0\}) = (\{r_0\}, [A]^{2k}\{r_0\})$$

$$= \{(b_1X'_1+b_2X'_2+\cdots+b_nX'_2), (a_1\lambda_1^{2k}X_1+a_2\lambda_2^{2k}X_2 + \cdots + a_n\lambda_n^{2k}X_n)\}$$

Using the known orthogonal properties of the eigenvectors $X_1, X_2, \cdots, X_n$ and those of $X'_1, X'_2, \cdots, X'_n$ of its transpose, it is seen from the above expression that

$$(r'_k, r_k) = (a_1b_1\lambda_1^{2k} + a_2b_2\lambda_2^{2k} + \cdots + a_nb_n\lambda_n^{2k})$$

$$= \lambda_1^{2k}(a_1b_1 + a_2b_2\left[\frac{\lambda_2}{\lambda_1}\right]^{2k} + \cdots) \rightarrow a_1b_1\lambda_1^{2k} + O\left[\frac{\lambda_2}{\lambda_1}\right]^{2k} \tag{6.136}$$

Similarly, for the vectors r'_{k-1} and r_k,

$$(r'_{k-1}, r_k) \rightarrow a_1b_1\lambda_1^{2k-1} + O\left[\frac{\lambda_2}{\lambda_1}\right]^{2k-1} \tag{6.137}$$

Hence, combining these equations, the approximation to the dominant eigenvalue λ_1 is

$$\lambda_1 = \frac{(r'_k, r_k)}{(r'_{k-1}, r_k)} + O\left[\frac{\lambda_2}{\lambda_1}\right]^{2k} \tag{6.138}$$

The process can be implemented without calculating powers of $[A]$ by generating the successive vectors,

$$\{r_{k-1}\}, \{r_k\}, \text{ and } \{r'_{k-1}\}, \{r'_k\}$$

Each step requiring the multiplication of a vector by a matrix. The last vector $\{r_k\}$ is an approximation to the mode shape.

In implementing the method it should be seen that if the matrix $[A]$ has not been normalized so that the maximum term in $[A]$ is 1.0, overflow will soon occur in calculating the iteration vectors. To overcome this problem, let $\alpha = $ maximum term in $[A]$, and generate $[C]$, such that

$$[C] = \frac{1}{\alpha}[A]$$

Then the approximation to λ_1 is

$$\lambda_1 \approx \frac{\alpha([C^T]^k\{r_0\}, [C]^k\{r_0\})}{([C^T]^{(k-1)}\{r_0\}, [C]^k\{r_0\})} \tag{6.139}$$

The method is highly convergent if λ_2 is not close to λ_1. If the number of degrees of freedom of the system is small, then perhaps it is convenient to calculate $([K^{-1}][G])$ and to use the above iteration scheme. However, since the first method given should be able to calculate the lowest critical load in approximately ten reductions of the stiffness matrix, it is difficult to see that the power method competes well with the pivot search method, unless $[F]$ has been formed directly.

6.6.3   Inverse iteration

Inverse iteration, as the name implies, is only a means by which the calculation of the inverse $[K]^{-1}$ in eqn (6.134) is circumvented. Clearly, in eqn (6.134), the inverse is not really required, but rather the solution of the equations for which the right–hand side is the vector $[K_G]\{r_k\}$, where $\{r_k\}$ is the kth iteration vector. The convergence of this method is examined in detail because being a practical method for obtaining critical loads, the difficulty which arises with pairs of roots $\pm\lambda$ must be overcome. To this end, consider the general equations,

$$[K]\{e_i\} = \lambda[G]\{e_i\} \tag{6.140}$$

Suppose that the lowest value of λ, λ_m is required. Again, as in the power method, a starting vector $\{x_0\}$ is generated, and the iteration scheme, based on the sequence is

$$[K]\{x_k\} = \{y_{k-1}\} \tag{6.141}$$

Solve eqn (6.141) for x_k, and generate y_k from

$$\{y_k\} = [G]\{x_k\} \tag{6.142}$$

On the first cycle,

$$[K]\{x_1\} = [G]\{x_0\} \tag{6.143}$$

From eqns (6.141) ff. it is seen that, symbolically,

$$\{x_{k+1}\} = [K]^{-1}[G]\{x_k\}$$

Proof of convergence of inverse iteration to the lowest eigenvalue of eqn (6.140)

As before, any vector x_0 in the n–dimensional space spanned by the eigenvectors may be expressed by coordinates in this vector space. That is, let

$$\{x_0\} = \sum a_i\{e_i\} \tag{6.144}$$

Then, from eqn (6.141),

$$\{x_1\} = [K]^{-1}[G]\{x_0\} = \sum_{i=1,n} a_i [K]^{-1}[G]\{e_i\}$$

However, from eqn (6.140),

$$[K]^{-1}[G]\{e_i\} = \frac{1}{\lambda_1}\{e_i\}$$

Therefore,

$$\{x_1\} = \sum_{i=1,n} a_i \left(\frac{1}{\lambda_i}\right)\{e_i\}$$

Repeated application of eqns (6.142) and (6.143) gives

$$\{x_k\} = \sum_{i=1,n} a_i \left(\frac{1}{\lambda_i}\right)^k \{e_i\}$$

That is, if λ_m is the lowest eigenvalue,

$$\{x_k\} = \left[\frac{1}{\lambda_m}\right]^k \left\{ a_m\{e_m\} + \sum_{i=1,\ n\ \neq m} a_i\{e_i\} \left[\frac{\lambda_m}{\lambda_i}\right]^k \right\} \tag{6.145}$$

so that,

$$\{x_k\} = \alpha_k\{e_m\} + \Delta_k$$

in which $\Delta_k \to 0$ as $k \to \infty$. Therefore,

$$\{x_k\} = \alpha_k\{e_m\}$$

With one more cycle of iteration

$$\{x_{k+1}\} = \alpha_{k+1}\{e_m\} = \frac{\alpha_k}{\lambda_m}\{e_m\}$$

To obtain an estimate of λ_m, the ratio of any two terms of $\{x_{k+1}\}$ and $\{x_k\}$ will suffice. That is, for example,

$$\frac{(x_k)_1}{(x_{k+1})_1} \approx \lambda_m$$

A better mathematical approach is, of course, to take the scalar products,

$$\{x_k\}^T\{x_k\} = \alpha_k^2\{e_m\}^T\{e_m\} = \alpha_k^2$$

and

$$\{x_{k+1}\}^T\{x_{k+1}\} = \left\{\frac{\alpha_k^2}{\lambda_m^2}\right\}\{e_m\}^T\{e_m\} = \frac{\alpha_k^2}{\lambda_m^2}$$

Thence, the best first approximation to λ_m is

$$\lambda_m^2 = \frac{\{x_k\}^T\{x_k\}}{\{x_{k+1}\}^T\{x_{k+1}\}} \tag{6.146}$$

6.6.4 Computational efficiencies in the calculation of λ_m

Eqn (6.140) is modified slightly so that it is unnecessary to form $\{x_{k+1}\}$ in order to obtain the estimate of λ_m. Firstly, it is seen from eqn (6.140), for the eigenvalue λ_m, vector $\{e_m\}$, that

$$\{e_m\}^T [K]\{e_m\} = \lambda_m \{e_m\}^T [G]\{e_m\} \tag{6.147}$$

Now the k th iteration vector can be expressed eqn (6.145):

$$\{x_k\} = \alpha_k \{e_m\} + \{\Delta_k\} \tag{6.148}$$

where $\{\Delta_k\}$ is the residual $\to 0$ as k becomes large. Then, from eqn (6.148), premultiplying both sides of the expression by $[K]$,

$$[K]\{x_k\} = \alpha_k [K]\{e_m\} + [K]\{\Delta_k\}$$

Thence,

$$([K]\{x_k\})^T \{x_k\} = (\alpha_k [K]\{e_m\} + [K]\{\Delta_k\})^T (\alpha_k \{e_m\} + \{\Delta_k\})$$

or,

$$\{x_k\}^T [K]\{x_k\} = \alpha_k^2 \{e_m\}^T [K]\{e_m\} + \{\Delta_k\}^T [K]\{\Delta_k\} + 2\alpha_k \{e_m\}^T [K]\{\Delta_k\}$$

$$= \alpha_k^2 \{e_m\}^T [K]\{e_m\} + (O\{\Delta_k^2\} \to 0)$$

From eqn (6.147), $$\{x_k\}^T [K]\{x_k\} \approx \alpha_k^2 \lambda_m \{e_m\}^T [G]\{e_m\}$$

By a similar process it is shown that

$$\{x_k\}^T [G]\{x_k\} \approx \alpha_k^2 \{e_m\}^T [G]\{e_m\}$$

So that, $$\{x_k\}^T [K]\{x_k\} \approx \lambda_m \{x_k\}^T [G]\{x_k\}$$

Hence, the approximation to λ_m is,

$$\lambda_m = \frac{\{x_k\}^T [K]\{x_k\}}{\{x_k\}^T [G]\{x_k\}} \tag{6.149}$$

However, in the iteration process, ((eqns (6.141) and (6.142)), expressions are given for $[K]\{x_k\}$ and $[G]\{x_k\}$, such that now eqn (6.149) is written

$$\lambda_m = \frac{\{x_k\}^T \{y_{k-1}\}}{\{x_k\}^T \{y_k\}} \tag{6.150}$$

Since the vectors $\{x_k\}$, $\{y_k\}$, $\{y_{k-1}\}$ are calculated during the iteration process, this represents little extra effort, and eqn (6.150) is prefered, to eqn (6.149). The above process breaks down when the eigenvalues occur in equal and opposite pairs, for example, $(\pm\lambda_m)$, as will be shown below. In buckling problems, this phenomena is a common occurrence for example, a simply supported plate subjected to constant in–plane shear stress can have the shear either positive or negative. Thus, suppose that $(+\lambda_m, \{e_m^+\})$ and $(-\lambda_m, \{e_m^-\})$ are an eigenpair. Then, from eqn (6.140),

$$[K]\{e_m^+\} = \lambda_m [G]\{e_m^+\}$$

and

$$[K]\{e_m^-\} = -\lambda_m [G]\{e_m^-\}$$

Alternatively, dividing both sides of these equations by λ_m,

$$\frac{1}{\lambda_m}\{e_m^+\} = [K]^{-1}[G]\{e_m^+\} \tag{6.151}$$

$$\frac{1}{\lambda_m}\{e_m^-\} = -[K]^{-1}[G]\{e_m^-\} \tag{6.152}$$

As before, suppose the iteration commences with a starting iteration vector x_0, with x_0 expressed as in eqn (6.144):

$$\{x_0\} = \sum a_i \{e_i\}$$

and, from eqn (6.143),

$$\{x_1\} = [K]^{-1}[G]\{x_0\} = \sum a_i [K]^{-1}[G]\{e_i\}$$

Using eqn (6.151), and assuming that λ_m, $-\lambda_m$ have coefficients a_1, a_2,

$$\{x_1\} = \frac{1}{\lambda_m}(a_1\{e_m^+\} - a_2\{e_m^-\}) + \sum_{i=3,n} \frac{a_i}{\lambda_i}\{e_i\}$$

After k cycles of iteration,

$$\{x_k\} = \frac{1}{\lambda_m^k}\left\{(a_1\{e_m^+\} + (-1)^k a_2\{e_m^-\}) + \sum_{i=3,n} a_i \left[\frac{\lambda_m}{\lambda_i}\right]^k \{e_i\}\right\}$$

That is, writing this equation in the same form as eqn (6.148),

$$\{x_k\} = \alpha_k(\{e_m^+\} + \beta_k\{e_m^-\}) + \{\Delta'_k\} \tag{6.153a}$$

Also, for the k 1 iteration,

$$\{x_{k+1}\} = \frac{\alpha_k}{\lambda_m}(\{e_m^+\} - \beta_k\{e_m^-\}) + \{\Delta'_k\} \tag{6.153b}$$

It is now clear that eqn (6.146) cannot be used to calculate λ_m^2. However, there is now the possibility of using $\{x_{k+2}\}$, because, from eqn (6.153),

$$\{x_{k+2}\} = \frac{\alpha_k}{\lambda_m^2}(\{e_m^+\} + \beta_k\{e_m^{-1}\}) + \{\Delta_k''\} \tag{6.154}$$

Using eqns (6.153) and (6.154),

$$\lambda_m^4 = \frac{(\{x_k\}^T\{x_k\})}{(\{x_{k+2}\}^T\{x_{k+2}\})} \tag{6.155}$$

Alternatively, proceeding in the same manner as to derive eqn (6.150), it is foun that,

$$\{x_k\}^T[K]\{x_k\} = \lambda_m \alpha_k^2[\{e_m^+\}^T[G]\{e_m^+\} - \beta_k^2\{e_m^-\}^T[G]\{e_m^-\}]$$

and

$$\{x_{k+1}\}^T [K]\{x_{k+1}\} = \frac{\alpha_k^2}{\lambda_m}[\{e_m^+\}^T [G]\{e_m^+\} - \beta_k^2\{e_m^-\}^T [G]\{e_m^-\}]$$

Thus,

$$\lambda_m^2 = \frac{\{x_k\}^T [K]\{x_k\}}{\{x_{k+1}\}^T [K]\{x_{k+1}\}} = \frac{\{x_k\}^T \{y_{k-1}\}}{\{x_{k+1}\}^T \{y_k\}} \tag{6.156}$$

Comparing eqns (6.156) and (6.150), it is seen that now $\{x_{k+1}\}$ is required. The presence of negative eigenvalues is easily detected during the iteration process, in which case the algorithm automatically uses eqn (6.156) rather than eqn (6.150). To obtain mode shapes, eqns (6.153) and (6.154) may be used. That is, solving these equations for $\{e_m^+\}$ and $\{e_m^-\}$, it is found that

$$\{e_m^+\} = \frac{1}{2\alpha_k}(\{x_k\} + \lambda_m \{x_{k+1}\})$$

and

$$\{e_m^-\} = \frac{1}{2\alpha_k \beta_k}(\{x_k\} - \lambda_m \{x_{k+1}\})$$

Now, since $\{e_m^+\}$ and $\{e_m^-\}$ are normalized, the factors $\dfrac{1}{2\alpha_k}$ and $\dfrac{1}{2\alpha_k \beta_k}$ can be neglected hence,

$$\{e_m^+\} = (\{x_k\} + \lambda_m \{x_{k+1}\}) \tag{6.157(a)}$$

and

$$\{e_m^-\} = (\{x_k\} - \lambda_m \{x_{k+1}\})$$

6.6.5 The subspace iteration method

The subspace iteration method allows the simultaneous calculation of the approximations to several eigenvalues and eigenvectors of eqn (6.125), using a modified version of the inverse iteration. Its advantage lies in the ability to calculate a few dominant eigenvalues of systems with very large numbers of degrees of freedom. Its disadvantage lies in the possibility that because the iteration is started with a set of trial vectors, eigenvalues may be missed. This condition can always be detected, however, with a Sturm sequence check. The method has been used successfully in the calculation both of buckling loads and natural frequencies. The ability to iterate simultaneously to eigenvalues other than the lowest one is based on the observation that if, in eqn (6.144), the coefficient a_m of the lowest eigenvector $\{e_m\}$ is zero, the iteration will converge to the lowest but one value of λ. A given vector $\{x'_0\}$ may always be chosen orthogonal to the approximation $\{x_0\}$ of $\{e_m\}$, by maintaining

$$\{x_0\}^T \{x'_0\} = 0$$

The space is reduced from the original n dimensions of $[K]$ and $[G]$ matrices to a subspace of dimension q, where q is the number of iteration vectors. It is suggested that if p accurate eigenvectors are required, then the subspace q should contain $2p$ vectors for $p \le 8$ and $(p + 8)$ vectors for $(p \ge 8$. Using a value of q higher than p gives improved convergence rates, especially to the higher modes. The actual eigenvalue and eigenvector determination is carried out in the q subspace. The theory is based on the knowledge that similarity transformations [5], are used to obtain the reduced $(q \times q)$ matrices $[\bar{K}]$ and $[\bar{G}]$.

Similarity transformations are of the type,

$$[P]^{-1}[A][P]$$

and, from eqn (6.125), it is evident that the eigenvalues of eqn (6.158),

$$([P]^{-1}[A][P])[P]^{-1}\{x\} = \lambda([P]^{-1}[B][P])[P]^{-1}\{x\} \tag{6.158}$$

are the same as for the original set. The eigenvectors $\{Y\}$ are related by the transformation,

$$\{Y\} = [P]^{-1}\{X\} \text{ or } \{X\} = [P]\{Y\} \tag{6.159}$$

Eqn (6.158) is written

$$[\bar{A}]\{Y\} = \lambda[\bar{B}]\{Y\} \tag{6.160}$$

where,

$$[\bar{A}] = [P]^{-1}[A][P], \quad [\bar{B}] = [P]^{-1}[B][P] \tag{6.161}$$

If the matrix $[P]$ is orthogonal, its transpose is also its inverse, so that, in this case,

$$[P]^T = [P]^{-1}$$

and

$$[P]^T[P] = [I] \tag{6.162}$$

However, this is the condition to be imposed on the iteration vectors in the subspace scheme. Hence, the eigenvalues of the subspace iteration are a subset of those of the total system. That is, if $[P]$ contains q vectors and eqn (6.162) still holds where $[I]$ is now $(q \times q)$, then $[\bar{A}]$ and $[\bar{B}]$ are $(q \times q)$ matrices, and the solution of eqn (6.160) produces q approximations to the corresponding eigenvalues of the complete system. The calculation of the eigenvalues of eqn (6.160) also provides a means of ensuring that the iteration vectors are orthogonal. Thus, with the notation used in the description of the inverse iteration, let the reduced problem be written

$$[\bar{K}][\bar{X}] = [\bar{G}][\bar{X}]\lfloor\Omega\rfloor \tag{6.163}$$

where $\lfloor\Omega\rfloor$ is a diagonal matrix of the eigenvalues and $[\bar{X}] = q \times q$ matrix whose columns are the eigenvalues of the system eqn (6.163). That is, if $[X_k]$ are a set of trial vectors of the complete system, then the vectors $[X'_k]$ given by the transformation in eqn (6.164) are orthogonal with respect to $[K]$,

$$[X_k^1] = [X_k][\bar{X}] \tag{6.164}$$

This is so because, $[X_k^1]^T[K][X_k^1] = [\bar{X}]^T[X_k]^T[K][X_k][\bar{X}] = [\bar{X}]^T[\bar{K}][\bar{X}] = [I]$

In the iteration cycle, $\{Y_k\}$ is calculated prior to $\{X_k^1\}$. This is because it is required to calculate $[\bar{G}]$. Hence, rather than form $\{X_k^1\}$ from eqn (6.164) and then $\{Y'_k\}$ from

$$\{Y'_k\} = [G]\{X_k^1\}$$

it is more efficient to form $\{Y'_k\}$ directly from $\{Y_k\}$ by

$$\{Y'_k\} = \{Y_k\}[\bar{X}]$$

Then $\{Y'_k\}$ is used to calculate $\{X_{k+1}\}$ from

$$[K]\{X_{k+1}\} = \{Y'_k\}$$

The flow chart of the subspace iteration scheme is set out below.

1.   Establish a set of starting vectors $\{Y'_0\}$.

2.   Calculate $\{X_1\}$, $\{Y_1\}$, $[\bar{K}_1]$ and $[\bar{G}_1]$ by the following steps

$$[K]\{X_1\} = \{Y'_0\}$$

$$\{Y_1\} = [G]\{X_1\}$$

$$[\bar{K}_1] = \{X_1\}^T\{Y'_0\}$$

$$[\bar{G}_1] = \{X_1\}^T\{Y_1\}$$

3.   Solve the system $[\bar{K}_1][\bar{X}] = \Omega[\bar{G}_1][\bar{X}]$ for all eigenpairs Ω and $[\bar{X}]$ using a generalized Jacobi method.

4.   Perform the transformation, $\{Y'_1\} = \{Y_1\}[\bar{X}]$, to orthogonalize the matrix $\{Y'_1\}$ with respect to $[K]$ and $[G]$.

5.   Repeat steps 2 to 4 with $\{Y'_1\}$ instead of $\{Y'_0\}$ until convergence is achieved.

6.7 NON–LINEAR TRANSFORMATION MATRICES

6.7.1   Plane frame elements

In the stiffness method of analysis of plane frames in small displacement theory, the transformation of displacement components of the joints required to calculate member deformations are

1. from global coordinates to local member coordinate axes;
2. from the local member axes to the straining modes of the member.

The first of these transformations is given by eqn (5.103). That is,

Displacements

$$\{r'\} = [L]\{r\}$$

Forces

$$\{R'\} = [L]\{R\} \tag{6.165}$$

where, in eqn (6.165), the transformation matrix $[L]$ is defined as

$$[L] = \begin{bmatrix} \cos\alpha & \sin\alpha & 0 \\ -\sin\alpha & \cos\alpha & 0 \\ 0 & 0 & 1 \end{bmatrix} \tag{6.166}$$

The second transformation, for small displacements, is, from eqn (5.100),

$$\begin{Bmatrix} \phi_i \\ \phi_j \\ \delta \end{Bmatrix} = \begin{bmatrix} 0 & \frac{1}{l} & 1 & 0 & -\frac{1}{l} & 0 \\ 0 & \frac{1}{l} & 0 & 0 & -\frac{1}{l} & 1 \\ -1 & 0 & 0 & 1 & 0 & 0 \end{bmatrix} \begin{Bmatrix} r'_i \\ r'_j \end{Bmatrix}$$ (6.167)

Eqn (6.165) is valid, regardless of the magnitude of the member displacements, if the original member position is always used as reference (and, as in two, dimensional problems, θ_z appears as a scalar, rather than as a vector). eqn (6.167), however, requires that the displacements be small for the approximation in order that the member chord rotation increment may be expressed as

$$\Delta\alpha \approx \frac{v_2 - v_1}{l} = \frac{r'_{2y} - r'_{1y}}{l}$$ (6.168)

Referring to Fig. 6.28, define the variables (u, w) as

$$u = r'_{jx} - r'_{ix}$$ (6.169a)

$$w = r'_{jy} - r'_{iy}$$ (6.169b)

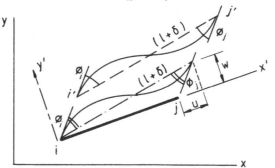

Fig. 6.28   Large displacements of member.

Then the deformation parameters (δ, ϕ_i, ϕ_j) are given

$$\delta = \sqrt{(l + u)^2 + w^2} - l$$ (6.170)

$$\phi_i = \theta_i - \tan^{-1}\frac{w}{l + u}$$ (6.171a)

$$\phi_j = \theta_j - \tan^{-1}\frac{w}{l + u}$$ (6.171b)

6.7.2   Incremental–force displacement relationships for a member

eqns (5.90) and (5.95) give, in turn, the non–linear relationship between the end rotations of a member and the corresponding moments for the 'exact' integration of the differential equation for the elastic curve of the member, on the one hand, and the value calculated, assuming a cubic approximation, on the other.

That is,

Transcendental functions, eqn (5.90),

$$\left\{\begin{array}{c} M_i \\ M_j \end{array}\right\} = \frac{2EI}{l} \left[\begin{array}{cc} 2\phi_3 & \phi_4 \\ \phi_4 & 2\phi_3 \end{array}\right] \left\{\begin{array}{c} \phi_i \\ \phi_j \end{array}\right\} \qquad (6.172)$$

Cubic approximation, eqn (6.55),

$$\left\{\begin{array}{c} M_i \\ M_j \end{array}\right\} = \left\{\frac{2EI}{l} \left[\begin{array}{cc} 2 & 1 \\ 1 & 2 \end{array}\right] + \frac{Pl}{30} \left[\begin{array}{cc} 4 & -1 \\ -1 & 4 \end{array}\right]\right\} \left\{\begin{array}{c} \phi_i \\ \phi_j \end{array}\right\} \qquad (6.173)$$

In these expressions, the change in bending stiffness is due to the presence of the axial force P. No account is taken, however, of the effect of the moments on the axial stiffness. To obtain the appropriate relationship, it is necessary first to obtain the change in length of the member due to the transverse deflection. For the cubic approximation this was calculated in eqn (5.93) as

$$\Delta l = \frac{l}{30}(2\phi_i^2 - \phi_i\phi_j + 2\phi_j^2) \qquad (6.174)$$

The corresponding expression for eqn (6.172) will be developed in the following section (eqn (6.177)). Using eqn (6.174), the distance between the nodes $i-j$ is $(l + \delta)$ (Fig. 6.26), and is given as the original length, less the effect of the bowing plus the extension of the member due to the axial force P. That is,

$$l - \Delta l + \frac{Pl}{AE} = l + \delta \qquad (6.175)$$

or, solving for P,

$$P = EA\left[\frac{\delta}{l} + \frac{\Delta l}{l}\right]$$

Writing this expression, using the value of Δl given in eqn (6.174),

$$P = EA\left[\frac{\delta}{l} + \frac{1}{30}(2\phi_i^2 - \phi_i\phi_j + 2\phi_j^2)\right] \qquad (6.176)$$

To obtain the incremental stiffness relationship it is necessary to differentiate eqns (6.173) and (6.176) partially, with respect to δ, ϕ_i and ϕ_j, in turn. Thence,

$$[\Delta S] = \left[\begin{array}{c} \Delta P \\ \Delta M_i \\ \Delta M_j \end{array}\right] = \left[\begin{array}{ccc} \dfrac{EA}{l} & \dfrac{EA}{30}(4\phi_i - \phi_j) & \dfrac{EA}{30}(-\phi_i + 4\phi_j) \\ \cdot & k_{\Delta ii} & k_{\Delta ij} \\ \cdot & symmetric & k_{\Delta jj} \end{array}\right] \left[\begin{array}{c} \Delta\delta \\ \Delta\phi_i \\ \Delta\phi_j \end{array}\right] \qquad (6.177)$$

In eqn (6.177),

$$k_{\Delta ii} = \frac{4EI}{l} + \frac{4EA}{30}\delta + \frac{EAl}{300}(8\phi_i^2 - 4\phi_i\phi_j + 3\phi_j^2) \qquad (6.178a)$$

$$k_{\Delta ij} = \frac{2EI}{l} - \frac{EA}{30}\delta + \frac{EAl}{300}(-2\phi_i^2 + 6\phi_i\phi_j - 2\phi_j^2) \tag{6.178b}$$

$$k_{\Delta jj} = \frac{4EI}{l} + \frac{4EA}{30}\delta + \frac{EAl}{300}(3\phi_i^2 - 4\phi_i\phi_j + 8\phi_j^2) \tag{6.178c}$$

The incremental stiffness eqn (6.177) will be written

$$\{\Delta S\} = [k_\Delta]\{\Delta v\} \tag{6.179}$$

The expression corresponding to eqn (6.177) containing the transcendental functions is obtained as follows. Let,

$$q = \frac{Pl^2}{\pi^2 EI} \; ; \quad \lambda^2 = \frac{Al^2}{I} \tag{6.180}$$

Firstly, write eqn (6.172) in a slightly modified form. That is,

$$\left\{\begin{matrix} M_i \\ M_j \end{matrix}\right\} = \frac{EI}{l}\begin{bmatrix} c_1 & c_2 \\ c_2 & c_1 \end{bmatrix}\left\{\begin{matrix} \phi_i \\ \phi_j \end{matrix}\right\} \tag{6.181}$$

$$P = EA\left[\frac{\delta}{l} - c_b\right] \tag{6.182}$$

In eqn (6.182), c_b is the effect of bowing of the member on the change in the distance between the nodes i and j. The functions c_1, c_2 and c_b are given as follows:

1.   q > 0 (compression) $\phi^2 = \pi^2 q$

$$c_1 = \frac{\phi(\sin\phi - \phi\cos\phi)}{2(1 - \cos\phi) - \phi\sin\phi} \tag{6.183a}$$

$$c_2 = \frac{\phi(\phi - \sin\phi)}{2(1 - \cos\phi) - \phi\sin\phi} \tag{6.183b}$$

2.   q < 0 (tension) $\psi^2 = -\pi^2 q$

$$c_1 = \frac{\psi(\psi\cosh\psi - \sinh\psi)}{2(1 - \cosh\psi) + \psi\sinh\psi} \tag{6.184a}$$

$$c_2 = \frac{\psi(\sinh\psi - \psi)}{2(1 - \cosh\psi) + \psi\sinh\psi} \tag{6.184b}$$

$$c_b = b_1(\phi_1 + \phi_2)^2 + b_2(\phi_1 - \phi_2)^2 \tag{6.185}$$

where

$$b_1 = \frac{(c_1 + c_2)(c_2 - 2)}{8\pi^2 q} \quad \text{and} \quad b_2 = \frac{c_2}{8(c_1 + c_2)} \tag{6.186}$$

Notice that in the calculation of c_b, the symmetric $(\phi_1 + \phi_2)$ and antisymmetric $(\phi_1 - \phi_2)$ modes of the bending deformation have been used. It can also be proven that

$$b_1 = \frac{-(c'_1 + c'_2)}{4\pi^2} \quad \text{and} \quad b_2 = \frac{-(c'_1 - c'_2)}{4\pi^2} \tag{6.187}$$

Thus, the derivatives c'_1, c'_2 are found to be

$$c'_1 = -2\pi^2(b_1 + b_2) ; \quad \text{and} \quad c'_2 = -2\pi^2(b_1 - b_2) \tag{6.188}$$

Note: $c'_1 = \dfrac{\partial c_1}{\partial q}$, etc. Finally, define G_1, G_2 and H as

$$G_1 = c'_1\phi_1 + c'_2\phi_2 \tag{6.189a}$$

$$G_2 = c'_2\phi_1 + c'_1\phi_2 \tag{6.189b}$$

$$H = \frac{\pi^2}{\lambda^2} + b'_1(\phi_1 + \phi_2)^2 + b'_2(\phi_1 - \phi_2)^2 \tag{6.189c}$$

Then, on differentiation of eqns (6.181) and (6.182) partially with respect to ϕ_1, ϕ_2 and δ, noting that

$$\frac{dF}{d\phi_1} = \frac{\partial F}{\partial \phi_1} + \frac{\partial F}{\partial q}\frac{\partial q}{\partial \phi_1}, \text{ etc.}$$

$$\left\{ \begin{array}{c} \Delta P \\ \Delta M_i \\ \Delta M_j \end{array} \right\} = \left[\begin{array}{ccc} \dfrac{\pi^2}{H} & \dfrac{G_1}{H} & \dfrac{G_2}{H} \\[2ex] \cdot & c_1 + \dfrac{G_1^2}{\pi^2 H} & c_2 + \dfrac{G_1 G_2}{\pi^2 H} \\[2ex] \cdot & symmetric & c_1 + \dfrac{G_1^2}{\pi^2 H} \end{array} \right] \left\{ \begin{array}{c} \Delta \delta \\ \Delta \phi_i \\ \Delta \phi_j \end{array} \right\} \tag{6.190}$$

The derivatives b'_1 and b'_2 are calculated to be

$$b'_1 = \frac{[-4\pi^2 b_1(c_2 - 2) - (c_1 + c_2)(b_1 - b_2)]\phi - (c_1 + c_2)(c_2 - 2)}{8\phi^2} \tag{6.191a}$$

and

$$b'_2 = \frac{\pi^2[-(b_1 - b_2)(c_1 + c_2) + 2c_2 b_1]}{4(c_1 + c_2)^2} \tag{6.191b}$$

for the case of q positive, with similar expressions for q negative (i.e. tensile).

6.8 LARGE DISPLACEMENT ANALYSIS OF FRAME STRUCTURES

6.8.1 Introduction

Thus far, the discussion of large displacement analysis has been confined, firstly, to tension structures which necessarily exhibit special characteristics and, secondly, to the calculation of bifurcation points of ideal structures whose performance can be assumed perfectly linear up to the bifurcation point. In this section the analysis techniques are extended so that the whole of the load/deflection path can be traced. This proves to be a daunting task, and it will be found that each structure presents its own particular difficulties. It is hoped, however, that the theory, together with the algorithm for tracing the equilibrium path presented here, is sufficiently general at least to suggest solutions to the problems encountered in most situations.

Because it is difficult to envisage situations in the postbifurcation range which are not three–dimensional in character, only the case of three dimensional structures will be considered, and then only those whose flexural properties can be specified in terms of the second moments of the area and the St Venant's torsion constant. Apart from the problems associated with tracing equilibrium paths, a further difficulty arises because joint rotations may become large and are no longer vector quantities, so that there is often a division of the topic into

1. large displacements with small to moderate rotations;

2. large displacements with large joint rotations.

An incremental approach presented by Oran [30] will be given to approximate the effects encountered in 2.

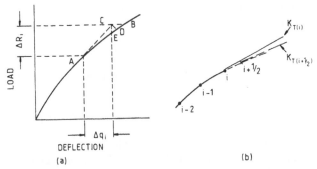

Fig. 6.29 Non–linear load deflection curve.

6.8.2 Basic theory for equilibrium iteration

As mentioned in section 6.1, there are several facets to the tracing of the non–linear equilibrium path. The simple situation is shown in Fig. 6.29(a). A load increment ΔR_i produces the linear step Δq_i, and in one dimensional space this represents the displacement from the equilibrium position A to C which is no longer on the equilibrium path.

The curve in Fig. 6.29 has been purposely drawn for a softening structure because the alternative stiffening structure has been discussed adequately in connection with tension nets. Examining Fig. 6.29(a), it is seen that it may be possible to return to the equilibrium path by iterating:

1. at constant load from C to B,

2. at constant displacement from C to E,

3. along a spherical path or arc in both load and displacements from C to D.

A simplistic view of Fig. 6.29 may lead to the conclusion that the shortest of the paths CB, CD and CE will provide the best strategy for iteration. If the path C to B is chosen, then the technique reduces to one of the Newton–Raphson iteration techniques. The basic incremental stiffness equations for the ith iteration are written

$$\{\Delta q_i\} = [K_{Ti}]^{-1}\{\Delta R_i\} = [K_{Ti}]^{-1}\psi\{q_{i-1}\} \tag{6.192}$$

and

$$\{q_i\} = \{q_{i-1}\} + \{\Delta q_i\} \tag{6.193}$$

$[K_{Ti}]$ is the tangent stiffness matrix and, hopefully, $\psi\{q_i\} \to 0$ as i increases. As written, eqn (6.192) is the Newton–Raphson method. If the evaluation of $[K_{Ti}]$ is considered expensive, a variation to eqn (6.192) is to use $[K_{T0}]$, the tangent stiffness at the beginning of the iteration cycle, i.e.

$$\{\Delta q_i\} = [K_{T0}]^{-1}\psi\{q_{i-1}\} \tag{6.194}$$

The calculation of $\{\Delta q_i\}$, then involves a back substitution only. The penalty is, however, that convergence will be much slower. Various methods have been proposed to obtain a secant method [31]. In the present applications, this secant Newton–Raphson method has not proved to be very beneficial, particularly in structures where wide variations in increments to individual terms of Δq_i occur. A simple and effective means of improving convergence is to use a technique of stiffness matrix extrapolation. The iteration points are shown, $i-2$, $i-1$, i, in Fig. 6.29(b). Evidently, from Fig. 6.29(b), if the stiffness at $(i + 1/2)$ is known, then a much better first iteration step is calculated towards B. That is,

$$\{\Delta q_i\} = [K_{T(i+1/2)}]^{-1}\{\Delta R_i\} \tag{6.195}$$

If equal load increments have been used, then the terms of $[K_T]$ are extrapolated through the expression,

$$[K_T]_{i+1/2} \approx 0.375[K_T]_{i-2} - 1.25[K_T]_{i-1} + 1.875[K_T]_i \tag{6.196}$$

In many instances iteration to the final value is achieved in two or three steps. Basic to the above theory is the calculation of both $[K_T]$ and $\psi\{q_i\}$. These problems have been discussed in relation to cable structures and frame members in sections 6.2 and 6.4. The above theory breaks down at limit points and bifurcation points, as shown in Fig. 6.30. Because $|K_T| \to 0$ at A or perhaps ∞ at C iterations will diverge at such points, when either zero or negative stiffness is encountered on the diagonal during the reduction of K_T. In the latter case the load should, of course, be decremented rather than incremented. A simple approach is to determine the onset of the zero pivot condition and, near A, simply stop iterations, allowing the solution to snap to B, where $[K_T]$ is again positive definite. Such a technique, then, fails to trace the path ACB, which may be important if some secondary bifurcation occurs in the region A-B.

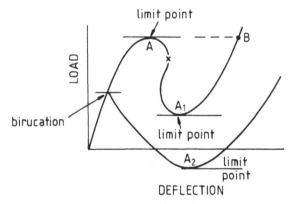

Fig. 6.30   Limit points.

Means are required to sense the approach of A. Two methods of doing this are by checking the magnitude of,

1.   the determinate of $[K_T]$;

2.   the current stiffness parameter.

The current stiffness parameter (CSP) [32] is simply an estimate of the structure stiffness in its current lowest eigenmode. Thus, if the load factor λ is close to an eigenvalue, the increment in the displacement vector approximates X_I the eigenmode. That is, let

$$\{\Delta q_i\} \approx \beta\{X_i\}$$

where β is a scalar multiplier. Then,

$$\{\Delta q_i\}^T[K_{Ti}]\{\Delta q_i\} \rightarrow \beta^2\{X_i\}^T\{K_{Ti}\}\{X_i\} \rightarrow \beta^2\alpha\{X_i\}^T\{X_i\} = \beta^2\alpha \qquad (6.197)$$

In eqn (6.197), $\{X_i^T X_i\}$ has been normalized to equal unity. Then, when the load is close to a limit point, $\beta^2 \rightarrow \infty$. Thus,

$$\text{CSP} = \frac{\{\Delta q_0\}^T[K_0]\{\Delta q_0\}}{\{\Delta q_i\}^T[K_{Ti}]\{\Delta q_i\}} \rightarrow 0 \qquad (6.198)$$

Thus, the plots of CSP and the stiffness matrix determinant vary with λ, as shown in Fig. 6.31.

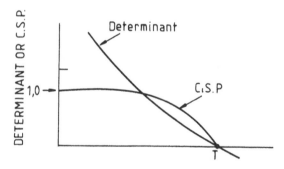

Fig. 6.31   Plots of CSP and determinant of stiffness matrix.

6.8.3   Tracing equilibrium paths – control of iteration steps

A superficially simple technique [33] to analyse snap–through is obtained by augmenting $[K_T]$ with artificial springs so that the stiffness of the modified structure is always positive definite. Such a system for a two–bar toggle mechanism is shown in Fig. 6.32.

The process of addition of the spring in one–dimensional load/deflection space is shown in Fig. 6.33 For a structure with many nodes and loads, many springs are required, and the augmented matrix must be singular. The spring matrix can be generated from the load vector $\{\theta\}$ as

$$[K_a] = \frac{k}{\theta^2}\{\theta\}\{\theta\}^T \qquad (6.199)$$

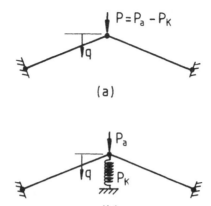

Fig. 6.32   Toggle mechanism–actual and modified structure.

In eqn (6.199), $\bar{\theta}$ is the norm of the load vector $\{\theta\}$, and k is a reference spring stiffness. The method is appealing for simple structures, but, on a little reflection, loses its attraction for complex structures because of the difficulty in handling bifurcation points as well as turning points.

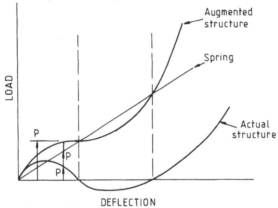

Fig. 6.33   Load deflection curves–actual and augmented structure.

6.8.3.1   Iteration on load and displacement increments

From Fig. 6.29(a) it is seen that the distances CE, CD from the linear increment step AC, can be considerably less that the distance CB at constant load. One of the first useful attempts at utilizing this fact was formulated by Batoz & Dhatt [34]. In their approach the tangent stiffness $[K_T]$ is used to calculate $\{\Delta q_r\}$, the incremental displacement due to the residual forces $\{\psi(q_{i-1})\}$, and also $\{\Delta q_e\}$, which is that displacement increment produced by the external load vector $\{Q\}$. That is,

$$\{\Delta q_r\} = [K_T]^{-1}\{\psi(q_{i-1})\} \tag{6.200}$$

and

$$\{\Delta q_e\} = [K_T]^{-1}\{Q\} \tag{6.201}$$

The actual applied load increment is $\Delta\lambda\{Q\}$, and $\Delta\lambda$ is evaluated by constraining the displacement increment at a particular degree of freedom k to be equal to $\overline{\delta}$. Thus,

$$\overline{\delta} = \Delta q_r^k + \Delta\lambda q_e^k \tag{6.202}$$

The value of $\Delta\lambda$ is given by

$$\Delta\lambda = \frac{\overline{\delta} - \Delta q_r^k}{\Delta q_e^k} \tag{6.203}$$

The increment of the displacement vector is then

$$\{\Delta q_i\} = \{\Delta q_r\} + \Delta\lambda\{\Delta q_e\} \tag{6.204}$$

and

$$\{q_i\} = \{q_{i-1}\} + \{\Delta q_i\} \tag{6.205}$$

Eqns (6.202) and (6.203) apply for the first iteration cycle, and if the iteration has converged at the $(i-1)$th load step, $\Delta q_r^k = 0$. For subsequent cycles, $\overline{\delta} = 0$, since the chosen point has fixed displacements during the equilibrium iterations. Evidently, this approach follows the path CE in Fig. 6.29(a). This algorithm works well if for the chosen degree of freedom, the load deflection path, is monotonically increasing. However, if the load deflection path exhibits snap–back characteristics, as illustrated by point C in Fig. 6.34, the displacement control method will diverge. The solution is to locate another point for which the equilibrium path is monotonically increasing.

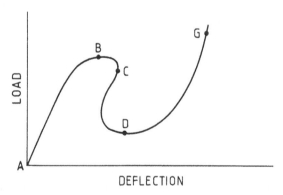

Fig. 6.34   Load deflection path with snap–back characteristics.

6.8.3.2   Iteration on load and displacements

The difficulty encountered in the Batoz and Dhatt algorithm suggests that instead of using the displacement at a single point as the control on the iteration on the load vector, it may be better to control the norm of the displacement vector in the same way. The method was suggested by Riks [35], Ramm [36] and Crisfield [37]. That is, the constraint equation employed is written

$$\{\Delta q\}^T\{\Delta q\} = \Delta l^2 \tag{6.206}$$

Thus, Δl is the magnitude of the multi–degree of freedom vector $\{\Delta q\}$. It may be expected that the use of this constraint equation should have a smoothing effect and thus help avoid some of the difficulties encountered in the Batoz and Dhatt method. The process now follows the same steps as in eqns (6.200) and (6.201). That is, at the ith iteration,

$$\{\Delta q_e^i\} = [K_T]^{-1}\{Q\} \tag{6.207}$$

and

$$\{\Delta q_r^i\} = [K_T]^{-1}\psi\{(q)^{i-1}\} \tag{6.208}$$

If $[K_T]$ is constant, eqn (6.207) is evaluated only once at the beginning of the load step. The actual iterative displacement is

$$\{\Delta q^i\} = \Delta\lambda^i\{\Delta q_e^i\} + \{\Delta_r^i\} \tag{6.209}$$

The total of all increments up to the ith iteration is, thus,

$$\{\delta q^i\} = \{\delta q^{i-1}\} + \{\Delta q^i\} \tag{6.210}$$

The constraint on the norm of the displacement vector is expressed as

$$\{\delta q^i\}^T\{\delta q^i\} = \Delta l^2 \tag{6.211}$$

That is, substituting eqn (6.210) into eqn (6.211),

$$[\{\delta q^{i-1}\} + \{\Delta q^i\}]^T[\{\delta q^{i-1}\} + \{\Delta q^i\}] = \Delta l^2 \tag{6.212}$$

The value of the load multiplier $\Delta\lambda^2$ in eqn (6.209) is now determined by substitution for $\{\Delta q^i\}$ in eqn (6.212). The constraint equation thus reduces to

$$a_1\lambda^2 + a_2\lambda + a_3 = 0 \tag{6.213}$$

where

$$a_1 = \{\Delta q_e^i\}^T\{\Delta q_e^i\} \tag{6.214}$$

$$a_2 = 2[\{\delta q^{i-1}\} + \{\Delta q_r^i\}]^T\{\Delta q_e^i\} \tag{6.215}$$

$$a_3 = [\{\delta q^{i-1}\} + \{\Delta q_r^i\}]^T[\{\delta q^{i-1}\} + \{\Delta q_r^{i-1}\}] - \Delta l^2 \tag{6.216}$$

There will be, in general, two real roots of eqn (6.213). The correct choice of $\Delta\lambda$ which will avoid doubling back on the solution path satisfies the condition,

$$\{\delta q^i\}^T\{\delta q^{i-1}\} > 0 \tag{6.217}$$

This expresses the fact that the projection of $\{\delta q^i\}$ on $\{\delta q^{i-1}\}$ should be positive. In the event that both roots satisfy eqn (6.217), the required root is the one nearest the linear solution,

$$\Delta\lambda = -\frac{a_3}{a_2} \tag{6.218}$$

The arc length strategy, as the above process is now called, is robust, but will fail to converge if eqn (6.213) has only imaginary roots. That is, if

$$a_2 - 4a_1a_3 < 0 \qquad\qquad (6.219)$$

This problem will be discussed in more detail subsequently. Eqs (6.213) or (6.218) determine the magnitude of the load increment. They do not, however, determine the sign of $\Delta\lambda$; that is, if the load is to be incremented or decremented. A load deflection curve with limit point and snap back phenomena in the unloading path is shown in Fig. 6.34. Let $\Delta\lambda$, $\Delta\delta$, and ΔW be the increments in the load factor λ, δ some measure of displacements, and W the work done. Then, for an increment in the path A to B,

$$\Delta\lambda = +ve, \quad \Delta\delta = +ve, \quad \Delta W = +ve$$

Also the determinant of $[K_T]$ is positive. Proceeding from B, on applying a positive load increment,

$$\Delta\lambda = +ve, \quad \Delta\delta = -ve, \quad \Delta W = -ve$$

Now, both the sign of the determinant of $[K_T]$, as measured by the sign count of negatives on the diagonal, and the sign of the incremental work have changed from those of the previous increment, and hence the actual values to be used are

$$\Delta\lambda = -ve, \quad \Delta\delta = +ve.$$

and the solution advances along the load path. At the increment from C to D, a negative incremental load was initially applied ($\Delta\lambda$ remaining -ve in the range B to C). Thence, in passing C,

$$\Delta\lambda = -ve, \quad \Delta\delta = -ve, \quad \Delta W = +ve. \qquad\qquad (6.220)$$

The sign of ΔW is different from that obtained for the previous increment. However, it will be found that the sign count of negatives on the diagonal of the reduced matrix will be the same as on that from B to C. Hence, the sign of $\Delta\lambda$ will not be changed. Similarly, past the lower limit point D, the stiffness matrix has again become positive definite and $\Delta\lambda$ must now be positive.

The above criterion is robust and serves many situations well. However, special attention must be given to the situation in eqn (6.213) when imaginary roots occur. Better control is achieved by reducing the arc length $\Delta\lambda$ and the arbitrary external load P proportionally. Thus, the imaginary roots occur at the jth increment, the tangent stiffness matrix is recomputed at the $(j-1)$th increment from the stored relevant parameters. Both the arc length and the external load increment are reduced by half and the iteration continued. Once a solution is determined, the arc length and the external load are reset to their original values. This modification has proved to be a more stable, efficient and automatic strategy. Even with all the above mentioned devices to improve convergence, it will not be found an easy task to trace the complete load deflection path. Oscillating solutions still occur at points just before limit points. When the pivots of the factorized tangent stiffness matrix are examined, it is found that multiple negatives occur, indicating the possible presence of an eigenvalue cluster. When a point where the solution oscillates is reached, it is necessary to revert to an imperfect approach. The previously converged load point is used, and the eigenvalues and eigenvectors calculated from this point. These correspond to the multiple negative pivots on the diagonal of the reduced stiffness equations. It is then necessary to activate some or all of the relevant eigenmodes. This is achieved by applying small forces at the appropriate nodes and directions. An example will be given when such conditions occur.

6.8.3.3   The variable arc length method

The idea of a variable arc length has been discussed in section 6.8.3.2. In this technique both the load reference vector and the length Δl have been varied. In the present context a variable length is achieved by relating the step length to the number of iterations used in the previous increment, $I^{(i-1)}$, to a desired value $I^{(d)}$. Following Ramm [36],

$$\Delta l^{(i)} = \left\{ \frac{I^{(d)}}{I^{(i-1)}} \right\}^{1/2} \Delta l^{(i-1)} \tag{6.221}$$

Where $\Delta l^{(i)}$ is the predicted increment and $\Delta l^{(i-1)}$ is the step length used in the previous increment. An alternative to this scheme can be obtained by utilizing the current stiffness parameter as the indication of the non–linearity of the system. Thence,

$$\Delta l^{(i)} = \frac{\Delta S_P^{(d)}}{\Delta S_P^{(i-1)}} \Delta l^{(i-1)} \tag{6.222}$$

In this equation, $\Delta S_P^{(i-1)} = S_P^{(i)} - S_P^{(i-1)}$, and $\Delta S_P^{(d)}$ is a desired change in S_P per increment. It is noted in passing that these methods appear to have beneficial effects when applied to simple problems. However, in more complex situations with many nodes and members, they have not always proved satisfactory in overcoming the problems outlined above.

6.8.4   Application of theory to large displacement analysis of 3–D frames

It is necessary first to develop the tangent stiffness expressions for the 3–D frame member in a manner similar to that for the two–dimensional frame in section 6.4. The basic member force and displacement quantities are shown in vector form in Fig. 6.35.

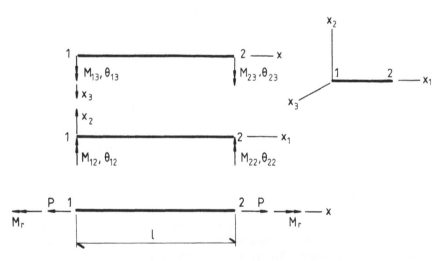

Fig. 6.35   Basic member forces–three dimensions.

The coordinate axes shown in Fig. 6.35 are uch that OX_1 coincides with the chord connecting the ends 1–2 of the member, and the OX_2, OX_3 axes are the principal axes of the cross section. For this situation the basic member force/distortion relationship for bending moments is expressed in eqn (6.223).

$$
\begin{Bmatrix} M_{13} \\ M_{23} \\ M_{12} \\ M_{22} \end{Bmatrix} = \frac{2E}{l} \begin{bmatrix} \begin{bmatrix} 2I_3 & I_3 & \cdot & \cdot \\ I_3 & 2I_3 & \cdot & \cdot \\ \cdot & \cdot & 2I_2 & I_2 \\ \cdot & \cdot & I_2 & 2I_2 \end{bmatrix} + \frac{Pl}{30} \begin{bmatrix} 4 & -1 & \cdot & \cdot \\ -1 & 4 & \cdot & \cdot \\ \cdot & \cdot & 4 & -1 \\ \cdot & \cdot & -1 & 4 \end{bmatrix} \end{bmatrix} \begin{Bmatrix} \theta_{13} \\ \theta_{23} \\ \theta_{12} \\ \theta_{22} \end{Bmatrix}
\tag{6.223}
$$

In eqn (6.223), I_2, I_3 are the second moments of area about the X_3, X_2 principal axes, respectively. The axial shortening due to bowing, assumed to be in the principal planes and to be calculated independently in each plane, is

$$
u_b = \frac{l}{30}\{(2\theta_{12}^2 - \theta_{12}\theta_{22} + 2\theta_{22}^2) + (2\theta_{13}^2 - \theta_{13}\theta_{23} + 2\theta_{23}^2)\}
\tag{6.224}
$$

Thence,

$$
P = \frac{EA}{l}(e + u_b)
\tag{6.225}
$$

where the distance between the nodes *1* and *2* after deformation is expressed as

$$
e = u_{21} - u_{11}
\tag{6.226}
$$

Finally, the torsion stiffness expressed simply in terms of J the St Venant's torsion constant is given by

$$
M_T = \frac{GJ}{l}\theta_T
\tag{6.227}
$$

Partial differentiation of eqn (6.223) with respect to member deformations gives the incremental stiffness relationship,

$$
\begin{bmatrix} \Delta M_{13} \\ \Delta M_{23} \\ \Delta M_{12} \\ \Delta M_{22} \\ \Delta M_T \\ \Delta P \end{bmatrix} = \begin{bmatrix} k_{11} & k_{12} & k_{13} & k_{14} & k_{15} & k_{16} \\ \cdot & k_{22} & k_{23} & k_{24} & k_{25} & k_{26} \\ \cdot & \cdot & k_{33} & k_{34} & k_{35} & k_{36} \\ \cdot & \cdot & \cdot & k_{44} & k_{45} & k_{46} \\ \cdot & \cdot & symmetric & \cdot & k_{55} & k_{56} \\ \cdot & \cdot & \cdot & \cdot & \cdot & k_{66} \end{bmatrix} \begin{bmatrix} \Delta\theta_{13} \\ \Delta\theta_{23} \\ \Delta\theta_{12} \\ \Delta\theta_{22} \\ \Delta\theta_T \\ \Delta e \end{bmatrix}
\tag{6.228}
$$

That is,

$$
\{\Delta F^e\} = [k^e]\{\Delta\delta^e\}
\tag{6.229}
$$

In eqn (6.228) the terms of the $[k^e]$ matrix are defined as follows:

$$k_{11} = \frac{EAl}{900}(8\theta_{12}^2 - 4\theta_{12}\theta_{22} + 8\theta_{22}^2)$$

$$k_{12} = \frac{2EI_3}{l} - \frac{EAe}{30} - \frac{EAl}{300}(2\theta_{13}^2 - 6\theta_{13}\theta_{23} + 2\theta_{23}^2)\frac{-EAl}{900}(2\theta_{12}^2 - \theta_{12}\theta_{22} + 2\theta_{22}^2)$$

$$k_{13} = \frac{EAl}{900}(16\theta_{13}\theta_{12} - 4\theta_{13}\theta_{22} - 4\theta_{12}\theta_{23} + \theta_{23}\theta_{22})$$

$$k_{14} = \frac{EAl}{900}(-4\theta_{13}\theta_{12} + 16\theta_{13}\theta_{22} + \theta_{23}\theta_{12} - 4\theta_{23}\theta_{22})$$

$$k_{15} = 0$$

$$k_{16} = \frac{EA}{30}(4\theta_{13} - \theta_{23})$$

$$k_{22} = \frac{4EI_3}{l} + \frac{EAe}{30} + \frac{EAl}{300}(3\theta_{13}^2 - 4\theta_{13}\theta_{23} + 8\theta_{23}^2) + \frac{EAl}{900}(8\theta_{12}^2 - 4\theta_{12}\theta_{22} + 8\theta_{22}^2)$$

$$k_{23} = \frac{EAl}{900}(-4\theta_{12}\theta_{13} + \theta_{13}\theta_{22} + 16\theta_{23}\theta_{12} - 4\theta_{23}\theta_{22})$$

$$k_{24} = \frac{EAl}{900}(\theta_{13}\theta_{12} - 4\theta_{23}\theta_{12} + 16\theta_{23}\theta_{22})$$

$$k_{25} = 0$$

$$k_{26} = \frac{EA}{30}(-\theta_{13} + 4\theta_{23})$$

$$k_{33} = \frac{4EI_2}{l} = \frac{4EAe}{30} + \frac{EAl}{900}(8\theta_{13}^2 - 4\theta_{13}\theta_{23} + 8\theta_{23}^2) + \frac{EAl}{300}(8\theta_{12}^2 - 4\theta_{12}\theta_{22} + 3\theta_{22}^2)$$

$$k_{34} = \frac{2EI_2}{l} - \frac{EAe}{30} - \frac{EAl}{900}(2\theta_{13}^2 - \theta_{13}\theta_{23} + 2\theta_{23}^2) - \frac{EAl}{300}(2\theta_{1}2^2 - 6\theta_{12}\theta_{22} + 2\theta_{22}^2)$$

$$k_{35} = 0$$

$$k_{36} = \frac{EA}{30}(4\theta_{12} - \theta_{22})$$

$$k_{44} = \frac{4EI_2}{l} + \frac{EAe}{30} + \frac{EAl}{900}(8\theta_{13}^2 - 4\theta_{13}\theta_{23} + 8\theta_{23}^2) + \frac{EAl}{300}(3\theta_{12}^2 - 4\theta_{12}\theta_{22} + 8\theta_{22}^2)$$

$$k_{45} = 0 \qquad k_{46} = \frac{EA}{30}(-\theta_{12} + 4\theta_{22})$$

$$k_{55} = \frac{GJ}{l} \qquad k_{56} = 0$$

$$k_{66} = \frac{EA}{l}$$

6.8.5 Transformation of member forces/displacements to nodal forces/displacements

From geometry and equilibrium considerations, the relationships between the local nodal forces and member basic forces are

$$
\begin{bmatrix} \hat{F}_1 \\ \hat{F}_2 \\ \hat{F}_3 \\ \hat{F}_4 \\ \hat{F}_5 \\ \hat{F}_6 \\ \hat{F}_7 \\ \hat{F}_8 \\ \hat{F}_9 \\ \hat{F}_{10} \\ \hat{F}_{11} \\ \hat{F}_{12} \end{bmatrix}
=
\begin{bmatrix}
0 & 0 & 0 & 0 & 0 & -1 \\
\frac{1}{l} & \frac{1}{l} & 0 & 0 & 0 & 0 \\
0 & 0 & -\frac{1}{l} & -\frac{1}{l} & 0 & 0 \\
0 & 0 & 0 & 0 & -1 & 0 \\
0 & 0 & 1 & 0 & 0 & 0 \\
1 & 0 & 0 & 0 & 0 & 0 \\
0 & 0 & 0 & 0 & 0 & 1 \\
-\frac{1}{l} & -\frac{1}{l} & 0 & 0 & 0 & 0 \\
0 & 0 & \frac{1}{l} & \frac{1}{l} & 0 & 0 \\
0 & 0 & 0 & 0 & 1 & 0 \\
0 & 0 & 0 & 1 & 0 & 0 \\
0 & 1 & 0 & 0 & 0 & 0
\end{bmatrix}
\begin{bmatrix} M_{13} \\ M_{23} \\ M_{12} \\ M_{22} \\ M_T \\ P \end{bmatrix}
\qquad (6.230)
$$

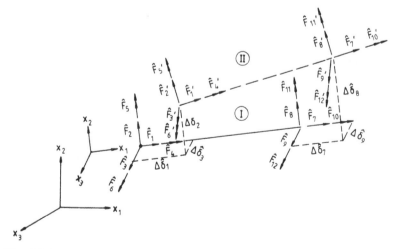

Fig. 6.36   Local and global coordinate force/displacement components.

That is,

$$\{\hat{F}\} = [M]\{F^e\} \tag{6.231}$$

The local nodal forces $\{\hat{F}\}$ and the incremental displacements $\{\Delta\hat{\delta}\}$ are defined in Fig. 6.36. Use of the contragredient principle results in the following relation:

$$\{\Delta\delta^e\} = [M]^T\{\Delta\hat{\delta}\} \tag{6.232}$$

The incremental local nodal forces $\{\Delta\hat{F}\}$ are obtained by differentiating equation (6.231) with respect to both $[M]$ and $\{F^e\}$. That is,

$$\{\Delta\hat{F}\} = [M]\{\Delta F^e\} + [\Delta M]\{F^e\} \tag{6.233}$$

It is seen that $[\Delta M]$ relates $\{\Delta\hat{F}\}$ to $\{F^e\}$ when the member has been given an increment of displacements. That is, the member has been rotated slightly and has suffered an extension, δ. Thus, for example, for small incremental displacements of the member $\Delta\delta_2$, $\Delta\delta_3$, $\Delta\delta_8$, $\Delta\delta_9$, as shown in Fig. 6.37

$$\rho_1 = 0 \; ; \quad \rho_2 = -\frac{(\Delta\delta_9 - \Delta\delta_3)}{l} \; ; \quad \rho_3 = \frac{(\Delta\delta_8 - \Delta\delta_2)}{l}$$

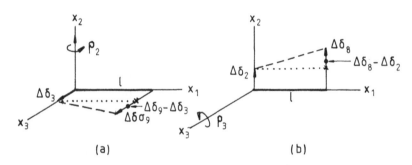

(a)   (b)

Fig. 6.37   Rigid–body rotations.

Thus, relative to the local coordinate axes, the displaced axes have the direction cosines,

$$[\rho] = \begin{bmatrix} 1 & -\rho_3 & \rho_2 \\ \rho_3 & 1 & 0 \\ -\rho_2 & 0 & 1 \end{bmatrix} \tag{6.234}$$

That is, if $\{\hat{F}\}$ components are unchanged in the rotated position, these same components will have changed in the initial member axes. That is,

$$\{F'\} = [N]\{F\} \; ; \quad [N] = \begin{bmatrix} [\rho]^T & \cdot & \cdot \\ \cdot & \cdot & \cdot \\ \cdot & \cdot & [\rho]^T & \cdot \\ \cdot & \cdot & \cdot \end{bmatrix} \tag{6.235}$$

The change in components is, thus,

$$\{\Delta \hat{F}'\} = \{F'\} - \{F\} \tag{6.236}$$

This gives

$$\{\Delta F'\} = [\Delta N]\{\Delta F\} \; ; \quad \text{with} \quad [\Delta \rho] = \begin{bmatrix} 0 & -\rho_3 & \rho_2 \\ \rho_3 & 0 & 0 \\ -\rho_2 & 0 & 0 \end{bmatrix} \tag{6.237}$$

Furthermore, it is assumed that if S_0 is the transverse shear, then

$$(S_0 + \Delta S_0)(l + e) = S_0 l$$

Thence,

$$\Delta S_0 = -\frac{S_0 e}{l} \tag{6.238}$$

where

$$e = \Delta \delta_7 - \Delta \delta_1 \tag{6.239}$$

Combining eqn (6.237) with the effect in eqn (6.238), define the transformation matrix $[\Delta \rho']$:

$$[\Delta \rho'] = \begin{bmatrix} 0 & -\rho_3 & \rho_2 \\ \rho_3 & -\dfrac{e}{l} & 0 \\ -\rho_2 & 0 & -\dfrac{e}{l} \end{bmatrix} \tag{6.240}$$

Then

$$\{\Delta F'\} = [\Delta N']\{F\} \tag{6.241}$$

Then, with,

$$[\Delta N'] = \begin{bmatrix} [\Delta \rho']^T & 0 & 0 & 0 \\ 0 & 0 & 0 & 0 \\ 0 & 0 & [\Delta \rho']^T & 0 \\ 0 & 0 & 0 & 0 \end{bmatrix}$$

it follows that

$$\{\Delta F'\} = [\Delta N'][M]\{F^e\} \tag{6.242}$$

The matrix $[\Delta M]$ in eqn (6.233) is given as

$$[\Delta M] = \begin{bmatrix} \bar{\rho} \\ 0 \\ -\bar{\rho} \\ 0 \end{bmatrix} \tag{6.243}$$

where

$$[\bar{\rho}] = \frac{1}{l^2} \begin{bmatrix} -l\rho_3 & -l\rho_3 & -l\rho_3 & -l\rho_2 & 0 & 0 \\ le & le & 0 & 0 & 0 & -l^2\rho_3 \\ 0 & 0 & l^2e & l^2 & 0 & l^2\rho_2 \end{bmatrix}$$

Now, because $[\Delta M]$ involves the displacement increments $\{\Delta\delta\}$, it follows that

$$[\Delta M]\{F^e\} \rightarrow [Q]\{\Delta\delta\} \tag{6.244}$$

where $[Q]$ takes the more usual form,

$$[Q] = \begin{bmatrix} Q_{11} & Q_{12} \\ Q_{21} & Q_{22} \end{bmatrix} \tag{6.245}$$

The submatrices on the right–hand side of eqn (6.245) are defined, with $SM\,1 = \dfrac{M_{13} + M_{23}}{l^2}$ and $SM\,2 = \dfrac{M_{12} + M_{22}}{l^2}$

$$[Q_{11}] = [Q_{22}] = \begin{bmatrix} 0 & SM\,1 & -SM\,2 & 0 & 0 \\ \cdot & P/l & 0 & 0 & 0 \\ \cdot & \cdot & P/l & 0 & 0 \\ \cdot & symmetric & \cdot & 0 & 0 \\ \cdot & \cdot & \cdot & \cdot & 0 \\ \cdot & \cdot & \cdot & \cdot & \cdot & 0 \end{bmatrix} \tag{6.246}$$

and with $SM\,3 = \dfrac{M_{12} + M_{23}}{l^2}$

$$[Q_{12}] = -[Q_{21}]^T = \begin{bmatrix} 0 & -SM\,3 & SM\,2 & 0 & 0 & 0 \\ -SM\,1 & -P/l & 0 & 0 & 0 & 0 \\ SM\,2 & 0 & -P/l & 0 & 0 & 0 \\ 0 & 0 & 0 & 0 & 0 & 0 \\ 0 & 0 & 0 & 0 & 0 & 0 \\ 0 & 0 & 0 & 0 & 0 & 0 \end{bmatrix} \tag{6.247}$$

Therefore, substituting eqns (6.229), (6.232) and (6.244) into eqn (6.233), the incremental equilibrium equations in the convected coordinate system are

$$\{\Delta F\} = [M][k^e]\{\Delta\delta^e\} + [Q]\{\Delta\delta\} = [M][k^e][M]^T\{\Delta\delta\} + [Q]\{\Delta\delta\}$$

$$= ([M][k^e][M]^T + [Q])\{\Delta\delta\} = [k]\{\Delta\delta\} \tag{6.248}$$

That is, the tangent stiffness relationship in local coordinates is

$$\{\Delta F\} = [K_T]\{\Delta\delta\} \tag{6.249}$$

The transformation from local to global components is now straightforward. To this end, define $[L]$ as in eqn (5.127) and form $[L_D]$,

$$[L_D] = \lceil [L] \rfloor_4$$

where $[L]$ is the direction cosine matrix of the local axes:

$$\{\Delta F^G\} = [L_D][K_T][L_D]^T\{\Delta\delta^G\} \tag{6.250}$$

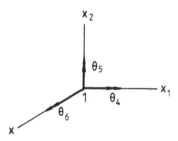

Fig. 6.38   Global rotation components.

It now remains to modify the analysis for large rotations. Consider end *1* of the member, Fig. 6.38. The increment in the rotation at the joint, is expressed as:

$$\Delta \begin{Bmatrix} \theta_4 \\ \theta_5 \\ \theta_6 \end{Bmatrix} = \begin{Bmatrix} \Delta\omega_1 \\ \Delta\omega_2 \\ \Delta\omega_3 \end{Bmatrix} \tag{6.251}$$

A set of axes (X_1, X_2, X_3) is rigidly attached to a joint and initially these coincide with the global axes, but change position, however, as the joint rotates. The incremental joint rotation matrix $[\Delta\omega]$ is written

$$[\Delta\omega] = \begin{bmatrix} 0 & -\Delta\omega_3 & \Delta\omega_2 \\ \Delta\omega_3 & 0 & -\Delta\omega_1 \\ -\Delta\omega_2 & \Delta\omega_1 & 0 \end{bmatrix} \tag{6.252}$$

Eqn (6.257) for the matrix $[r]$ for principal axes direction cosines is developed as follows. Consider the incremental rotations, and because these are still small, superposition still applies. Hence apply the three rotations independently, as shown in Fig. 6.39(a) to (c). From Fig. 6.39, (X_1, X_2, X_3) axes $\rightarrow (X'_1, X'_2, X_3')$ axes such that

$$X'_1 \approx (1 \quad \Delta\omega_3 \quad -\Delta\omega_2)$$

$$X'_2 \approx (-\Delta\omega_3 \quad 1 \quad \Delta\omega_1)$$

$$X'_3 \approx (\Delta\omega_2 \quad -\Delta\omega_1 \quad 1)$$

These give

$$[\Delta\omega]^T = \begin{bmatrix} 0 & \Delta\omega_3 & -\Delta\omega_2 \\ -\Delta\omega_3 & 0 & \Delta\omega_1 \\ \Delta\omega_2 & -\Delta\omega_1 & 0 \end{bmatrix} \qquad (6.253)$$

This matrix is now used to transform components of a vector in a set of axes, $[\alpha_-]$ already rotated from the global set.

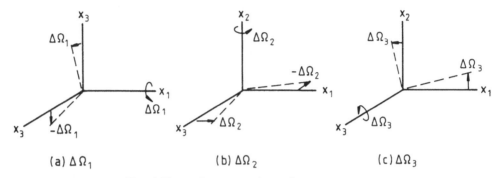

(a) $\Delta\Omega_1$ (b) $\Delta\Omega_2$ (c) $\Delta\Omega_3$

Fig. 6.39   Incremental rotation components.

The change in the global components of these axes themselves will be

$$[\Delta\alpha] = [\Delta\omega][\alpha_-]$$

giving the new position direction cosines,

$$[\alpha_+] = [\alpha_-] + [\Delta\alpha] \qquad (6.254)$$

The next concept concerns the principal axes of the cross section. The initial orientation of the member axes $[r] = [L]^T => [r_{-1}]$. These axes maintain a constant relationship to the $[\alpha_+]$ axes, except as modified below. The global components of $[r_-]$ for the end i, $(i = 1$ or $2)$, will now be different. That is, the principal axes no longer coincide with the member axes. Define the 'end section orientation matrices' $[p^{(1)}]$, $[p^{(2)}]$ such that

$$[p^{(i)}] = [\alpha^{(i)}][r_0] \qquad (6.255)$$

That is, $[p^{(i)}]$ are global components of the principal axes for the matrix $[r]$ in the deformed position, and the first column must represent the direction cosines of the chord. Thus, $[p^{(i)}]^T \{r_1\}$ are the direction cosines of X_1, X_2 in rotated member coordinates. Now, for increments $\{r_1\}$, close to $1'$ axis, so that it may be considered to be obtained from the (X'_1) axes by a rotation about X'_2, X'_3 (Fig. 6.40). That is,

$$[p^{(i)}]^T \{r_1\} = \left\{ \begin{array}{c} 1 \\ \theta_{13} \\ -\theta_{12} \end{array} \right\} \qquad (6.256)$$

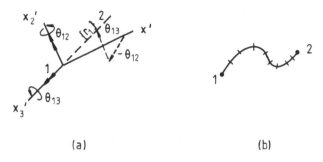

(a)   (b)

Fig. 6.40   Rotation about 2′ and 3′ axes.

Define the rotation matrix $[e_1]$:

$$[e_1] = \begin{bmatrix} 1 & -\theta_{13} & \theta_{12} \\ \theta_{13} & 1 & 0 \\ -\theta_{12} & 0 & 1 \end{bmatrix} \tag{6.257}$$

Then the new average $[r]$ is given by

$$[r] = \frac{1}{2}\{[p^{(1)}][e_1] + [p^{(2)}][e_2]\} \tag{6.258}$$

This theory will apply if the member is not excessively distorted. For example, as is shown in Fig. 6.39(b), it will be necessary to subdivide into shorter segments. Finally, then,

$$[L] = [r]^T \tag{6.259}$$

6.8.6   Examples of large displacement analyses

In this section two–examples are given. The first is a two hinged deep arch, which illustrates the point that unless a plane frame is restrained to prevent out–of–plane displacements, these mode shapes can and do occur. The second example, a shallow trussed dome, exhibits many of the problems mentioned above, before complete snap–through has taken place.

Example 6.9 *Two–hinged deep arch*

The hinged arch is shown in Fig. 6.40, and, is analysed using eight frame elements. The complete structure has been analysed because of the obvious possibility of unsymmetric bifurcation modes. In this example, both the perfect and imperfect approaches have been used to trace the primary and secondary bifurcation paths. The results from the perfect approach are shown in Fig. 6.41, and are labelled (1). On tracing the primary bifurcation path, the number of negatives on the diagonal in the factorized tangent stiffness matrix changes from zero to two at the limit point. It is noted in the analysis that if the determinant criterion only is employed in determining the direction of the arc length increment, the solution strategy will fail to converge and will not be able to continue the rising fundamental path. The incremental work concept overcomes this problem. After the occurrence of the symmetric snap–through point (or limit point), the number of negative pivots in the reduced stiffness matrix is three. The two negative pivots in the rising fundamental path show that two secondary path branches are possible.

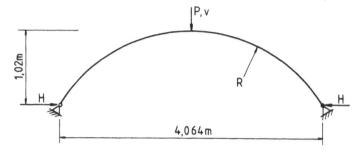

Radius, R = 2,54 m
Thickness, t = 25,4 mm
Area, A = 645 mm²
E = 0,0688 m Pa

Fig. 6.41   Deep arch–dimensions, loads and properties.

The out–of–plane secondary path, labelled (2, 3, 4 or 5) in Fig. 6.41, was obtained by including a small imperfection in geometry (2), or, applying a small force in the out–of–plane direction. The direction of this force (or the imperfection) can be obtained by an eigenvalue analysis just below the bifurcation point, and, using the resulting mode shape as an indication of the secondary path to be activated. Finally, the in–plane non–linear asymmetric buckling path was obtained by applying an out–of–plane moment at the crown. The curves so obtained for various values of this moment are shown in Fig. 6.41 nos. (6, 7, 8).

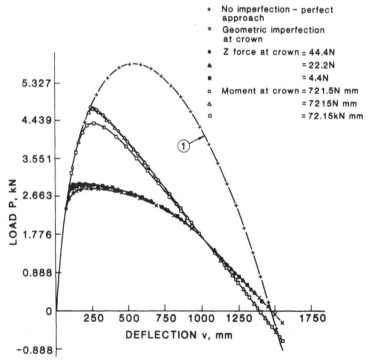

Fig. 6.42   Load deflection curves for two–hinged deep arch.

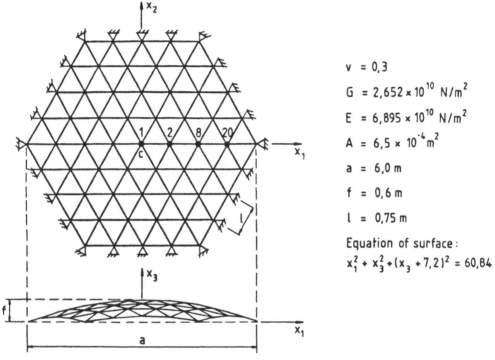

$v = 0,3$

$G = 2,652 \times 10^{10} \text{ N/m}^2$

$E = 6,895 \times 10^{10} \text{ N/m}^2$

$A = 6,5 \times 10^{-4} \text{m}^2$

$a = 6,0 \text{ m}$

$f = 0,6 \text{ m}$

$l = 0,75 \text{ m}$

Equation of surface:
$$x_1^2 + x_3^2 + (x_3 + 7,2)^2 = 60,84$$

Fig. 6.43   Geometry of shallow geodesic dome.

Example 6.10 *Shallow geodesic dome*

The dome topology and geometry are shown in Fig. 6.43. This dome with a 1:10 rise to span ratio has been analysed by Noor and Peters [38] as a truss structure. In the present analysis the dome is idealized, with 156 frame elements with a concentrated load acting at its apex. When analysing this structure, the various numerical difficulties mentioned in the previous section were encountered. Oscillating solutions occurred either just before or just after the limit points. The load/vertical deflection curves for nodes *1*, *2*, *8* and *20* are plotted in Fig. 6.44, (a) to (d). The displacement pattern is symmetrical about the apex and, multiple snap–through phenomena occur as the various symmetric rings of the structure snap–through. Convergence difficulties occurred at or near points such as *1*, *2* and *3* in Fig. 6.44(a). These were relatively easy to overcome using the technique in the arch length strategy of simply decreasing the arc length and initial load step size. However, at points such as near *5* and *6*, difficulties arose which were unresolved by this strategy. It was then necessary to revert to the imperfect approach mentioned in section 6.8.4, and to calculate the eigenvalues at the geometric configuration just prior to the point of difficulty. This, of course, necessitates the continuous storage of the current stiffness as the solution proceeds. Then the appropriate mode can be activated with a small force imperfection, and the solution continues without oscillation. It was found necessary to determine a sign reversal for the arc length increment based on both the change in sign of the determinant and incremental work sign change from the previous increment. The deflected shapes of the structure at points *1*, *2*, *3*, *4*, *5*, *6* and *7* of the load/deflection curve, (Fig. 6.44(a)), are shown in Fig. 6.45(a) to (g). It is seen that the major computational difficulty associated with this structure occurs with the snap–through of the outer ring of nodes.

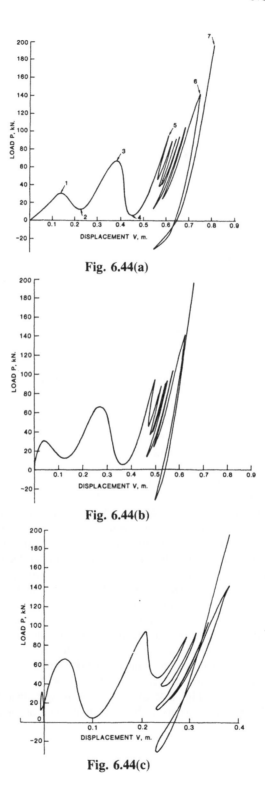

Fig. 6.44(a)

Fig. 6.44(b)

Fig. 6.44(c)

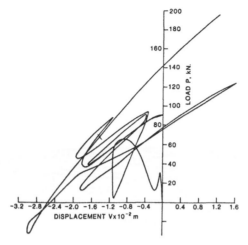

Fig. 6.44(d)   Load deflection curves.

Fig. 6.45(a)

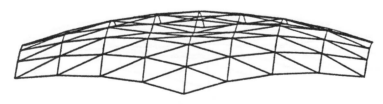

Fig. 6.45(b)

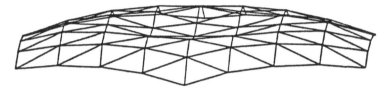

Fig. 6.45(c)

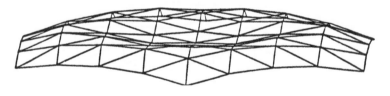

Fig. 6.45(d)

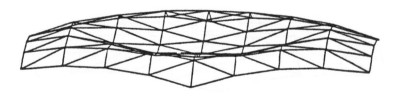

Fig. 6.45(e)

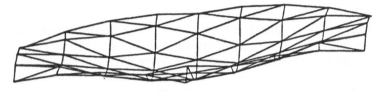

Fig. 6.45(f)

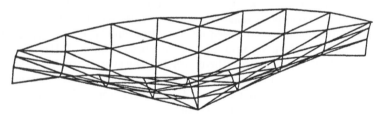

Fig. 6.45(g)   Structure deflected shapes.

CHAPTER 7

The solid continuum

7.1 PLANE ELASTICITY

7.1.1 Introduction

Thus far, the development of the theory in Chapters 4 to 6 has been confined to line element systems (trusses, frames and grids). These are restricted classes of structure in which, for the purpose of simplicity, the assumption of beam bending, that is, plane sections remain plane in the deformed state, has been invoked, so that the generalized forces of axial load, moment and torque can be used to describe the load deformation characteristics of the member. This is essentially a strength–of–materials approach. In their immediate use (that is, for elementary small deflection analyses), the assumption appears to be a simplification; however, in the event of complications in the spatial and deflection geometry, it may have to be modified extensively, or the general theory of continuum mechanics may have to be resorted to. It was shown, for example, that if warping is included in the behaviour of cross sections of thin walled members (section 6.7), the concept of the bimoment as a generalized force becomes necessary. The present chapter develops the general theory for the approximate solution to elasticity problems involving either plane stress, plane strain, axisymmetry or general three–dimensional elasticity. The St Venant's torsion problem is also discussed as a useful special class of restricted elasticity problem. The chapter concludes with a brief introduction to heat transfer because the topic is intimately connected with thermal stress analysis. This, in turn, introduces the concept of finite elements in space and time, that is, transient analysis. The basic mathematical preliminaries have been given in detail in Chapters 1 and 2, and in Chapter 3 the various approaches to the finite element method have been delineated. Herein, only the compatible displacement models will be studied. There will be, however, a parallel development of the conventional theory (using the contragredient principle and orthogonal interpolation functions (section 3.3), and the natural mode technique (section 3.4). Elements can be broadly classified (in so far as computations are concerned), into the categories of

1. those which can have their stiffness matrices calculated explicitly;
2. those which require numerical integration.

These various domains were illustrated in some detail in sections 1.4.4 and 1.4.6. It is seen that it is only for simple domains, that is, straight–sided triangles, tetrahedra, and rectangles, that the integration of the interpolation polynomials is of sufficient simplicity to warrant pursuing the task of explicit integration. In such cases the integrals of the products of the interpolation polynomials are given in eqns (1.152) to (1.154). The fundamental domains to be studied are shown in Chapter 1, Fig. 1.7.

7.1.2 Plane stress–plane strain and axisymmetry

These three situations are grouped together because, with minor modifications, software written for one situation can be adapted for the other two. The three physical representations are shown in Fig. 7.1.

The first is a thin disc loaded in its plane; the second, the slice of an infinitely long prism whose load conditions do not vary in the longitudinal direction. The axisymmetric case, in Fig. 7.1(c), is such that there is no stress or strain variation in the circumferential direction, and strains in this direction are then related to the radial displacements (eqn (7.1)). The problem remains as to how to incorporate the volume of revolution in the integration.

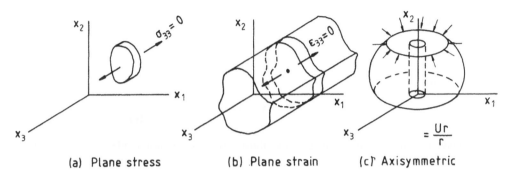

(a) Plane stress   (b) Plane strain   (c) Axisymmetric

Fig. 7.1   Plane stress–plane strain–axisymmetry.

It is found that in all situations certain recurring themes arise, namely

1.   the expression for the constitutive law for the material;

2.   the integration of the work equations to generate the element stiffness matrix relating nodal quantities, here generali:ed forces and nodal displacements;

3.   the calculation of statica.lly equivalent nodal forces, for applied surface pressures and body forces and the inclusion o.f thermal or shrinkage initial stresses.

7.1.3 Constitutive laws

The constitutive laws were given in section 2.7. It has been shown there that for the three situations shown in Fig. 7.1, the relevant tensor quantities for stress and strain are as follows:

	plane stress	plane strain	axisymmetry
stress:	$\begin{bmatrix} \sigma_{11} & \sigma_{12} & 0 \\ \sigma_{21} & \sigma_{22} & 0 \\ 0 & 0 & 0 \end{bmatrix}$	$\begin{bmatrix} \sigma_{11} & \sigma_{12} & 0 \\ \sigma_{21} & \sigma_{22} & 0 \\ 0 & 0 & \sigma_{33} \end{bmatrix}$	$\begin{bmatrix} \sigma_{11} & \sigma_{12} & 0 \\ \sigma_{21} & \sigma_{22} & 0 \\ 0 & 0 & \sigma_{\theta\theta} \end{bmatrix}$
strain:	$\begin{bmatrix} \varepsilon_{11} & \varepsilon_{12} & 0 \\ \varepsilon_{21} & \varepsilon_{22} & 0 \\ 0 & 0 & \varepsilon_{33} \end{bmatrix}$	$\begin{bmatrix} \varepsilon_{11} & \varepsilon_{12} & 0 \\ \varepsilon_{21} & \varepsilon_{22} & 0 \\ 0 & 0 & 0 \end{bmatrix}$	$\begin{bmatrix} \varepsilon_{11} & \varepsilon_{12} & 0 \\ \varepsilon_{21} & \varepsilon_{22} & 0 \\ 0 & 0 & \varepsilon_{\theta\theta} \end{bmatrix}$

It is seen that for the axisymmetric case, both $\sigma_{\theta\theta}$ and $\varepsilon_{\theta\theta}$ appear. However, $\varepsilon_{\theta\theta}$, is simply related to the radial displacements u_r, by the expression,

$$\varepsilon_{\theta\theta} = \frac{u_r}{r} \tag{7.1}$$

In eqn (7.1), r is the radial distance from the axis of symmetry. For both plane stress and strain, the constitutive law, relating stress and strain, is given (eqn (2.135)):

$$\begin{bmatrix} \sigma_{11} \\ \sigma_{22} \\ \sigma_{21} \end{bmatrix} = E_r \begin{bmatrix} \tau & \nu & 0 \\ \nu & \tau & 0 \\ 0 & 0 & \lambda \end{bmatrix} \begin{bmatrix} \varepsilon_{11} \\ \varepsilon_{22} \\ 2\varepsilon_{21} \end{bmatrix} \tag{7.2}$$

where, for plane stress,

$$\tau = 1, \quad \lambda = \frac{1 - \nu}{2}, \quad E_r = \frac{E}{1 - \nu^2} \tag{7.3}$$

and for plane strain,

$$\tau = 1 - \nu, \quad \lambda = 0.5 - \nu, \quad E_r = \frac{E}{(1 + \nu)(1 - 2\nu)} \tag{7.4}$$

The axisymmetric case must include the third direct stress–strain relationship, so that, from eqn (2.138),

$$\begin{bmatrix} \sigma_{rr} \\ \sigma_{22} \\ \sigma_{\theta\theta} \\ \sigma_{r2} \end{bmatrix} = E^* \begin{bmatrix} 1 - \nu & \nu & \nu & 0 \\ \nu & 1 - \nu & \nu & 0 \\ \nu & \nu & 1 - \nu & 0 \\ 0 & 0 & 0 & \frac{1 - 2\nu}{2} \end{bmatrix} \begin{bmatrix} \varepsilon_{rr} \\ \varepsilon_{22} \\ \varepsilon_{\theta\theta} \\ 2\varepsilon_{r2} \end{bmatrix} \tag{7.5}$$

In the plane stress (strain) material model, the description given in eqn (7.2) may not be adequate when anisotropic material properties are encountered. The axes associated with these properties may be inclined at an angle θ with the global X_1, X_2 axes. Eqn (7.2) can be written for the material constitutive relationship in the X'_1, X'_2 axes as

$$\begin{bmatrix} \sigma'_{11} \\ \sigma'_{22} \\ \sigma'_{21} \end{bmatrix} = \begin{bmatrix} c'_{11} & c'_{12} & c'_{13} \\ c'_{21} & c'_{22} & c'_{23} \\ c'_{31} & c'_{32} & c'_{33} \end{bmatrix} \begin{bmatrix} \varepsilon'_{11} \\ \varepsilon'_{22} \\ 2\varepsilon'_{12} \end{bmatrix} \tag{7.6}$$

That is, symbolically,

$$\{\sigma'\} = [C']\{\varepsilon'\} \tag{7.7}$$

The corresponding relationship in the global X_1, X_2 coordinates is obtained from eqns (2.30) and (2.42). From eqn (2.30), it can be shown that

$$\begin{bmatrix} \sigma_{11} \\ \sigma_{22} \\ \sigma_{12} \end{bmatrix} = \begin{bmatrix} c^2 & s^2 & -2cs \\ s^2 & c^2 & 2cs \\ cs & -cs & c^2 - s^2 \end{bmatrix} \begin{bmatrix} \sigma'_{11} \\ \sigma'_{22} \\ \sigma'_{12} \end{bmatrix} \tag{7.8}$$

That is,

$$\{\sigma\} = [T]\{\sigma\}' \tag{7.9}$$

and, furthermore, from eqn (2.113), for the strain components,

$$\{\varepsilon\}' = [T]^T\{\varepsilon\} \tag{7.10}$$

Substitution of eqn (7.10) in eqn (7.7) and premultiplication by $[T]$ gives

$$\{\sigma\} = [T][C'][T]^T\{\varepsilon\} \tag{7.11}$$

That is, the constitutive matrix in the X_1, X_2 global coordinate system is given by

$$[C] = [T][C'][T]^T \tag{7.12}$$

Presumably, the coefficients in the $[C']$ matrix will be found by experiment.

7.2.1 The triangle element family – straight sides

The two elements considered here in detail are the CST (constant strain triangle) and the LST (linear strain triangle). These have three and six nodes, respectively. They are shown in Fig. 7.2 together with the QST (quadratic strain triangle), which is given for reference only.

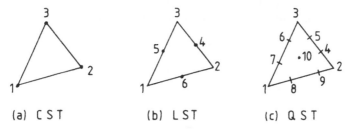

(a) C S T   (b) L S T   (c) Q S T

Fig. 7.2   Triangular elements.

The interpolation polynomials for the two elements are simply ϕ_1 and ϕ_2, as given in eqns (1.141) and (1.143), respectively. That is,

$$\{\phi_1\}^T = \{\zeta_1 \ \zeta_2 \ \zeta_3\} \tag{7.13a}$$

and

$$\{\phi_2\}^T = \{\zeta_1(2\zeta_1 - 1) \ \zeta_2(2\zeta_2 - 1) \ \zeta_3(2\zeta_3 - 1) \ 4\zeta_2\zeta_3 \ 4\zeta_3\zeta_1 \ 4\zeta_1\zeta_2\} \tag{7.13b}$$

The theory is developed generally and then applied, in turn, to both the CST and the LST. It is convenient for the integration of the interpolation functions to group nodal values of each variable (displacements in this case) together. Thus,

$$\{u\}^T = \{u_{1i}^T \ u_{2i}^T\}$$

is the vector of the nodal displacements, $i = 1$, to the number of nodes, and

$$\{\tilde{u}\}^T = \{\tilde{u}_1 \ \tilde{u}_2\}$$

is the displacement vector at any point within the triangle. Then, because $\tilde{u}_1$ and $\tilde{u}_2$ components are interpolated independently,

$$\{\tilde{u}\} = [\phi_u]\{u\} \tag{7.14}$$

where

$$[\phi_u] = \left[\phi_u^T\right]_2 \tag{7.15}$$

is a (2×2) supermatrix with the $\{\phi_u\}^T$ matrices as diagonal submatrix blocks. Because the strains are obtained from the displacement functions as first derivatives, they also can be interpolated in terms of their own nodal values, by polynomial functions of one degree lower than those for the displacements. That is, for strains, write

$$\{\varepsilon\} = [\phi_\varepsilon]\{\varepsilon_c\} \tag{7.16}$$

For example, for the LST,

$$[\phi_\varepsilon] = \left[\phi_1\right]_3 \tag{7.17}$$

and, is a (3×9) matrix, the values of $\{\varepsilon_c\}$ being $\{\varepsilon_{x1}\ \varepsilon_{x2}\ \varepsilon_{x3}\}$, etc. at the apices *1, 2, 3* of the triangle (Fig. 7.2 (b)). To obtain $\{\varepsilon_c\}$, the nodal strains, it is necessary simply to look for the nodal points of one lower order than those for the displacement polynomial interpolation. Thus, the nodal points for the strain interpolation for the LST are the nodal points for, displacement interpolation of the CST. The nodal strains are now related to the nodal displacements via the transformation,

$$\{\varepsilon_c\} = [\Phi_c]\{u\} \tag{7.18}$$

The matrix $[\Phi_c]$ is obtained by differentiating eqn (7.14), with respect to x_1, x_2, and remembering, from eqn (2.4), that

$$\varepsilon_{ij} = \frac{1}{2}(u_{i,j} + u_{j,i}) \tag{7.19}$$

The generalized stiffness matrix $[k_c]$ corresponding to the nodal strains is evaluated, using section 3.3, as

$$[k_c] = \int_{vol} [\phi_\varepsilon]^T [k][\phi_\varepsilon]\, dV \tag{7.20}$$

Finally, again applying the contragredient principle and using eqn (7.18), the element stiffness matrix, relating nodal displacements to nodal generalized forces, is given by

$$[k_e] = [\Phi_c]^T [k_c][\Phi_c] \tag{7.21}$$

For the CST, these matrices have already been evaluated, (eqns (3.19) and (3.22)). For the LST, from eqn (7.16),

$$\{\varepsilon_c\}^T = \{\varepsilon_{x1}\ \varepsilon_{x2}\ \varepsilon_{x3}\ \varepsilon_{y1}\ \varepsilon_{y2}\ \varepsilon_{y3}\ \gamma_{xy1}\ \gamma_{xy2}\ \gamma_{xy3}\} \tag{7.22}$$

Integrating eqn (7.20), using the expression for the integral of ζ functions over the triangle given in eqn (1.152),

$$[k_c] = \frac{AtE_r}{12}\begin{bmatrix} \Lambda\tau & \Lambda\nu & 0 \\ \Lambda\nu & \Lambda\tau & 0 \\ 0 & 0 & \Lambda\lambda \end{bmatrix} \tag{7.23}$$

In eqn (7.23), the submatrix Λ is defined by

$$[\Lambda] = \begin{bmatrix} 2 & 1 & 1 \\ 1 & 2 & 1 \\ 1 & 1 & 2 \end{bmatrix} \tag{7.24}$$

Now, the expression for $\{\varepsilon_c\}$ in terms of $\{u\}$, given in eqn (7.18), is written

$$[\Phi_c] = \begin{bmatrix} T_B & 0 \\ 0 & T_A \\ T_A & T_B \end{bmatrix} \tag{7.25}$$

where $[T_A]$ and $[T_B]$ are defined as follows, using the general matrix: $[T_t]$

$$[T_t] = \frac{1}{2A} \begin{bmatrix} 3t_1 & -t_2 & -t_3 & 4t_2 & 0 & 4t_3 \\ -t_1 & 3t_2 & -t_3 & 4t_1 & 4t_3 & 0 \\ -t_1 & -t_2 & 3t_3 & 0 & 4t_2 & 4t_1 \end{bmatrix} \tag{7.26}$$

To obtain $[T_A]$ and $[T_B]$ simply substitute a's and b's for t, as defined in eqn (1.117). It has thus been demonstrated that for the straight–sided triangular element, the element stiffness matrices for the planar elasticity can be integrated explicitly. These provide a useful function, not only for incorporation in computer software, but also as a check on the numerical accuracy of the numerically integrated elements.

7.2.2 Numerically integrated elements

The elements developed in section 7.2.1 have the advantage of simplicity of formulation and explicit integration. However, in the more general situation the element may have curved sides, so that, not only the displacement fields, but also the spatial coordinates must be interpolated over the element domain. It is, in general, better practice to use straight–sided elements wherever possible because excessive deviation from linearity along an element side will lead to negative stiffness and loss of element stiffness matrix rank, with consequently disastrous results. In practice, a solution algorithm which aborts when a negative pivot is encountered during the reduction process soon isolates the offending node, and a correction to its coordinates is easily made. In the formulation of the computer coding for the calculation of the element stiffness matrices it is most likely that the practical expediency is to integrate all elements numerically, simply to avoid the necessity of having to distinguish between possible alternative element types.

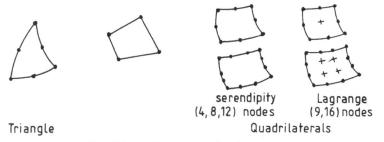

serendipity
(4, 8,12) nodes

Lagrange
(9,16) nodes

Triangle

Quadrilaterals

Fig. 7.3   Isoparametric element types.

The basic elements discussed herein are shown in Fig. 7.3, where it is seen that the four–node quadrilateral with straight sides falls into the isoparametric family of numerically integrated elements. The basis of the mathematical formulation for the numerical integration of these elements has been laid down in section 1.4.5. The isoparametric and Lagrange elements are considered first because the triangular domain has certain special features worthy of individual attention. From eqn (1.98), the coordinates are interpolated:

$$\{\tilde{x}\}^T = \{\phi\}^T [X]$$ (7.27)

The displacement vector $\{\tilde{u}\}$ is similarly interpolated, so that

$$\{\tilde{u}\}^T = \{\phi\}^T [\bar{U}]$$ (7.28)

The Jacobian of the transformation is given (eqn (1.104)) as

$$[J] = [\phi_{,\xi}]^T [X]$$ (7.29)

Thence, using the chain rule of differentiation, the derivatives of the displacements with respect to x_1, x_2 coordinates are given by

$$\{u_{,x}\}^T = [J]^{-1}[u_{,\xi}]^T = [J]^{-1}[\phi_{,\xi}]^T [\bar{U}] = [\phi_{,x}]^T [\bar{U}]$$ (7.30)

The strain transformation is given by

$$\{\varepsilon\} = \begin{bmatrix} \varepsilon_{x1} \\ \varepsilon_{x2} \\ \gamma_{x_1 x_2} \end{bmatrix} = \begin{bmatrix} \dfrac{\partial \tilde{u}}{\partial x_1} \\[2mm] \dfrac{\partial \tilde{v}}{\partial x_2} \\[2mm] \dfrac{\partial \tilde{u}}{\partial x_2} - \dfrac{\partial \tilde{v}}{\partial x_1} \end{bmatrix} = \begin{bmatrix} \dfrac{\partial \phi^T}{\partial x_1} & 0 \\[2mm] 0 & \dfrac{\partial \phi^T}{\partial x_2} \\[2mm] \dfrac{\partial \phi^T}{\partial x_2} & \dfrac{\partial \phi^T}{\partial x_2} \end{bmatrix} \begin{Bmatrix} U \\ V \end{Bmatrix}$$ (7.31)

In eqn (7.31), $\dfrac{\partial \phi^T}{\partial x_1}$ is the first row of $[\phi_{,x}^T]$, and $\dfrac{\partial \phi^T}{\partial x_2}$ the second. It must be noted that, whereas in eqn (7.30), the nodal displacement variables are grouped in matrix form, $[\bar{U}] \equiv [U\ V]$, in eqn (7.31), they are now grouped in vector form, again with all x_1 values together in $[U]$, and all x_2 values in $[V]$. Eqn (7.31) is, thus, written

$$\{\varepsilon\} = [a]\{r\}$$ (7.32)

The grouping of the terms in $\{r\}$ in eqn (7.32) allows the $[a]$ matrix to be arranged in submatrix form, which greatly facilitates the calculation of the element stiffness matrix. The evaluation of the element stiffness matrix is now straightforward and follows directly from section 3.3. Thus, if the stress state in the element is given by

$$\{\sigma\} = [C]\{\varepsilon\} + \{\sigma_0\}$$ (7.33)

using eqn (3.21), the nodal forces are given by

$$\{R\} = \int_{area} [a]^T [C][a]\ dA\ \{r\} + \int_{area} [a]^T \{\sigma_0\}\ dA$$ (7.34)

The first term on the right–hand side of eqn (7.34) gives the stiffness matrix, which, integrated numerically, is approximated

$$[k] = \sum\sum w_i w_j [a_{ij}]^T [C_{ij}][a_{ij}]\det J_{ij} \tag{7.35}$$

The number of integration points for the elements shown in Fig. 7.3(b) is given in Table 7.1.

Table 7.1

Number of nodes	Integration order
4	2×2
8 or 9	3×3
12 or 16	4×4

The Gauss quadrature points at which the function has to be evaluated, and the corresponding weight functions w_{ij}, are given in Chapter 1 for orders up to 4. By careful programming of the summations in eqn 7.35, making use of repetition of submatrix terms, constant multipliers taken outside the integration loops and so forth, there can be a significant reduction in the CPU time necessary to form the stiffness matrix as compared to that taken simply by programming the matrix multiplications of eqn (7.35) in full. This is discussed in further detail for three–dimensional elements. The numerical integration scheme can easily be extended to accommodate the axisymmetric situation shown in Fig. 7.1(c). In this case, the element actually forms a toroid, as shown in Fig. 7.4.

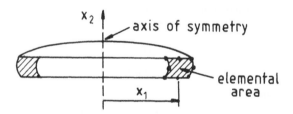

Fig. 7.4 Axisymmetric element volume.

To adapt the planar elasticity type of analysis to axisymmetry, the following modifications are necessary:

1. Include in the strain transformation the term for hoop strains,

$$\varepsilon_\theta = \frac{\tilde{u}}{\tilde{x}_1} = \frac{\{\phi\}^T \{u_1\}}{\{\phi\}^T \{X_1\}} \tag{7.36}$$

In eqn (7.36), $\tilde{u}$ is the x_1 component of displacements at the radial distance $\tilde{x}_1$.

2. Use the constitutive equation, as given in eqn (7.5), for isotropic material.

3. Replace the volume of the element, $detJ\ d\xi\ d\eta\ t$, by

$$dV = 2\pi \tilde{x}_1 detJ\ d\xi d\eta$$

That is,

$$dV = 2\pi\{\phi\}^T\{X_1\}detJ\ d\xi d\eta \qquad (7.37)$$

4, Replace the nodal forces, $\{R\}$ by the generalized forces which have been integrated over the surface area by including the term,

$$2\pi\tilde{x}_1 = 2\pi\{\phi\}^T\{X_1\}$$

Thus, using eqn (7.36) for the hoop strain, the strain transformation matrix (eqn (7.31)) now becomes

$$\{\varepsilon\} = \begin{bmatrix} \varepsilon_{x_1} \\ \varepsilon_{x_2} \\ \varepsilon_\theta \\ \gamma_{x_1,x_2} \end{bmatrix} = \begin{bmatrix} \dfrac{\partial \tilde{u}}{\partial x_1} \\ \dfrac{\partial \tilde{v}}{\partial x_2} \\ \dfrac{\tilde{u}}{x_1} \\ \dfrac{\partial \tilde{v}}{\partial \tilde{x}_1}+\dfrac{\partial \tilde{u}}{\partial x_2} \end{bmatrix} = \begin{bmatrix} \dfrac{\partial \phi^T}{\partial x_1} & 0 \\ 0 & \dfrac{\partial \phi^T}{\partial x_2} \\ \dfrac{\phi^T}{\phi^T X_1} & 0 \\ \dfrac{\partial \phi^T}{\partial x_2} & \dfrac{\partial \phi^T}{\partial x_1} \end{bmatrix} \begin{Bmatrix} U \\ V \end{Bmatrix} \qquad (7.38)$$

The coefficient matrix on the right–hand side of eqn (7.38) now defines the $[a]$ matrix in the axisymmetric element. Thence, eqn (7.35) now becomes

$$[k] = 2\pi\sum\sum w_i w_j [a_{ij}]^T [C_{ij}][a_{ij}]\{\phi_{ij}\}^T\{X\}detJ_{ij} \qquad (7.39)$$

The calculation of generalized nodal forces is left until section 7.3.2.

7.2.3   The triangular domain

The triangular domain, its interpolation functions, differentiation and integration thereof, has been treated in sections 1.4.6, 1.4.7 and 1.4.8. It was explained therein that special questions arise because of the use of the $(\zeta_1, \zeta_2, \zeta_3)$ coordinates. From section 1.4.7, the Jacobian of the transformation between the (x_1, x_2) and the $(\zeta_1, \zeta_2, \zeta_3)$ coordinates is given by eqn (1.159),

$$[J_A] = [e_3|\frac{\partial \phi^T}{\partial \zeta}[X]] \qquad (7.40)$$

The infinitesimal area is then given by

$$d\Omega = detJ_A d\zeta_i d\zeta_j \qquad (7.41)$$

Finally, from eqn (1.165), the x_1, x_2 derivatives are given by

$$\{u_{,x}\}^T = [L_R][D][U] \qquad (7.42)$$

It is noted that the form is similar to that of eqn (7.30); however, now the matrices $[L_R]$ and $[D]$ require special definition. In the notation of eqn (7.30),

$$[D] = [\phi_{,\zeta}]^T = \begin{bmatrix} \dfrac{\partial \phi^T}{\partial \zeta_1} \\ \dfrac{\partial \phi^T}{\partial \zeta_2} \\ \dfrac{\partial \phi^T}{\partial \zeta_3} \end{bmatrix} \tag{7.43}$$

and from eqn (1.170),

$$\begin{bmatrix} 0 \\ L_R \end{bmatrix} = [J_A]^{-1} \tag{7.44}$$

This expression shows that $[L_R]$ is the lower (2×3) submatrix of the inverse of the Jacobian matrix. Given the x_1, x_2 derivatives of the displacements, the procedure now follows directly that given in eqn (7.35) (or (7.36) for the axisymmetric case), except that now one of the integration schemes suitable for the triangular domain given in Table 1.3 (page 48), should be used. Thence,

$$[k] = \sum w_i [a_i]^T [C_i][a_i] detJ_i \tag{7.45}$$

For the calculation of the elastic stiffness of the LST, the seven–point integration scheme is adopted.

7.2.4   The natural mode method–constant strain triangle

In section 3.4 it was explained how the displacements of an element domain can be separated into,

1.   rigid body motion;

2.   natural modes describing the deformation of the element.

In Chapter 3 this idea was applied to a straight beam segment. The concept is now further developed for the triangular element in planar elasticity. The discussion is introduced with reference to the three–node CST, Fig. 7.5(a).

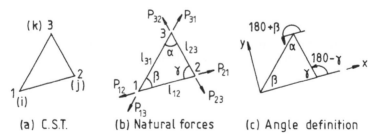

(a) C.S.T.   (b) Natural forces   (c) Angle definition

Fig. 7.5   Triangular element–natural mode method.

Using the notation of section 3.2.4, for ρ, choose

$$\{\rho\}^T = \{\rho_{xi} \ \rho_{yi} \ \rho_{xj} \ \rho_{yj} \ \rho_{xk} \ \rho_{yk} \} \tag{7.46}$$

Now, there are three rigid body modes in the plane, leaving three natural straining modes. For these natural modes, the changes in the lengths of the triangle sides are used. The strain field in the element is constant, and so also is the stress field. The stress field chosen is designated $\{\sigma_T\}$, defined by

$$\{\sigma_T\}^T = <\sigma_{1c} \ \sigma_{2c} \ \sigma_{3c}> \tag{7.47}$$

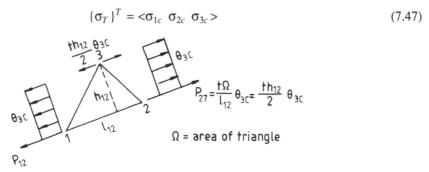

Fig. 7.6   Stress field σ_{3c}.

These stress fields are parallel to the sides of the triangle. It is important to realize that these are natural stresses and not total stresses. The equivalent nodal forces are shown in Fig. 7.5(b). These are the equal and opposite pairs (P_{12}, P_{21}), etc. If the length matrix $[l_D]$ is defined

$$[l_D] = \left| l_{23} \ l_{31} \ l_{12} \right| \tag{7.48}$$

then, the nodal forces P_{32}, P_{13}, P_{21} are defined by Fig. 7.6,

$$\{P_T\} = \begin{bmatrix} P_{32} \\ P_{13} \\ P_{21} \end{bmatrix} = -\begin{bmatrix} P_{23} \\ P_{31} \\ P_{12} \end{bmatrix} = \Omega t \, [l_D]^{-1}\{\sigma_T\} \tag{7.49}$$

It is now necessary to derive the constitutive law relating the total strains parallel to the element sides with the natural stress fields $\{\sigma_T\}$. To do this, use the second–order tensor transformation eqn (2.30). That is,

$$[\varepsilon'] = [L][\varepsilon][L]^T \tag{2.30}$$

The transformation matrix $[L]$ is defined by

$$[L] = \begin{bmatrix} \cos\theta & \sin\theta \\ -\sin\theta & \cos\theta \end{bmatrix}$$

and the values of θ, from Fig. 7.5 (c), are $(180 - \gamma)$, $(180 + \beta)$, and 0, for the sides 23, 31 and 12, respectively for the local coordinate axes shown. Thence, from eqn (2.30), the strain parallel with a triangle side is given by

$$\varepsilon'_{x_1} = \cos{}^2\theta \ \varepsilon_{x_1} + \sin{}^2\theta \ \varepsilon_{x_2} + \cos\theta \sin\theta \ \gamma_{x_1 x_2}$$

giving finally, for all three sides,

$$\begin{bmatrix} \varepsilon_{1c} \\ \varepsilon_{2c} \\ \varepsilon_{3c} \end{bmatrix} \equiv \begin{bmatrix} \varepsilon_{\beta} \\ \varepsilon_{\gamma} \\ \varepsilon_{\alpha} \end{bmatrix} = \begin{bmatrix} \cos^2\gamma & \sin^2\gamma & -\sin\gamma\cos\gamma \\ \cos^2\beta & \sin^2\beta & \sin\beta\cos\beta \\ 1 & 0 & 0 \end{bmatrix} \begin{bmatrix} \varepsilon_{x_1} \\ \varepsilon_{x_2} \\ \gamma_{x_1 x_2} \end{bmatrix} \tag{7.50}$$

Write this equation as

$$\{\varepsilon_T\} = [T]\{\varepsilon_{xy}\} \tag{7.51}$$

Contragredience gives the corresponding stress transformation as

$$\{\sigma_{xy}\} = [T]^T\{\sigma_T\} \tag{7.52}$$

Now, for the local *XY* coordinate system,

$$\{\varepsilon_{xy}\} = \frac{1}{E}[f]\{\sigma_{xy}\} \tag{7.53}$$

Then

$$\{\varepsilon_T\} = [T](\frac{1}{E}[f])[T]^T\{\sigma_T\} \tag{7.54}$$

The flexibility matrix in the natural coordinate system is thus given by

$$[f_T] = [T](\frac{1}{E}[f])[T]^T \tag{7.55}$$

Multiplication of the matrices in eqn (7.55) for the plane stress, isotropic elasticity gives

$$[f_T] = \frac{1}{E}\begin{bmatrix} 1 & \cos^2\alpha - v\sin^2\alpha & \cos^2\gamma - v\sin^2\gamma \\ \cos^2\alpha - v\sin^2\alpha & 1 & \cos^2\beta - v\sin^2\beta \\ \cos^2\gamma - v\sin^2\gamma & \cos^2\beta - v\sin^2\beta & 1 \end{bmatrix} \tag{7.56}$$

To obtain the stiffness relationship, it is necessary to invert the matrix expression in eqn (7.56):

$$[k_T] = [f_T]^{-1} \tag{7.57}$$

Thus, the constitutive law becomes

$$\{\sigma_T\} = [k_T]\{\varepsilon_T\} \tag{7.58}$$

It now remains to calculate the relationship between $\{P_T\}$ and $\{\rho_T\}$, and finally that between the global force and displacement nodal values. To this end, first define the changes in the lengths of the element sides,

$$\{\rho_T\}^T = \Delta\{l_{23}\ l_{31}\ l_{12}\} \tag{7.59}$$

Then, because the strains are constant along the element sides, it follows that

$$\{\varepsilon_T\} = [l_D]^{-1}\{\rho_T\} \tag{7.60}$$

Combining eqns (7.49), (7.58) and (7.60) gives

$$\{P_T\} = \Omega t\,[l_D]^{-1}[k_T][l_D]^{-1}\{\rho_T\} \tag{7.61}$$

The natural (3×3) stiffness matrix for the triangle is thus given by

$$[k_N] = \Omega t\,[l_D]^{-1}[k_T][l_D]^{-1} \tag{7.62}$$

so that eqn (7.61) is written

$$\{P_T\} = [k_N]\{\rho_T\} \tag{7.63}$$

It now only remains, to relate the natural $\{P_T\}$ forces to the nodal global forces, $\{P_{xy}\}$ so that the nodal global equations of equilibrium can be formed. The various nodal force components are shown in Fig. 7.7.

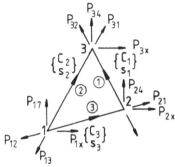

Fig. 7.7   Natural and global forces.

From Fig. 7.7, the direction cosines of side i are given by $\{c_i \ s_i\}$, and, thence, by projection and using eqn (7.49),

$$
\begin{bmatrix} P_{1x} \\ P_{1y} \\ P_{2x} \\ P_{2y} \\ P_{3x} \\ P_{3y} \end{bmatrix}
=
\begin{bmatrix}
0 & c_2 & -c_3 \\
0 & s_2 & -s_3 \\
-c_1 & 0 & c_3 \\
-s_1 & 0 & s_3 \\
c_1 & -c_2 & 0 \\
s_1 & -s_2 & 0
\end{bmatrix}
\begin{bmatrix} P_{32} \\ P_{13} \\ P_{21} \end{bmatrix}
\tag{7.64}
$$

Write this equation symbolically as,

$$\{P_{xy}\} = [a_N]^T \{P_T\} \tag{7.65}$$

Thence, again using contragredience, the corresponding displacement transformation is given

$$\{\rho_T\} = [a_N]\{\rho_{xy}\} \tag{7.66}$$

Combining eqns (7.63), (7.65) and (7.66),

$$\{P_{xy}\} = [a_N]^T [k_N][a_N]\{\rho_{xy}\} \tag{7.67}$$

The global stiffness matrix is thus obtained as,

$$[k_{xy}] = [a_N]^T [k_N][a_N] \tag{7.68}$$

Before discussing the merits of the natural approach, it is worthwhile to reflect on the physical significance of the above formulation. Examination of the three sets of stress fields shown in Figs 7.5(b) and 7.6 shows that each is equivalent to the stress in a simple link member. However, the evaluation of the deformation of the element is complicated by two facts:

1.   the material is in a state of plane stress (or strain);

2.   the cross section of the element is variable because of the triangular shape.

In addition, there are three force components instead of one, as is the case for the two–node link element. All these considerations lead to the derivation given above. The formulation may, at first sight, appear to be cumbersome. However, it should be noted that in forming $[k_N]$, only (3×3) matrices have to be manipulated, with the expansion to the (6×6) global stiffness matrix taking place in the last step. The advantages of the natural approach become more apparent and compelling when large displacement analyses are undertaken. An examination of an element in its undeformed and deformed configurations Fig. 7.8 shows that filaments parallel to the sides of the element can have their natural stress components added directly, even though displacements are large. This circumvents the problems associated with having to transform stress components.

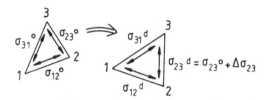

Fig. 7.8   Large displacements–addition of stress components.

Before proceeding to develop the ideas of the natural mode technique further to accommodate higher–order elements with curved sides, it is worth examining the stress fields in Fig. 7.5(b) to show how easily they may be used to generate the geometric stiffness matrix $[k_G]$ for the CST, from the corresponding theory for the two–node link element (section 6.2, eqn (6.22)). In eqn (6.22), it was shown that the increments in force components in the global coordinate axes for given node displacements are

$$\Delta \begin{bmatrix} \vec{P}_I \\ \vec{P}_J \end{bmatrix}_G = \frac{P_N}{l} \begin{bmatrix} A & -A \\ -A & A \end{bmatrix} \begin{bmatrix} \vec{v}_I \\ \vec{v}_J \end{bmatrix} \tag{6.22}$$

The submatrix $[A]$ is defined

$$[A] = [I_3] - \{c\}\{c\}^T \tag{7.69}$$

In the present situation, for the triangle in Fig. 7.5(b), it is seen that $\{P_{32}, P_{23}\}$, etc. form three sets of such forces, so that the geometric stiffness matrix for the triangle is given by the summation,

$$[K_G] = \sum_{i=1,3} [M_{ji}]^T [K_{Gij}][M_{ji}] \tag{7.70}$$

where

$$[K_{Gij}] = \frac{P_N}{l_{ij}} \begin{bmatrix} A_{ij} & -A_{ij} \\ -A_{ij} & A_{ij} \end{bmatrix} \tag{7.71}$$

and, for $j = 2$, $i = 1$,

$$[M_{ji}] = \begin{bmatrix} I_3 & 0 & 0 \\ 0 & I_3 & 0 \end{bmatrix} \tag{7.72}$$

In eqn (7.72), $[M_{ji}]$ varies cyclically for the permutations of i and j. In eqns (7.70) and (7.72), it has been assumed that the triangular element is in three–dimensional space. The restriction to planar elasticity is simply one of detail.

7.2.5   Higher order triangular elements – natural formulation – ur–elements

For the straight–sided triangular elements, it is a relatively simple task to extend the reasoning in section 7.2.4 to higher–order interpolation, for example, the LST. In this element there will be 12 nodal degrees of freedom, and, with three rigid body modes, the number of natural straining modes is equal to nine. The three straining modes parallel to side 1–2 of the element are shown in Fig. 7.9.

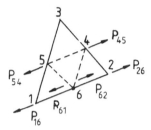

Fig. 7.9   Straining modes in LST parallel to side 1–2.

The derivation of the (9×9) natural stiffness matrix is relatively straightforward and is left to the reader as an exercise. It is of considerable interest, however, to extend the numerical integration techniques developed in section 7.2.4 for use with the natural mode method. To this end, the concept of the ur–element is introduced [40]. In this application of planar elasticity, the ur–element is simply a CST as given in section 7.2.4, with its sides now oriented parallel with the triangle filament directions, $\zeta_1 = \text{constant}$, $\zeta_2 = \text{constant}$, $\zeta_3 = \text{constant}$, as shown in Fig. 7.10.

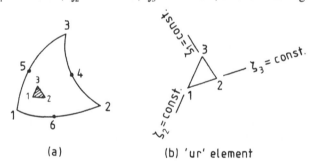

(a)   (b) 'ur' element

Fig. 7.10   Curved triangle–local element.

As mentioned, the ur–element is simply a CST, and hence, although its dimensions are infinitesimally small, the derivation of its stiffness matrix must follow directly from section 7.2.4. Herein this is modified slightly to follow the derivation in [40]. Firstly, the Cartesian flexibility matrix is written

$$
\begin{bmatrix} \varepsilon_x \\ \varepsilon_y \\ \dfrac{\gamma_{xy}}{\sqrt{2}} \end{bmatrix} = \frac{1}{E} \begin{bmatrix} 1 & -v & 0 \\ -v & 1 & 0 \\ 0 & 0 & 1+v \end{bmatrix} \begin{bmatrix} \sigma_x \\ \sigma_y \\ \sqrt{2}\tau_{xy} \end{bmatrix}
\tag{7.73}
$$

This is simply eqn (7.53), written here as

$$\{\varepsilon_{xy}\} = \frac{1}{E}[f]\{\sigma_{xy}\} \qquad (7.74)$$

The transformation between global and natural strains at any point in the triangle (i.e. for the ur–element) is given

$$\begin{bmatrix} \varepsilon_{23} \\ \varepsilon_{31} \\ \varepsilon_{12} \end{bmatrix} = [m]^{-2} \begin{bmatrix} (x_{,23}^1)^2 & (x_{,23}^2)^2 & \sqrt{2}(x_{,23}^1)(x_{,23}^2) \\ (x_{,31}^1)^2 & (x_{,31}^2)^2 & \sqrt{2}(x_{,31}^1)(x_{,31}^2) \\ (x_{,12}^1)^2 & (x_{,12}^2)^2 & \sqrt{2}(x_{,12}^1)(x_{,12}^2) \end{bmatrix} \begin{bmatrix} \varepsilon_x \\ \varepsilon_y \\ \dfrac{\gamma_{xy}}{\sqrt{2}} \end{bmatrix} \qquad (7.75)$$

This equation is obtained by application of eqn (2.30) to each of the directions, $\zeta_i = $ constant. The matrix $[m]$ is simply the matrix $[l_D]$ for the local triangle, and is written

$$[m] = \begin{bmatrix} l_{23} \ l_{31} \ l_{12} \end{bmatrix} \qquad (7.76)$$

The value l_{ij} is referred to the side i - j of the local triangle, and will be defined below. The transformation in eqn (7.75) is written

$$\{\varepsilon_N\} = [B]^T\{\varepsilon_{xy}\} \qquad (7.77)$$

Then, the natural flexibility matrix for the local triangle is calculated as

$$[f_N] = [B]^T[f][B] \qquad (7.78)$$

and, thence,

$$\{\varepsilon_N\} = [f_N]\{\sigma_N\} \qquad (7.79)$$

The stiffness matrix $[E_N] = [f_N]^{-1}$, so that

$$\{\sigma_N\} = [E_N]\{\varepsilon_N\} = E[\kappa_N]\{\varepsilon_N\} \qquad (7.80)$$

The operator $\dfrac{\partial}{\partial\zeta_{ij}}$ has been defined in section 1.4.8, eqn (1.157), as

$$\frac{\partial}{\partial\zeta_{ij}} = -\frac{\partial}{\partial\zeta_i} + \frac{\partial}{\partial\zeta_j} \qquad (7.81)$$

Thence, the local derivative vector a_{ij} is given by

$$\{a_{ij}\} = \{x_{,ij}\} = \frac{\partial\{\tilde{x}\}}{\partial\zeta_{ij}} \qquad (7.82)$$

and, from eqns (1.158) and (7.27),

$$\{a_{ij}\}^T = \{x_{,ij}\}^T = \frac{\partial\tilde{\tilde{x}}}{\partial\zeta_{ij}} = \frac{\partial\{\phi\}^T}{\partial\zeta_{ij}}[\bar{X}] = \{\phi_{,ij}\}^T[\bar{X}] \qquad (7.83)$$

Since a_{ij} is the local derivative in the direction i, j, it follows that the square of its magnitude is calculated from

$$m_{ij}^2 = \{a_{ij}\}^T\{a_{ij}\} = l_{ij}^2 \qquad (7.84)$$

This defines the terms of the matrix $[m]$. As before, from eqn (1.160), the Jacobian is given by

$$[J_A] = [e_3 \mid D\bar{X}]$$ (7.85)

and the infinitesimal ur–element area is

$$d\Omega = \det J_A \, d\zeta_i d\zeta_j$$ (7.86)

Similarly, the displacement vector in the filament direction along the element side $i-j$ is given by

$$\{u_{,ij}\} = \frac{\partial\{u\}}{\partial\zeta_{ij}}$$ (7.87)

The terms of $\{u_{,ij}\}$ are collected into a (6×1) supervector:

$$\{u_{,N}\} = \frac{\partial\{u\}}{\partial\bar{\zeta}_N} = \begin{bmatrix} u_{,23} \\ u_{,31} \\ u_{,12} \end{bmatrix}$$ (7.88)

Then the natural deformations, as defined in eqn (7.59), are given by

$$\{\rho_N\} = [m]^{-1}[A_N]\{u_{,N}\}$$ (7.89)

The matrix $[m]$ is defined in eqn (7.76), and $[A_N]$ is the superdiagonal matrix, obtained, using eqn (7.82), as

$$[A_N] = \begin{bmatrix} a_{23}^T & a_{31}^T & a_{12}^T \end{bmatrix}$$ (7.90)

Note that

$$[m]^2 = [A_N][A_N]^T$$

For example, then,

$$\rho_{N3} = m_{12}^{-1}\{a_{12}\}^T\{u_{,12}\}$$

expresses the projection of the Cartesian vector components $\{u_{,12}\}$ onto a_{12}. Now form the column matrix,

$$[A_N^c] = \begin{bmatrix} a_{23}^T \\ a_{31}^T \\ a_{12}^T \end{bmatrix} = \frac{\partial\{\phi\}^T}{\partial\bar{\zeta}_N}[\bar{X}] = [D][\bar{X}]$$ (7.91)

The matrix $[D_e]$ is defined, via the resh operator $\overline{2}$ such that

$$[D_e] = \overline{2}[D]$$ (7.92)

The resh operator $\overline{2}$ makes a (2×2) supermatrix in $[D_e]$ from the corresponding single term of $[D]$. If $\{u_I\}$ is the vector of node point displacements grouped in the order $\{u_{ix}, u_{iy}\}$ at each node, then, from eqn 7.88,

$$\{u_{,N}\} = [D_e]\{u_I\}$$ (7.93)

and, thence, from eqn (7.89),

$$\{\rho_N\} = [a_N]\{u_I\} = [m]^{-1}[A_N][D_e]\{u_I\}$$ (7.94)

Finally, the natural strains are given by

$$\{\varepsilon_N\} = [m]^{-1}\{\rho_N\} = [m]^{-2}[A_N][D_e]\{u_I\} \tag{7.95}$$

Thus the nodal stiffness matrix is calculated:

$$[k_{xy}] = \int_{area} [D_e]^T [A_N]^T [m]^{-2} [E_N][m]^{-2}[A_N][D_e]d\Omega \tag{7.96}$$

Now, the infinitesimal area is given, by

$$d\Omega = detJ_A \, d\zeta_i \, d\zeta_j = \Omega_N d\zeta_i \, d\zeta_j \tag{7.97}$$

Defining,

$$[k_N] = E\,\Omega_N [m]^{-1}[\kappa_N][m]^{-1}$$

and

$$[a_N] = [m]^{-1}[A_N][D_e]$$

then, eqn (7.96) is written simply as

$$[k_{xy}] = \int_{area} [a_N]^T [k_N][a_N]d\zeta_i \, d\zeta_j \tag{7.98}$$

The elegance of the above formulation lies in the fact that it can be adapted so readily to large displacement analysis, with the direct addition of the increments of the $\{\sigma_N\}$ filament stress components at any point. These natural stresses may be transformed to Cauchy stresses by using contragredience in conjunction with eqn (7.75)

$$\{\sigma_{xy}\} = [B]\{\sigma_N\} \tag{7.99}$$

7.2.6 Geometric stiffness

It remains now to follow the ur–element concept for the geometric stiffness matrix given in section 7.2.4. Thus, from eqn (6.22), the geometric stiffness for the component σ_{ij} of the ur–element stresses is given by

$$\Delta \begin{bmatrix} u_i \\ u_j \end{bmatrix} = \frac{\sigma_{ij} V_N}{m_{ij}^2} \begin{bmatrix} A & -A \\ -A & A \end{bmatrix} \begin{bmatrix} u_i \\ u_j \end{bmatrix} \tag{7.100}$$

In eqn (7.100), $V_N = t\,\Omega_N$ is the infinitesimal area of the ur–element, so that

$$\frac{\sigma_{ij} V_N}{m_{ij}^2} = \frac{\sigma_{ij} t \, h_{ij} m_{ij}}{2m_{ij}^2} = \frac{P_N}{m_{ij}} \quad \text{(see eqn (7.71))}$$

From eqn (7.100),

$$\Delta u_j = \frac{\sigma_{ij}}{m_{ij}^2}[A][u_j - u_i]V_N \tag{7.101}$$

It is seen that

$$\frac{\partial u}{\partial s_{ij}} = \frac{1}{m_{ij}}\left\{\frac{\partial u}{\partial \zeta_j} - \frac{\partial u}{\partial \zeta_i}\right\}$$

and, hence,

$$\delta u = u_j - u_i = \frac{ds_{ij}}{m_{ij}}\left\{\frac{\partial u}{\partial \zeta_j} - \frac{\partial u}{\partial \zeta_i}\right\}$$

In the limit, $ds_{ij} \equiv m_{ij}$ and $V_N = d\Omega$, so that

$$\Delta u_j = \frac{\sigma_{ij}}{m_{ij}^2}[A]\frac{\partial \bar{u}}{\partial \zeta_{ij}}d\Omega \qquad (7.102)$$

Now, from eqn (7.93),

$$\frac{\partial \bar{u}}{\partial \zeta_{ij}} = [D_{ije}]\{u_I\} \qquad (7.103)$$

so that applying the contragredient principle gives the increment in the nodal force components $\delta \Delta R_{Iij}$ as,

$$\{\delta \Delta R_{Iij}\} = [D_{ije}]^T\{\Delta u_j\} = [D_{ije}]^T\left[\sigma_{ij}m_{ij}^{-2}[A]\right][D_{ije}]d\Omega\{u_I\} \qquad (7.104)$$

To obtain the effects of all three components of the ur–element stresses, define the matrix $[S]$

$$[S] = \left[s_{23}\ s_{23}\ s_{31}\ s_{31}\ s_{12}\ s_{12}\right] \qquad (7.105)$$

where

$$\{s_{ij}\} = (\sigma_{ij}m_{ij}^{-2})[A] \qquad (7.106)$$

Thence,

$$\{\delta R_I\} = [D_{23e}^T\ D_{31e}^T\ D_{12e}^T][S]\begin{bmatrix}D_{23e}\\D_{31e}\\D_{12e}\end{bmatrix}d\Omega\{U_i\} = [D_e]^T[S][D_e]d\Omega\{u_I\}$$

Integrating over the area of the triangular element gives

$$\{\Delta R_I\} = \int[D_e]^T[S][D_e]\Omega_N d\zeta_i d\zeta_j \qquad (7.107)$$

Thus, the elegant expression has been obtained for the element geometric stiffness matrix $[k_G]$

$$[k_G] = \int[D_e]^T[S][D_e]\Omega_N d\zeta_i d\zeta_j \approx \sum w_i[D_{ei}]^T[S_i][D_{ei}]\Omega_{N_i} \qquad (7.108)$$

It should be noted that stresses may be calculated at the midside nodes, and three–point integration used in the calculation of $[k_G]$.

7.3 GENERALIZED (KINEMATICALLY EQUIVALENT) FORCE SYSTEMS

7.3.1 Introduction

Thus far, the theory has been concerned only with the generation of element stiffness matrices which, when assembled, form the global stiffness matrix $[K]$ which enters into the nodal force displacement relationship, $[K]\{r\} = \{R\}$.

In practice, the forces applied to the body will usually be distributed in nature, either on the surface or throughout the body (gravity and initial strain restraint forces), and it is necessary to determine the equivalent concentrated, generalized nodal forces. The most common effects to be considered are

1. distributed surface forces, both pressures and tractions;

2. distributed body forces, which may be due to gravity or, in dynamics, to the D'Alembert inertia forces;

3. restraint release forces due to the temperature variation in the kinematically determinate state $r = 0$.

For all these force systems, the basic concept for determining the nodal forces is that the two force systems are both kinematically and statically equivalent. That is, the work done by the generalized nodal force for a corresponding perturbation of its shape function must equal the integral over the element domain (surface or volume) of the corresponding distributed force and the interpolated shape function. Again, as with the stiffness matrix formulation, the discussion can be divided into those domains or surfaces with regular geometry which is amenable to direct integration and those which are not and thus require numerical integration. Each will be considered separately, although in computer software it is probably convenient to rely simply on numerical integration to avoid repetitive coding.

7.3.2 Surface forces

The inplane surface loading is the variable $\tilde{p}(\zeta_b)$ acting on a given boundary (B), and is considered to have been integrated through the element thickness (t) so that it is a force per unit length. Let $\{R_P\}$ be the node forces which are both statically and kinematically equivalent to $\tilde{p}(\zeta_b)$. Then, according to the principle of virtual displacements, the work done by these node forces must be equal to that done by the distributed surface forces $\tilde{p}(\zeta_B)$. This, of course, leads naturally to a direct application of the contragredient principle, as is demonstrated in this section.

Thus, let the distributed force $\tilde{p}(\zeta_B)$ vary as a pth–order polynomial. The number of node values needed to specify $\tilde{p}(\zeta_B)$ will be (section 1.4.2)

$$m_p = \frac{(n_p + 1)!}{n_p!} = n_p + 1$$

since, along the boundary, the variable $\tilde{p}(\zeta_B)$ is a function of a single variable.
Then

$$\tilde{p}(\zeta_B) = \begin{bmatrix} \tilde{p}_x(\zeta_B) \\ \tilde{p}_y(\zeta_B) \end{bmatrix} = \begin{bmatrix} \phi_P^T & \phi_P^T \end{bmatrix} \begin{bmatrix} p_x \\ p_y \end{bmatrix} = [\psi]^T \{p\} \qquad (7.109)$$

In eqn (7.109), $\tilde{p}_x(\zeta_B)$, $\tilde{p}_y(\zeta_B)$ are the X and Y components of the distributed force, and p_x and p_y the values specified at the $(n_p + 1)$ node points. For convenience it is usual to restrict the order of m_p to the same as that of the number of node points in the element side. Thus, for end points only, a linear variation of $\tilde{p}(\zeta_B)$ is possible, while the addition of a mid–side node allows for quadratic variation. The displacement of the boundary B_P for a compatible displacement model depends only on the nodal points located within that boundary, and hence the boundary displacement is given by

$$\tilde{r}_B(\zeta_B) = \begin{bmatrix} u(\zeta_B) \\ v(\zeta_B) \end{bmatrix} = \begin{bmatrix} \phi_B^T & \phi_B^T \end{bmatrix} \begin{bmatrix} U_B \\ V_B \end{bmatrix} = [\Phi_B]^T \{r_B\} \qquad (7.110)$$

Applying the contragredient principle, the contribution $\Delta\{R_B\}$ to the nodal forces statically equivalent to $\tilde{p}(\zeta_B)$ is given by

$$\{\Delta R_B\} = [\Phi_B]\tilde{p}(\zeta_b)ds = [\Phi_B][\psi_B]^T ds\{p\}$$

Integrating over B_P,

$$\{R_B\} = \int\limits_{B_P} [\Phi_B][\psi_B]^T ds\{p\} \tag{7.111}$$

Substitution for $[\Phi_B]$, $[\psi_B]$ and $\{p\}$ gives the components,

$$\{R_{Bx}\} = \int\limits_{B_P} [\phi_B][\phi_P]^T ds\{p_x\} ; \quad \text{and} \quad \{R_{By}\} = \int\limits_{B_P} [\phi_B][\phi_P]^T ds\{p_y\} \tag{7.112}$$

Example 7.1 *Linear and parabolic variation*

I.   Suppose that both $[\phi_B]$ and $[\phi_P]$ are linear functions for the surface shown in Fig. 7.11(a).

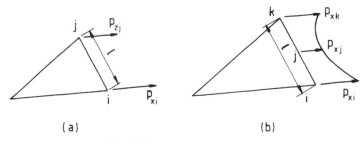

Fig. 7.11   Surface forces.

Then,

$$[\phi_B]^T = [\phi_P]^T = [\frac{1}{2}(1-\zeta) \quad \frac{1}{2}(1+\zeta)] ; \quad \text{with} \quad \zeta = \frac{2s}{l}, \quad ds = \frac{l}{2}d\zeta$$

Thence,

$$\begin{bmatrix} R_{xi} \\ R_{xj} \end{bmatrix} = \frac{l}{8}\int\limits_{-1}^{+1} \begin{bmatrix} (1-\zeta) \\ (1+\zeta) \end{bmatrix}[(1-\zeta) \ (1+\zeta)]\ d\zeta \begin{bmatrix} p_{xi} \\ p_{xj} \end{bmatrix} = \frac{l}{6}\begin{bmatrix} 2 & 1 \\ 1 & 2 \end{bmatrix}\begin{bmatrix} p_{xi} \\ p_{xj} \end{bmatrix} \tag{i}$$

II.   The surface in Fig. 7.11(b) has a parabolic variation in $\tilde{p}(\zeta_B)$, for an element whose polynomial variation for displacement interpolation in the side is quadratic (three nodes, $i-j-k$). Using the polynomial ζ_2 as for the triangular element, varying from 0 at i to 1 at k, the interpolation polynomials are given by

$$[\phi_B] = [\phi_P] = \begin{bmatrix} \zeta_2(2\zeta_2 - 1) \\ 4\zeta_2(1 - \zeta_2) \\ (1 - \zeta_2)(1 - 2\zeta_2) \end{bmatrix}$$

Substituting into eqn (7.103) and noting that

$$\int_0^l ds = l \int_0^1 d\zeta_2$$

gives, finally,

$$\begin{bmatrix} R_{xi} \\ R_{xj} \\ R_{xk} \end{bmatrix} = \frac{l}{30} \begin{bmatrix} 4 & 2 & -1 \\ 2 & 16 & 2 \\ -1 & 2 & 4 \end{bmatrix} \begin{bmatrix} p_{xi} \\ p_{xj} \\ p_{xk} \end{bmatrix} \qquad \text{(ii)}$$

In both no. I. and II., similar expressions apply for p_y. It is seen from these examples that for a straight–sided element, it is a relatively simple task to calculate the statically equivalent nodal forces for the distributed pressure p_x or p_y. It must be noted, however, that the specification for $\tilde{p}_x(\tilde{p}_y)$ in Fig. 7.11 is a little unusual, being specified as a force per unit length on a skew dimension s. It is more usual to specify surface forces in terms of pressure and traction; that is, in terms of normal and tangential components in the local coordinate axes associated with the surface. This definition of surface forces is used in the following derivation for the numerical evaluation of the node forces. A surface element $i-j-k$ is shown in Fig. 7.12 and is assumed to be curved, being specified in this instance by the three node points $i-j-k$. A single coordinate, η, is used to measure distance along the curve. Both ds and $d\eta$ are measured as positive in the anticlockwise sense. Pressures are taken to be positive when in the coordinate directions shown in Fig. 7.12. The interpolation for the (x, y) coordinates in the side will be

$$\{\tilde{x}\ \tilde{y}\} = \{N\}^T[X\ Y] \qquad (7.113)$$

where $\{N\}^T$ are the shape functions, here given, in terms of the single parameter η, as

$$\{N\}^T = \{\frac{1}{2}\eta(\eta - 1)\ (1 - \eta^2)\ \frac{1}{2}\eta(\eta + 1)\} \qquad (7.114)$$

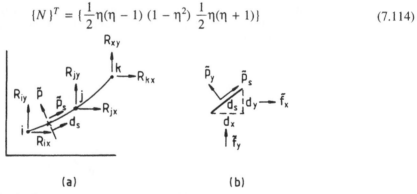

(a)   (b)

Fig. 7.12   Surface pressures–local coordinate systems.

The length vector $\vec{ds}$ along the side is given, in terms of its x and y components, as

$$\vec{ds} = \begin{bmatrix} dx \\ dy \end{bmatrix} = \begin{bmatrix} \dfrac{\partial \tilde{x}}{\partial \eta} \\[2mm] \dfrac{\partial \tilde{y}}{\partial \eta} \end{bmatrix} d\eta = \begin{bmatrix} \dfrac{\partial N^T}{\partial \eta} X \\[2mm] \dfrac{\partial N^T}{\partial \eta} Y \end{bmatrix} d\eta \qquad (7.115)$$

The transformation matrix $[L]$ relating the (x,y) and (s,n) coordinate systems can now be calculated from eqn (7.115):

$$[L] = \frac{1}{\sqrt{detJ}} \begin{bmatrix} \dfrac{\partial N^T}{\partial \eta}X & \dfrac{\partial N^T}{\partial \eta}Y \\[2ex] -\dfrac{\partial N^T}{\partial \eta}Y & \dfrac{\partial N^T}{\partial \eta}X \end{bmatrix} \tag{7.116}$$

In eqn (7.116), the rows of the matrix on the right–hand side are the direction numbers of the (s,n) axes, and $detJ$ is the determinant of this matrix, or alternatively ds^2, so that the rows of the $[L]$ matrix are the direction cosines of the (s,n) axes, (eqn (1.38)). Thence, the x,y components of the $\{\tilde{p}_x, \tilde{p}_y\}$ pressures at the points i, j and k can be expressed as

$$\begin{bmatrix} p_x \\ p_y \end{bmatrix}_{i,j \ or \ k} = [L]^T \begin{bmatrix} p_s \\ p_n \end{bmatrix}_{i,j \ or \ k} \tag{7.117}$$

The stress resultants in the x and y directions are then given by

$$\tilde{f}_x = \tilde{p}_x \, |dy| = \tilde{p}_x \left| \frac{\partial N^T}{\partial \eta} Y \right| d\eta \tag{7.118a}$$

and

$$\tilde{f}_y = \tilde{p}_y \, |dx| = \tilde{p}_y \left| \frac{\partial N^T}{\partial \eta} X \right| d\eta \tag{7.118b}$$

In eqn (7.118) it is seen that the infinitesimal lengths $|dy|$, $|dx|$ are unsigned values, so that the sense of $\tilde{f}_x$ or $\tilde{f}_y$ is obtained from that of $\tilde{p}_x$ or $\tilde{p}_y$, respectively. Applying the contragredient principle to eqn (7.118), with the displacements being interpolated using the same functions as in eqn (7.113), gives the nodal force components,

$$\{\Delta R_x\} = [N]\tilde{f}_x = [N]\tilde{p}_x \left| \frac{\partial N^T}{\partial \eta} Y \right| d\eta$$

and

$$\{\Delta R_y\} = [N]\tilde{f}_y = [N]\tilde{p}_y \left| \frac{\partial N^T}{\partial \eta} X \right| d\eta$$

Interpolating $\tilde{p}_x$ and $\tilde{p}_y$ in terms of their nodal values,

$$\{\tilde{p}_x \ \tilde{p}_y\} = [N]^T \{p_x \ p_y\}$$

and integrating from -1 to $+1$ on η gives, finally, the expressions for the nodal forces as

$$\{R_x\} = \int_{-1}^{+1} [N][N]^T \left| \frac{\partial N^T}{\partial \eta} Y \right| d\eta \{p_x\} \tag{7.119a}$$

and

$$\{R_y\} = \int_{-1}^{+1} [N][N]^T \left| \frac{\partial N^T}{\partial \eta} X \right| d\eta \{p_y\} \tag{7.119b}$$

The similarity should be noticed between eqns (7.119) and (7.113). In eqn (7.119), the Jacobian of the transformation varies along the surface arc, and thus necessitates the use of numerical integration. Alternatively, $\{\tilde{p}_s, \tilde{p}_n\}$ may be interpolated::

$$\{\tilde{p}_s, \tilde{p}_n\} = [N]^T \{p_s, p_n\}$$

Then, calculating the transformation matrix $[L]$ at the integration point give

$$\begin{bmatrix} \tilde{p}_x \\ \tilde{p}_y \end{bmatrix} = [L]^T \begin{bmatrix} \tilde{p}_s \\ \tilde{p}_n \end{bmatrix}$$

The expressions for the numerical integration of eqn (7.119) then become

$$\{R_x\} = \sum w_i [N_i] \left| \frac{\partial N_i^T}{\partial \eta} Y \right| \tilde{p}_{xi} \tag{7.120a}$$

and

$$\{R_y\} = \sum w_i [N_i] \left| \frac{\partial N_i^T}{\partial \eta} X \right| \tilde{p}_{yi} \tag{7.120b}$$

The three–point Gauss integration rule will suffice in this instance.

Example 7.2 *Surface force calculation*

Eqn (7.120) has been incorporated in a planar elasticity finite element program, and the results printed for the calculation of the equivalent nodal forces on the semicircular opening shown in Fig. 7.13. Two surface load conditions are given:

1. internal pressure of 10 units;

2. clockwise traction of 10 units.

Four surface elements of equal length have been used, with the node numbering 1 to 9, as shown in Fig. 7.13. From the summation of nodal forces, it is seen that the resultants are (0, 100) for the internal pressure, and (100, 0) for the traction.

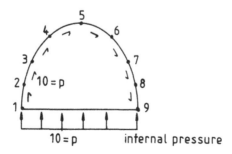

Fig. 7.13   Surface loading–pressure and traction.

Table 7.2

	Node	Load case 1		Load case 2	
		R_x	R_y	R_x	R_y
side 1	1	−6.864	0.096	0.096	6.864
	2	−23.570	9.763	9.763	23.570
	3	−4.922	4.785	4.785	4.992
side 2	3	−4.785	4.922	4.922	4.785
	4	−9.763	23.570	23.570	9.763
	5	−0.096	6.864	6.864	0.096
		Antisymmetric about 5	Symmetric about 5	Symmetric about 5	Antisymmetric about 5

7.3.3 Body forces

The calculation of the generalized nodal forces to replace distributed body forces is similar to that for surface forces, except that now the volume of the element must be considered rather than the surface area. Thus, define the nodal system N_f for the body force interpolation. That is, body force components at the nodes are

$$[f] = [f_x \ f_y] \tag{7.121}$$

Then, at any point ζ_i in the body, the value of the body force $\tilde{f}(\zeta_i)$ is given by the interpolation formula,

$$[\tilde{f}_x \ \tilde{f}_y] = \tilde{h}\,[\phi_f]^T [f_x \ f_y] \tag{7.122}$$

In eqn (7.122), $\tilde{h}$ is the variable thickness, which itself may be interpolated in terms of nodal values:

$$\tilde{h} = [\phi_h]^T \{h\} \tag{7.123}$$

Then, again assuming that the element displacement functions are given by the interpolation similar to eqn (7.113),, allows the application of the contragredient principle to the force system, giving the increment to the nodal forces,

$$\Delta[R_{xf} \ R_{yf}] = \{N\}[\tilde{f}_x \ \tilde{f}_y]dA = \tilde{h}\{N\}[\phi_f]^T dA\,[\tilde{f}_x \ \tilde{f}_y]$$

Integrating over the area,

$$[R_{xf} \ R_{yf}] = \int_{area} \tilde{h}\{N\}[\phi_f]^T dA\,[f_x \ f_y] \tag{7.124}$$

For the axisymmetric case, $\tilde{h}$ is replaced by the volume of revolution $2\pi\tilde{x}$; that is, $2\pi[N]^T\{X\}$.

Thence, for axisymmetry,

$$[R_{xf} \ R_{yf}] = 2\pi \int_{area} [N][N]^T\{X\}[\phi_f]^T dA\,[f_x \ f_y] \tag{7.125}$$

Example 7.3 *Body forces CST*

Calculate the equivalent nodal forces for the CST with constant body forces and constant thickness h. In this case,

$$[N]^T = [\zeta_1 \ \zeta_2 \ \zeta_3] \quad \text{and} \quad [\phi_f] = 1$$

so that f_x, f_y will be integrated at a single point, for example, the centroid with values f_{xc}, f_{yc}. Substitution into eqn (7.124) gives

$$[R_{xf} \ R_{yf}] = \frac{hA}{3} \begin{bmatrix} 1 \\ 1 \\ 1 \end{bmatrix} [f_{xc} \ f_{yc}] \tag{i}$$

That is, one–third of the body force is distributed to each of the apex nodes, an expected result.

Example 7.4 *Body forces LST*

Calculate the equivalent nodal forces for the LST with constant body force and constant thickness h. In this case,

$$[N]^T = [\zeta_1(2\zeta_1 - 1) \ \zeta_2(2\zeta_2 - 1) \ \zeta_3(2\zeta_3 - 1) \ 4\zeta_3\zeta_2 \ 4\zeta_1\zeta_3 \ 4\zeta_2\zeta_1]$$

Again, as in example 7.3, $\phi_f = 1$ and substitution into eqn (7.124), using eqn (1.152) to evaluate the integrals, gives the interesting result:

$$[R_{xf} \ R_{yf}] = \frac{hA}{3} \begin{bmatrix} 0 \\ 0 \\ 0 \\ 1 \\ 1 \\ 1 \end{bmatrix} [f_{xc} \ f_{yc}] \tag{ii}$$

In this case one–third of the total body force is concentrated at each of the mid–side nodes of the triangle.

7.3.4 Initial strains

Initial strains within the continuum arise from a variety of causes, the most common being temperature variation from the ambient state, and the volume changes due to change in moisture content in materials such as timber, soil and concrete. These strains are, in general, transient in nature, and if the time span is short, it may be necessary to consider the time domain in the analysis. Herein, only the steady state will be considered. Temperature strains have been discussed in some detail for framed structure members in section 5.2, and the basic theory developed therein applies in the present discussion of the two–dimensional continuum. The concept of initial stresses was also mentioned briefly in section 3.3. The fundamental concept is that the analysis must start from the kinematically determinate strain state, which implies that $\{\varepsilon_i\}$ the strain vector, must everywhere be equal to zero. Now, since, from eqn (7.32),

$$\{\varepsilon\} = [a]\{r\}$$

strains may be induced by nodal displacements, according to the assumed displacement fields within the element. If, then, $\{\varepsilon_i\}$ is the initial strain field at any point within the element, the restraint stress necessary to prevent this strain is

$$\{\sigma_i\} = -[k]\{\varepsilon_i\} \tag{7.126}$$

In eqn (7.126), $[k]$ is, of course, simply the constitutive matrix for the material, and, should include the third direct stress component for plane strain problems. Using eqn (7.32), together with the contragredient principle, gives the restraint nodal forces as

$$\{R_i\} = \int_{vol} [a]^T \{\sigma_i\} dV \qquad (7.127)$$

The forces which must be applied to release these nodal restraints are the negative of $\{R_i\}$, so that, if $\{R_r\}$ are the release forces,

$$\{R_r\} = \int_{vol} [a]^T [k] \{\varepsilon_i\} dV \qquad (7.128)$$

The analysis now proceeds in the usual way. Finally, when the stress distribution due to $\{R_i\}$ has been determined, it must be remembered that the initial state given by eqn (7.126) must be included in the solution.

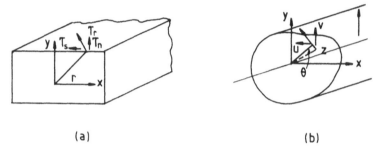

(a)   (b)

Fig. 7.14   Torsion of prismatic shaft.

7.4 TORSION OF PRISMATIC SHAFTS

7.4.1   Introduction –St Venant's theory

The St Venant's torsion solution is included at this stage because the planar discretization of the member cross section has many similarities to the planar elasticity problem, in so far as the development of its computer software is concerned. It is well known from the elementary strength of materials that, in the torsion of a prismatic shaft of circular cross section, the assumption that cross sections remain plane and rotate rigidly with respect to one another leads to a valid solution for the shear stress distribution. The shear stress at any point of the cross section is then directed at right angles to the radius vector from the centroid of the cross section. It is easily shown that this assumption is invalid for members of non–circular cross section (such as the rectangle, shown in Fig. 7.14). From Fig. 7.14(a), the shear stress τ_r has components τ_s and τ_n, and, since τ_n is non–zero, for equilibrium at the surface, longitudinal traction forces are necessary. The St Venant's torsion problem can, however, be solved as a type of initial strain situation, by first making the assumption for the calculation of the displacements (u, v) that the elementary assumption remains valid, and that the surface stresses so induced are removed by allowing the cross sections to warp in the z direction. All sections are assumed to warp alike, so that longitudinal σ_z stresses are not induced. Thus, if θ is the twist per unit length of the shaft at the cross sectional distance z from the fixed end, the three components of the displacements are given by

$$u = -y\theta z \qquad v = x\theta z; \quad \text{and} \quad w = \theta\psi(x, y) \qquad (7.129)$$

The function $\psi(x, y)$ must be determined by the analysis. Given the displacement assumptions in eqn (7.129), the in–plane shear strains are

$$\gamma_{xz} = w_{,x} + u_{,z} = \tilde{w}_{,x} - \tilde{y}\theta \qquad (7.130a)$$

$$\gamma_{yz} = w_{,y} + v_{,z} = \tilde{w}_{,y} + \tilde{x}\theta \qquad (7.130b)$$

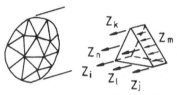

(a) Cross section $\qquad$ (b) Nodal forces on triangular element

Fig. 7.15 $\qquad$ Cross section idealization.

7.4.2 Torsion–finite element approximation

The cross section is now subdivided into finite elements, shown in Fig. 7.15 (in this case of triangular shape). Within the finite elements as shown in Fig. 7.15(a), express the longitudinal warping function $\tilde{w}$ in terms of its nodal values, using ϕ , the interpolation polynomial,

$$\tilde{w} = \{\phi\}^T \{w\} \qquad (7.131)$$

Then the x and y derivatives are given by

$$\begin{bmatrix} \tilde{w}_{,x} \\ \tilde{w}_{,y} \end{bmatrix} = \begin{bmatrix} \phi_{,x}^T \\ \phi_{,y}^T \end{bmatrix} \{w\} \qquad (7.132)$$

Thence, from eqn (7.130), the shear strains are expressed as

$$\begin{bmatrix} \gamma_{xz} \\ \gamma_{yz} \end{bmatrix} = \begin{bmatrix} \phi_{,x}^T \\ \phi_{,y}^T \end{bmatrix} \{w\} + \begin{bmatrix} 0 & -\theta \\ \theta & 0 \end{bmatrix} \begin{bmatrix} \tilde{x} \\ \tilde{y} \end{bmatrix} \qquad (7.133)$$

Evidently, $\tilde{x}, \tilde{y}$ can also be interpolated in terms of nodal values, and for the straight–sided triangle,

$$\tilde{x} = [\zeta_1 \ \zeta_2 \ \zeta_3] \begin{bmatrix} x_1 \\ x_2 \\ x_3 \end{bmatrix} = \{\phi_1\}^T \{X\} \qquad (7.134)$$

and, similarly, for $\tilde{y} = \{\phi_1\}^T \{Y\}$. Substituting these values in eqn (7.133) gives

$$\begin{bmatrix} \gamma_{xz} \\ \gamma_{yz} \end{bmatrix} = \begin{bmatrix} \phi_{,x}^T \\ \phi_{,y}^T \end{bmatrix} \{w\} + \begin{bmatrix} 0 & -\theta\phi_1^T \\ \theta\phi_1^T & 0 \end{bmatrix} \begin{bmatrix} X \\ Y \end{bmatrix} \qquad (7.135)$$

Eqn (7.135) is written

$$\tilde{\varepsilon} = [T]\{w\} + \{\varepsilon_0\} \qquad (7.136)$$

It is noted that the second term on the right–hand side of this equation is independent of the warped shape and that ε_0 is the initial strain produced by the distortions which violate the stress boundary conditions. The constitutive equation relating shear stresses and strains is, simply,

$$\begin{bmatrix} \tau_{xz} \\ \tau_{yz} \end{bmatrix} = \begin{bmatrix} G & 0 \\ 0 & G \end{bmatrix} \begin{bmatrix} \gamma_{xz} \\ \gamma_{yz} \end{bmatrix} \tag{7.137}$$

or, symbolically,

$$\{\tau\} = [G]\{\overline{\varepsilon}\} \tag{7.138}$$

Thence, combining eqns (7.136) and (7.138),

$$\{\tau\} = [G][T]\{w\} + [G]\{\varepsilon_0\} \tag{7.139}$$

The nodal forces which are to be considered are the shears complementary to those on the XY plane of the cross section, and, for a single element, they are shown in Fig. 7.15(b). The contribution to these nodal forces for an infinitesimal area is (applying contragredience)

$$\Delta\{R\} = [T]^T\{\tau\} \tag{7.140}$$

For the whole element,

$$\{R\} = \int\limits_{area} [T]^T[G][T]dA\,\{w\} + \int\limits_{area} [T]^T[G]\{\varepsilon_0\}dA \tag{7.141}$$

For the whole finite element configuration, all contributions to the node forces from each triangle must be added. At an internal node or at a free boundary, R_z forces must vanish. The solution of eqn (7.141) for all elements thus takes the form,

$$[K]\{w\} = -\{R_0\} \tag{7.142}$$

where

$$[K] = \sum\int [T]^T[G][T]dA \tag{7.143a}$$

and

$$\{R_0\} = \sum\int [T]^T[G][\varepsilon_0]dA \tag{7.143b}$$

For the general cross section without axes of symmetry, one value of $\{w\}$ is set to zero. On axes of symmetry $\{w\} = 0$ and only a portion of the cross section need be considered.

The above theory is based on the approximation to the Laplace equation,

$$\nabla^2 w = -2G\theta \tag{i}$$

An alternative approach to the solution to the St Venant's torsion problem is to use a stress function, ϕ, which leads to Poisson's equation,

$$\nabla^2\phi = 0 \tag{ii}$$

In the finite element analysis, the solution to (i) is much to be preferred to (ii) because, in the latter case, voids must be discretized, whereas, in the former, they do not.

7.4.3   Calculation of the shear constant and shear stress distribution

From eqn (7.142) written for all node points, the solution is obtained for w, the warping function, in terms of the twist per unit length θ. It is seen from eqn (7.129c), that $\psi(x, y) = w$ for $\theta = 1$. Once w has been approximated, the element shear stresses are calculated from eqn (7.139). That is,

$$\begin{bmatrix} \tau_{xz} \\ \tau_{yz} \end{bmatrix} = G\theta \left(\begin{bmatrix} \phi_{,x}^T \\ \phi_{,y}^T \end{bmatrix} \{w_1\} + \begin{bmatrix} 0 & -\theta\phi_1^T \\ \theta\phi_1^T & 0 \end{bmatrix} \begin{Bmatrix} X \\ Y \end{Bmatrix} \right) \tag{7.144}$$

The twisting moment M_T may be obtained as a function of θ, since

$$M_T = \int_{area} (\tilde{x}\tau_{yz} - \tilde{y}\tau_{xz})dA \tag{7.145}$$

For a given finite element the contribution to M_T is

$$\Delta M_T = \int_{area} [\tilde{y} \ \tilde{x}] \begin{bmatrix} -\tau_{xz} \\ \tau_{yz} \end{bmatrix} dA$$

Now,

$$\begin{bmatrix} \tilde{y} \\ \tilde{x} \end{bmatrix} = \begin{bmatrix} 0 & \phi_1^T \\ \phi_1^T & 0 \end{bmatrix} \begin{bmatrix} X \\ Y \end{bmatrix}$$

and the shear stresses have been calculated in eqn (7.144) thence,

$$\Delta M_T = [X^T \ Y^T]G \int_{area} \begin{bmatrix} 0 & \phi_1^T \\ \phi_1^T & 0 \end{bmatrix} \left(\begin{bmatrix} -\phi_{,x}^T \\ \phi_{,y}^T \end{bmatrix} \theta \{w_1\} + \begin{bmatrix} 0 & \phi_1^T \\ \phi_1^T & 0 \end{bmatrix} \begin{Bmatrix} X \\ Y \end{Bmatrix} \theta \right) dA \tag{7.146}$$

In eqn (7.146), w_1 is the warping surface calculated for $\theta = 1$. For all triangles, eqn (7.146) becomes

$$M_T = K_s\theta \tag{7.147}$$

so that, for a given torque M_T,

$$\theta = \frac{M_T}{K_s} \tag{7.148}$$

Thus, if M_T is given, θ is calculated from eqn (7.148) and, thence, eqn (7.144) gives the elemental stress values.

Example 7.5 *Cross section modelled with three–node triangles*

In this case,

$$\phi^T = \phi_1^T = [\zeta_1 \ \zeta_2 \ \zeta_3] \tag{i}$$

Hence the x and y derivatives are given by

$$\{\phi_{1,x}\}^T = \frac{1}{2A}\{b_1 \ b_2 \ b_3\} \equiv \frac{1}{2A}\{b\}^T$$

$$\{\phi_{1,y}\}^T = \frac{1}{2A}\{a_1 \ a_2 \ a_3\} \equiv \frac{1}{2A}\{a\}^T$$

and, thence,

$$[T] = \frac{1}{2A}\begin{bmatrix} b_1 & b_2 & b_3 \\ a_1 & a_2 & a_3 \end{bmatrix} \tag{ii}$$

and

$$[T]^T[G][T] = \frac{G}{4A^2}\begin{bmatrix} b_1 & a_1 \\ b_2 & a_2 \\ b_3 & a_3 \end{bmatrix}\begin{bmatrix} b_1 & b_2 & b_3 \\ a_1 & a_2 & a_3 \end{bmatrix}$$

Integrating over the area of the triangle,

$$\int_{area} [T]^T[G][T]\, dA = \frac{G}{4A}[b\ a]\begin{bmatrix} b^T \\ a^T \end{bmatrix} = \frac{G}{4A}[bb^T + aa^T] \tag{iii}$$

The term, $\int[T]^T[G]\{\varepsilon_0\}dA$, is given by

$$\int_{area} [T]^T[G]\{\varepsilon_0\}dA = \frac{G}{2A}\int_{area}[b\ a]\begin{bmatrix} 0 & -\phi_1^T \\ \phi_1^T & 0 \end{bmatrix}\begin{bmatrix} X \\ Y \end{bmatrix}dA\,\theta = \frac{G}{2A}[b\ a]\begin{bmatrix} -\bar{y} \\ \bar{x} \end{bmatrix}\theta \tag{iv}$$

In eqn (iv),

$$\bar{y} = \frac{1}{3}(y_1 + y_2 + y_3) \quad \text{and} \quad \bar{x} = \frac{1}{3}(x_1 + x_2 + x_3)$$

are the coordinates of the triangle centroid. Thence, the nodal force vector $\{R\}$ is given by

$$\{R\} = \begin{bmatrix} R_{zi} \\ R_{zj} \\ R_{zk} \end{bmatrix} = \frac{G}{4A}[bb^T + aa^T]\{w_1\}\theta + \frac{G\theta}{2}[b\ a]\begin{bmatrix} -\bar{y} \\ \bar{x} \end{bmatrix} \tag{v}$$

Example 7.6 *Cross section modelled with six–node triangles*

The St Venant's torsion problem responds well to analysis using the six–node triangle, and accurate results are obtained for quite coarse element subdivision. In this case,

$$\{\phi\}^T = \{\phi_2\}^T = [\zeta_1(2\zeta_1 - 1)\ \zeta_2(2\zeta_2 - 1)\ \zeta_3(2\zeta_3 - 1)\ 4\zeta_2\zeta_3\ 4\zeta_3\zeta_1\ 4\zeta_1\zeta_2] \tag{i}$$

Then define the matrix $[V]$, such that

$$[V] = \begin{bmatrix} 3v_1^2 & -v_1v_2 & -v_1v_3 & 0 & 4v_1v_3 & 4v_1v_2 \\ -v_1v_2 & 3v_2^2 & -v_2v_3 & 4v_2v_3 & 0 & 4v_1v_2 \\ -v_1v_3 & -v_2v_3 & 3v_3^2 & 4v_2v_3 & 4v_1v_3 & 0 \\ 0 & 4v_2v_3 & 4v_2v_3 & 8\alpha_1 & 8v_1v_2 & 8v_1v_3 \\ 4v_1v_3 & 0 & 4v_1v_3 & 8v_1v_2 & 8\alpha_2 & 8v_2v_3 \\ 4v_1v_2 & 4v_1v_2 & 0 & 8v_1v_3 & 8v_2v_3 & 8\alpha_3 \end{bmatrix} \tag{ii}$$

When i, j, k are cyclic permutations of 1, 2, 3, α_i is defined by

$$\alpha_i = v_j^2 + v_k^2 + v_j v_k \tag{iii}$$

Then, from eqn (7.143),

$$[k_{elt}] = \int\limits_{area} [T]^T [G][T]\, dA = \frac{G}{12A}[V_a + V_b] \tag{iv}$$

where V_a and V_b are obtained from (ii) by substituting $v_i = a_i$ and $v_i = b_i$, respectively.
Similarly, define the matrix $[E]$,

$$[E] = \begin{bmatrix} e_1 & 0 & 0 \\ 0 & e_2 & 0 \\ 0 & 0 & e_3 \\ e_2 + e_3 & e_2 + 2e_3 & e_3 + 2e_2 \\ 2e_3 + e_1 & e_3 + e_1 & e_3 + 2e_1 \\ e_1 + 2e_2 & 2e_1 + e_2 & e_1 + e_2 \end{bmatrix} \tag{v}$$

Then,

$$\int\limits_{area} [T]^T [G]\{\varepsilon_0\} dA = \frac{G\theta}{6}[-E_b Y + E_a X] \tag{vi}$$

where $[E_a]$ and $[E_b]$ are obtained by substituting $e_i = a_i$ and $e_i = b_i$, respectively, in eqn (v).
   In eqn (7.147), the first term now become

$$\Delta M_T = [X^T\ Y^T]G \int\limits_{area} \begin{bmatrix} \phi_1\phi_{,y}^T \\ -\phi_1\phi_{,x}^T \end{bmatrix} dA\,\{w_1\}\theta = \frac{G}{6}[X^T\ Y^T]\begin{bmatrix} E_a^T \\ -E_b^T \end{bmatrix}\{w_1\}\theta \tag{vii}$$

   The program strategy is to set up the stiffness matrix and the $[E]$ matrices, element by element. The initial strain matrices with sign changed are loaded into the load vector. The torsion constant is obtained from eqn (7.148) as the applied torque for $\theta = 1$. The stress distribution for $M_T = 1$ is obtained, element, by element using eqns (7.144) and (7.149).

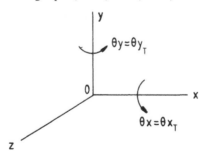

Fig. 7.16   Rotations of XY plane.

7.4.4   Centre of twist–shear centre

The displacements of the member cross section were calculated (eqn (7.129)):

$$u = -\theta zy \quad \text{and} \quad v = \theta zx$$

In these equations the assumption was made that the displacement of the origin was zero. In fact, the displacement is zero at the shear center or center of twist, (coordinates (x_T, y_T)). Then the cross section displacements are given by

$$u = -\theta z (y - y_T) \quad \text{and} \quad v = \theta z (x - x_T) \tag{7.149}$$

Now examine the expressions, see (Fig. 7.16),

$$u = \theta z y_T \quad \text{and} \quad v = -\theta z x_T$$

From Fig. 7.16, for θ_x,

$$v = -\theta_x z$$

and, for θ_y,

$$u = \theta_y z$$

Thus, the above terms represent rotations about the $y-$ and $x-$ axes respectively. Thus, write the warping displacement w

$$w = \theta \psi(x, y) + \theta x_T y - \theta y_T x \tag{7.150}$$

To locate x_T, y_T, reason as follows. The w function should be such that if a σ_z stress exists representing pure bending about either the $x-$ or the $y-$ axis, the generalized force corresponding to the w displacement produced by pure twist should be zero. For example, for an M_x generalized force (i.e. a moment), the stress distribution is

$$\sigma_z = cy$$

and the only generalized force is M_x. Thus, give the cross section the generalized displacement of eqn (7.150). Then,

$$R_z = \int \sigma_z w dA = \int cy (\theta \psi(x, y) + \theta x_T y - \theta y_T x) \, dA \tag{7.151}$$

If the origin is at the centroid and principal axes are used,

$$\int y^2 dA = I_x \quad \text{and} \quad \int xy dA = 0$$

and eqn (7.151) locates x_T, since, then,

$$x_T = -\frac{\int y \, \psi(x, y) dA}{I_x} \tag{7.152a}$$

Similarly,

$$y_T = \frac{\int x \, \psi(x, y) dA}{I_y} \tag{7.152b}$$

If non–principal axes are used, then eqn (7.151) and the corresponding equation for y_T result in two simultaneous equations to be solved for x_T, y_T in terms of $\psi(x,y)$, I_x, I_y, and I_{xy}. The function $\psi(x,y)$ has been approximated in the finite element analysis as the warped shape for $\theta = 1$. In this case, and for principal axes, let

$$\psi(x, y) = \{\phi\}^T \{w\}$$

and, as before,

$$y = \{\phi_1\}^T \{Y\} \quad \text{and} \quad x = \{\phi_1\}^T \{X\}$$

It follows that

$$x_T \approx -\frac{\sum\{Y\}^T \int \phi_1 \phi^T \, dA \{w\}}{\sum\{Y\}^T \int \phi_1^T \phi \, dA \{Y\}} \tag{7.153(a)}$$

and

$$y_T \approx \frac{\sum\{X\}^T \int \phi_1 \phi^T \, dA \{w\}}{\sum\{X\}^T \int \phi_1^T \phi_1 \, dA \{X\}} \tag{7.153(b)}$$

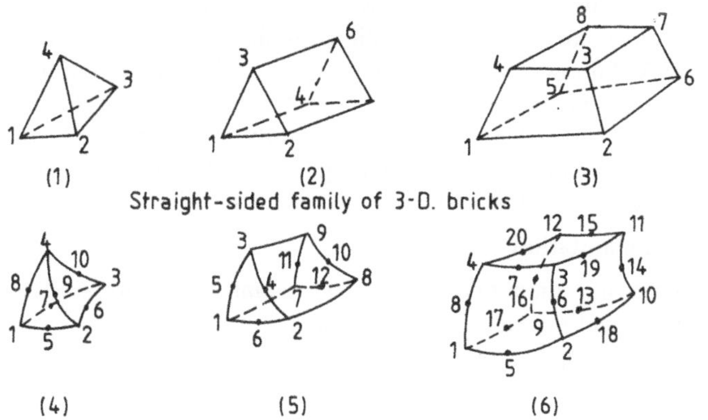

Fig. 7.17   Curved–sided family of three–dimensional bricks.

7.5 THE THREE–DIMENSIONAL SOLID

7.5.1   Introduction

The various element configurations are shown in Fig. 7.17. The element family in Fig. 7.17(a) has apex nodes only, and so their sides are straight, whereas, in Fig. 7.17(b), the additional mid–side nodes have been added. The element number (6) in Fig. 7.17(b) is the 20–node isoparametric brick. There are other versions possible here, for example, a 27–node Lagrange type element with a centre node and nodes in the centre of each face. The details of such elements are not provided because they require simple routine modifications of the element given herein. In three–dimensional problems, it is found that the magnitude of solving the nodal equations of equilibrium quickly becomes an overwhelming task as the mesh dimensions are increased. For example, a cube of 10 eight–node bricks contains 1331 nodes or 3993 degrees of freedom, and the corresponding mesh of 20–node bricks contains 5082 nodes or 15246 degrees of freedom.

Clearly, the size of the problem has become nearly intractable if all three dimensions require the same degree of mesh fineness. Considerable attention to programming design, both in element stiffness generation and the out–of–core equation solver is necessary to avoid excessive computer CPU times. Nevertheless, it may become apparent that for such three–dimensional solids, the boundary element method becomes a viable alternative to the finite element method, because, with this technique, the discretization process is confined to the surface of the domain so that the number of node points (and hence the number of equations to be solved) increases only as the square of the dimensions of the problem, and not as the cube.

With the finite element method in three dimensions, data preparation becomes a serious problem, and automatic mesh generation with graphical display of data, containing such features as partial mesh display and the option for hidden line removal, becomes an absolute necessity if error–free data files are to be produced. Complex structural shapes may be difficult to model with the isoparametric bricks alone, and the element library should contain both the tetrahedron and triangular prismatic bricks (1, 2, 3, 4 in Fig. 7.17). The calculation of element stiffness matrices is along lines very similar to those for the planar element counterparts. For example, for the straight–sided tetrahedra (4, 6, or 8) nodes, the integration of the zeta functions over the volume can be obtained explicitly, whereas for the isoparametric bricks (or any of the elements with curved sides), numerical integration is again required. Because the integration extends over the third dimension, care must be taken to use the minimum order sufficient for the accurate calculation of the stiffness matrix. The minimum requirements are second–order integration for the 8–node brick, and third–order integration for the 20–node brick. Following the same principles as for the two–dimensional elements, it is seen that stress values may be calculated at the centre point of the 8–node brick, and at the $(2 \times 2 \times 2)$ Gauss points for the 20–node brick. Evidently, the display of stress values and stress contouring will require considerable ingenuity, and again it will be demonstrated how well the boundary element formulation addresses this problem.

7.5.2 Stiffness matrix formulation

The various element types, like their two–dimensional counterparts, are sufficiently distinct in their properties to warrant individual development. The tetrahedron family is discussed first, because, like the triangle, it allows a very simple explicit formulation.

The tetrahedron

Constant strain

The stiffness matrix for the tetrahedron can be obtained by using the interpolation functions $(\zeta_1, \zeta_2, \zeta_3, \zeta_4)$. The transformations for the derivatives of the displacement functions expressed in these coordinates and the global (x_i) set are obtained in a similar way as for the triangular element. Alternatively, the natural mode technique can be used again, as for the triangle, adopting the concept of the ur–element [39] (which in this case is an elementary tetrahedron) and all the ramifications thereof. The non–dimensional coordinates (ζ_i) are defined in terms of the volume ratios V_i/V, where V_i is the volume contained between the side area whose opposite apex is node i and the interior point. The apex nodes thus have the non–dimensional coordinates (1, 0, 0, 0), (0, 1, 0, 0), etc., and

$$\zeta_1 + \zeta_2 + \zeta_3 + \zeta_4 = 1 \tag{7.154}$$

The integral of the ζ polynomials over the volume of the tetrahedron has been given in eqn (1.153)

$$I_3 = \int_V \zeta_1^p \zeta_2^q \zeta_3^r \zeta_4^s \, dV = 6V \frac{p\,!q\,!r\,!s\,!}{(p + q + r + s + 3)!}$$

The coordinate transformations are established by noting that $\tilde{x}_i$ within the element can be interpolated by

$$\tilde{x}_i = \{\phi_1\}^T \{x_i\} \tag{7.155}$$

where

$$\{\phi_1\}^T = [\zeta_1 \ \zeta_2 \ \zeta_3 \ \zeta_4]$$

That is, for all three coordinates,

$$
\begin{bmatrix} 1 \\ \tilde{x}_1 \\ \tilde{x}_2 \\ \tilde{x}_3 \end{bmatrix}
=
\begin{bmatrix}
1 & 1 & 1 & 1 \\
x_{11} & x_{12} & x_{13} & x_{14} \\
x_{21} & x_{22} & x_{23} & x_{24} \\
x_{31} & x_{32} & x_{33} & x_{34}
\end{bmatrix}
\begin{bmatrix} \zeta_1 \\ \zeta_2 \\ \zeta_3 \\ \zeta_4 \end{bmatrix}
\tag{7.156}
$$

The determinant of this coefficient matrix is easily calculated to be $6V$. This can be proven by calculating the volume V of the tetrahedron as the scalar triple product,

$$6V = \vec{A} \cdot \vec{B} \times \vec{C} \tag{7.157}$$

In this equation the vectors $\vec{A}, \vec{B}$ and $\vec{C}$ correspond to three sides of the tetrahedron, such that

$$\vec{A} = \{x_{i4} - x_{i1}\} \ ; \quad \vec{B} = \{x_{i2} - x_{i1}\} \ ; \quad \vec{C} = \{x_{i3} - x_{i1}\}$$

This result will be found to agree with the determinant of the coefficient matrix in eqn (7.156). The scalar triple product in eqn (7.157) gives the Jacobian of the transformation between the coordinate systems as

$$
[J] = 6V =
\begin{bmatrix}
x_{41} - x_{11} & x_{42} - x_{12} & x_{43} - x_{13} \\
x_{21} - x_{11} & x_{22} - x_{12} & x_{23} - x_{13} \\
x_{31} - x_{11} & x_{32} - x_{12} & x_{33} - x_{13}
\end{bmatrix}
\tag{7.158}
$$

The inverse transformation to eqn (7.156) relates ζ coordinates to the global coordinates within the tetrahedron and is written

$$
\begin{bmatrix} \zeta_1 \\ \zeta_2 \\ \zeta_3 \\ \zeta_4 \end{bmatrix}
=
\frac{1}{6V}
\begin{bmatrix}
A_{11} & b_1 & c_1 & d_1 \\
A_{21} & b_2 & c_2 & d_2 \\
A_{31} & b_3 & c_3 & d_3 \\
A_{41} & b_4 & c_4 & d_4
\end{bmatrix}
\begin{bmatrix} 1 \\ \tilde{x}_1 \\ \tilde{x}_2 \\ \tilde{x}_3 \end{bmatrix}
\tag{7.159}
$$

Note that

$$\sum_{i=1,4} b_i = \sum_{i=1,4} c_i = \sum_{i=1,4} d_i = 0 \tag{7.160}$$

The terms b_i, c_i, d_i are the cofactors of the second, third and fourth columns of the transposed coefficient matrix in eqn (7.156).

Then the derivatives with respect to to the global coordinates are given by the chain rule as

$$f_{,x_1} = \frac{1}{6V} b_i f_{,\zeta_i}, \text{ sum } i = 1,4$$

$$= \frac{1}{6V} \{b\}^T \{f_{,\zeta}\} \tag{7.161}$$

with similar expressions for the derivatives with respect to x_2, x_3 replacing $\{b\}$ by $\{c\}$ and $\{d\}$, respectively. The strain displacement transformation is now given in eqn (7.162) and is constant throughout the tetrahedron. That is, it is a constant stress element.

$$
\begin{bmatrix} \varepsilon_{x_1} \\ \varepsilon_{x_2} \\ \varepsilon_{x_3} \\ \gamma_{x_1 x_2} \\ \gamma_{x_2 x_3} \\ \gamma_{x_3 x_1} \end{bmatrix} = \frac{1}{6V} \begin{bmatrix} b^T & 0 & 0 \\ 0 & c^T & 0 \\ 0 & 0 & d^T \\ c^T & b^T & 0 \\ 0 & d^T & c^T \\ d^T & 0 & b^T \end{bmatrix} \begin{bmatrix} U_1 \\ U_2 \\ U_3 \end{bmatrix} \tag{7.162}
$$

This is the expression:

$$\{\varepsilon\} = [a]\{r\}$$

The constitutive equation for homogeneous isotropic material is, from eqn (2.132), expressed by

$$\{\sigma\} = [k]\{\varepsilon\} \tag{7.163}$$

Using eqn (3.21), the element stiffness matrix is calculated:

$$[k_e] = \int_{vol} [a]^T [k][a] \, dV = \frac{1}{36V} [a]^T [k][a] \tag{7.164}$$

The linear strain tetrahedron

The linear strain tetrahedron has quadratic displacement interpolation, and, using the nodal point system in Fig. 7.17(b), the interpolation polynomial $\{\phi_2\}$ is given by

$$\{\phi_2\} = <\zeta_1(2\zeta_1 - 1) \cdots \zeta_4(2\zeta_4 - 1), 4\zeta_1\zeta_2, 4\zeta_2\zeta_3, 4\zeta_3\zeta_1, 4\zeta_2\zeta_4, 4\zeta_3\zeta_4> \tag{7.165}$$

Again, the similarity with $\{\phi_2\}$ for the LST should be noted. The strains are interpolated linearly, being first derivatives of the displacements, so that, if $\{\varepsilon_c\}$ is the vector of nodal strains,

$$\{\varepsilon_c\}^T = [\bar{\varepsilon}_{x_1} \ \bar{\varepsilon}_{x_2} \ \bar{\varepsilon}_{x_3} \ \bar{\gamma}_{x_1 x_2} \ \bar{\gamma}_{x_2 x_3} \ \bar{\gamma}_{x_3 x_1}]^T \tag{7.166}$$

where

$$\{\bar{\varepsilon}_{x_1}\}^T = [\varepsilon_{x_1}1 \ \varepsilon_{x_1}2 \ \varepsilon_{x_1}3 \ \varepsilon_{x_1}4]^T, \text{ etc.} \tag{7.167}$$

Then,

$$\{\varepsilon\} = \left[\phi_1^T\right]_6 \{\varepsilon_c\} = [\phi]\{\varepsilon_c\} \tag{7.168}$$

The notation $\lceil-\rfloor_6$ implies a matrix with six diagonal blocks each equal to the enclosed matrix. The generalized stiffness matrix $[k_c]$ relating the nodal stresses to the nodal strains is calculated as

$$[k_c] = \int_{vol} [\phi]^T [k][\phi] \, dV \tag{7.169}$$

noting that

$$\int_V \{\phi_1\}\{\phi_1\}^T \, dV = \frac{V}{20} \begin{bmatrix} 2 & 1 & 1 & 1 \\ 1 & 2 & 1 & 1 \\ 1 & 1 & 2 & 1 \\ 1 & 1 & 1 & 2 \end{bmatrix} = \frac{V}{20}[\Lambda]$$

The matrix $[k_c]$ is written

$$[k_c] = \frac{VE^*}{20}[k]^*[\Lambda] \tag{7.170}$$

The terms in $[k]$ and E^* are defined in eqn (2.132), and the notation $[k]^*[\Lambda]$ implies that each term of $[k]$ is to be replaced by the 4×4 submatrix $k_{ij} \times [\Lambda]$, thus making $[k_c]$ into a 24×24 matrix. The nodal strains are expressed in terms of the nodal displacements, using the transformation matrix $[\Phi_c]$. To this end, the derivative transformation matrix $[T_t]$ is defined by

$$[T_t] = \frac{1}{6V} \begin{bmatrix} 3t_1 & -t_2 & -t_3 & -t_4 & 4t_2 & 0 & 4t_3 & 4t_4 & 0 & 0 \\ -t_1 & 3t_2 & -t_3 & -t_4 & 4t_1 & 4t_3 & 0 & 0 & 4t_4 & 0 \\ -t_1 & -t_2 & 3t_3 & -t_4 & 0 & 4t_2 & 4t_1 & 0 & 0 & 4t_4 \\ -t_1 & -t_2 & -t_3 & 3t_4 & 0 & 0 & 0 & 4t_1 & 4t_2 & 4t_3 \end{bmatrix} \tag{7.171}$$

Then, for example, the relationship between the nodal x_1 strains $\{\bar{\varepsilon}_{x_1}\}$ and nodal displacements x_1 is given, using replacement in $[T_t]$, $t \to b$:

$$\{\bar{\varepsilon}_{x_1}\} = [T_b]\{U_1\} \tag{7.172}$$

For all components,

$$\{\varepsilon_c\} = [\Phi_c]\{r\} \tag{7.173}$$

The matrix $[\Phi_c]$ is a 24×30 matrix similar in form to that in eqn (7.162). Now, however, $\{b\}^T$ is replaced by $[T_b]$ etc. The nodal displacement vector $\{r\}$ has been grouped as

$$\{r\}^T = <U_1^T \ U_2^T \ U_3^T>$$

where

$$\{U_1\}^T = <U_{x_11} \ U_{x_12} \cdots \ U_{x_110}> \tag{7.174}$$

Thence (as in eqn (7.21) for the triangular element), the 30×30 linear strain tetrahedron stiffness matrix is given by

$$[k_e] = [\Phi_c]^T [k_c][\Phi_c] \tag{7.175}$$

7.5.3 The three–dimensional isoparametric family

The three–dimensional isoparametric family is given in Fig. 7.17, element numbers 3 and 6, representing the 8– and 20–node bricks, respectively, whose shape functions have been given in section 1.4.4. The theory for these elements is identical to that given in section 7.2.3 except that now integration must be performed over the three dimensions x_1, x_2, x_3. The equation corresponding to the two–dimensional case (eqn 7.31), relating strains at a point within the element to the nodal displacements, is written

$$\{\varepsilon\} = \begin{bmatrix} \varepsilon_{11} \\ \varepsilon_{22} \\ \varepsilon_{33} \\ \gamma_{12} \\ \gamma_{23} \\ \gamma_{31} \end{bmatrix} = \begin{bmatrix} \tilde{u}_{1,1} \\ \tilde{u}_{2,2} \\ \tilde{u}_{3,3} \\ \tilde{u}_{1,2} + \tilde{u}_{2,1} \\ \tilde{u}_{2,3} + \tilde{u}_{3,2} \\ \tilde{u}_{3,1} + \tilde{u}_{1,3} \end{bmatrix} = \begin{bmatrix} \phi_{,1}^T & 0 & 0 \\ 0 & \phi_{,2}^T & 0 \\ 0 & 0 & \phi_{,3}^T \\ \phi_{,2}^T & \phi_{,1}^T & 0 \\ 0 & \phi_{,3}^T & \phi_{,2}^T \\ \phi_{,3}^T & 0 & \phi_{,1}^T \end{bmatrix} \begin{bmatrix} U_1 \\ U_2 \\ U_3 \end{bmatrix} \tag{7.176}$$

where $\tilde{u}_{1,i}$, $\phi_{,i}$ etc., denote differentiation with respect to x_i. This is the equation:

$$\{\varepsilon\} = [a]\{r\}$$

The constitutive relationship eqn (2.132) is written

$$\{\sigma\} = [k]\{\varepsilon\}$$

Thence the element stiffness matrix is calculated as

$$[k_e] = \sum_i \sum_j \sum_k w_i w_j w_k [a_{ijk}]^T [k][a_{ijk}] \det J_{ijk} \tag{7.177}$$

If the matrix multiplication in eqn (7.177), $[a]^T[k][a]$, is carried out for the 20–node brick and a 27 point integration rule used, then 27 (60×60) matrices will be calculated in order to generate $[k_e]$. Approximately three–quarters of a million multiplications are required. However, if the $[a]$ matrix in eqn (7.176) is written

$$[a] = \begin{bmatrix} A & 0 & 0 \\ 0 & B & 0 \\ 0 & 0 & C \\ B & A & 0 \\ 0 & C & B \\ C & 0 & A \end{bmatrix}$$

The matrix product $[a]^T[k][a]$ is written, using the notation $v1 = (1 + v)$ and $v2 = (1 - 2v)/2$,

$$E^* \begin{bmatrix} v1A^TA + v2(B^TB + C^TC) & vA^TB + v2B^TA & vA^TC + v2C^TA \\ & v1B^TB + v2(A^TA + C^TC) & vB^TC + v2C^TB \\ \text{symmetric} & & v1C^TC + v2(B^TB + A^TA) \end{bmatrix} \tag{7.178}$$

Now it is seen that integration involves the (20×1) vectors $\{A\}$, $\{B\}$ and $\{C\}$, and that their products, $\{A\}^T\{A\}$, $\{B\}^T\{B\}$, $\{C\}^T\{C\}$, $\{A\}^T\{B\}$, $\{A\}^T\{C\}$, $\{B\}^T\{C\}$, $\{C\}^T\{A\}$, $\{C\}^T\{B\}$; are required; in all only eight, 20×20 matrices.

It is found that only about one–tenth of the number of operations are required as in the direct use of eqn (7.177). This is reflected in a corresponding reduction in CPU time necessary to calculate an element stiffness matrix. It should be noted that a reduction in time can be achieved by similar programming in the two–dimensional case. However, the result there is nowhere as spectacular as for the threedimensional case because the problem of CPU time is not as critical.

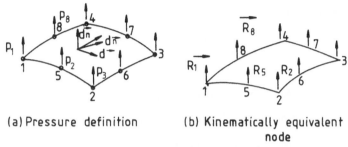

(a) Pressure definition   (b) Kinematically equivalent
                                                node

Fig. 7.18   Surface force definition.

7.5.4   Surface forces

The simple case of normal or hydrostatic pressure loading acting on the surface is considered. The nodal pressures p_i are given at the the surface can be interpolated, using eqn (1.108):

$$\tilde{p}(\zeta) = \{\phi\}^T \{\bar{p}(\zeta)\} \tag{7.179}$$

To obtain the direction of the normal, and hence its x_i components at (ζ, η), the vectors in the isoparametric directions (ζ, η) must first be calculated. Thus, let

$$[X] = [\{\bar{x}_1\} \ \{\bar{x}_2\} \ \{\bar{x}_3\}] \tag{7.180}$$

be the matrix with nodal coordinate components as columns. Then the vectors in the parametric directions are

$$\vec{ds}_\zeta = [X]^T \{\phi_{,\zeta}\} \tag{7.181}$$

$$\vec{ds}_\eta = [X]^T \{\phi_{,\eta}\} \tag{7.182}$$

Hence the normal is obtained by the cross product,

$$\vec{ds}_\xi = \vec{ds}_\zeta \times \vec{ds}_\eta$$

and the unit normal is

$$\{n\} = \frac{\vec{ds}_\xi}{|ds_\xi|} \tag{7.183}$$

Thus, a normal pressure $\tilde{p}(\zeta)$ on the surface acting on the elemental area $ds_\zeta \, ds_\eta$, has force components, in the x_i coordinate directions,

$$\{\Delta p\} = \tilde{p}(\zeta)\{n\}|ds|d\zeta d\eta = \{n\}\{\phi\}^T \{\tilde{p}(\zeta)\}|ds|d\zeta d\eta \tag{7.184}$$

Now the displacements at a point on the surface are interpolated from nodal values, using the same interpolation polynomial, as

$$\tilde{u}_i = \{\phi\}^T \{\bar{u}_i\} \tag{7.185}$$

Contragredience thus gives the contribution to the kinematically equivalent node forces as

$$\{\Delta R_i\} = \{\phi\}\Delta p_i = \{\phi\}\{\phi\}^T n_i |ds| d\zeta d\eta \{\bar{p}_\zeta\}$$

Integration over the surface gives

$$\{R_i\} = \int_{area} \{\phi\}\{\phi\}^T n_i |ds| d\zeta d\eta \{\bar{p}_\zeta\} \tag{7.186}$$

Using numerical quadrature,

$$\{R_i\} = \sum\sum w_\alpha w_\beta \{\phi_{\alpha\beta}\}\{\phi_{\alpha\beta}\}^T n_{i\alpha\beta} |ds_{\alpha\beta}| \{\bar{p}_\zeta\} \tag{7.187}$$

In this case a (3×3) integration order will be adequate to generate the nodal loads.

7.6 HEAT TRANSFER–FINITE ELEMENT APPROXIMATION

7.6.1 Introduction

It has been shown (section 7.3.4) that initial strains within an elastic body, in general, produce internal stresses and perhaps external reactions. These external reactions will necessarily be self-equilibrating and so have zero external resultant force. Initial strains may occur from a variety of causes, such as shrinkage and creep in concrete. However, perhaps the most common occurrence is due to a variation in temperature from the ambient condition. Temperature variation is caused by many effects, the most obvious of which is the daily and seasonal thermal cycles produced by the sun. Other potentially more significant effects can be produced by heat of hydration in concrete construction, local heat sources in welding, skin friction heat in high–speed aircraft, and a multitude of other sources associated with manufacturing processes. All of these effects will cause internal strains and hence stresses in the structure in which they occur, thus having an influence on the performance and integrity thereof. There are fundamentally two types of heat transfer problem:

1.   the steady state, in which conditions are considered stable and the temperature at any point in the body is constant;

2.   the transient state, in which conditions are time dependent, and the temperature at any point in the body is changing.

7.6.2 Fourier's law for heat conduction

The basis for the calculation of temperature throughout a body is the conduction ent eof heat law, which is a statement of the observation that the rate of flow of heat depends on the inverse temperature gradient; that is, heat flows from a high–temperature region to a lower–temperature region. The rate of flow, $Q_{,t}$, is given by the equation,

$$Q_{,t} = -kAT_{,l} \tag{7.188}$$

From Fig. 7.19 it is seen that A is the area through which the heat energy flows, l is the direction of its normal, Q is the quantity of heat, and k is a constant for the material called the thermal conductivity.

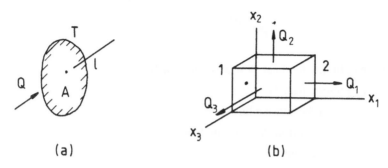

Fig. 7.19   Heat flow–global components.

The expression in eqn (7.188) is generalized for heat flow within a three–dimensional body, and the differential equation for heat transfer by conduction is obtained. From Fig. 7.19(b), the heat flows in the x_1, x_2, x_3 coordinate directions are Q_1, Q_2, Q_3, respectively, and the body is considered to be homogeneous and isotropic, with the conductivity k the same in all directions. Considering first the x_1 direction, the heat flow on the (1) face is

$$dQ_{11,t} = -kdx_2 dx_3 \frac{\partial T}{\partial x_1} dt$$

and, on the (2) face,

$$dQ_{12,t} = -kdx_2 dx_3 \left[\frac{\partial T}{\partial x_1} + \frac{\partial^2 T}{\partial x_1^2} dx_1 \right] dt$$

Similar expressions may be written for the x_2 and x_3 directions. Heat may be stored or taken from the infinitesimal element, depending on the temperature change, and the change in stored energy is given by

$$dQ_v = \rho C_P dx_1 dx_2 dx_3 T_{,t} dt$$

where ρ is the material density, C_P is the specific heat for the material, and $T_{,t} dt$, the temperature change in the time interval dt. (Note the subscript $,t$ denotes differentiation with respect to time.) The balance of energy is expressed as

heat stored = heat inflow

That is,

$$\rho C_P dVT_{,t} dt = \sum_{i=1,3} (dQ_{i1,t} - dQ_{i2,t})$$

Substituting for $dQ_{i1,t}$ etc. gives the differential equation,

$$\rho C_P \frac{\partial T}{\partial t} = k \frac{\partial^2 T}{\partial x_i \partial x_i} \tag{7.189}$$

In this equation the summation convention is used on the repeated index.

The solution of eqn (7.189) is controlled by the boundary conditions of the problem. If $k(t)$ varies with temperature, then the expression corresponding to eqn (7.189) is

$$\rho C_P \frac{\partial T}{\partial t} = \frac{\partial}{\partial x_i} \left[k(t) \frac{\partial T}{\partial x_i} \right] \tag{7.190}$$

7.6.3 Boundary conditions for heat flow at a free surface

Three effects may be present, namely:

1. convection to or from the surrounding fluid medium;

2. radiation to or absorption from the surrounding bodies;

3. heat input from a given source.

The heat flow will be in the direction $\hat{n}$, the normal to the surface. Each of these conditions will be considered in turn.

1. Convection

The expression for q_c, the heat flux per unit area across the boundary due to convection, is

$$q_c = h_c (T - T_a) \tag{7.191}$$

where (T, T_a) are the temperatures of the body surface and the surrounding fluid, and h_c is the convection coefficient of heat transfer.

2, heat radiation

The heat flux per unit area q_r for radiation to the surrounding environment is again expressed by

$$q_r = h_r (T - T_a) \tag{7.192}$$

However, now the radiation constant h_r is expressed

$$h_r = \varepsilon\sigma(T_{ks}^2 + T_{ka}^2)(T_{ks} + T_{ka}) \tag{7.193}$$

In eqn (7.193), T_{ka}, T_{ks} are the temperatures of the body surface and the surrounding environment, respectively, measured in degrees Kelvin, and σ is the Stefan–Boltzmann constant.

3. Heat input

This will be a given parameter, a heat flux q_i per unit area. Thus, for all three effects,

$$-k\frac{\partial T}{\partial n} = q_i + q_c + q_r \tag{7.194}$$

That is,

$$q_i + k\frac{\partial T}{\partial n} + (T - T_a)(h_r + h_c) = 0 \tag{7.195}$$

This equation may be written in simplified form:

$$k\frac{\partial T}{\partial n} + hT + q = 0 \tag{7.196}$$

where

$$q = q_i - hT_a \tag{7.197}$$

and

$$h = h_r + h_c \tag{7.198}$$

7.6.4 Finite element approximation

The solution to a given problem must satisfy the differential equation (eqn (7.189)) subject to the boundary conditions eqns (7.196) to (7.198). The method of weighted residuals, as outlined in section 3.3, will be used herein to make the finite element approximation. For convenience, only the two–dimensional, case is considered, as the extension to three dimensions is one of detail and not of concept. Thus, if $\bar{T}$ is an approximate solution to eqn (7.189), it is required that the residual $\bar{e}$ be orthogonal to the shape functions N_i within the element. That is,

$$\int_{area} (k \nabla^2 \bar{T} - \rho C_P \bar{T}_{,t}) N_i \, dA = 0 \tag{7.199}$$

The approximate solution $\bar{T}$ is interpolated in terms of the element nodal temperatures by the expression,

$$\bar{T} = \{N\}^T \{T\} \tag{7.200}$$

Using Gauss' theorem (eqn (1.82)),

$$\int_{area} \frac{\partial}{\partial x_k} \left[N_i \frac{\partial \bar{T}}{\partial x_k} \right] dA = \int C N_i \frac{\partial \bar{T}}{\partial x_k} n_k \, dS$$

Differentiating the left–hand side of this equation, and transposing terms gives

$$\int_{area} k N_i \frac{\partial^2 \bar{T}}{\partial x_k \partial x_k} dA = \int_C k N_i \frac{\partial \bar{T}}{\partial x_k} n_k \, dS - \int_{area} k \frac{\partial N_i}{\partial x_k} \frac{\partial \bar{T}}{\partial x_k} dA$$

$$= \int_C k N_i \frac{\partial \bar{T}}{\partial n} dS - \int_{area} k \frac{\partial N_i}{\partial x_k} \frac{\partial \bar{T}}{\partial x_k} dA \tag{7.201}$$

Now, from eqn (7.196), on the boundary,

$$k \frac{\partial \bar{T}}{\partial n} = -(h\bar{T} + q) = -(h \{N\}^T \{T\} + q) \tag{7.202}$$

Hence substitution in eqn (7.201) gives (using eqn (7.200))

$$\int_{area} k \frac{\partial N_i}{\partial x_k} \frac{\partial \{N\}^T}{\partial x_k} dV \{T\} + \int_{area} \rho C_P N_i \{N\}^T dA \frac{\partial \{T\}}{\partial t} + \int_C k N_i \{N\}^T ds \{T\} + \int_C q N_i \, dS = 0 \tag{7.203}$$

For all shape functions, N_i $i = 1$ to n,

$$\int_{area} k \frac{\partial \{N\}}{\partial x_k} \frac{\partial \{N\}^T}{\partial x_k} dV \{T\} + \int_{area} \rho C_P \{N\}\{N\}^T dA \frac{\partial \{T\}}{\partial t} + \int_C k \{N\}\{N\}^T dS + \int_C q \{N\} dS = 0 \tag{7.204}$$

That is, in symbolic matrix notation,

$$[k_e]\{T\} + [C_e]\frac{\partial\{T\}}{\partial t} + \{F_e\} = 0 \tag{7.205}$$

where

$$[k_e] = \int\limits_{area} k\frac{\partial\{N\}}{\partial x_k}\frac{\partial\{N\}^T}{\partial x_k}dA + \int\limits_{C} k\{N\}\{N\}^T dS \tag{7.206}$$

and

$$[C_e] = \int\limits_{area} \rho C_P \{N\}\{N\}^T dA \tag{7.207}$$

and

$$[F_e] = \int\limits_{C} q\{N\}dS \tag{7.208}$$

The boundary conditions in the heat transfer problem are many and varied and quite demanding to satisfy exactly. Some typical situations are given:

1. $h = 0$ on a non–conducting surface. An axis of symmetry both of the body and of the heat boundary conditions is a non–conducting surface.

2. $q_i = 0$ on a boundary not subjected to any heat input.

3. $T = T_0$ on a boundary on which the temperature is prescribed.

In thin plates, two special situations must also be considered:

1. Heat flow only in the x_1, x_2 plane (the plane of the plate). This is a true two–dimensional case, and surface integrals become line integrals on the boundary.

2. The thin plate problem in which T is constant in the x_3 direction, but the surface has heat transference in the x_3 direction.

    Terms in $\dfrac{\partial T}{\partial x_3}$ are deleted from the volume integral. It is seen from Fig. 7.20 that heat flow now occurs across the x_2, x_3 surfaces of the plate.

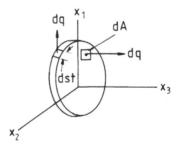

Fig. 7.20   Heat flow thin disc boundaries.

If the steady state condition is being analysed in eqn (7.205), the time derivatives $\dfrac{\partial \{T\}}{\partial t}$ are zero by virtue of the fact that the only boundary conditions are the temperatures specified on the surface, and the problem reverts to a simple finite element analysis involving only the spatial coordinates (x_1, x_2) as the independent variables. However, the transient state is of more interest because it allows the calculation of strains which vary with time in problems where the heat source is not constant. A typical practical example is in the welding of structural elements for which an intense local heat source obviously produces large local strains and distortions. For this transient type of problem it is thus necessary to approximate in the time domain as well as the spatial domain.

7.6.5   Solution in the time domain

From eqn (7.205) it is seen that in the time domain only a first–order differential equation has to be solved, and, this is readily amenable to approximate solutions, here developed in terms of the time shape functions. Thus, suppose, as shown in Fig. 7.21, that the solution in the time domain interval Δt is T_a, given as a linear function,

$$\{T_a\} = [(1 - \zeta)\ \zeta]\begin{bmatrix} T_0 \\ T_1 \end{bmatrix} = [N_0\ N_1]\begin{bmatrix} T_0 \\ T_1 \end{bmatrix} \tag{7.209}$$

and

$$\{F_a\} = \{F_0\} + \zeta\{\Delta F_1\} \tag{7.210}$$

From Fig 7.21,

$$\zeta = \frac{t}{\Delta t}; \quad d\zeta = \frac{1}{\Delta t}dt$$

so that,

$$\frac{\partial T_a}{\partial t} = \frac{\partial T_a}{\partial \zeta}\frac{\partial \zeta}{\partial t} = \frac{1}{\Delta t}[-1\ 1\ ]\begin{bmatrix} T_0 \\ T_1 \end{bmatrix} \tag{7.211}$$

This temperature function does not satisfy eqn (7.205) exactly, so that, now, not only is the spatial distribution approximated, but also that with respect to time. There are many time–marching algorithms which may be used to solve the problem. For example, it can be assumed that the integral of the weighted residual error of the differential equation is zero.

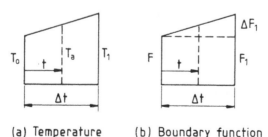

(a) Temperature   (b) Boundary function

Fig. 7.21   Interpolation in the time domain.

That is

$$\int_0^{\Delta t} wD(\bar{T}_a)dt = 0 \tag{7.212}$$

Several weight factors are possible and have been used. For example:

1. $w = 1$ at some point and zero elsewhere;
2. $w = 1$ over the whole time sub–domain;
3. $w = N_1$ (the Galerkin method applied in the time domain).

These three strategies will be developed.

1. Suppose, in the first instance, $w = 1$ at the beginning of the interval, and $t = 0$, and is zero elsewhere. Then, substitution in eqn (7.203) gives

$$[K]\{T_0\} + \frac{1}{\Delta}t[C](\{T_1\} - \{T_0\}) + \{F_0\} = 0 \tag{7.213}$$

Given the conditions $\{T_0\}$ at the beginning of the interval, those $\{T_1\}$ at the end may be determined.

2. Suppose $w = 1$ over the whole time subdomain. Then, using eqn (7.213),

$$\int_0^1 \left\{ [k][1-\zeta \; \zeta] \begin{bmatrix} T_0 \\ T_1 \end{bmatrix} + \frac{1}{\Delta t}[C][-1 \; 1]\begin{bmatrix} T_0 \\ T_1 \end{bmatrix} + \{F_0\} + \zeta\{\Delta F_1\} \right\} \Delta t d\zeta = 0 \tag{7.214}$$

In eqn (7.214), let

$$\{\Delta t\} = \{T_1\} - \{T_0\}$$

Then, integrating eqn (7.214),

$$\left[[k] - \frac{2}{\Delta t}[C]\right]\{\Delta T\} + 2[k]\{T_0\} + 2\{F_0\} + \{\Delta F_1\} = 0 \tag{7.215}$$

This equation may be solved directly for $\{\Delta T\}$, except that $\{\Delta F_1\}$ is dependent on $\{T_1\}$. It may be approximated by setting $\{\Delta F_1\} = 0$.

3. Finally, the third approximation, the Galerkin approach, assumes that $w = N_1$ over the interval 0 to 1, so that eqn (7.205) becomes

$$\int_0^1 \zeta \left\{ [k][(1-\zeta) \; \zeta]\begin{bmatrix} T_0 \\ T_1 \end{bmatrix} + \frac{1}{\Delta t}[C][-1 \; 1]\begin{bmatrix} T_0 \\ T_1 \end{bmatrix} + \{F\} \right\} \Delta t d\zeta = 0$$

On integration, this gives

$$[k](\frac{1}{6}\{T_0\} + \frac{1}{3}\{T_1\}) + [C]\frac{1}{2\Delta t}(\{T_1\} - \{T_0\}) + \frac{1}{2}\{F_0\} + \frac{1}{3}\{\Delta F_1\} = 0$$

Finally, using the same form as in eqn (7.215),

$$([k] + [C]\frac{3}{2\Delta t})\{\frac{\Delta T}{3}\} + \frac{1}{2}[k]\{T_0\} + \frac{1}{2}\{F_0\} + \frac{1}{3}\{\Delta F_1\} = 0 \tag{7.216}$$

Again, this may be solved for $\{\Delta T\}$, but the method is again no longer explicit, in that $\{\Delta F_1\}$ depends on the temperatures $\{T_1\}$.

CHAPTER 8

Plate bending

8.1 INTRODUCTION TO PLATE THEORY

8.1.1 Curvature tensor of plate surface

The assumption is made that distortion of a thin plate loaded transversely to the plane of the middle surface of the plate can be described in terms of the curvatures of the middle surface. These curvatures are given as second derivatives of the transverse deflection u_3, and it is thus important to the understanding of plate theory to establish the transformation laws for the components of the curvature tensor under rotation of the axes about the x_3–axis. The axes chosen are shown in Fig. 8.1. The x_1, x_2–axes lie in the middle plane of the plate, and the (x'_1, x'_2) set are obtained from (x_1, x_2) by a positive rotation about x_3.

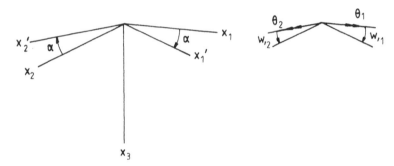

Fig. 8.1   Coordinate axes for plate middle surface.

The radii of curvature r_{11}, r_{22}, r_{12} and the curvature and the twist of the surface χ_{11}, χ_{22} and χ_{12}, respectively, are defined such that

$$\chi_{11} = \frac{1}{r_{11}} = -u_{3,11} ; \quad \chi_{22} = \frac{1}{r_{22}} = -u_{3,22} ; \quad \chi_{12} = \frac{1}{r_{12}} = -u_{3,12} \tag{8.1}$$

It is seen from these definitions that the curvatures are positive when the plate surface is concave when looked at from above (that is, from the negative x_3 direction). For small rotations, the rotation vector $<\theta_1, \theta_2>$ obeys the law of vector transformation of components, such that

$$\begin{Bmatrix} \theta'_1 \\ \theta'_2 \end{Bmatrix} = \begin{bmatrix} \cos \alpha & \sin \alpha \\ -\sin \alpha & \cos \alpha \end{bmatrix} \begin{Bmatrix} \theta_1 \\ \theta_2 \end{Bmatrix} \tag{8.2}$$

That is, following eqn (1.38),

$$\{\theta'\} = [L]\{\theta\} \tag{8.3}$$

358

The slopes are related to positive rotations (Fig. 8.1(b)) by the expressions,

$$\theta_1 = u_{3,2} \; ; \quad \theta_2 = -u_{3,1} \tag{8.4}$$

With similar expressions for the primed axes. Substituting these values in eqn (8.2) and interchanging rows and columns gives

$$\left\{ \begin{matrix} u'_{3,1} \\ u'_{3,2} \end{matrix} \right\} = \left[\begin{matrix} \cos \alpha & \sin \alpha \\ -\sin \alpha & \cos \alpha \end{matrix} \right] \left\{ \begin{matrix} u_{3,1} \\ u_{3,2} \end{matrix} \right\} \tag{8.5}$$

That is,

$$\{u'_{3,}\} = [L]\{u_{3,}\} \tag{8.6}$$

where the comma subscript $u_{3,}$ implies differentiation with respect to x_1 and x_2 of the displacement u_3. The curvature tensor whose components are the second derivatives of $u_{3,}$ is defined by

$$[\chi] = -\left[\begin{matrix} u_{3,11} & u_{3,12} \\ u_{3,21} & u_{3,22} \end{matrix} \right] = -\left[\begin{matrix} \dfrac{1}{r_{11}} & \dfrac{1}{r_{12}} \\ \dfrac{1}{r_{21}} & \dfrac{1}{r_{22}} \end{matrix} \right] = -\left\{ \begin{matrix} \dfrac{\partial}{\partial x_1} \\ \dfrac{\partial}{\partial x_2} \end{matrix} \right\} [u_{3,1} \; u_{3,2}] = -\nabla \nabla^T u_3 \tag{8.7}$$

The operator ∇ is defined by

$$\nabla = \left\{ \begin{matrix} \dfrac{\partial}{\partial x_1} \\ \dfrac{\partial}{\partial x_2} \end{matrix} \right\} \tag{8.8}$$

Alternatively, in index notation,

$$\chi_{ij} = -u_{3,ij} \tag{8.9}$$

In the x'_1, x'_2, axes, the same definition applies, so that

$$[\chi'] = -\nabla' \nabla'^T u_3 \tag{8.10}$$

Then, from eqn (8.5),

$$\{\nabla'\} = [L]\{\nabla\} \tag{8.11}$$

so that, substituting eqns (8.1) and (8.7) in eqn (8.10),

$$[\chi'] = [L]\{\nabla\}\{\nabla\}^T u_3[L]^T = [L][\chi][L]^T \tag{8.12}$$

This establishes the tensor character of the $[\chi]$ matrix. The inverse transformation is given by

$$[\chi] = [L]^T [\chi'][L] \tag{8.13}$$

If eqn (8.12) is expanded, the Mohr's circle type of expressions is obtained, relating individual components. That is, for example,

$$\chi'_{11} = u'_{3,11} = \chi_{11} \cos^2 \alpha + \chi_{22} \sin^2 \alpha - 2\chi_{12} \cos \alpha \sin \alpha \tag{8.14}$$

$$\chi'_{12} = u'_{3,12} = -(\chi_{22} - \chi_{11})\sin \alpha \cos \alpha + \chi_{12}(\cos^2 \alpha - \sin^2 \alpha) \qquad (8.15)$$

The principal directions at a point on the surface are the directions for which there is no twist, and the corresponding curvatures are the principal curvatures. The principal directions are obtained by setting $\chi'_{12} = 0$ in eqn (8.15), so that, if the angle is α_o,

$$\tan 2\alpha_o = \frac{-2\chi_{12}}{\chi_{22} - \chi_{11}} \qquad (8.16)$$

The magnitudes of χ_{11o}, χ_{22o}, the principal curvatures, can be obtained by substituting the value of α_o in eqn (8.12).

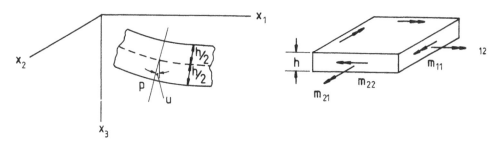

Fig. 8.2   Plate displacements and stress resultants.

8.1.2   Small deflection plate theory

The Kirchoff hypothesis

The Kirchoff hypothesis for plates is similar to the Navier hypothesis for the bending of beams (plane sections remain plane), and simply states that normals to the undeformed middle surface remain normal in the bent plate. Thence, from Fig. 8.2(a), it is seen that the displacements (u_1, u_2) of the point P, a distance x_3 from the middle surface of the plate, are

$$u_i = -x_3 u_{3,1} \quad i = 1 \text{ or } 2 \qquad (8.17)$$

From Chapter 2, eqn (2.4), it follows that the strain tensor at the point P in the bent plate (based on the Kirchoff hypothesis) is given in index notation as

$$\varepsilon_{ij} = -\frac{1}{2}x_3(u_{3,ij} + u_{3,ji}) = -x_3 u_{3,ij} \qquad (8.18)$$

or, in matrix notation, as

$$[\varepsilon] = \begin{bmatrix} \varepsilon_{11} & \varepsilon_{12} \\ \varepsilon_{21} & \varepsilon_{22} \end{bmatrix} = -x_3(\nabla\nabla^T u_3) \qquad (8.19)$$

Since the assumption is made that there is no stress in the normal direction of the plate, and that transverse shear strains are zero, it follows that a state of plane stress exists at the point P.

Stress resultants

The moment tensor $[M]$ for the plate is defined by

$$[M] = \begin{bmatrix} M_{11} & M_{12} \\ M_{21} & M_{22} \end{bmatrix} = \begin{bmatrix} m_{11} & -m_{12} \\ m_{21} & m_{22} \end{bmatrix} \tag{8.20}$$

The negative sign occurs in the term $-m_{12}$, because of the nature of the twist moments being caused by the complementary shears, τ_{12}, τ_{21}. It is seen that $[M]$ is related to the stress tensor $[\sigma]$ in the plate by the integral expression,

$$[M] = \int_{-\frac{h}{2}}^{\frac{h}{2}} x_3 [\sigma] dx_3 \tag{8.21}$$

or, in index notation,

$$M_{ij} = \int_{-\frac{h}{2}}^{\frac{h}{2}} x_3 \sigma_{ij} dx_3 \tag{8.22}$$

Constitutive laws

The constitutive law for the homogeneous isotropic material in a state of plane stress is

$$\sigma_{ij} = \frac{E}{1-v^2} \{(1-v)\varepsilon_{ij} + v\varepsilon_{kk}\delta_{ij}\} \tag{8.23}$$

and, substituting eqns (8.21) and (8.19) in eqn (8.22), the expression for the moments in terms of derivatives of the transverse deflection is obtained as

$$M_{ij} = -D\{(1-v)u_{3,ij} + vu_{3,kk}\delta_{ij}\} \int_{-\frac{h}{2}}^{\frac{h}{2}} x_3^2 \, dx_3$$

That is,

$$M_{ij} = -D\{(1-v)u_{3,ij} + vu_{3,kk}\delta_{ij}\} \tag{8.24}$$

where D is the flexural rigidity of the plate defined as

$$D = \frac{Eh^3}{12(1-v^2)} \tag{8.25}$$

Since $u_{3,ij}$ is a rank–two tensor, so also is M_{ij}, so that the components of $[M]$, in eqn (8.20), transform in a way identical to those of $[\chi]$. Thus,

$$[M'] = [L][M][L]^T \tag{8.26}$$

For the finite element formulation, it is convenient to express the constitutive eqn (8.24) in matrix form, and this gives

$$\begin{Bmatrix} M_{11} \\ M_{22} \\ M_{21} \end{Bmatrix} = D \begin{bmatrix} 1 & v & 0 \\ v & 1 & 0 \\ 0 & 0 & \dfrac{1-v}{2} \end{bmatrix} \begin{Bmatrix} -u_{3,11} \\ -u_{3,22} \\ -2u_{3,21} \end{Bmatrix} \tag{8.27}$$

In symbolic form, eqn (8.27) will be written

$$\{M\} = [k]\{\chi\} \tag{8.28}$$

where

$$\{\chi\} = \begin{Bmatrix} \chi_{11} \\ \chi_{22} \\ 2\chi_{12} \end{Bmatrix}$$

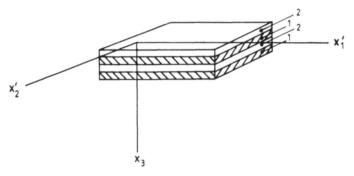

Fig. 8.3   Laminate plate.

Example 8.1 *Constitutive law for laminated plates*

An important class of plates is that of laminated plywood, in which the properties vary from lamination to lamination. It is usually assumed, however, that in a plywood plate the properties of layers alternate, odd layers having one value and even layers a second value, as shown in Fig. 8.3. The axes in Fig. 8.3 are labelled (x'_1, x'_2) since they correspond to the material axes of the laminations, which may not correspond to the global axes of the plate. Within a layer of the material the constitutive law of plane stress for the material with orthotropic properties will be given by

$$\begin{Bmatrix} \sigma'_{11} \\ \sigma'_{22} \\ \sigma'_{12} \end{Bmatrix} = \begin{bmatrix} c_{11} & c_{12} & 0 \\ c_{21} & c_{22} & 0 \\ 0 & 0 & c_{33} \end{bmatrix} \begin{Bmatrix} \varepsilon'_{11} \\ \varepsilon'_{22} \\ 2\varepsilon'_{12} \end{Bmatrix} \tag{i}$$

That is,

$$\{\sigma'\} = [C]\{\varepsilon'\} \tag{ii}$$

where the c_{ij} terms are defined by

$$c_{11} = \frac{E_1}{1 - v_1 v_2} \; ; \quad c_{22} = \frac{E_2}{1 - v_1 v_2}$$

$$c_{12} = \frac{E_2 v_1}{1 - v_1 v_2} = \frac{E_1 v_2}{1 - v_1 v_2} = c_{21} \; ; \quad c_{33} = G \tag{iii}$$

The global (x_1, x_2) axes are related to the material (x'_1, x'_2) axes by a rotation θ, so that

$$\{\sigma'\} = [T]\{\sigma\} \tag{iv}$$

where

$$[T] = \begin{bmatrix} \cos^2 \theta & \sin^2 \theta & 2 \sin \theta \cos \theta \\ \sin^2 \theta & \cos^2 \theta & -2 \sin \theta \cos \theta \\ 2 \sin \theta \cos \theta & -2 \sin \theta \cos \theta & \cos^2 \theta - \sin^2 \theta \end{bmatrix} \tag{v}$$

and similarly, for strains,

$$\{\varepsilon'\} = [T]^T \{\varepsilon\} = [T]\{\varepsilon\} \tag{vi}$$

Thence, in the (x_1, x_2) axes,

$$\{\sigma\} = [T]^{-1}[C][T]\{\varepsilon\} = [\bar{C}]\{\varepsilon\} \tag{vii}$$

For n layers, layer k of thickness h_k, the stress resultants are given

$$\begin{Bmatrix} M_{11} \\ M_{22} \\ M_{21} \end{Bmatrix} = \sum_{k=1}^{n} \int_{h_{k-1}}^{h_k} \begin{Bmatrix} \sigma_{11} \\ \sigma_{22} \\ \sigma_{21} \end{Bmatrix} x_3 dx_3 \tag{viii}$$

Using the Kirchoff assumption, eqn (8.19), and the constitutive equations (vii) gives these moment resultants as

$$\begin{Bmatrix} M_{11} \\ M_{22} \\ M_{21} \end{Bmatrix} = [k] \begin{Bmatrix} \sigma_{11} \\ \sigma_{22} \\ \sigma_{21} \end{Bmatrix} \tag{ix}$$

where the general term k_{ij} of $[k]$ is defined as

$$k_{ij} = \sum_{h=1}^{n} \int_{h_{k-1}}^{h_k} \bar{c}_{ij}^{(k)} x_3^2 dx_3 = \frac{1}{3} \sum_{k=1}^{n} \bar{c}_{ij}^{(k)} (h_k^3 - h_{k-1}^3) \tag{x}$$

If the layers of the plywood are symmetrically placed in pairs around the centre layer so that each pair is placed with its grain perpendicular to the previous layer, a standard plywood configuration results. If the coordinates are chosen so that the x_1–axis is parallel to the exterior grain and the x_2–axis perpendicular to it, then $\theta = 0°$ for odd layers and $90°$ for the even layers. This gives $\bar{c}_{13} = \bar{c}_{23} = 0$ in eqn (vii), with the corresponding simplification in (x).

8.1.3   The differential equation of equilibrium for the bent plate

The differential equation for the deflected shape u_3 of the bent plate is obtained by substituting the compatibility conditions (Kirchoff assumptions) and the constitutive equations in the differential equation of equilibrium for transverse load and moment stress resultants. The stress resultants and their variations are shown in Fig. 8.4.

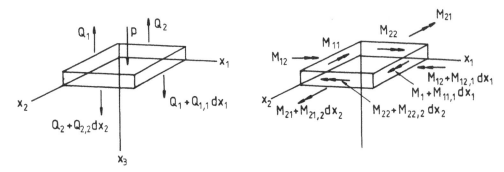

Fig. 8.4   Plate stress resultants.

From Fig. 8.4(a) and (b) it follows that for moment equilibrium about the x_1, x_2–axes,

$$M_{ij,i} = Q_j \quad (\textstyle\sum \text{ on } j = 1, 2) \tag{8.29}$$

and, for transverse force equilibrium,

$$Q_{j,j} = -p \quad (\textstyle\sum \text{ on } j = 1, 2) \tag{8.30}$$

Differentiating eqn (8.29), and substituting in eqn (8.30),

$$M_{ij,ij} = -p \quad (\textstyle\sum \text{ on } j = 1, 2) \tag{8.31}$$

Expanding eqn (8.31) gives

$$M_{11,11} + M_{12,12} + M_{21,21} + M_{22,22} = -p \tag{8.32}$$

Since $M_{12} = M_{21}$ and the order of the differentiation can be reversed, this equation may be written

$$M_{11,11} + 2M_{21,21} + M_{22,22} = -p \tag{8.33}$$

To obtain eqn (8.31) in terms of displacement derivatives, it is only necessary to substitute eqn (8.24) in eqn (8.31). That is,

$$(1 - v)u_{3,iijj} + vu_{3,kkjj} = \frac{p}{D} \tag{8.34}$$

Replacing dummy index k by i in the second term gives

$$u_{3,iijj} = \frac{p}{D} \tag{8.35}$$

On expanding eqn (8.35),

$$\frac{\partial^4 u_3}{\partial x_1^4} + \frac{2\partial^4 u_3}{\partial x_1^2 \partial x_2^2} + \frac{\partial^4 u_3}{\partial x_2^4} = \frac{p}{D} \tag{8.36}$$

Alternatively, defining the operator ∇ as in eqn (8.8)/eqn (8.35) (or (8.36)) may be written

$$\nabla^T \nabla \nabla^T \nabla u_3 = \frac{p}{D} \tag{8.37}$$

8.1.4   The Mindlin plate theory

The Mindlin theory of plate bending is a plate theory which allows for shear distortions in a manner that is applicable to moderately thick plates. As will be shown later in this chapter, the theory has application in some finite element formulations. The Mindlin theory makes the assumption that the slope at a point on the middle surface of the plate arises from two effects, namely:

1.   The rotation caused by the bending distortions;

2.   The transverse shear distortions which violate the plane section (or Kirchoff hypothesis) on the normals.

These effects are shown in Fig. 8.4, for the distortion in the x_1 direction.

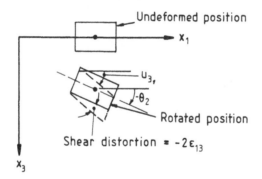

Fig. 8.5   Rotation and shear distortion of plate.

From Fig. 8.5 it is seen that the direct and shear strains ε_{11}, ε_{22} and ε_{12} are still given by eqn (8.17). The transverse shear distortions ε_{13} and ε_{23} are given, from Fig. 8.5:

$$\varepsilon_{13} = \frac{1}{2}\{\theta_2 + u_{3,1}\}$$

$$\varepsilon_{23} = \frac{1}{2}\{-\theta_1 + u_{3,2}\} \tag{8.38}$$

The transverse stress resultants Q_1 and Q_2 are calculated from the transverse shear stresses:

$$\begin{Bmatrix} Q_1 \\ Q_2 \end{Bmatrix} = \int_{-\frac{h}{2}}^{\frac{h}{2}} \begin{Bmatrix} \tau_{13} \\ \tau_{23} \end{Bmatrix} dx_3$$

and, in terms of the shear strains,

$$\left\{ \begin{array}{c} \tau_{13} \\ \tau_{23} \end{array} \right\} = \alpha G \begin{bmatrix} 1 & 0 \\ 0 & 1 \end{bmatrix} \left\{ \begin{array}{c} 2\varepsilon_{13} \\ 2\varepsilon_{23} \end{array} \right\} \tag{8.39}$$

The coefficient $\alpha \approx 5/6$ is used to allow for the parabolic distribution of shear stress through the plate thickness. Thence, finally, using eqn (8.38),

$$\left\{ \begin{array}{c} Q_1 \\ Q_2 \end{array} \right\} = \alpha h G \begin{bmatrix} 1 & 0 \\ 0 & 1 \end{bmatrix} \left\{ \begin{array}{c} \theta_2 + u_{3,1} \\ -\theta_1 + u_{3,2} \end{array} \right\} \tag{8.40}$$

It should be noted that in the Kirchoff assumptions Q_1, Q_2 follow from eqn (8.29). The Mindlin theory is used in the present text for plate elements where the plate rotations θ_1, θ_2 are interpolated independently to the transverse deflection u_3.

Example 8.2 *Constitutive equations for sandwich plates*

A section of a sandwich plate is shown in Fig. 8.6.

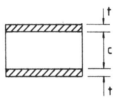

Fig. 8.6   Sandwich plate cross section.

The constitutive equations are now given from eqn (8.28):

$$[M] = [k]\{\chi\}$$

where the stiffness of the plate element is given by

$$[k] = \frac{Et(c+t)^2}{2(1-v^2)} \begin{bmatrix} 1 & v & 0 \\ v & 1 & 0 \\ 0 & 0 & \dfrac{1-v}{2} \end{bmatrix} \tag{i}$$

Here, in eqn (i), E, v are the properties of the sandwich plates. If the core has the elastic constant G, the shear force resultants are given in terms of the shear strains:

$$\left\{ \begin{array}{c} Q_1 \\ Q_2 \end{array} \right\} = \frac{G(c+t)^2}{c} \begin{bmatrix} 1 & 0 \\ 0 & 1 \end{bmatrix} \left\{ \begin{array}{c} \gamma_1 \\ \gamma_2 \end{array} \right\} \tag{ii}$$

8.2 THE FINITE ELEMENT FORMULATION

Plate bending problems lend themselves to a great variety of elements and methods of solution. Displacement, equilibrium and hybrid elements have all been used with varying degrees of success.

Results for any particular element should be examined carefully because many give good displacement results (particularly for distributed loads), but poor moment fields (i.e. second derivatives). The complexity of the plate bending problem arises from various causes, for example:

1. It would appear that C_1 as well as C_0 compatibility is required for elements to be conforming displacement models. This C_1 compatibility between two elements is shown in Fig. 8.7 where the normals n_1, n_2 to the surface must coincide.

2. Finite element configurations made up of plates are highly redundant compared with planar elasticity elements, and hence there is a great variety of ways in which it is possible to select these redundants.

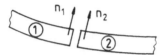

Fig. 8.7 $\quad C_1$ compatibility of matching normals.

8.2.1 Displacement models–triangular elements

As shown in section 8.1.2, bending moments in a plate are related to curvatures given as second derivatives of deflections. Hence, if the constant strain (curvature) condition is to be satisfied, the deflection function must be at least cubic. The complete cubic in two–dimensional space contains ten terms. If this is used for the triangle, the term $\zeta_1 \zeta_2 \zeta_3$ occurs in the interpolation polynomial, and this must be activated in some other way than by the deflections of node points located on the sides of the triangle. Possible combinations are shown in Fig. 8.8, and all look unpromising at first appraisal.

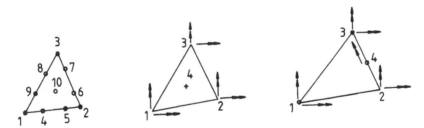

Fig. 8.8 $\quad$ Triangular elements node configurations.

It would appear that:

1. The Fig. 8.8(a) element is unsatisfactory because no rotation or moment terms appear, and hence there is no possibility of C_1 compatibility.

2.   If the element in Fig. 8.8(b) is used, it is found to be too flexible, because C_1 compatibility is not achieved. That is, rotations of normals are not compatible between adjacent elements.

3.   The element in Fig. 8.8(c) has an odd node (4), with one rotation degree of freedom in side 1.

All these elements do not give C_1 compatibility measured here by normal slopes along adjoining element sides. For example, in Fig. 8.8(b) along side 2–3, the deflection is a cubic curve and is completely defined by $u_{3(2)}$, $\theta_{s(2)}$, $u_{3(3)}$, $\theta_{s(3)}$. However, the normal slope $u_{3,n}$, being a first derivative of u_3, is a quadratic curve and cannot be completely specified by $\theta_{s(2)}$, $\theta_{s(3)}$. Hence C_1 compatibility is violated, and the element is too flexible. If a quintic polynomial is used, it is found that both C_0 and C_1 compatibilities can be satisfied. However, a heavy penalty is incurred, in that second derivatives must be used as nodal parameters. See Fig 8.9 for the nodal variable configuration.

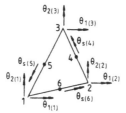

Fig. 8.9   Quintic element nodal parameters.

In two dimensions the quintic polynomial has 21 terms and is interpolated using the following function values:

at nodes *1, 2, 3* u_3, $u_{3,1}$, $u_{3,2}$, $u_{3,11}$, $u_{3,22}$, $u_{3,12}$
at nodes *4, 6* θ_s (normal slope)

It is possible to delete the mid–side nodes by using constraint equations, such as eqn (8.41) which force a linear variation of slope along the sides.

$$\theta_{s(4)} = \frac{\theta_{s(2)} + \theta_{s(3)}}{2} \tag{8.41}$$

This gives an element with 18 degrees of freedom without the mid–side nodes. The simplest of the three triangular elements have six moment values, defining three linear varying moment fields in M_{11}, M_{22}, M_{12}. These must be chosen in such a way that constant strain terms are included. This may be achieved in at least two ways, which will be described in the following sections.

8.2.2   The Bazeley–Irons element

The deformation of the triangular element [74] can be derived in terms of six relative nodal slopes, which, in turn, can be expressed in terms of nine nodal values, namely:

$$(u_{3i} \ u_{3j} \ u_{3k}) \qquad \text{deflections}$$
$$(\theta_{1i} \ \theta_{2i} \ \theta_{1j} \ \theta_{2j} \ \theta_{1k} \ \theta_{2k}) \quad \text{slopes}$$

The deformed position is shown in Fig. 8.10. The plate distortion u_3^* is measured relative to the plane through the nodes i, j, k. That is, the displacement of the plate from the unloaded position i_0, j_0, k_0, to i, j, k is a rigid body motion only.

The deflection to the plane $i\,j\,k$ will be thus denoted u_3^R and,

$$u_3^R = \{\phi_1\}^T \{u_{3c}\} = [\zeta_1 \ \zeta_2 \ \zeta_3] \begin{Bmatrix} u_{3i} \\ u_{3j} \\ u_{3k} \end{Bmatrix}$$

(8.42)

Thus, the deflection at any point on the middle surface of the plate will be

$$u_3 = u_3^R + u_3^*$$

(8.43)

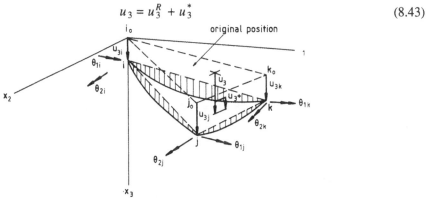

Fig. 8.10   Plate deformations—Bazeley–Irons element.

The six nodal values of rotation $\{\theta_1^* \ \theta_2^*\}_l$, $l = i,\, j,\, k$ are required to specify the relative displacement field. Assuming a cubic interpolation function but omitting the $\zeta_1 \ \zeta_2 \ \zeta_3$ term in the first instance, relative rotations are given by

$$\theta_1^* = u_{3,2}^* = \theta_1 - u_{3,2}^R \ ; \quad \text{and} \quad \theta_2^* = -u_{3,1}^* = \theta_2 + u_{3,1}^R$$

(8.44)

From eqn 8.42 (using eqns 1.136 and (1.137)) to obtain global coordinate derivatives,

$$\begin{Bmatrix} -u_{3,2}^R \\ u_{3,1}^R \end{Bmatrix} = \frac{1}{2A} \begin{bmatrix} -a_1 & -a_2 & -a_3 \\ b_1 & b_2 & b_3 \end{bmatrix} \begin{Bmatrix} u_{3i} \\ u_{3j} \\ u_{3k} \end{Bmatrix}$$

(8.45)

Define the displacement vector $\{r^*\}$ such that

$$\{r^*\}^T = <\theta_{1i}^* \ \theta_{2i}^* \ \theta_{1j}^* \ \theta_{2j}^* \ \theta_{1k}^* \ \theta_{2k}^*>$$

Thence from eqns (8.43), (8.44),

$$\{r^*\} = \frac{1}{2A} \begin{bmatrix} -a_1 & 2A & 0 & -a_2 & 0 & 0 & -a_3 & 0 & 0 \\ b_1 & 0 & 2A & b_2 & 0 & 0 & b_3 & 0 & 0 \\ -a_1 & 0 & 0 & -a_2 & 2A & 0 & -a_3 & 0 & 0 \\ b_1 & 0 & 0 & b_2 & 0 & 2A & b_3 & 0 & 0 \\ -a_1 & 0 & 0 & -a_2 & 0 & 0 & -a_3 & 2A & 0 \\ b_1 & 0 & 0 & b_2 & 0 & 0 & b_3 & 0 & 2A \end{bmatrix} \begin{Bmatrix} r_i \\ r_j \\ r_k \end{Bmatrix}$$

(8.46)

where the displacement subvectors on the right–hand side are

$$\{r_l\} = \begin{Bmatrix} u_{3l} \\ \theta_{1l} \\ \theta_{2l} \end{Bmatrix} \quad l = i, j, k \tag{8.47}$$

Write eqn 8.46,

$$\{r^*\} = [T]\{r\} \tag{8.48}$$

Nodal forces obey the contragredient transformation, so that

$$\{R\} = [T]^T \{R^*\} \tag{8.49}$$

It is evident that

$$\{R_l^*\}^T = <M_{1l} \ M_{2l}> \quad l = i, j, k$$

and

$$\{R_l\}^T = <F_{3l} \ M_{1l} \ M_{2l}> \quad l = i, j, k \tag{8.50}$$

If the stiffness relationship,

$$\{R^*\} = \{k^*\}\{r^*\} \tag{8.51}$$

is known, it follows that

$$\{R\} = [T]^T [k^*][T]\{r\} \tag{8.52}$$

and

$$[k] = [T]^T \{k^*\}[T] \tag{8.53}$$

is a (9×9) symmetric matrix. To determine $\{r^*\}$, proceed as follows. Write the interpolation for u_3^*

$$u_3^* = \{\phi_3\}^T \{r_*\} \tag{8.54}$$

where $\{\phi_3\}$ is as yet undetermined. Choose the non–orthogonal interpolation polynomial

$$\{\psi\}^T = <\zeta_1^2\zeta_2 \ \zeta_1^2\zeta_3 \ \zeta_2^2\zeta_1 \ \zeta_2^2\zeta_3 \ \zeta_3^2\zeta_1 \ \zeta_3^2\zeta_2> \tag{8.55}$$

Then the orthogonal polynomial for node slopes is obtained by using section 1.4.2:

$$\{\phi_3\} = \begin{bmatrix} \zeta_1^2(-\zeta_2 b_3 + \zeta_3 b_2) \\ \zeta_1^2(-\zeta_2 a_3 + \zeta_3 a_2) \\ \zeta_2^2(\zeta_1 b_3 - \zeta_3 b_1) \\ \zeta_2^2(\zeta_1 a_3 - \zeta_3 a_1) \\ \zeta_3^2(-\zeta_1 b_2 + \zeta_2 b_1) \\ \zeta_3^2(-\zeta_1 a_2 + \zeta_2 a_1) \end{bmatrix} \tag{8.56}$$

Unfortunately, the function u_3^* thus obtained does not satisfy the constant strain criterion necessary for convergence. It may be modified, however, by adding the function $\beta \zeta_1 \zeta_2 \zeta_3$ to each term. It can be proven that $\beta = 1/2$. For example, with $\beta = 1/2$ the first term in eqn (8.56) becomes

$$\phi_{31} = -(\zeta_1^2 \zeta_2 + \frac{1}{2}\zeta_1 \zeta_2 \zeta_3)b_3 + (\zeta_1^2 \zeta_3 + \frac{1}{2}\zeta_1 \zeta_2 \zeta_3)b_2 \tag{8.57}$$

Thus, given $\{\phi_3\}$, the generation of $[k_c]$ follows the same procedure as for any finite element analysis. Thus, differentiate $\{\phi_3\}$ twice with respect to x_1, x_2 and once with respect to (x_1, x_2) and obtain nodal values $\{\varepsilon_c\}$, expressed as

$$\{\varepsilon_c\} = [\Phi_c]\{r^*\} \tag{8.58}$$

Then,

$$[k_c] = \int_{area} \{\phi_\varepsilon\}[k]\{\phi_\varepsilon\}^T dA \tag{8.59}$$

and

$$[k^*] = [\Phi_c]^T [k_c][\Phi_c] \tag{8.60}$$

In eqn (8.59), $\{\phi_\varepsilon\} \equiv \{\phi_1\}$, so that for an element of constant thickness t,

$$[k_c] = \frac{Et^3A}{144(1-v^2)} \begin{bmatrix} \Lambda & v\Lambda & 0 \\ v\Lambda & \Lambda & 0 \\ 0 & 0 & \dfrac{1-v}{2}\Lambda \end{bmatrix} \tag{8.61}$$

Thus, $[k_c^*]$ is a 9×9 matrix with the submatrix $[\Lambda]$ defined as

$$[\Lambda] = \begin{bmatrix} 2 & 1 & 1 \\ 1 & 2 & 1 \\ 1 & 1 & 2 \end{bmatrix} \tag{8.62}$$

Examine now why $\beta = 1/2$ in eqn (8.57); That is

$$\phi_{31} = \{-(\zeta_1^2 \zeta_2 + \beta \zeta_1 \zeta_2 \zeta_3)b_3 + (\zeta_1^2 \zeta_2 + \beta \zeta_1 \zeta_2 \zeta_3)b_2\} \tag{8.63}$$

To include constant curvature terms, that is, second derivatives of u_3^*, it should be possible to express u_3^* as

$$u_3^* = [\zeta_2 \zeta_1 \quad \zeta_3 \zeta_1 \quad \zeta_1 \zeta_2] \begin{Bmatrix} A_1 \\ A_2 \\ A_3 \end{Bmatrix} \tag{8.64}$$

Then,

$$2A \begin{Bmatrix} -u_{3,1}^* \\ u_{3,2}^* \end{Bmatrix} = \begin{bmatrix} -(b_2\zeta_3 + b_3\zeta_2) & -(b_3\zeta_1 + b_1\zeta_3) & -(b_1\zeta_3 + b_2\zeta_1) \\ (a_2\zeta_3 + a_3\zeta_2) & (a_3\zeta_1 + a_1\zeta_3) & (a_1\zeta_2 + a_2\zeta_1) \end{bmatrix} \begin{Bmatrix} A_1 \\ A_2 \\ A_3 \end{Bmatrix} \tag{8.65}$$

Thence,

$$
2A\,\{r^*\} = 2A\begin{Bmatrix}\theta_{1i}^* \\ \theta_{2i}^* \\ \theta_{1j}^* \\ \theta_{2j}^* \\ \theta_{1k}^* \\ \theta_{2k}^*\end{Bmatrix} = \begin{bmatrix} 0 & a_3 & a_2 \\ 0 & -b_3 & -b_2 \\ a_3 & 0 & a_1 \\ -b_3 & 0 & -b_1 \\ a_2 & a_1 & 0 \\ -b_2 & -b_1 & 0 \end{bmatrix}\begin{Bmatrix} A_1 \\ A_2 \\ A_3 \end{Bmatrix} = [U]\{A\} \qquad (8.66)
$$

Therefore,

$$
2Au_3^* = \{\phi_3\}^T [U]\{A\} = \begin{bmatrix}\phi_{33}a_3 - \phi_{34}b_3 + \phi_{35}a_2 - \phi_{36}b_2 \\ \cdot \\ \cdot\end{bmatrix}^T \{A\}
$$

Now, for the first term,

$$
\phi_{33}a_3 - \phi_{34}b_3 + \phi_{35}a_2 - \phi_{36}b_2 = 2A\,\zeta_2\zeta_3(\zeta_3 + \zeta_2 + 2\beta\zeta_1)
$$

If $\beta = 1/2$, this gives, as required,

$$
2A\,\zeta_2\zeta_3(\zeta_1 + \zeta_2 + \zeta_3) = 2A\,\zeta_2\zeta_3
$$

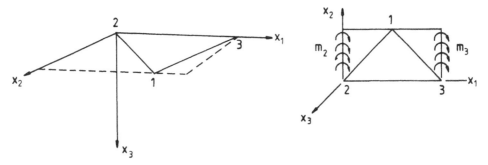

Fig. 8.11   Natural mode moment field.

8.2.3   The natural mode approach

Non-compatible plate bending element (Argyris [1])

The element is considered to have three linear moment fields parallel with the sides 1, 2, 3 of the triangle. Obviously, then, $\langle m_{11}\ m_{22}\ m_{12}\rangle$ vary linearly over the triangle, as do $\langle m_{\zeta_1}\ m_{\zeta_2}\ m_{\zeta_3}\rangle$. The element is shown in Fig. 8.11 with a set of local Cartesian axes, such that OX_1 is coincident with side 1 of the triangle. The bending moment field shown in Fig. 8.11(a) or (b) is consistent with the convention concave positive when viewed from the negative x_3-axis. n keeping with the philosophy of the natural mode method, the moment field $\langle m_2\ m_3\rangle$, will be expressed in terms of its symmetric and antisymmetric components $\langle m_{S1}\ m_{A1}\rangle$ respectively. These moment fields together with their displacement modes are shown in Fig. 8.12.

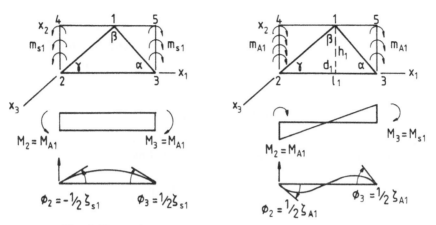

Fig. 8.12 Symmetric and antisymmetric moment fields.

From Fig. 8.12 the rotations $<\phi_2\ \phi_3>$ are expressed in terms of the symmetric and antisymmetric components as

$$\left\{\begin{matrix}\phi_2\\\phi_3\end{matrix}\right\}=\frac{1}{2}\begin{bmatrix}-1&1\\1&1\end{bmatrix}\left\{\begin{matrix}\zeta_{S1}\\\zeta_{A1}\end{matrix}\right\} \tag{8.67}$$

Solving for $<\zeta_{S1}\ \zeta_{A1}>$,

$$\left\{\begin{matrix}\zeta_{S1}\\\zeta_{A1}\end{matrix}\right\}=\begin{bmatrix}-1&1\\1&1\end{bmatrix}\left\{\begin{matrix}\phi_2\\\phi_3\end{matrix}\right\} \tag{8.68}$$

The relationship between the moments $<M_2\ M_3>$ and the symmetric and antisymmetric components $<M_{S1}\ M_{A1}>$ is found, by contragredience from eqn (8.68), to be

$$\left\{\begin{matrix}M_2\\M_3\end{matrix}\right\}=\begin{bmatrix}-1&1\\1&1\end{bmatrix}\left\{\begin{matrix}M_{S1}\\M_{A1}\end{matrix}\right\} \tag{8.69}$$

Within the element the bending moment at any point may be described by

1.   the moment tensor;

$$[M]=\begin{bmatrix}m_{11}&-m_{12}\\m_{21}&m_{22}\end{bmatrix}$$

In this case 1 and 2 axes are related to side 1 of the triangle.

2.   by moments .

$$\{m_\zeta\}^T=<m_1\ m_2\ m_3>$$

These are the components of the linear moment fields parallel with the sides 1, 2 and 3 of the triangle. Following section 1.4.6 and using contragredience, it can be shown that

$$\{v_\zeta\}=[f_\zeta]\{m_\zeta\} \tag{8.70}$$

where

$$\{f_\zeta\} = \frac{12}{12Eh^3}\begin{bmatrix} 1 & \cos^2\alpha - v\sin^2\alpha & \cos^2\gamma - v\sin^2\gamma \\ \cos^2\alpha - v\sin^2\alpha & 1 & \cos^2\beta - v\sin^2\beta \\ \cos^2\gamma - v\sin^2\gamma & \cos^2\beta - v\sin^2\beta & 1 \end{bmatrix} \tag{8.71}$$

In eqn (8.70),

$$\{v_\zeta\} = <v_1, \ v_2, \ v_3,> \tag{8.72}$$

are the curvatures along sides 1, 2 and 3 of the triangle. The moment vector, $\{m_\zeta\}^T = <m_1 \ m_2 \ m_3>$, is now expressed in terms of the natural modes M_S, M_A, so that

$$\{m_\zeta\} = \{M_S\} + \xi\{M_A\} = [I_3 \ \xi]\begin{Bmatrix} M_S \\ M_A \end{Bmatrix} \tag{8.73}$$

The matrix $[\xi]$ is calculated from the modes shown in Fig. 8.12, using the three sets of local coordinates (s_i, n_i), $i = 1, 2, 3$. For example, consider side 1, and express local coordinates, in terms of natural coordinates, as

$$\begin{Bmatrix} 1 \\ s_1 \\ n_1 \end{Bmatrix} = \begin{bmatrix} 1 & 1 & 1 \\ 0 & l_1 & d_1 \\ 0 & 0 & h_1 \end{bmatrix}\begin{Bmatrix} \zeta_2 \\ \zeta_3 \\ \zeta_1 \end{Bmatrix} \tag{8.74}$$

Then,

$$m_{A1} = M_{A1}\begin{Bmatrix} \dfrac{2s_1}{l_1} - 1 \end{Bmatrix}$$

Substituting for s_1, from eqn (8.74), gives

$$m_{A1} = \frac{M_{A1}}{l_1}[-l_1 \ l_1 \ (2d_1 - l_1)]\begin{Bmatrix} \zeta_2 \\ \zeta_3 \\ \zeta_1 \end{Bmatrix}$$

Similar expressions for the other two sides give the expression for $[\xi]$,

$$[\xi] = \begin{bmatrix} \{(\dfrac{2d_1}{l_1} - 1)\zeta_1 - \zeta_2 + \zeta_3\} & \{\zeta_1 + (2\dfrac{d_2}{l_2} - 1)\zeta_2 - \zeta_3\} & \{-\zeta_1 + \zeta_2 + (\dfrac{2d_3}{l_3} - 1)\zeta_3\} \end{bmatrix}\begin{Bmatrix} M_{A1} \\ M_{A2} \\ M_{A3} \end{Bmatrix} \tag{8.75}$$

Statically equivalent forces at the nodes to the distributed moment fields $<M_{S1}M_{A1}>$ may be calculated to be

$$\begin{Bmatrix} T_{S1} \\ T_{A1} \end{Bmatrix} = \frac{h_1}{2}\begin{Bmatrix} M_{S1} \\ M_{A1} \end{Bmatrix} = \frac{A}{l_1}\begin{Bmatrix} M_{S1} \\ M_{A1} \end{Bmatrix} \tag{8.76}$$

For all three components,

$$\begin{Bmatrix} T_{S1} \\ T_{S2} \\ T_{S3} \\ T_{A1} \\ T_{A2} \\ T_{A3} \end{Bmatrix} = A \begin{bmatrix} l^{-1} & 0 \\ 0 & l^{-1} \end{bmatrix} \begin{Bmatrix} M_{S1} \\ M_{S2} \\ M_{S3} \\ M_{A1} \\ M_{A2} \\ M_{A3} \end{Bmatrix} \qquad (8.77)$$

Inverting this expression,

$$\begin{Bmatrix} M_S \\ M_A \end{Bmatrix} = A \begin{bmatrix} l & 0 \\ 0 & l \end{bmatrix} \{P_{NB}\} \qquad (8.78)$$

where the diagonal matrix $[l] = \lceil l_1 l_2 l_3 \rfloor$, and

$$\{P_{NB}\}^T = <T_{S1} T_{S2} T_{S3} T_{A1} T_{A2} T_{A3}> \qquad (8.79)$$

Corresponding to the nodal forces $\{P_{NB}\}$ are the natural displacements,

$$\{\rho_{NB}\}^T = <\rho_{S1} \rho_{S2} \rho_{S3} \rho_{A1} \rho_{A2} \rho_{A3}> \qquad (8.80)$$

The flexibility is calculated to be

$$\{\rho_{NB}\} = [f_{NB}]\{P_{NB}\} \qquad (8.81)$$

where

$$[f_{NB}] = \int_{area} [b]^T [f_\zeta][b]dA \qquad (8.82)$$

It is seen from eqns (8.73) and (8.78) that

$$[b] = \frac{1}{A}[I_3 \, \xi]\lceil l \; l \rfloor \qquad (8.83)$$

For constant thickness,

$$\int_{area} [b]^T [f_\zeta][b]dA = \frac{12}{Eh^3}\frac{1}{A^2}\begin{bmatrix} l & 0 \\ 0 & l \end{bmatrix}\int_{area}\begin{bmatrix} I_3 \\ \xi^T \end{bmatrix}[f^*][I_3 \, \xi]dA\begin{bmatrix} l & 0 \\ 0 & l \end{bmatrix} \qquad (8.84)$$

The integrals are given by

1. $$\int_{area} [f^*]dA = A[f^*]$$

2. $$\int_{area} [f^*][\xi]dA = \frac{[f^*]A}{3}\begin{bmatrix} 2d_1 - l_1 & 0 & 0 \\ 0 & 2d_2 - l_2 & 0 \\ 0 & 0 & 2d_3 - l_3 \end{bmatrix}[l]^{-1} = \frac{[f^*]A[V][l]^{-1}}{3} \qquad (8.85)$$

where $[V]$ is defined as the diagonal matrix,

$$[V] = \lceil (2d_1 - l_1)\ (2d_2 - l_2)\ (2d_3 - l_3) \rfloor \qquad (8.86)$$

3. $\int\limits_{area}^{T} [f^*][\xi]dA$

To calculate this integral, write $[\xi] = \begin{vmatrix} t_1 & t_2 & t_3 \end{vmatrix}$. Then,

$$\int\limits_{area} [\xi]^T [f^*][\xi]dA = [f^*]* \int\limits_{area} \{t\}\{t\}^T dA \tag{8.87}$$

where '*' implies term–by–term multiplication, and

$$\{t\}^T = <t_1 \ t_2 \ t_3> \tag{8.88}$$

On integration of the terms in eqn (8.87), it is found that

$$\int\{t\}\{t\}^T dA = \frac{A[l]^{-1}}{3} \begin{bmatrix} (d_1-l_1)^2+d_1^2 & d_2(d_1-l_1) & d_1(d_3-l_3) \\ d_2(d_1-l_1) & (d_2-l_2)^2+d_2^2 & d_3(d_2-l_2) \\ d_1(d_3-l_3) & d_3(d_2-l_2) & (d_3-l_3)^2+d_3^2 \end{bmatrix} [l]^{-1}$$

$$= \frac{A[l]^{-1}[U][l]^{-1}}{3} \tag{8.89}$$

Thence,

$$[f_{NB}] = \int\limits_{area} [b]^T [f_\zeta][b]dA = \frac{12}{Eh^3 A} \begin{bmatrix} [l][f^*] & \frac{1}{3}[l][f^*][V] \\ \frac{1}{3}[V]^T[f^*][l] & \frac{1}{3}[f^*]*[U] \end{bmatrix} \tag{8.90}$$

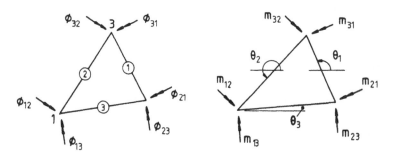

Fig. 8.13   Nodal rotations and moments.

Given $[f_{NB}]$, a 6×6 flexibility matrix, the stiffness matrix is obtained by inversion; that is

$$[k_{NB}] = [f_{NB}]^{-1} \tag{8.91}$$

and

$$\{P_{NB}\} = [k_{NB}]\{\rho_{NB}\} \tag{8.92}$$

The stiffness matrix in global (rotation and moment) components is given

$$[\bar{k}] = [a_N]^T [k_{NB}][a_N] \tag{8.93}$$

where $[a_N]$ is calculated from the following:

$$\{\rho_{NB}\} = [a_N]\{\rho_B\} \quad \text{and} \quad \{P_B\} = [a_N]^T \{P_{NB}\}$$

From Fig. 8.13,

$$\begin{Bmatrix} \rho_{S1} \\ \rho_{S2} \\ \rho_{S3} \\ \rho_{A1} \\ \rho_{A2} \\ \rho_{A3} \end{Bmatrix} = \begin{bmatrix} 0 & -1 & 0 & 0 & 1 & 0 \\ 0 & 0 & -1 & 0 & 0 & 1 \\ -1 & 0 & 0 & 1 & 0 & 0 \\ 0 & 1 & 0 & 0 & 1 & 0 \\ 0 & 0 & 1 & 0 & 0 & 1 \\ 1 & 0 & 0 & 1 & 0 & 0 \end{bmatrix} \begin{Bmatrix} \phi_{13} \\ \phi_{12} \\ \phi_{21} \\ \phi_{23} \\ \phi_{32} \\ \phi_{31} \end{Bmatrix} \tag{8.94}$$

That is,

$$\{\rho_{NB}\} = [C]\{\phi_{ij}\}$$

and, hence, using contragredience,

$$\{M_{ij}\} = [C]^T \{\rho_{NB}\} \tag{8.95}$$

Projection of the moments $\{M_{ij}\}$ onto the x_1, x_2 coordinate axes in Fig. 8.13(b) gives

$$\begin{Bmatrix} M_{11} \\ M_{12} \\ M_{21} \\ M_{22} \\ M_{31} \\ M_{32} \end{Bmatrix} = \begin{bmatrix} -\sin\theta_3 & -\sin\theta_2 & 0 & 0 & 0 & 0 \\ \cos\theta_3 & \cos\theta_2 & 0 & 0 & 0 & 0 \\ 0 & 0 & -\sin\theta_1 & -\sin\theta_3 & 0 & 0 \\ 0 & 0 & \cos\theta_1 & \cos\theta_3 & 0 & 0 \\ 0 & 0 & 0 & 0 & -\sin\theta_2 & -\sin\theta_3 \\ 0 & 0 & 0 & 0 & \cos\theta_2 & \cos\theta_3 \end{bmatrix} \begin{Bmatrix} m_{13} \\ m_{12} \\ m_{21} \\ m_{22} \\ m_{31} \\ m_{32} \end{Bmatrix} \tag{8.96}$$

That is,

$$\{M_{12}\} = [B]\{m_{12}\} \tag{8.97}$$

and, hence, by contragredience,

$$\{\phi_{12}\} = [B]^T \{\theta_{12}\} \tag{8.98}$$

Thence, finally,

$$[\bar{k}] = [B][C]^T [k_{NB}][C][B]^T \tag{8.99}$$

The stiffness $[\bar{k}]$ is still only a 6×6 matrix. To expand to a 9×9 matrix including reactions, use eqn (8.100), in which $[T]$ has been defined in eqns (8.46) and (8.48):

$$\{\delta_e^*\} = [T]\{\delta_e\} \tag{8.100}$$

The two examples given so far for the simple three–node triangle show an important difference in the philosophy of finite element applications. The first, using the global coordinate system in its formulation, requires considerable adaptation if large displacements are to be considered. The second, when combined with the natural mode formulation for the plane stress triangle, easily adapts to large displacements since the linear moment fields are associated with the deformed directions of the triangle sides. The natural mode formulation highlights a second interesting feature of plate bending element formulation, the moment fields being parallel to the element sides. Since it is the intention to attempt to enforce C_1 compatibility on elements, it would appear to be a superior strategy to attempt slope connections through θ_s type variables located in the element sides rather than θ_n type variables. Examples are given later of these types of elements.

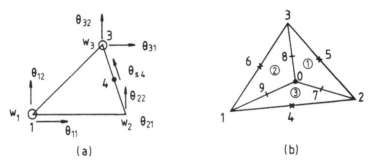

Fig. 8.14   The C_1 compatible plate bending triangular element.

8.2.4   The 12– d.o.f. compatible plate bending triangular element

8.2.4.1   Element configuration

The previous two elements, as already mentioned, are C_0 compatible only. As such, their deflection calculations converge from above; that is, they are too flexible. If attempts are made to enforce C_1 compatibility on such elements, they tend to become too stiff. They also have another defect, in that the second derivatives (and, hence, moment fields) tend to have a saw–tooth pattern which, unfortunately, does not disappear as the mesh is refined. The best estimate of the moment fields in an element is probably that at the centroid. The 12–d.o.f. element first presented by Felippa and Clough [40] overcomes both of these deficiencies. That is, it is a C_1–compatible element and possesses much better–controlled second derivatives of the displacements. The basis of the element is the complete ten–term cubic interpolation polynomial in which the nodal parameters are taken as (w, θ_1, θ_2) at nodes $1, 2, 3$ and θ_s the normal slope at node 4 as shown in Fig. 8.14(a). This element alone appears to be an unpromising candidate for a plate bending element. However, if the single triangular element is divided into three subtriangles with their common apex at the centroid, the (θ_{s4}) variable can then be employed in each of the external triangle sides to produce an element which becomes C_1 compatible, as shown in Fig. 8.14(b). These sub–triangles also have the internal nodes $0, 7, 8, 9$ as shown in Fig. 8.14(b). If the triangle in Fig. 8.14(b) is examined in detail it is seen (excluding, for the present, the nodes $7, 8, 9$) that the three cubic polynomials spanning the subtriangles have 15 degrees of freedom. They are evidently, as yet, C_1 incompatible along sides 1-$0, 2$-0 and 3-0. However, the nodes $7, 8, 9$, are introduced to enforce this internal normal slope compatibility. This provides three equations of constraint from which the three degrees of freedom at the centroid can be expressed in terms of the 12 external degrees of freedom. This will be explained in further detail below.

To this end the geometry of the subtriangle (2) is shown in Fig. 8.15. Thus the triangle *1, 2, 3* is subdivided into three elements, *1-0-3*, *2-0-1*, and *3-0-1* where *0* is the internal point. The midpoints of the external sides are numbered *4, 5, 6*, and the midpoints of the internal sides *7, 8, 9*. The subelement number corresponds to the opposite corner number of the main triangle.

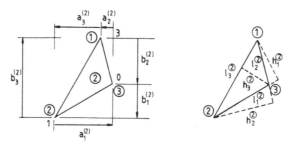

Fig. 8.15   Subtriangle geometry.

Quantities associated with a specific subelement are identified by a superscript in parentheses. The usual ζ notation applies to the whole triangle. For each subtriangle, the same conventions hold if its three apex nodes are renumbered *1-2-3* with node *3* always being the internal point *0*. Thus, for the subelements, the global dimensions are given by

$$a_1 = a_3^{(1)}; \quad a_1^{(1)} + a_2^{(2)} = 0$$

$$a_2 = a_3^{(2)}; \quad a_1^{(2)} + a_2^{(3)} = 0$$

$$a_3 = a_3^{(3)}; \quad a_1^{(3)} + a_2^{(1)} = 0 \tag{8.101}$$

and similarly for the *b* dimensions.

For the whole triangle, the intrinsic dimensions and their ratios are related to the global dimensions a_i, b_i (section 1.4.6) as follows:

$$l_i^2 = a_i^2 + b_i^2 \tag{8.102a}$$

$$h_i = \frac{2A}{l_i} \tag{8.102b}$$

$$d_i = -\frac{1}{l_i}(a_i a_k + b_i b_k) \tag{8.102c}$$

$$\lambda_i = \frac{d_i}{l_i} = -\frac{(a_i a_k + b_i b_k)}{a_i^2 + b_i^2} \tag{8.102d}$$

$$\mu_i = 1 - \lambda_i = \frac{-a_i a_j + b_i b_j}{a_i^2 + b_i^2} \tag{8.102e}$$

In eqn (8.102), i, j, k permute 1, 2, 3 in cyclic order. For each subtriangle, simply substitute the appropriate subtriangle dimensions. With 0 at the centroid of the triangle,

$$A^{(1)} = A^{(2)} = A^{(3)} = \frac{A}{3} \tag{8.103}$$

Many simple relationships exist between the whole triangle and its three subtriangles. Thus,

$$a_1^{(i)} = -a_2^{(j)} = -\frac{(a_i - a_j)}{3} \tag{8.104a}$$

$$b_1^{(i)} = -b_2(j) = \frac{(b_i - b_j)}{3} \tag{8.104b}$$

also,

$$\mu_1^{(i)} + \lambda_2^{(k)} = -1 \tag{8.105a}$$

$$\mu_2^{(i)} + \lambda_1^{(k)} = 3 \tag{8.105b}$$

$$\lambda_3^{(i)} = \frac{1}{3}(1 + \lambda_i) \tag{8.105c}$$

$$\mu_3^{(i)} = \frac{1}{3}(1 + \mu_i) \tag{8.105d}$$

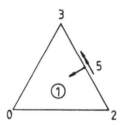

Fig. 8.16   Subtriangle nodes.

The subelement nodal system (N_1) is given in Fig. 8.16. The nodal variables consist of the transverse deflection u_3 and global rotations θ_1, θ_2 at each corner, plus the side rotation θ_s at the midpoint of the external side. For subelement (1):

$$\{r^{(1)}\}^T = \langle u_{32}\ \theta_{12}\ \theta_{22}\ u_{33}\ \theta_{13}\ \theta_{23}\ u_{30}\ \theta_{10}\ \theta_{20}\ \theta_{s5} \rangle \tag{8.106}$$

The deflection of its surface is given

$$u_3^{(1)} = \{\phi_{(3)}\}^T \{r^{(1)}\} \tag{8.107}$$

The interpolation polynomial is given in eqn (8.108), and for this the dimensions and triangular coordinates of the subelement must be used.

$$
\{\bar{\phi}^{(1)}\} = \begin{bmatrix}
0 \\
0 \\
0 \\
\zeta_1^2(3 - 2\zeta_1) + 6\mu_3^{(1)}\zeta_1\zeta_2\zeta_3 \\
\zeta_1^2(b_2^{(1)}\zeta_3 - b_3^{(1)}\zeta_2) - (b_3^{(1)}\mu_3^{(1)} - b_1^{(1)})\zeta_1\zeta_2\zeta_3 \\
\zeta_1^2(a_2^{(1)}\zeta_3 - a_3^{(1)}\zeta_2) - (a_3^{(1)}\mu_3^{(1)} - a_1^{(1)})\zeta_1\zeta_2\zeta_3 \\
\zeta_1^2(3 - 2\zeta_2) + 6\lambda_3^{(1)}\zeta_1\zeta_2\zeta_3 \\
\zeta_2^{(2)}(b_3^{(1)}\zeta_1 - b_1^{(1)}\zeta_3) - (b_2^{(1)} - b_3^{(1)}\lambda_3^{(1)})\zeta_1\zeta_2\zeta_3 \\
\zeta_2^2(a_3^{(1)}\zeta_1 - a_1^{(1)}\zeta_3) - (a_2^{(1)} - a_3^{(1)}\lambda_3^{(1)})\zeta_1\zeta_2\zeta_3 \\
0 \\
4h_3^{(1)}\zeta_1\zeta_2\zeta_3 \\
0 \\
\zeta_3^2(3 - 2\zeta_3) \\
\zeta_3^2(b_1^{(1)}\zeta_2 - b_2^{(1)}\zeta_1) \\
\zeta_3^2(a_1^{(1)}\zeta_2 - a_2^{(1)}\zeta_1)
\end{bmatrix}
\tag{8.108}
$$

It follows that the complete kinematic field for triangle *1-2-3* can be specified in terms of 15 degrees of freedom arranged as follows:

$$
\{\bar{r}\}^T = <u_{31}\ \theta_{11}\ \theta_{21}\ u_{32}\ \theta_{12}\ \theta_{22}\ u_{33}\ \theta_{13}\ \theta_{23}\ \theta_{s4}\ \theta_{s5}\ \theta_{s6}\ |u_{30}\ \theta_{10}\ \theta_{20}> = <r^T\ |r_0^T>
\tag{8.109}
$$

For subelement k $(k = 1, 2, 3)$

$$
u_3^{(k)} = \{\bar{\phi}^{(k)}\}^T\{\bar{r}\} = <\{\phi_e^{(k)}\}^T\ |\{\phi_0^{(k)}\}^T> \begin{bmatrix} r \\ r_0 \end{bmatrix}
\tag{8.110}
$$

where $\{\bar{\phi}^{(k)}\}$ has the same structure as in eqn (8.108), as follows:

for (1), nodes *1, 4, 6*, not present
for (2), nodes *2, 4, 5*, not present
for (3), nodes *3, 5, 6*, not present

For all three triangles, the interpolation polynomials form a 15×15 matrix,

$$
\begin{bmatrix}
0 & 2 & 1 \\
1 & 0 & 2 \\
2 & 1 & 0 \\
0 & 0 & 4 \\
4 & 0 & 0 \\
0 & 4 & 0 \\
3 & 3 & 3
\end{bmatrix}
\tag{8.111}
$$

In eqn (8.111) each column is calculated by using the global dimensions of the corresponding triangle.

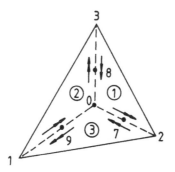

Fig. 8.17   Matching internal subtriangle normal slopes.

8.2.4.2   Internal compatibility conditions

As mentioned previously, the normal slopes (side rotations) of the adjacent subelements are matched at the nodal points 7-8-9. This provides three equations of constraint from which the three degrees of freedom at the centroid are expressed, in terms of the 12 external degrees of freedom, as

$$\{r_0\} = [G]\{r\} \tag{8.112}$$

To achieve this use is made of the expression for the derivative of $f(\zeta_1, \zeta_2, \zeta_3)$ in the direction of the internal normals (section 1.4.6):

$$f_{,n_1} = \frac{1}{h_1}(f_{,\zeta_1} - f_{,\zeta_2}\mu_1 - f_{,\zeta_3}\lambda_1)$$

and

$$f_{,n_2} = \frac{1}{h_2}(-f_{,\zeta_1}\lambda_2 + f_{,\zeta_2} - f_{,\zeta_3}\mu_3)$$

where the subscripted variable $f_{,\zeta_i}$ implies differentiation with respect to ζ_i. Consider, for example, subelement (3). Differentiate the interpolation vector in eqn (8.110) with respect to $\zeta_1^{(3)}, \zeta_2^{(3)}, \zeta_3^{(3)}$, and evaluate at the midpoints:

$$\zeta_2^{(3)} = \zeta_3^{(3)} = 1/2, \ \zeta_1^{(3)} = 0$$

and

$$\zeta_1^{(3)} = \zeta_3^{(3)} = 1/2, \ \zeta_2^{(3)} = 0$$

to obtain

$$\theta_7^{(3)} = (u_{3,n_1}^{(3)})_7 = <r^T \ r_0^T> \begin{bmatrix} q_7^{(3)} \\ q_{70}^{(3)} \end{bmatrix} \tag{8.113a}$$

and

$$\theta_9^{(3)} = (u_{3,n_2}^{(3)})_9 = <r^T \ r_0^T> \begin{bmatrix} q_9^{(3)} \\ q_{90}^{(3)} \end{bmatrix} \tag{8.113b}$$

Similarly, $\theta_7^{(1)}$, $\theta_8^{(1)}$, $\theta_8^{(2)}$, $\theta_9^{(2)}$ can be evaluated. The positive sense of normal rotations is shown in Fig. 8.16. Thence, from Fig. 8.17,

$$\theta_7^{(3)} = -\theta_7^{(1)} \tag{8.114a}$$

$$\theta_8^{(1)} = -\theta_8^{(2)} \tag{8.114b}$$

$$\theta_9^{(2)} = -\theta_9^{(3)} \tag{8.114c}$$

Using eqn (8.113),

$$\begin{bmatrix} q_{07}^{(3)} + q_{07}^{(1)} \\ q_{08}^{(1)} + q_{08}^{(2)} \\ q_{09}^{(2)} + q_{09}^{(3)} \end{bmatrix} \{r_0\} + \begin{bmatrix} q_7^{(3)} + q_7^{(1)} \\ q_8^{(1)} + q_8^{(2)} \\ q_9^{(2)} + q_9^{(3)} \end{bmatrix} \{r\} = \{0\} \tag{8.115}$$

This equation may be written

$$[Q_0]\{r_0\} + [Q]\{r\} = \{0\} \tag{8.116}$$

Thence,

$$\{r_0\} = [G]\{r\} \tag{8.117}$$

so that

$$[G] = -[Q_0]^{-1}[Q] \tag{8.118}$$

Thus, the interpolation functions eqn (8.110) plus the kinematic constraints eqn (8.116) uniquely define the displacement field of the flat triangular element with 12 external degrees of freedom satisfying both external and internal C_1 compatibility. Symbolically,

$$\{u_3\} = \{\phi\}^T \{r\}$$

where $\{\phi\}$ has a different expression over each subtriangle. That is,

$$u_3^{(k)} = \{\phi^{(k)}\}^T \{r\}$$

where

$$\{\phi^{(k)}\}^T = \{\phi_e^{(k)}\}^T + \{\phi_0^k\}^T [G] \tag{8.119}$$

Felippa has obtained the following closed form expression for $\{\phi^{(k)}\}$ by given in eqn (8.120). For sublement 3, the expression for $\{\phi^{(3)}\}$ is given b

$$\{\phi^{(3)}\} = \{\phi_{31}^{(3)} \ \phi_{\theta11}^{(3)} \ \phi_{\theta21}^{(3)} \ \phi_{32}^{(3)} \ \phi_{\theta12}^{(3)} \ \phi_{\theta22}^{(3)} \ \phi_{33}^{(3)} \ \phi_{\theta13}^{(3)} \ \phi_{\theta32}^{(3)} \ \phi_{\theta4}^{(3)} \ \phi_{\theta5}^{(3)} \ \phi_{\theta6}^{(3)} \} \tag{8.120}$$

in which (in terms of the dimensions and coordinates of the complete triangle)

$$\phi_{31}^{(3)} = \zeta_1^2(3 - 2\zeta_1) + 6\mu_3\zeta_1\zeta_2\zeta_3 + \zeta_3^2[3(\lambda_2 - \mu_3)\zeta_1 + (2\mu_3 - \lambda_2)\zeta_3 - 3\mu_3\zeta_2] \tag{8.121}$$

$$\phi_{\theta11}^{(3)} = \zeta_1^2(b_2\zeta_3 - b_3\zeta_2) + (b_1 - b_3\mu_3)\zeta_1\zeta_2\zeta_3$$

$$+ \frac{1}{6}\zeta_3^2[3(b_2\lambda_2 + b_3\mu_3 - 2b_1)\zeta_1 + 3(b_3\mu_3 - b_1)\zeta_2 + (3b_1 - b_2\lambda_2 - 2b_3\mu_3)\zeta_3] \tag{8.122}$$

$$\phi_{32}^{(3)} = \zeta_2^2(3 - 2\zeta_2) + 6\lambda_3\zeta_1\zeta_2\zeta_3 + \zeta_3^2[3(\mu_1 - \lambda_3)\zeta_2 + (2\lambda_3 - \mu_1)\zeta_3 - 3\lambda_3\zeta_1] \tag{8.123}$$

$$\phi_{\theta12}^{(3)} = \zeta_2^2(b_3\zeta_1 - b_1\zeta_2) + (b_3\lambda_3 - b_2)\zeta_1\zeta_2\zeta_3 +$$

$$\frac{1}{6}\zeta_3^2[3(2b_2 - b_3\lambda_3 - b_1\mu_1)\zeta_2 + 3(b_2 - b_3\lambda_3)\zeta_1 + (-3b_2 + b_1\mu_1 + 2b_3\lambda_3)\zeta_3] \tag{8.124}$$

$$\phi_{33}^{(3)} = \zeta_3^2[3(1 + \mu_2)\zeta_1 + 3(1 + \lambda_1)\zeta_2 + (1 - \mu_2 - \lambda_1)\zeta_3] \tag{8.125}$$

$$\phi_{\theta13}^{(3)} = \frac{1}{6}\zeta_3^2[3(3b_1 + b_2 + b_1\lambda_1)\zeta_2 + (b_2\mu_2 - b_1\lambda_1)\zeta_3 - 3(b_1 + 3b_2 + b_2\mu_2)\zeta_1] \tag{8.126}$$

$$\phi_{\theta4}^{(3)} = \frac{4A}{3l_3}[6\zeta_1\zeta_2\zeta_3 + \zeta_3^2(5\zeta_3 - 3)] \tag{8.127}$$

$$\phi_{\theta5}^{(3)} = \frac{4A}{3l_1}[\zeta_3^2(3\zeta_2 - \zeta_3)] \tag{8.128}$$

$$\phi_{\theta6}^{(3)} = \frac{4A}{3l_2}[\zeta_3^2(3\zeta_1 - \zeta_3)] \tag{8.129}$$

For $\phi_{\theta2i}^{(3)}$, change all b_i in $\phi_{\theta1i}^{(3)}$ to a_i. The above expressions apply to subelement 3 where ($\zeta_1 \geq \zeta_3$, $\zeta_2 \geq \zeta_3$). In subelement 1 (where $\zeta_2 \geq \zeta_1$, $\zeta_3 \geq \zeta_1$) and in subelement 2 (where $\zeta_3 \geq \zeta_2$, $\zeta_1 \geq \zeta_2$) permute cyclically all subscripts and superscripts (*1-2-3* permutes to *2-3-1*, and *4-5-6* to *5-6-4*). For example,

$$\phi_{11}^{(1)} = \zeta_1^2[3(1 + \mu_3)\zeta_2 + 3(1 + \lambda_2)\zeta_3 + (1 - \mu_3 - \lambda_2)\zeta_1] \tag{8.130}$$

8.2.4.3   Generation of the element stiffness matrix

Curvature displacement relationships

Since $u_3^{(k)}$ is expressed in terms of a cubic polynomial, the three curvature components vary linearly over each subregion. This variation may be specified by the curvatures at the three subtriangle apices, and nine values would thus be required for each subelement (27 for the complete triangle). However the internal compatibility requirements impose equal curvatures at the centroid for the three adjacent subelements. Therefore, only seven values are required to specify fully the variation of a curvature component. For example,

$$\{\chi_{11}\}^T = <\chi_{11i}^{(3)} \ \chi_{11j}^{(3)} \ \chi_{11j}^{(1)} \ \chi_{11k}^{(1)} \ \chi_{11i}^{(2)} \ \chi_{11i}^{(2)} \ \chi_{110}> \tag{8.131}$$

In eqn (8.131), *i, j, k* are the whole triangle nodes *1, 2, 3*. In a similar way the curvature terms $\{\chi_{22}\}$ and $\{\chi_{12}\}$, may be expressed in terms of their nodal values.

That is, 21 values are sufficient to determine the complete curvature field.

$$\{\chi\}^T = <\{\chi_{11}\}^T \ \{\chi_{22}\}^T \ \{2\chi_{21}\}^T> \tag{8.132}$$

The curvature variation in terms of the nodal displacements can be obtained by differentiating the interpolation vector with respect to the global coordinates:

$$\{\chi_{11}^{(k)}\} = \{\bar{\phi}_{,11}^{(k)}\}^T \{\bar{r}\} \tag{8.133}$$

and, similarly, for $\{\chi_{22}^{(k)}\}$, $\chi_{21}^{(k)}\}$. *Note:*

$$\{\bar{\phi}_{,11}^{(k)}\} = \{\bar{\phi}_{,\zeta_m \zeta_n}\} b_m^{(k)} b_n^{(k)}, \quad \text{sum } m, n = 1, 2, 3 \tag{8.134}$$

On substituting the triangular coordinates of the subtriangle apices, the nodal curvatures are obtained

$$\{\chi\} = [\bar{B}]\{\bar{r}\} = [B_e \ | B_0] \begin{Bmatrix} r \\ r_0 \end{Bmatrix} \tag{8.135}$$

Finally, using the constraint condition eqn (8.117),

$$\{\chi\} = [B_e + B_0 G]\{r\} = [B]\{r\} \tag{8.136}$$

Note: It is important to realize that $\{\chi_0\}$ should be evaluated once only, because it is constant in all the triangles. The remainder of the stiffness matrix generation follows the standard procedure using orthogonal interpolation functions, here linear $\{\phi_1\}$ functions, for the curvatures. Thus, for subelement (k),

$$\{\chi^{(k)}\}^T = <\{\chi_{11}^{(k)}\}^T \ \{\chi_{22}^{(k)}\}^T \ 2\{\chi_{21}^{(k)}\}^T> $$

where

$$\{\chi_{11}^{(k)}\}^T = <\chi_{11i}^{(k)} \ \chi_{11j}^{(k)} \ \chi_{110}^{(k)}>, \text{ etc.} \tag{8.137}$$

In eqn (8.137), i, j, k are cyclic permutations of 1, 2, 3. Because of the linear variation of each component $\chi^{(k)}$, write

$$\{\bar{\chi}^{(k)}\} = \begin{Bmatrix} \bar{\chi}_{11}^{(k)} \\ \bar{\chi}_{22}^{(k)} \\ 2\bar{\chi}_{21}^{(k)} \end{Bmatrix} = \left[\{\phi_1\}^T \ \{\phi_1\}^T \ \{\phi_1\}^T \right] \{\chi^{(k)}\} \tag{8.138}$$

The constitutive matrix $[C]$ is assumed constant over the whole triangle, see eqn (8.27), while the thickness may vary according to the interpolation,

$$h = \{\phi_\xi\}\{\xi\}h_r \tag{8.139}$$

The reference thickness is h_r, $\{\xi\}$ is the vector of relative thickness nodal values, and $\{\phi_\xi\}$ is the interpolation polynomial. For the triangle apex nodes *1, 2, 3*, $\{\phi_\xi\}$ will usually be linear; that is, $\{\phi_\xi\} = \{\phi_1\}$. Thence, using the equation,

$$[k_\chi] = \int\limits_{area} [a_\chi]^T [C][a_\chi] dA \tag{8.140}$$

it is found that

$$[N^{(k)}] = \frac{A^{(k)}h_r^{\,3}}{12} \begin{bmatrix} c_{11}\bar{Q} & c_{12}\bar{Q} & c_{13}\bar{Q} \\ c_{21}\bar{Q} & c_{22}\bar{Q} & c_{23}\bar{Q} \\ c_{31}\bar{Q} & c_{32}\bar{Q} & c_{33}\bar{Q} \end{bmatrix} \qquad (8.141)$$

In eqn (8.141),

$$\bar{Q} = \frac{1}{A^{(k)}} \int\limits_{area} \{\phi_1\}[\{\{\phi_\xi\}^T\{\xi\}]^3\{\phi_1\}^T dA = \frac{1}{12}\begin{bmatrix} 2 & 1 & 1 \\ 1 & 2 & 1 \\ 1 & 1 & 2 \end{bmatrix} \quad \text{for constant thickness} \quad (8.142)$$

Once $[N^{(k)}]$ has been calculated from eqn (8.145), it can be assembled into the 21×21 nodal curvature stiffness matrix, $[k_\chi]$. Then to obtain the 12×12 nodal stiffness matrix $[k_r]$, the $[B]$ matrix of eqn (8.141) is used:

$$[k_r] = [B]^T[k_\chi][B] \qquad (8.143)$$

The 12 d.o.f. compatible triangular plate element is a highly successful element, producing curvature fields which tend to be free of the saw–tooth effects experienced by the incompatible elements discussed previously. Evidently, the combination of the three cubic interpolation functions together with the C_1 compatibility has beneficial effects. It may possibly be considered a disadvantage to have to accomodate the midside rotational degree of freedom nodes. These nodes can be constrained out by imposing a linear variation of normal slopes. For example, along side *2-3* of the triangle (Fig. 8.14(b)), set

$$\theta_{s5} = \frac{1}{2}(\theta_{s2} + \theta_{s3}) \qquad (8.144)$$

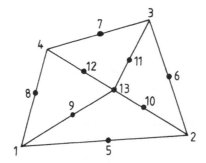

Fig. 8.18   Q19 arbitrary quadrilateral.

This may have the effect of stiffening the element excessively. A better approach is to construct an arbitrary quadrilateral of four triangular elements, as shown in Fig. 8.18. In the quadrilateral nodes *9* to *13* are internal nodes, and, can be removed from the global stiffness matrix by static condensation before assembly. Nodes *5* to *8* have θ_s degrees of freedom only and can be condensed out by the same process as in eqn (8.144). That is, for example,

$$\theta_{s5} = \frac{1}{2}(\theta_{s1} + \theta_{s2}) \quad \text{etc.} \tag{8.145}$$

The resulting element is a quadrilateral with corner nodes only, and with 12 d.o.f. It is possibly the best plate element available for incorporation in standard finite element packages. It would thus appear that the discussion of triangular plate bending elements has been exhausted from the point of view of utility. It will be shown that such is, however, not the case, and that there are at least two strong contenders which are useful not only as plate elements, but also for their incorporation in flat facit shell elements. Before discussing these elements, the hybrid approach and an equilibrium element are outlined.

8.2.5 A hybrid triangular element

Remember, from section 3.6, the basic steps of the hybrid element approach. Firstly, the stress assumption is made:

$$\{\sigma\} = [P]\{\beta\} \tag{8.146}$$

where $\{\beta\}$ are a set of generalized forces with no direct physical interpretation. In the plate problem,

$$\{\sigma\} = \begin{Bmatrix} M_{11} \\ M_{22} \\ M_{21} \\ Q_1 \\ Q_2 \end{Bmatrix} \tag{8.147}$$

The deformation (strains) in the plate are calculated as

$$\begin{Bmatrix} \chi_{,11} \\ \chi_{,22} \\ 2\chi_{,21} \\ \gamma_1 \\ \gamma_2 \end{Bmatrix} = \begin{bmatrix} C_B & 0 \\ 0 & C_S \end{bmatrix} \{\sigma\} \tag{8.148}$$

or, in shorthand notation,

$$\{\varepsilon\} = [f]\{\sigma\} \tag{8.149}$$

Where, for sandwich plates,

$$[C_B] = \frac{1}{B} \begin{bmatrix} 1 & -\nu & 0 \\ -\nu & 1 & 0 \\ 0 & 0 & 2(1+\nu) \end{bmatrix} ; \quad B = \frac{Et(c+t)^2}{2} \tag{8.150}$$

and, (example 8.2 and Fig. 8.6).

$$[C_S] = \frac{1}{S} \begin{bmatrix} 1 & 0 \\ 0 & 1 \end{bmatrix} ; \quad S = \frac{G(c+t)^2}{c} \tag{8.151}$$

Applying contragredience to eqn (8.146) gives the generalized displacements corresponding to $\{\beta\}$ as

$$\{r_\beta\} = \int\limits_{area} [P]^T\{\varepsilon\}dA = \int\limits_{area} [P]^T[f][P]dA\,\{\beta\} \tag{8.152}$$

Write this equation as

$$\{r_\beta\} = [F]\{\beta\} \tag{8.153}$$

Inversion gives the equivalent stiffness relationship as

$$\{\beta\} = [k_\beta]\{r_\beta\} \tag{8.154}$$

It now remains to obtain nodal forces which are statically equivalent to the moment and shear fields along the element sides. The surface forces are calculated from the stress field,

$$\{G_s\} = [R]\{\beta\} \tag{8.155}$$

The nodal displacements $\{u\}$ are chosen to give displacements along the sides corresponding to $\{G_s\}$

$$\{\tilde{u}\} = [L]\{u\} \tag{8.156}$$

Using contragredience, the nodal forces are:

$$\{R_u\} = \int\limits_s [L]^T\{G_s\}ds = \int\limits_s [L]^T[R]ds\,\{\beta\} \tag{8.157}$$

Define the matrix $[T]$

$$[T] = \int\limits_s [R]^T[L]ds \tag{8.158}$$

Again applying contragredience,

$$\{r_\beta\} = [T]\{u\} \tag{8.159}$$

Thence, finally,

$$\{R_u\} = [T][k_\beta][T]\{u\} \tag{8.160}$$

The stiffness matrix for the nodal forces is, thus,

$$[K] = [T]^T[k_\beta][T] \tag{8.161}$$

These equations (8.148) to (8.154) are, of course, the standard application of the hybrid element method. Various moment and shear fields can be chosen to satisfy the differential equations of equilibrium eqn (8.29). For example, a nine, stress mode element is given by

$$\begin{Bmatrix} M_1 \\ M_2 \\ M_{12} \\ Q_1 \\ Q_2 \end{Bmatrix} = \begin{bmatrix} 1 & 0 & 0 & x_1 & 0 & 0 & x_2 & 0 & 0 \\ 0 & 1 & 0 & 0 & x_1 & 0 & 0 & x_2 & 0 \\ 0 & 0 & 1 & 0 & 0 & x_1 & 0 & 0 & x_2 \\ 0 & 0 & 0 & 1 & 0 & 0 & 0 & 0 & 1 \\ 0 & 0 & 0 & 0 & 0 & 1 & 0 & 1 & 0 \end{bmatrix} \begin{Bmatrix} \beta_1 \\ \beta_2 \\ \cdot \\ \cdot \\ \beta_9 \end{Bmatrix} \tag{8.162}$$

Alternatively, a five, stress mode element is obtained by simply dropping columns 5, 6, 7 and 9 from eqn (8.162) and retaining only five β terms. In this case the matrix multiplication in eqn (8.151) gives the integrand of $[F]$ as

$$[F] = \begin{bmatrix} & & x_1 c_{11} & x_2 c_{12} \\ & [C] & x_1 c_{21} & x_2 c_{22} \\ & & x_1 c_{31} & x_2 c_{32} \\ x_1 c_{11} & x_1 c_{21} & x_1 c_{31} & (x_1^2 c_{11} + d_{11}) & (x_1 x_2 c_{12} + d_{12}) \\ x_2 c_{12} & x_2 c_{22} & x_3 c_{32} & (x_1 x_2 c_{12} + d_{12}) & (x_2^2 c_{22} + d_{22}) \end{bmatrix} \qquad (8.163)$$

Constant terms are integrated by multiplying them by the triangle area. Linear terms in x_1, x_2 disappear if, for the local coordinates, the origin at the triangle centroid is used, and

$$\int_{area} x_1^2 dA = \frac{(x_{1i}^2 + x_{1j}^2 + x_{1k}^2)A}{12} \quad \text{etc.} \qquad (8.164)$$

Probably the most intricate part of the hybrid formulation is in the calculation of the $[L]$ and $[G]$ matrices and the subsequent line integration around the element perimeter. Herein, (Q_1, Q_2) are chosen as linear functions, whereas u_3 is chosen to have a linear term from the end displacements of a side plus a quadratic variation due to the end rotations.

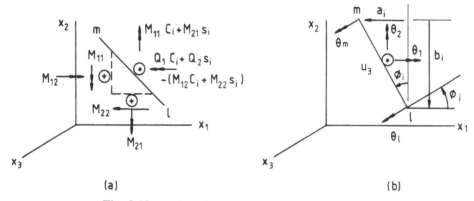

(a)   (b)

Fig. 8.19   Hybrid moment element side forces.

The force fields on side $(i\text{–}m)$ are shown in Fig. 8.19(a), and the corresponding displacement fields in Fig. 8.19(b).

$$a_i = x_{1m} - x_{1l}$$

$$b_i = x_{2l} - x_{2m} \qquad (8.165)$$

$$c_i = \cos \phi_i = \frac{x_{2m} - x_{2l}}{L_i} = \frac{-b_i}{L_i}$$

$$s_i = \sin\phi_i = \frac{x_{1l} - x_{1m}}{L_i} = \frac{-a_i}{L_i} \tag{8.166}$$

Thus, choose

$$\left\{ \begin{matrix} \theta_1 \\ \theta_2 \end{matrix} \right\} = \begin{bmatrix} (1 - \xi) & \xi & 0 & 0 \\ 0 & 0 & (1 - \xi) & \xi \end{bmatrix} \left\{ \begin{matrix} \theta_{1l} \\ \theta_{1m} \\ \theta_{2l} \\ \theta_{2m} \end{matrix} \right\} \tag{8.167a}$$

and

$$u_3 = [(1 - \xi) \ \xi] \begin{bmatrix} u_{3l} \\ u_{3m} \end{bmatrix} + \frac{L_i \xi (1 - \xi)}{2} (\theta_m - \theta_l)$$

That is,

$$u_3 = [(1 - \xi) \ \xi] \begin{bmatrix} u_{3l} \\ u_{3m} \end{bmatrix} + \frac{\xi(1 - \xi)}{2} [-b_i \ b_i \ -a_i \ a_i] \left\{ \begin{matrix} \theta_{1l} \\ \theta_{1m} \\ \theta_{2l} \\ \theta_{2m} \end{matrix} \right\} \tag{8.167b}$$

That is, for all three components,

$$\begin{bmatrix} \theta_1 \\ \theta_2 \\ u_3 \end{bmatrix} = \begin{bmatrix} 0 & (1 - \xi) & 0 & 0 & \xi & 0 \\ 0 & 0 & (1 - \xi) & 0 & 0 & \xi \\ (1 - \xi) & -\dfrac{b_i}{2}\xi(1 - \xi) & -\dfrac{a_i}{2}\xi(1 - \xi) & \xi & \dfrac{b_i}{2}\xi(1 - \xi) & \dfrac{a_i}{2}\xi(1 - \xi) \end{bmatrix} \begin{bmatrix} u_{3l} \\ \theta_{1l} \\ \theta_{2l} \\ u_{3m} \\ \theta_{1m} \\ \theta_{2m} \end{bmatrix} \tag{8.168}$$

This is eqn (8.156). The generalized forces corresponding to $<\theta_1 \ \theta_2 \ u_3>$ on side l, m will be designated $<F_1 \ F_2 \ F_3>$. From Fig. 8.19(a),

$$\left\{ \begin{matrix} F_1 \\ F_2 \\ F_3 \end{matrix} \right\} = \begin{bmatrix} 0 & a_i & b_i & 0 & 0 \\ -b_i & 0 & -a_i & 0 & 0 \\ 0 & 0 & 0 & -b_i & -a_i \end{bmatrix} \left\{ \begin{matrix} M_{11} \\ M_{22} \\ M_{21} \\ Q_1 \\ Q_2 \end{matrix} \right\} \tag{8.169}$$

Since $<M_1\, M_2\, M_{21}>$ vary linearly on the side l, m, whereas $<Q_1\, Q_2>$ are constant, it follows that

$$
\begin{bmatrix} M_{11} \\ M_{22} \\ M_{21} \\ Q_1 \\ Q_2 \end{bmatrix}
=
\begin{bmatrix}
(1-\xi) & 0 & 0 & \xi & 0 & 0 & 0 & 0 \\
0 & (1-\xi) & 0 & 0 & \xi & 0 & 0 & 0 \\
0 & 0 & (1-\xi) & 0 & 0 & \xi & 0 & 0 \\
0 & 0 & 0 & 0 & 0 & 0 & 1 & 0 \\
0 & 0 & 0 & 0 & 0 & 0 & 0 & 1
\end{bmatrix}
\begin{bmatrix} M_{11l} \\ M_{22l} \\ M_{21l} \\ M_{11m} \\ M_{22m} \\ M_{21m} \\ Q_1 \\ Q_2 \end{bmatrix}
\tag{8.170}
$$

Thus, combining eqns (8.169) and (8.170),

$$
\begin{Bmatrix} F_1 \\ F_2 \\ F_3 \end{Bmatrix}
=
\begin{bmatrix}
0 & a_i(1-\xi) & b_i(1-\xi) & 0 & a_i\xi & b_i\xi & 0 & 0 \\
-b_i(1-\xi) & 0 & -a_i(1-\xi) & -b_i\xi & 0 & -a_i\xi & 0 & 0 \\
0 & 0 & 0 & 0 & 0 & 0 & -b_i & -a_i
\end{bmatrix}
\begin{Bmatrix} M_{11l} \\ M_{22l} \\ M_{21l} \\ M_{11m} \\ M_{22m} \\ M_{21m} \\ Q_1 \\ Q_2 \end{Bmatrix}
\tag{8.171}
$$

Now, from eqn (8.163), for five β terms,

$$
\begin{Bmatrix} M_{11l} \\ M_{22l} \\ M_{21l} \\ M_{11m} \\ M_{22m} \\ M_{21m} \\ Q_1 \\ Q_2 \end{Bmatrix}
=
\begin{bmatrix}
 & & & x_{1l} & 0 \\
I_3 & & & 0 & x_{2l} \\
 & & & 0 & 0 \\
 & & & x_{1m} & 0 \\
I_3 & & & 0 & x_{2m} \\
 & & & 0 & 0 \\
0 & 0 & 0 & 1 & 0 \\
0 & 0 & 0 & 0 & 1
\end{bmatrix}
\{\beta\}
\tag{8.172}
$$

Hence, on multiplication of the matrices in eqns (8.170) and (8.177), the matrix $[R]$ in eqn (8.156) has been obtained for side l, m of the triangle. That is,

$$
[R] =
\begin{bmatrix}
0 & a_i & b_i & 0 & a_i(x_{2l} - b_i\xi) \\
-b_i & 0 & -a_i & -b_i(x_{1l} + \xi a_i) & 0 \\
0 & 0 & 0 & -b_i & -a_i
\end{bmatrix}
\tag{8.173}
$$

To form the $[T]$ matrix, it is necessary to calculate $[R]^T[L]$ and integrate from 0 to l, and then sum the contributions for the three sides of the triangle as follows, (Fig. 8.20)

$$[T] = \begin{bmatrix} (j)n & (i)l & (j)m \\ + & + & + \\ (k)n & (k)l & (i)l \end{bmatrix} \tag{8.174}$$

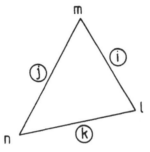

Fig. 8.20   Triangle node numbers.

The computer coding for this process is given in [41], for either five or nine β terms. The above hybrid element is of interest because of the insight it gives into this type of formulation. It remains now to look at an equilibrium element formulation for the triangle, in order to examine this potential avenue for the solution to the plate problem.

8.2.6   An equilibrium plate bending element

The following element was formulated by Anderhagen [42]. The basic element is the six–node triangle. However, four of these may be combined to make an arbitrary quadrilateral, similar to the 12–d.o.f. displacement element. The element node system is shown in Fig. 8.21.

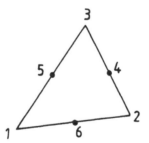

Fig. 8.21   Triangle node numbering.

The equilibrium element formulation starts by specifying the moment field, which varies quadratically, thus requiring the six node points shown in Fig. 8.21. That is,

$$\{\tilde{m}\} = \begin{bmatrix} \tilde{m}_{11} \\ \tilde{m}_{22} \\ \tilde{m}_{21} \end{bmatrix} = \begin{bmatrix} \phi_2 & \phi_2 & \phi_2 \end{bmatrix} \begin{Bmatrix} m_{11} \\ m_{22} \\ m_{21} \end{Bmatrix} \tag{8.175}$$

where the interpolation polynomial is given by

$$\{\phi_2\}^T = <\zeta_1(2\zeta_1 - 1) \quad \zeta_2(2\zeta_2 - 1) \quad \zeta_3(2\zeta_3 - 1) \quad 4\zeta_2\zeta_3 \quad 4\zeta_3\zeta_1 \quad 4\zeta_1\zeta_2> \tag{8.176}$$

Eqn (8.175) will be written

$$\{\tilde{m}\} = [\Phi]\{m\} \tag{8.177}$$

Let $\{\tilde{\chi}\}$ be the plate curvatures, related to the moments $\{\tilde{m}\}$ through the constitutive relationship:

$$\{\tilde{\chi}\} = \left\{ \begin{array}{c} \tilde{\chi}_{11} \\ \tilde{\chi}_{22} \\ 2\tilde{\chi}_{21} \end{array} \right\} = \frac{12}{Et^3} \begin{bmatrix} 1 & -v & 0 \\ -v & 1 & 0 \\ 0 & 0 & 2(1+v) \end{bmatrix} \left\{ \begin{array}{c} \tilde{m}_{11} \\ \tilde{m}_{22} \\ \tilde{m}_{21} \end{array} \right\} \tag{8.178}$$

That is,

$$\{\chi\} = [f]\{\tilde{m}\} \tag{8.179}$$

Then, if $\{\chi_m\}$ are the generalized curvatures corresponding to $\{m\}$,

$$\{\chi_m\} = \int_{area} [\Phi]^T [f][\Phi] dA \tag{8.180}$$

Let the integral of the product of the interpolation polynomial be expressed by

$$\int_{area} \{\phi_2\}^T \{\phi_2\} dA = \frac{A}{180} [\Psi] \tag{8.181}$$

where

$$[\Psi] = \begin{bmatrix} 6 & -1 & -1 & 4 & \cdot & \cdot \\ -1 & 6 & -1 & \cdot & 4 & \cdot \\ -1 & -1 & 6 & \cdot & \cdot & 4 \\ 4 & \cdot & \cdot & 32 & 16 & 16 \\ \cdot & 4 & \cdot & 16 & 32 & 16 \\ \cdot & \cdot & 4 & 16 & 16 & 32 \end{bmatrix} \tag{8.182}$$

Then write eqn (8.180) as

$$\{\chi_{mi}\} = [F_i]\{m_i\} \tag{8.183}$$

where, now,

$$[F_i] = \frac{A_i}{15Et^3} \begin{bmatrix} \Psi & -v\Psi & 0 \\ -v\Psi & \Psi & 0 \\ 0 & 0 & 2(1+v)\Psi \end{bmatrix} \tag{8.184}$$

The flexibility of the whole system can be assembled in the usual way, noting that at any node point the moments are the same for all elements incident thereon. That is,

$$\{\chi\} = [F]\{M\} \tag{8.185}$$

Now, however, the nodal values of the moments $\{m_{11}, m_{22}, m_{21}\}$ which compose $\{M\}$ are not free variables. They may be chosen only in such a way that equilibrium conditions are satisfied. These equilibrium equations are written

$$[U]\{M\} - \{P\} = 0 \tag{8.186}$$

The matrix $[U]$ is a rectangular equilibrium matrix with more columns than rows for all statically indeterminate structures, and $\{P\}$ is a vector of applied loads.

There are two types of equilibrium conditions. The first of these involves the differential equation of equilibrium, and the second the moment and shear equilibrium across element interfaces.

1.   Differential equation of equilibrium

From eqn (8.33),

$$M_{11,11} + 2M_{21,21} + M_{22,22} = p \tag{8.187}$$

Thus, for a specific element, i, using eqn (8.177),

$$\left\langle \frac{\partial^2 \phi_2^T}{\partial x_1^2} \quad \frac{\partial^2 \phi_2^T}{\partial x_2^2} \quad \frac{2\partial^2 \phi_2^T}{\partial x_1 \partial x_2} \right\rangle \left\{ \begin{array}{c} m_{11} \\ m_{22} \\ m_{21} \end{array} \right\}_i - \tilde{p}_i = 0$$

Write this equation as

$$\langle v_i \rangle \{m_i\} - \tilde{p}_i = 0 \tag{8.188}$$

Now, from section 1.4.6, the second derivatives are expressed

$$\frac{\partial^2 \phi_2^T}{\partial x_1^2} = \frac{1}{A^2} [b_1^2 \ b_2^2 \ b_3^2 \ 2b_2 b_3 \ 2b_3 b_1 \ 2b_1 b_2] \tag{8.189}$$

$$\frac{\partial^2 \phi_2^T}{\partial x_2^2} = \frac{1}{A^2} [a_1^2 \ a_2^2 \ a_3^2 \ 2a_2 a_3 \ 2a_3 a_1 \ 2a_1 a_2] \tag{8.190}$$

$$\frac{2\partial^2 \phi_2^T}{\partial x_1 \partial x_2} = \frac{2}{A^2} [a_1 b_1 \ a_2 b_2 \ a_3 b_3 \ (a_2 b_3 + a_3 b_2) \ (a_3 b_1 + a_1 b_3) \ (a_1 b_2 + a_2 b_1)] \tag{8.191}$$

It is seen that $\langle v_i \rangle \{m_i\}$ has a constant value, so that for the chosen interpolation of the moment fields only a constant $\tilde{p}_i$ can be accommodated over the element.

2.   Moment and shear continuity across element interfaces

Moment continuity is satisfied by choice of the quadratic functions and node points. Shear force continuity must still be satisfied. Shear forces are obtained as first derivatives of moments, and so they vary linearly. Continuity between adjacent elements is assured when, at the end points along the common side, the shear forces acting on the interface are the same for both elements. A given side l–m is shown in Fig. 8.22.

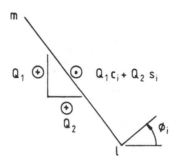

Fig. 8.22   Shear in element side *l-m*.

Then, from the equation of equilibrium (eqn (8.29)),

$$\begin{Bmatrix} Q_1 \\ Q_2 \end{Bmatrix} = \begin{bmatrix} \dfrac{\partial}{\partial x_1} & 0 & \dfrac{\partial}{\partial x_2} \\[2mm] 0 & \dfrac{\partial}{\partial x_2} & \dfrac{\partial}{\partial x_1} \end{bmatrix} \begin{Bmatrix} M_{11} \\ M_{22} \\ M_{21} \end{Bmatrix} \tag{8.192}$$

Then,

$$Q_n = [c_{i,}] \begin{Bmatrix} Q_1 \\ Q_2 \end{Bmatrix} = -\frac{1}{l_i}[b_i \ a_i] \begin{Bmatrix} Q_1 \\ Q_2 \end{Bmatrix}$$

$$= -\frac{1}{l_i}[b_i \ a_i] \begin{bmatrix} \dfrac{\partial}{\partial x_i} & 0 & \dfrac{\partial}{\partial x_2} \\[2mm] 0 & \dfrac{\partial}{\partial x_2} & \dfrac{\partial}{\partial x_1} \end{bmatrix} \begin{Bmatrix} m_{11} \\ m_{22} \\ m_{21} \end{Bmatrix} = -\frac{1}{L_i}[b_i \ a_i] \begin{bmatrix} T_b & 0 & T_a \\ 0 & T_a & T_b \end{bmatrix} \begin{Bmatrix} m_{11} \\ m_{22} \\ m_{21} \end{Bmatrix} \tag{8.193}$$

To calculate $[T_b]$, $[T_a]$, substitute $t = b$, $t = a$, respectively, in the expression:

$$[T_t] = \frac{1}{A}[(2\zeta_1-1)t_1 \ (2\zeta_2-1)t_2 \ (2\zeta_3-1)t_3 \ 4(t_3\zeta_2+t_2\zeta_3) \ 4(t_1\zeta_3+t_3\zeta_1) \ 4(t_1\zeta_2+t_2\zeta_1)] \tag{8.194}$$

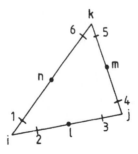

Fig. 8.23   Element–side shears.

Thus, for nodes i-j-k in Fig. 8.22,

$$\begin{Bmatrix} q_1 \\ q_2 \\ q_3 \\ q_4 \\ q_5 \\ q_6 \end{Bmatrix} = [q] \begin{Bmatrix} m_{11} \\ m_{22} \\ m_{21} \end{Bmatrix} \tag{8.195}$$

To ensure shear force continuity, two equilibrium equations for each side, relating the shear forces of the element on the right with those of the element on the left, must be written:

$$q^i_{left} + q^i_{right} = 0 \tag{8.196}$$

$$q^j_{left} + q^j_{right} = 0 \tag{8.197}$$

The '+' sign occurs as a consequence of the direction of the normal n for the side being reversed in the two triangles. It should also be evident that the zeros on the right–hand side can be replaced with a linearly varing line load. The curvature continuity is now enforced as follows. Suppose that λ_p, λ_V are the average element displacement and the average side of element displacement respectively. Then eqns (8.188) and (8.196) give, applying the contragredient principle,

$$\{\chi_{p_i}\} = [v_i]^T \{\lambda_{p_i}\} \tag{8.198}$$

$$\{\chi_{V_i} = [Q_i]^T \{\lambda_{V_i}\} \tag{8.199}$$

The continuity requirement is, thus,

$$\chi_m + \chi_P + \chi_V = 0$$

That is,

$$[F]\{m\} + [v]^T \lambda_P + [Q]^T \lambda_V = 0 \tag{8.200}$$

This equation must be solved subject to the equilibrium constraints,

$$[v]\{m\} = \{P\} \tag{8.201}$$

and

$$[Q]\{m\} = 0 \tag{8.202}$$

These equations clearly can be rearranged in partitioned form

$$\begin{bmatrix} [F] & [v]^T & [Q]^T \\ [v] & 0 & 0 \\ [Q] & 0 & 0 \end{bmatrix} \begin{Bmatrix} m \\ \lambda_P \\ \lambda_V \end{Bmatrix} = \begin{Bmatrix} 0 \\ p \\ 0 \end{Bmatrix} \tag{8.203}$$

Obviously, this is a mixed system of equations in which moments and displacements are evaluated simultaneously. It is interesting to note the various boundary conditions which arise, and a summary is shown in Table 8.1.

Table 8.1

Boundary condition	Variable Fixity				
	M_n	M_s	M_{ns}	λ_V	λ_Q
Free	×		×		
Simply supported	×				×
Clamped					×
Symmetric			×		
Continuous support					×
Column				×	

It would appear that, with so many examples thus given, the quest for the most useful plate triangular element must have been exhausted. Because of the peculiar nature of the plate bending problem with its C_1 compatibility requirements, and also the role these elements play in facet shell element combinations, such is not the case. It will be observed that the C_1 compatibility requirement involves the θ_n variables, in the element sides, and it would thus seem appropriate to make better use of these variables as was done in the 12–d.o.f. compatible cubic triangle element.

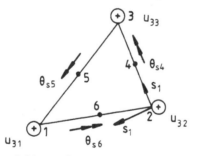

Fig. 8.24   Constant moment element.

The two further elements are thus reviewed: the Morley constant moment element, and the DKL (discrete Kirchoff with loof nodes) element family. Both of these are strong contenders, not only for small deflection plate analysis, but also for small and large deflection analysis of shells.

8.2.7   The Morley constant moment element

This is a non–conforming element [48] derived from quadratic variation of the transverse deflection expressed in terms of apex deflections and mid–side normal rotations. The nodal configuration is thus shown in Fig. 8.23. As with the previous elements, the transverse deflections are expressed by

$$u_3 = u_3^r + u_3^R \tag{8.204}$$

The superscripts r, R refer to the relative and rigid–body deflections, respectively. The interpolation for u_3^R is linear and is given simply by

$$u_3^R \{\phi_1\}^T \{u_e\}$$

(8.205)

where

$$\{u_e\}^T = <u_{31} \ u_{32} \ u_{33}>$$

(8.206)

The local side coordinates are (s_i, n_i) $i = 1, 2, 3$ and the derivatives of the rigid–body deflections are given by

$$u_{3,n_i}^R = \{\phi_{1,n_i}\}^T \{u_e\}$$

(8.207)

Using the expression for the normal derivatives,

$$\frac{\partial f}{\partial n_i} = \frac{1}{2A} \{ \frac{\partial f}{\partial \zeta_i} l_i + \frac{\partial f}{\partial \zeta_j}(d_i - l_i) - \frac{\partial f}{\partial \zeta_k} d_i \}$$

(8.208)

the derivatives of the relative deflections are given by

$$\left\{ \begin{array}{c} u_{34,n_1}^r \\ u_{35,n_2}^r \\ u_{36,n_3}^r \end{array} \right\} = [G] \left\{ \begin{array}{c} \{u_e\} \\ u_{34,n_1} \\ u_{35,n_2} \\ u_{36,n_3} \end{array} \right\}$$

(8.209)

where, from eqns (8.204) and (8.208),

$$[G] = \frac{1}{2A} \begin{bmatrix} -l_1 & l_1 - d_1 & d_1 & 2A & 0 & 0 \\ d_2 & -l_2 & l_2 - d_2 & 0 & 2A & 0 \\ l_3 - d_3 & d_3 & -l_3 & 0 & 0 & 2A \end{bmatrix}$$

(8.210)

The quadratic function for u_3^r is given by

$$\{u_3^r\} = [\zeta_2\zeta_3 \ \zeta_3\zeta_1 \ \zeta_1\zeta_2] \left\{ \begin{array}{c} c_1 \\ c_2 \\ c_3 \end{array} \right\} = \{\psi\}^T \{c\}$$

(8.211)

The vector $\{c\}$ is composed of undetermined generalized displacements. It is thus convenient to develop from eqn (8.211), a set of orthogonal interpolation functions involving u_{34,n_1} etc. Thus, evaluate u_{3,n_i}^r at nodes $4, 5, 6$, using eqn (8.207). This gives

$$\{c\} = [\Phi]^{-1} \begin{bmatrix} u_{34,n_1}^r \\ u_{35,n_2}^r \\ u_{36,n_3}^r \end{bmatrix}$$

(8.212)

where

$$[\Phi] = \frac{1}{4A} \lceil l \rfloor [I_{-1,1}] \tag{8.213}$$

In eqn (8.211),

$$\lceil l \rfloor = \text{diagonal}(l_1 \ l_2 \ l_3) \tag{8.214}$$

and

$$[I_{a,1}] = \begin{bmatrix} a & 1 & 1 \\ 1 & a & 1 \\ 1 & 1 & a \end{bmatrix} \tag{8.215}$$

The inverse of $[\Phi]$ is then calculated to be

$$[\Phi]^{-1} = 2A \ [I_{0,1}] \lceil l \rfloor^{-1} \tag{8.216}$$

Substitution of eqns (8.215) and (8.212) in eqn (8.211) gives the required orthogonal functions:

$$\{u_3^r\} = 2A \ <(\zeta_1(1 - \zeta_1) \ \zeta_2(1 - \zeta_2) \ \zeta_3(1 - \zeta_3)> \lceil l \rfloor^{-1} \{u_{3,n}^r\} \tag{8.217}$$

Thence, the curvature matrix $\{\chi_b\}$ is calculated by differentiating eqn (8.217). Thus,

$$\{\chi_b\} = \begin{Bmatrix} -u_{3,11}^r \\ -u_{3,22}^r \\ -2u_{3,12}^r \end{Bmatrix} = [B]\{u_{3,n}^r\} \tag{8.218}$$

where the $[B]$ matrix has been calculated to be

$$[B] = \frac{1}{A} \begin{bmatrix} b_1^2 & b_2^2 & b_3^2 \\ a_1^2 & a_2^2 & a_3^2 \\ a_1 b_1 & a_2 b_2 & a_3 b_3 \end{bmatrix} \lceil l \rfloor^{-1} \tag{8.219}$$

The constitutive equation is, as before, in eqn (8.27),

$$[D] = \frac{Et^3}{12(1 - v^2)} \begin{bmatrix} 1 & v & 0 \\ v & 1 & 0 \\ 0 & 0 & \dfrac{1 - v}{2} \end{bmatrix} \tag{8.220}$$

Thus, firstly,

$$[k_n] = \int_{area} [B]^T [D][B] dA \tag{8.221}$$

and, finally,

$$[k] = [G]^T [k_n][G] \tag{8.222}$$

In eqn (8.221), all matrices are constant, so that the integration gives, simply,

$$[k_n] = A [B]^T [D][B] \tag{8.223}$$

It is seen that $[k_n]$ is a (6×6) matrix, and the structure stiffness matrix must be assembled with apex nodes for transverse deflection, and mid–side nodes for normal (θ_n) moments. The above simple element passes the patch test, and is a suitable element to combine with the CST elasticity element for a flat facit shell formulation. Its stiffness matrix has been derived in at least two other independent ways:

1. By using a mixed variational principle (Herrmann [43]);

2. By the DKL (discrete Kirchoff with Loof nodes) approach, which is developed in the next section.

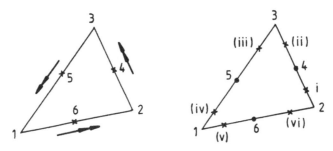

Fig. 8.25   The DKL family.

8.2.8   The DKL family of plate elements

The presentation of triangular element stiffness matrices concludes with that for the DKL family of elements, the two members of which are shown in Fig. 8.24. These elements, like the Morley constant moment element, start from the premise that to satisfy moment continuity across element boundaries it is better to use the normal rotations, (θ_s) variables, (as was, for example, the approach with the natural mode element (Fig. 8.11). The method of formulation is, however, quite distinct from all elements discussed previously, in that deflections and rotations are interpolated independently (as in the Ahmad type of isoparametric plate and shell elements). The Mindlin plate theory is used, and, to connect the deflections and rotations, the Kirchoff assumption of zero shear strain is invoked. (See section 8.1.4 for further details of the Mindlin plate theory.) The positive sense of the rotation and slope terms is shown in Fig. 8.26.

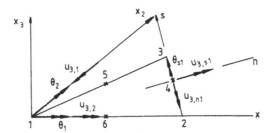

Fig. 8.26   Coordinate systems–signs of rotations and slopes.

Thus following the Mindlin theory, the deflections of the point P, distance x_3 from the middle surface, are given by

$$u_1 = x_3 \theta_2 ; \qquad u_2 = -x_3 \theta_1 \tag{8.224}$$

Thence the strains are calculated to be

$$\varepsilon_{11} = u_{1,1} = x_3\theta_{2,1} = x_3\chi_{11}$$

$$\varepsilon_{22} = u_{2,2} = -x_3\theta_{1,2} = x_3\chi_{22}$$

$$\gamma_{12} = x_3\{\theta_{2,2} - \theta_{1,1}\} = x_3\chi_{12} \tag{8.225}$$

with these definitions of rotations,

$$\{\varepsilon_b\} = x_3\{\chi\} \tag{8.226}$$

The transverse shear strains are (eqn (8.38)),

$$\{\varepsilon_s\} = \begin{Bmatrix} \gamma_{13} \\ \gamma_{23} \end{Bmatrix} = \begin{Bmatrix} u_{3,1} + \theta_2 \\ -\theta_1 + u_{3,2} \end{Bmatrix} \tag{8.227}$$

The plane strain assumption gives

$$\{\sigma_b\} = [D]\{\varepsilon_b\} \tag{8.228}$$

$$\{\sigma_s\} = [G]\{\varepsilon_s\} \tag{8.229}$$

Integrating eqns (8.226) and (8.228) through the plate thickness gives the usual assumption:

$$\{M\} = \frac{t^3}{12}[D]\{\chi\} \tag{8.230}$$

and, for transverse shear strains

$$\{Q\} = \begin{Bmatrix} Q_1 \\ Q_2 \end{Bmatrix} = t[G]\begin{Bmatrix} \gamma_{13} \\ \gamma_{23} \end{Bmatrix} \tag{8.231}$$

To formulate the element stiffness matrix, a linear basis is employed to interpolate both the transverse deflections u_3 and the normal rotations (θ_1, θ_2). Thus, transverse deflections are

$$\tilde{u}_3 = \{\phi_1\}^T\{u_3\} \tag{8.232}$$

where

$$\{u_3\}^T = \langle u_{31}\ u_{32}\ u_{33}\rangle \tag{8.233}$$

The shape functions for the rotations $\langle\theta_1\ \theta_2\rangle$ at the mid–side nodes are

$$\begin{Bmatrix} \tilde{\theta}_1 \\ \tilde{\theta}_2 \end{Bmatrix} = \begin{bmatrix} \bar{\phi}_1^T & \bar{\phi}_1^T \end{bmatrix}\begin{Bmatrix} \theta_1 \\ \theta_2 \end{Bmatrix} \tag{8.234}$$

In this case,

$$\{\bar{\phi}_1\}^T = \langle(-\zeta_1 + \zeta_2 + \zeta_3)\ (\zeta_1 - \zeta_2 + \zeta_3)\ (\zeta_1 + \zeta_2 - \zeta_3)\rangle \tag{8.235}$$

where, in eqn (8.231),

$$\{\theta_i\}^T = \langle\theta_{i4}\ \theta_{i5}\ \theta_{i6}\rangle\ i = 1, 2 \tag{8.236}$$

To link the transverse displacements to the side rotations, as well as constrain the element to model thin plate behaviour, the Kirchoff hypothesis is invoked at the mid–side nodes. That is, $\gamma_{s3} = 0$, or, using the second of eqns (8.227) in (n, s) axes,

$$-\theta_{nj} + u_{3,s_j} = 0 \tag{8.237}$$

These provide three constraint equations which reduce the original nine d.o.f. to six, resulting in the (6×6) stiffness matrix with nodal displacements:

(u_{3i}), $i = 1, 2, 3$

(θ_{sj}), $j = 4, 5, 6$

With the (n, s) axes chosen in Fig. 8.26 the following transformations hold:

1. global to local

$$\begin{bmatrix} n_i \\ s_i \end{bmatrix} = \frac{1}{l_i} \begin{bmatrix} -b_i & -a_i \\ a_i & -b_i \end{bmatrix} \begin{Bmatrix} x_1 - x_{1j} \\ x_2 - x_{2j} \end{Bmatrix} \tag{8.238}$$

2. local to natural

$$\begin{Bmatrix} \zeta_j \\ \zeta_k \\ \zeta_i \end{Bmatrix} = \frac{1}{2A} \begin{bmatrix} 2A & -(d_i - l_i) & -h_i \\ 0 & d_i & h_i \\ 0 & -l_i & 0 \end{bmatrix} \begin{Bmatrix} 1 \\ n_i \\ s_i \end{Bmatrix} \tag{8.239}$$

Thence the side derivatives are given by

$$f_{,s_i} = \frac{1}{l_i} \{ f_{,\zeta_k} - f_{,\zeta_j} \} \tag{8.240}$$

Using eqn (8.232), with $\lceil l \rfloor = \lceil l_1 l_2 l_3 \rfloor$,

$$\begin{Bmatrix} \theta_{n(4)} \\ \theta_{n(5)} \\ \theta_{n(6)} \end{Bmatrix} = \lceil l \rfloor^{-1} \begin{bmatrix} 0 & -1 & 1 \\ 1 & 0 & -1 \\ -1 & 1 & 0 \end{bmatrix} \{ u_3 \} = \lceil l \rfloor^{-1} [\Omega] \{ u_3 \} \tag{8.241}$$

Now, using eqn (8.238) to express local rotations in terms of global rotations, it follows, for example at node 4, that

$$\theta_{n(4)} = \frac{1}{l_1} [-b_1 \theta_{1(4)} - a_1 \theta_{2(4)}] \tag{8.242}$$

Thus, using the two similar expressions for $\theta_{n(5)}$, $\theta_{n(6)}$, the following relationship relating θ_2, θ_1 at nodes $4, 5, 6$, is obtained:

$$\{\theta_2\} = -[I_a]^{-1}[I_b]\{\theta_1\} - [I_a]^{-1}[\Omega]\{u_3\} \tag{8.243}$$

or, alternatively,

$$\{\theta_1\} = -[I_b]^{-1}[I_a]\{\theta_2\} - [I_b]^{-1}[\Omega]\{u_3\} \tag{8.244}$$

where

$$[I_p] = \begin{bmatrix} p_1 & p_2 & p_3 \end{bmatrix}$$

with $p = a$ or b, and

$$\Omega = \begin{bmatrix} 0 & -1 & 1 \\ 1 & 0 & -1 \\ -1 & 1 & 0 \end{bmatrix} \tag{8.245}$$

The side rotations are given from the second of eqns (8.238):

$$\{\theta_s\} = \lceil l \rfloor^{-1}[I_a \ -I_b] \begin{Bmatrix} \theta_1 \\ \theta_2 \end{Bmatrix} \tag{8.246}$$

Substitute for $\{\theta_2\}$, from eqn (8.243), to give

$$\{\theta_1\} = \lceil l \rfloor^{-1}[-\lceil l \rfloor^{-1}I_b[\Omega] \ \ I_a] \begin{Bmatrix} u_3 \\ \theta_s \end{Bmatrix} = [C_x]\{\delta^e\} \tag{8.247}$$

where

$$[C_x] = \lceil l \rfloor^{-1}[-\lceil l \rfloor^{-1}I_b[\Omega] \ \ I_a] \tag{8.248}$$

Alternatively, substitute for $\{\theta_1\}$ from eqn (8.244) in eqn (8.246); thence,

$$\{\theta_2\} = -\lceil l \rfloor^{-1}[\lceil l \rfloor^{-1}I_a[\Omega] \ \ I_b] \begin{bmatrix} u_3 \\ \theta_s \end{bmatrix} = [C_y]\{\delta^e\} \tag{8.249}$$

where

$$[C_y] = -\lceil l \rfloor^{-1}[\lceil l \rfloor^{-1}I_a[\Omega] \ \ I_b] \tag{8.250}$$

The interpolation functions for $\tilde{\theta}_1$, $\tilde{\theta}_2$ in eqn (8.234) have now been obtained, since

$$\begin{bmatrix} \tilde{\theta}_1 \\ \tilde{\theta}_2 \end{bmatrix} = \begin{bmatrix} \bar{\phi}_1^T & \bar{\phi}_1^T \end{bmatrix} \begin{bmatrix} C_x \\ C_y \end{bmatrix} \{\delta^e\} \tag{8.251}$$

Now, from eqn (8.225),

$$\{\varepsilon_b\} = x_3 \begin{bmatrix} \{\phi_{1,1}\}^T[C_y] \\ -\{\phi_{1,2}\}^T[C_x] \\ \{\phi_{1,2}\}^T[C_y]-\{\phi_{1,1}\}^T[C_x] \end{bmatrix} \{\delta^e\} = x_3[B_b]\{\delta^e\} \tag{8.252}$$

Thence the element stiffness matrix is given (integrating first through the thickness):

$$[k_e] = \frac{t^3}{12}\int_A [B_b]^T[D][B_b]dA \tag{8.253}$$

8.2.9   The six–node DKL plate bending element

The element is shown in its local coordinate system in Fig. 8.27. It is referred to as a six–node element, because the rotational (or bending) nodes (i) to (vi) can be grouped with the mid–side displacement nodes 4 to 6 in so far as the assembly of the global stiffness matrix is concerned. This gives nodes with 1 and 3 degrees of freedom at the apices and mid–sides, respectively. It will be seen that for plate bending analysis no transformation is required from local to global coordinates, except for the rotation terms, which will be multiplied by ±1 according to the apex nodes (*I-J*) in the side being in ascending or descending order as the nodes are traversed in an anticlockwise direction.

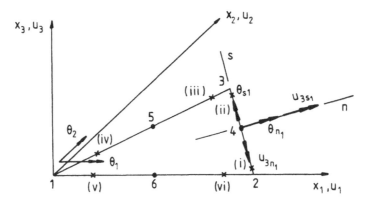

Fig. 8.27   The six–node DKL plate element.

In this case the interpolation function for u_3 displacements is given by $\{\phi_2\}$, in section 1.4.6. That is,

$$\{u_3\} = \{\phi_2\}^T \{u_3\} \tag{8.254}$$

In eqn (8.254), $\{u_3\}$ is the 6×1 vector of nodal point displacements. Then the x_1 and x_2 derivatives are, simply,

$$\begin{bmatrix} \tilde{u}_{3,1} \\ \tilde{u}_{3,2} \end{bmatrix} = \begin{bmatrix} \phi_{2,1}^T & \phi_{2,2}^T \end{bmatrix} \{u_3\} \tag{8.255}$$

In evaluating the shape functions for the normal rotations $\{\theta_1, \theta_2\}$, it is found that a quadratic basis cannot be formed from the nodal values at (i) to (vi). This is due to the presence of neutral functions [44]. The problem is overcome by adding an additional higher–order term to the quadratic basis, necessitating the inclusion of the centroidal node (c). Thus, in evaluating the shape functions for β_1 and β_2, the basis used is

$$\{L\}^T = \{1, \zeta_1, \zeta_2, \zeta_1^2, \zeta_1\zeta_2, \zeta_2^2, (2\zeta^3 + 3\zeta_1^2\zeta_2 - 3\zeta_1\zeta_2^2 - 2\zeta_2^3)\} \tag{8.256}$$

If $\{a\}$ is the (7×1) vector of generalized displacements, the rotation of the normal about the local x_1–axis is

$$\theta_1 = \{L\}^T \{a\} \tag{8.257}$$

On substituting the area coordinates of the Loof and centroidal nodes into eqn (8.257), the following transformation results:

$$
\begin{bmatrix} \theta_{1(i)} \\ \theta_{1(ii)} \\ \theta_{1(iii)} \\ \theta_{1(iv)} \\ \theta_{1(v)} \\ \theta_{1(vi)} \\ \theta_{1(c)} \end{bmatrix} = \begin{bmatrix}
1.0 & 0.00000 & 0.78867 & 0.00000 & 0.00000 & 0.62201 & -0.98112 \\
1.0 & 0.00000 & 0.21132 & 0.00000 & 0.00000 & 0.04466 & -0.01887 \\
1.0 & 0.21132 & 0.00000 & 0.04465 & 0.00000 & 0.00000 & 0.01887 \\
1.0 & 0.78867 & 0.00000 & 0.62201 & 0.00000 & 0.00000 & 0.98112 \\
1.0 & 0.78867 & 0.21132 & 0.62201 & 0.16667 & 0.04466 & 1.25095 \\
1.0 & 0.21132 & 0.78867 & 0.04466 & 0.16667 & 0.62201 & -1.25095 \\
1.0 & 0.33333 & 0.33333 & 0.11111 & 0.11111 & 0.11111 & 0.00000
\end{bmatrix} \begin{bmatrix} a_1 \\ a_2 \\ a_3 \\ a_4 \\ a_5 \\ a_6 \\ a_7 \end{bmatrix}
\qquad (8.258)
$$

Write this eqn (8.258) as,

$$\{\theta\} = [A]\{a\} \qquad (8.259)$$

Inverting,

$$\{a\} = [A]^{-1}\{\theta\} \qquad (8.260)$$

The inverse of $[A]$, $[A]^{-1}$ is evaluated as

$$
[A]^{-1} = \begin{bmatrix}
-0.24402 & 0.91068 & 0.91068 & -0.24402 & 0.33333 & 0.33333 & -1.0 \\
2.04145 & -6.04145 & 1.57735 & 0.42265 & -1.42265 & -2.57735 & 6.0 \\
0.42265 & 1.57735 & -6.04105 & 2.04145 & -2.57735 & -1.42265 & 6.0 \\
-4.92820 & 8.92820 & -6.19615 & 4.19615 & -1.46410 & 5.46410 & -6.0 \\
-4.46410 & 2.46410 & 2.46410 & -4.46410 & 5.00000 & 5.00000 & -6.0 \\
4.19615 & 8.92820 & 8.92820 & -4.92820 & 5.46410 & -1.46410 & -6.0 \\
1.73205 & 1.73205 & 1.73205 & -1.73205 & -1.73205 & -1.73205 & 0.0
\end{bmatrix}
\qquad (8.261)
$$

Note that $[A]^{-1}$ is invariant and may be formed in the computer software as a data statement. Substituting eqn (8.260) into eqn (8.257) gives the orthogonal interpolation functions as

$$\{\Phi\}^T = \{L\}^T [A]^{-1} \qquad (8.262)$$

so that

$$
\begin{bmatrix} \tilde{\theta}_1 \\ \tilde{\theta}_2 \end{bmatrix} = \begin{bmatrix} \Phi^T & \Phi^T \end{bmatrix} \begin{bmatrix} \theta_1 \\ \theta_2 \end{bmatrix}
\qquad (8.263)
$$

It is now necessary to apply the shear constraints so that $\theta_{n(i)}$ to $\theta_{n(vi)}$ can be expressed in terms of $\theta_{s(i)}$ to $\theta_{s(vi)}$. The Kirchoff assumptions are imposed at the discrete points (i) to (vi). That is,

$$\gamma_{s3} = 0 \quad (i) \text{ to } (vi)$$

This is the same condition as that in eqn (8.233). That is,

$$-\theta_{n(j)} + u_{3,s_j} = 0 \quad j = (i) \text{ to } (vi) \qquad (8.264)$$

Again, as with the first member of the DKL family, the choice of the constraints in eqn (8.264) enables the rotations parallel to the sides of the element to be retained as nodal parameters. Various possibilities exist for eliminating the two centroidal rotations.

It is found that the application of the areal constraints (eqn (8.265)) produces the best results:

$$\int_{area} \gamma_{13} = 0 \; ; \qquad \int_{area} \gamma_{23} = 0 \tag{8.265}$$

The first of these yields the condition,

$$\int_{area} (\theta_2 + u_{3,1}) \, dA = 0 \tag{8.266}$$

and, the second,

$$\int_{area} (-\theta_2 + u_{3,2}) \, dA = 0 \tag{8.267}$$

On application of eqns (8.264), (8.266) and (8.267), the original d.o.f. of the bending element are reduced to 12. That is,

$$\left\{ \begin{array}{c} \theta_1 \\ \theta_2 \end{array} \right\} = \left\{ \begin{array}{c} C_1 \\ C_2 \end{array} \right\} \{\delta^b\} \tag{8.268}$$

where, now,

$$\{\delta^b\}^T = \{u_{31}, u_{32}, \cdots, u_{36}, \theta_{s(i)}, \theta_{s(ii)}, \cdots, \theta_{s(vi)}\} \tag{8.269}$$

The matrices $[C_1]$, $[C_2]$, are as follows:

$$[C_1] = \begin{bmatrix} 0 & -c_1\dfrac{x_{23}}{l_{23}^2} & -c_2\dfrac{x_{23}}{l_{23}^2} & 0 & c_3\dfrac{x_{23}}{l_{23}^2} & 0 & -\dfrac{y_{23}}{l_{23}} & 0 & 0 & 0 & 0 & 0 \\[2mm] 0 & c_2\dfrac{x_{23}}{l_{23}^2} & c_1\dfrac{x_{23}}{l_{23}^2} & 0 & -c_3\dfrac{x_{23}}{l_{23}^2} & 0 & 0 & -\dfrac{y_{23}}{l_{23}} & 0 & 0 & 0 & 0 \\[2mm] c_2\dfrac{x_{13}}{l_{13}^2} & 0 & c_1\dfrac{x_{13}}{l_{13}^2} & 0 & 0 & -c_3\dfrac{x_{13}}{l_{13}^2} & 0 & 0 & \dfrac{y_{13}}{l_{13}} & 0 & 0 & 0 \\[2mm] -c_1\dfrac{x_{13}}{l_{13}^2} & 0 & -c_2\dfrac{x_{13}}{l_{13}^2} & 0 & 0 & c_3\dfrac{x_{13}}{l_{13}^2} & 0 & 0 & 0 & \dfrac{y_{13}}{l_{13}} & 0 & 0 \\[2mm] -c_1\dfrac{x_{12}}{l_{12}^2} & -c_2\dfrac{x_{12}}{l_{12}^2} & 0 & c_3\dfrac{x_{12}}{l_{12}^2} & 0 & 0 & 0 & 0 & 0 & 0 & -\dfrac{y_{12}}{l_{12}} & 0 \\[2mm] c_2\dfrac{x_{12}}{l_{12}^2} & c_1\dfrac{x_{12}}{l_{12}^2} & 0 & -c_3\dfrac{x_{12}}{l_{12}^2} & 0 & 0 & 0 & 0 & 0 & 0 & 0 & -\dfrac{y_{12}}{l_{12}} \\[2mm] \alpha_1 & \alpha_2 & \alpha_3 & c_4\dfrac{y_{12}}{A} & c_4\dfrac{y_{12}}{A} & -c_4\dfrac{y_{13}}{A} & c_5\dfrac{y_{23}}{l_{23}} & c_5\dfrac{y_{23}}{l_{23}} & -c_5\dfrac{y_{13}}{l_{13}} & -c_5\dfrac{y_{13}}{l_{13}} & c_5\dfrac{y_{12}}{l_{12}} & c_5\dfrac{y_{12}}{l_{12}} \end{bmatrix}$$

where,

$$\alpha_1 = \frac{1}{3}\left[\frac{x_{12}}{l_{12}^2} + \frac{x_{13}}{l_{13}^2} - \frac{y_{23}}{A}\right]; \quad \alpha_2 = \frac{1}{3}\left[-\frac{x_{12}}{l_{12}^2} + \frac{x_{23}}{l_{23}^2} + \frac{y_{13}}{A}\right]; \quad \alpha_3 = -\frac{1}{3}\left[\frac{x_{12}}{l_{12}^2} + \frac{x_{13}}{l_{13}^2} + \frac{y_{12}}{A}\right]$$

$$c_1 = 2.1547004 \; ; \quad c_2 = 0.1547004 \; ; \quad c_3 = 2.3094008 \; ; \quad c_4 = \frac{4}{3} \; ; \quad c_5 = \frac{2}{3}$$

$$[C_2] = \begin{bmatrix} 0 & c_1\dfrac{y_{23}}{l_{23}^2} & c_2\dfrac{y_{23}}{l_{23}^2} & 0 & -c_3\dfrac{y_{23}}{l_{23}^2} & 0 & -\dfrac{x_{23}}{l_{23}} & 0 & 0 & 0 & 0 & 0 \\[2mm] 0 & -c_2\dfrac{y_{23}}{l_{23}^2} & -c_1\dfrac{y_{23}}{l_{23}^2} & 0 & c_3\dfrac{y_{23}}{l_{23}^2} & 0 & 0 & -\dfrac{x_{23}}{l_{23}} & 0 & 0 & 0 & 0 \\[2mm] -c_2\dfrac{y_{13}}{l_{13}^2} & 0 & -c_1\dfrac{y_{13}}{l_{13}^2} & 0 & 0 & c_3\dfrac{y_{13}}{l_{13}^2} & 0 & 0 & \dfrac{x_{13}}{l_{13}} & 0 & 0 & 0 \\[2mm] c_1\dfrac{y_{13}}{l_{13}^2} & 0 & c_2\dfrac{y_{13}}{l_{13}^2} & 0 & 0 & -c_3\dfrac{y_{13}}{l_{13}^2} & 0 & 0 & 0 & \dfrac{x_{13}}{l_{13}} & 0 & 0 \\[2mm] c_1\dfrac{y_{12}}{l_{12}^2} & c_2\dfrac{y_{12}}{l_{12}^2} & 0 & -c_3\dfrac{y_{12}}{l_{12}^2} & 0 & 0 & 0 & 0 & 0 & 0 & -\dfrac{x_{12}}{l_{12}} & 0 \\[2mm] -c_2\dfrac{y_{12}}{l_{12}^2} & -c_1\dfrac{y_{12}}{l_{12}^2} & 0 & c_3\dfrac{y_{12}}{l_{12}^2} & 0 & 0 & 0 & 0 & 0 & 0 & 0 & -\dfrac{x_{12}}{l_{12}} \\[2mm] \alpha_4 & \alpha_5 & \alpha_6 & c_4\dfrac{x_{12}}{A} & c_4\dfrac{x_{12}}{A} & -c_4\dfrac{x_{13}}{A} & c_5\dfrac{x_{23}}{l_{23}} & c_5\dfrac{x_{23}}{l_{23}} & -c_5\dfrac{x_{13}}{l_{13}} & -c_5\dfrac{x_{13}}{l_{13}} & c_5\dfrac{x_{12}}{l_{12}} & c_5\dfrac{x_{12}}{l_{12}} \end{bmatrix}$$

where,

$$\alpha_4 = \frac{1}{3}\left[\frac{y_{12}}{l_{12}^2} + \frac{y_{13}}{l_{13}^2} - \frac{x_{23}}{A}\right] ; \quad \alpha_5 = \frac{1}{3}\left[-\frac{y_{12}}{l_{12}^2} + \frac{y_{23}}{l_{23}^2} + \frac{x_{13}}{A}\right] ; \quad \alpha_6 = -\frac{1}{3}\left[\frac{y_{12}}{l_{12}^2} + \frac{y_{13}}{l_{13}^2} + \frac{x_{12}}{A}\right]$$

$$c_1 = 2.1547004 ; \quad c_2 = 0.1547004 ; \quad c_3 = 2.3094008 ; \quad c_4 = \frac{4}{3} ; \quad c_5 = \frac{2}{3}$$

Thus, finally, the normal rotations $\{\tilde{\theta}_1, \tilde{\theta}_2\}$ can be interpolated in terms of $\{\delta^b\}$, using eqns (8.263) and (8.268). That is,

$$\begin{Bmatrix} \tilde{\theta}_1 \\ \tilde{\theta}_2 \end{Bmatrix} = \begin{bmatrix} \Phi^T C_1 \\ \Phi^T C_2 \end{bmatrix}\{\delta^b\} \tag{8.270}$$

It is seen in eqn (8.270):

$$\{\Phi\}^T[C_1] = \{L\}^T[A]^{-1}[C_1] \text{ etc.} \tag{8.271}$$

The matrix $[A]^{-1}$ is constant, and $[C_1]$, $[C_2]$ are constant for each element. The strain nodal displacement relationship is given (eqn (8.225)) by

$$\{\varepsilon\} = x_3 \begin{bmatrix} \{L_{,1}\}^T[A]^{-1}[C_2] \\[2mm] -\{L_{,2}\}^T[A]^{-1}[C_1] \\[2mm] \{L_{,2}\}^T[A]^{-1}[C_2] - \{L_{,2}\}[A]^{-1}[C_1] \end{bmatrix} \tag{8.272}$$

That is,

$$\{\varepsilon\} = x_3[B_b]\{\delta^e\} \tag{8.273}$$

In eqn (8.272),

$$\{L_{,1}\}^T = \frac{1}{2A}[0 \quad b_1 \quad b_2 \quad 2b_2\delta_1 \quad (b_1\delta_2 + b_2\delta_1) \quad 2b_2\delta_2 \quad L_{7,1}] \tag{8.274}$$

where,

$$L_{7,1} = b_1(6\zeta_1^2 + 6\zeta_1\zeta_2 - 3\zeta_2^2) + b_2(3\zeta_1^2 - 6\zeta_1\zeta_2 - 6\zeta_2^2) \tag{8.275}$$

Similarly,

$$\{L_{,2}\}^T = \frac{1}{2A}[0 \quad a_1 \quad a_2 \quad 2a_1\delta_1 \quad (a_1\zeta_2 + a_2\zeta_1) \quad 2a_2\zeta_2 \quad L_{7,2}] \qquad (8.276)$$

where

$$L_{7,2} = a_1(6\zeta_1^2 + 6\zeta_1\zeta_2 - 3\zeta_2^2) + a_2(3\zeta_1^2 - 6\zeta_1\zeta_2 - 6\zeta_2^2) \qquad (8.277)$$

Finally, the element stiffness matrix is calculated, using eqn (8.278)

$$[k_e] = \int_{area} \int_{-t/2}^{t/2} \frac{x_3^2}{4A^2}[B_b]^T[D][B_b]dA = \frac{t^3}{24A}\int_0^1 \int_0^{1-\zeta_1} [B_b]^T[D][B_b]d\zeta_1\zeta_2 \qquad (8.278)$$

The presence of fourth–order terms in eqn (8.278) entails the use of a seven–point quadrature rule (Chapter 1, Table 1.2), to integrate numerically the stiffness matrix.

8.3 QUADRILATERAL PLATE BENDING ELEMENTS

In this final section on plate bending, two element types are detailed. The first of these, the rectangular element family, obviously has wide application in regular structures. When combined with a planar elasticity element to form a flat facet shell element, it has obvious application for box type structures. The second group, quadrilateral elements, are based on isoparametric concepts and form an introduction to the curved Ahmad shell elements.

8.3.1   Rectangular plate elements

Because the polynomials used in the interpolation of the transverse displacement u_3 are incomplete, a variety of formulations is possible. Some of these are detailed in what follows.

Type 1 Elements

These elements are based on three d.o.f. per node, namely $(u_3, \theta_{x1}, \theta_{x2})$, giving a 12×12 stiffness matrix. Elements with common sides have compatible displacements expressed in their common interface as a cubic function. The linear cubic requires four terms, and the constants along an element side are supplied by the displacement and slope terms in that side. If, for example, side i–j is parallel with the OX_1–axis, the displacements (u_{3i}, u_{3j}) and slopes $(\theta_{2i}, \theta_{2j})$ Fig. 8.28, are used in the expression,

$$u_3 = A_1 + A_2x_1 + A_3x_1^2 + A_4x_1^3 \qquad (8.279)$$

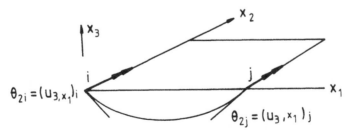

Fig. 8.28   Rectangular element–slopes along side i–j.

For the rectangle with $(u_3, \theta_1, \theta_2)$ specified at each node, a total of 12 displacement unknowns occur. Since the complete cubic function in (x_1, x_2) has ten terms, two additional quartic terms are used which must:

1. retain x_1, x_2 symmetry;
2. reduce to cubic expressions along the lines, $x_1 = $ constant, $x_2 = $ constant.

Suitable terms are obviously $x_1^3 x_2$, $x_1 x_2^3$. For example, the term $\alpha x_1^3 x_2$ is cubic in x_1 along lines $x_2 = $ constant, and, similarly, $\beta x_1 x_2^3$ is linear along the same line. A variety of elements can be chosen to fulfil the above requirements.

1. Based on modes derived from the polynomial,

$$u_3 = a_1 + a_2 x_1 + a_4 x_1^2 + a_5 x_1 x_2 + a_6 x_2^2 + a_7 x_1^2 x_2$$

$$+ a_8 x_1 x_2^2 + a_9 x_2^3 + a_{10} x_2^3 + a_{11} x_1^3 x_2 + a_{12} x_1 x_2^3 \tag{8.280}$$

2. Based on Hermitian polynomials [11] denoted H_{0i}, H_{1i}, etc. , defined such that

$$H_{0i}^{(N)}(s_j) = \delta_{ij} \tag{8.281}$$

$$\frac{d^N H_{0i}^{(N)}(s_j)}{ds_j} = 0 \tag{8.282}$$

where (N) is the degree of the derivative up to which the function can interpolate. Similarly,

$$H_{1i}^{(N)}(s_j) = 0 \tag{8.283}$$

$$\frac{dH_{1i}^{(1)}(s_j)}{ds_j} = \delta_{ij} \tag{8.284}$$

3. Based on mode functions such as those given by Melosh [45], Argyris [1], Dawe [46], Holand [47], and others. For example, assume a cubic function along one side with linear variation to zero on the opposite edge as shown in Fig. 8.29.

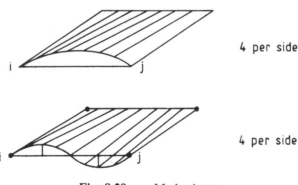

4 per side

4 per side

Fig. 8.29   Mode shapes.

In addition, a twist term, (eqn (8.285)), is added:

$$u_3 = \alpha_1 + \alpha_2 x_1 + \alpha_3 x_2 + \alpha_4 x_1 x_2 \tag{8.285}$$

This ensures that constant twist modes are included in the displacement patterns for the element. Thus nine straining modes give the required element deformation. Rigid–body motions expand the stiffness matrix to 12×12. It has been found that type (1) and (3) elements give results which converge reasonably well to analytic (or Fourier series) solutions. Element (2) does not converge to the correct result, however, because the Hermitian polynomials, as given, are unable to represent the constant strain state. In all these elements it is seen that although four constants are available in a side to specify displacements completely, only two are available to specify normal slopes (C_1 compatibility) whose variation is quadratic, and an extra displacement parameter is required in each side. These four parameters must either

1. give the normal slope at another point in the side; or

2. give the normal slope derivative at one of the apices, e.g. the twist terms ($u_{3,12}$) may be supplied as extra nodal parameters.

Then in the side $i-j$, parallel with the x_1–axis, the quadratic variation of normal slope is specified as

$$(\theta_{2i}, \theta_{2j}, \theta_{12i})$$

These functions, depending on the twist terms, may be added to any one of the above sets, e.g. extra Hermitian polynomials in no. 2. When this is done, the Hermitian elements give convergent results. These will be termed type II elements.

Type II elements

These have 16 degrees of freedom, 4 per node, and results give better convergence for a given number of elements than the 12–d.o.f. elements. Two possibilities for type II elements are shown in Fig. 8.30.

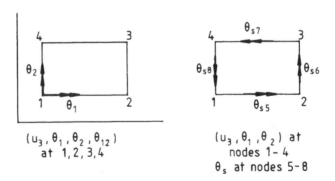

Fig. 8.30   Sixteen–d.o.f. rectangular element node configuration.

It is doubtful if the improvement in the convergence characteristics of the type II elements warrents the additional complexity of their formulation or their inclusion in finite element codes. In this text only one version of the simple 12×12 element stiffness matrix is reproduced. In this element, orthogonal interpolation functions are obtained by first defining a set of simple mode shapes. The rectangular element dimensions (2a, 2b) are shown in Fig. 8.31.

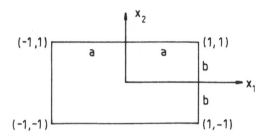

Fig. 8.31   Rectangular element–dimensions.

From Fig. 8.31, non–dimensional coordinates are, simply,

$$\xi = \frac{x_1}{a}, \quad \eta = \frac{x_2}{b} \tag{8.286}$$

Thence, first derivatives of the function $f(x_1, x_2)$ are given by

$$f_{,1} = \frac{1}{a} f_{,\xi}; \quad f_{,2} = \frac{1}{b} f_{,\eta} \tag{8.287}$$

As noted previously, 12 shape functions are chosen, with two fourth order–terms, isotropic with respect to the coordinate axes. Write the plate deflection as

$$\tilde{u}_3 = \{\phi\}^T \{\alpha\} \tag{8.288}$$

In eqn (8.286) the α parameters are, as yet, undefined, generalized coordinates. First the displacements of the nodes are grouped:

$$\{u_3\} = \begin{Bmatrix} u_{3(1)} \\ u_{3(2)} \\ u_{3(3)} \\ u_{3(4)} \end{Bmatrix}; \quad \{\theta_1\} = \begin{Bmatrix} \theta_{1(1)} \\ \theta_{1(2)} \\ \theta_{1(3)} \\ \theta_{1(4)} \end{Bmatrix}; \quad \{\theta_2\} = \begin{Bmatrix} \theta_{2(1)} \\ \theta_{2(2)} \\ \theta_{2(3)} \\ \theta_{2(4)} \end{Bmatrix} \tag{8.289}$$

The procedure outlined in section 1.4.2, eqn (1.93), is used to obtain a set of orthogonal interpolation functions. Thus, if $\{r\}$ is the vector of nodal displacements, evaluate $\{r\}$ in terms of $\{\alpha\}$ from eqn (8.288). That is,

$$\{r\} = [A]\{\alpha\} \tag{8.290}$$

Thence,

$$\{\alpha\} = [A]^{-1}\{r\} \tag{8.291}$$

That is, from eqn (8.286),

$$\tilde{u}_3 = \{\phi\}^T [A]^{-1}\{r\} \tag{8.292}$$

so that the orthogonal functions $\{\Phi\}$ are given by

$$\{\Phi\} = [A^{-1}]^T \{\phi\} = [B]\{\phi\} \tag{8.293}$$

In the present instance, choose $\{\phi\}$:

$$\{\phi\} = \frac{1}{4} \begin{bmatrix} 2 \\ \xi(3 - \xi^2) \\ \eta(3 - \eta^2) \\ \xi\eta(4 - \xi^2 - \eta^2) \\ a\xi(\xi^2 - 1) \\ a(\xi^2 - 1) \\ a\xi\eta(\xi^2 - 1) \\ a\eta(\xi^2 - 1) \\ b\eta(\eta^2 - 1) \\ b\xi\eta(\eta^2 - 1) \\ b(\eta^2 - 1) \\ b\xi(\eta^2 - 1) \end{bmatrix} \tag{8.294}$$

These functions have been chosen by drawing independent shape functions, as shown in Fig. 8.32. Thus, for example, consider the group (II) displacements and note

$$\phi_{i,1} = \frac{1}{a}\phi_{i,\xi} \tag{8.295}$$

Thence,

$$\{\phi_{,1}\}_{II} = \frac{1}{4} \begin{bmatrix} 3\xi^2 - 1 \\ 2\xi \\ \eta(3\xi^2 - 1) \\ 2\eta\xi \end{bmatrix} \tag{8.296}$$

Evaluate these slopes at the nodes:

1.   (1,1) $\dfrac{1}{2}\begin{bmatrix} 1 \\ 1 \\ 1 \\ 1 \end{bmatrix}$

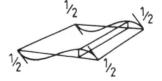

2. (1,−1) $\dfrac{1}{2}\begin{bmatrix} 1 \\ 1 \\ -1 \\ -1 \end{bmatrix}$

3. (−1,1) $\dfrac{1}{2}\begin{bmatrix} 1 \\ -1 \\ 1 \\ -1 \end{bmatrix}$

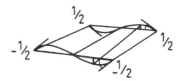

4. (−1,−1) $\dfrac{1}{2}\begin{bmatrix} 1 \\ -1 \\ -1 \\ 1 \end{bmatrix}$

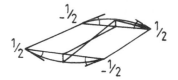

Fig. 8.32   Rectangular element–mode shapes.

When $[A^{-1}]^T$ is evaluated, it is found to have the very simple form,

$$[B] = [A^{-1}]^T = \begin{bmatrix} \omega & 0 & 0 \\ 0 & \omega & 0 \\ 0 & 0 & \omega \end{bmatrix} \qquad (8.297)$$

where $[\omega]$ is defined,

$$[\omega] = \frac{1}{2}\begin{bmatrix} 1 & 1 & 1 & 1 \\ 1 & 1 & -1 & -1 \\ 1 & -1 & 1 & -1 \\ 1 & -1 & -1 & 1 \end{bmatrix} \qquad (8.298)$$

The constitutive equation is, as in eqn (8.27), $\{\tilde{m}\} = [k]\{\varepsilon\}$, where, now,

$$\{\varepsilon\} = [a]\{r\} = -\begin{bmatrix} \Phi^T_{,11} \\ \Phi^T_{,22} \\ \Phi^T_{,12} \end{bmatrix}\{r\} = -\begin{bmatrix} \phi^T_{,11} \\ \phi^T_{,22} \\ 2\phi^T_{,12} \end{bmatrix}[B]^T\{r\} = \{\phi_\varepsilon\}[B]^T\{r\} \qquad (8.299)$$

The element stiffness matrix follows from

$$[K] = \int_{area} [a]^T [k][a] dV = [B] \int_{area} \{\phi_\varepsilon\}^T [k] \{\phi_\varepsilon\} dA \, [B]^T \tag{8.300}$$

In eqn (8.300),

$$2\{\phi_\varepsilon\} = \begin{bmatrix} 0 & -\dfrac{3\xi}{a^2} & 0 & -\dfrac{3\xi\eta}{a^2} & \dfrac{3\xi}{a} & \dfrac{1}{a} & \dfrac{3\xi\eta}{a} & \dfrac{\eta}{a} & 0 & 0 & 0 \\[2mm] 0 & 0 & -\dfrac{3\eta}{b^2} & -\dfrac{3\xi\eta}{b^2} & 0 & 0 & 0 & 0 & \dfrac{3\eta}{b} & \dfrac{1}{b} & \dfrac{\xi}{b} \\[2mm] 0 & 0 & 0 & \dfrac{(4-3\xi-3\eta^2)}{ab} & 0 & 0 & \dfrac{(3\xi^2-1)}{a} & \dfrac{3\xi}{b} & \dfrac{(3\eta^2-1)}{a} & 0 & \dfrac{3\eta}{a} \end{bmatrix} \tag{8.301}$$

Let $[k_c]$ be defined such that

$$[k_c] = \int_{area} \{\phi_\varepsilon\}^T [k] \{\phi_\varepsilon\} dA \quad \text{and} \quad D = \frac{Et^3}{12(1-v^2)} \tag{8.302}$$

Then the integration in eqn (8.302) is easily carried out, giving $[k_c]$, with the following non–zero terms (upper triangle only is given; the remaining terms are obtained from symmetry):

$$k_{11} = 0$$

$$k_{22} = \frac{3b}{a^3} \; ; \; k_{25} = -\frac{3b}{a^3} \; ; \; k_{2,12} = \frac{-v}{a}$$

$$k_{33} = \frac{3a}{b^3} \; ; \; k_{38} = \frac{-v}{b} \; ; \; k_{39} = -\frac{3a}{b^2}$$

$$k_{44} = \frac{1}{5a^3b^3}(5a^4 + 5b^4 + 14a^2b^2 - 4va^2b^2)$$

$$k_{47} = -\frac{1}{5a^2b}(5b^2 + 2a^2 + 3va^2)$$

$$k_{4,10} = -\frac{1}{5a^2b}(5a^2 + 2b^2 + 3vb^2)$$

$$k_{55} = \frac{3b}{a} \; ; \; k_{5,12} = v$$

$$k_{66} = \frac{b}{a} \; ; \; k_{6,11} = v$$

$$k_{77} = \frac{1}{5ab}(5b^2 + 2a^2 - 2va^2) \; ; \quad k_{7,10} = v$$

$$k_{88} = \frac{1}{3ab}(2a^2 + b^2 - 2va^2) \; ; \quad k_{89} = v$$

$$k_{99} = \frac{3a}{b}$$

$$k_{10,10} = \frac{1}{5ab}(5a^2 + 2b^2 - 2vb^2)$$

$$k_{11,11} = \frac{a}{b}$$

$$k_{12,12} = \frac{1}{3ab}(2b^2 + a^2 - 2vb^2) \tag{8.303}$$

8.3.2 Quadrilateral elements based on the isoparametric concept

The term isoparametric is used here loosely to mean those elements for which the same interpolation for the spatial coordinates within the element is used for the displacement (and/or) rotation functions. These elements also form an introduction to the isoparametric family of curved–shell elements. The isoparametric plate elements attempt to model the plate theory in a way similar to the DKL elements except that the shear constraints are not applied. Other means must be used, (e.g. reduced integration or the 'heterosis concept'), aimed at reducing the consequences of shear locking, which occurs in these elements when the thickness becomes small relative to the span of the plate. Although this may be of minor consequence in plates, because deflection control will usually prevent plate thicknesses becoming excessively thin, such is not the case with shells, where thickness–to–span ratios of 1/200 are often encountered. In such situations care must be taken to avoid locking. The concept of shear locking in plates is discussed further in this section. In the element it is assumed that at any point the displacement vector is controlled by three parameters, namely, $(\tilde{u}_3, \tilde{\theta}_1, \tilde{\theta}_2)$, such that

$$\{\tilde{r}\} = \begin{bmatrix} \tilde{u}_3 \\ \tilde{\theta}_1 \\ \tilde{\theta}_2 \end{bmatrix} = \begin{bmatrix} N^T \end{bmatrix}_3 \{r_n\} \tag{8.304}$$

In eqn (8.304), $\{r_n\}$ is a $(3 \times n)$ vector, where n is the number of node points in the element grouped with all $\tilde{u}_3$ nodal values together followed by all θ_1 values etc., and $\begin{bmatrix} N^T \end{bmatrix}_3$ is the matrix with the shape functions $\{N^T\}$ as diagonal submatrices. As with the elasticity isoparametric elements, the grouping of variables in eqn (8.304) expedites the computation of the element stiffness matrix. The basic force displacement stiffness relationships have been derived in eqns (8.27) and (8.40), where now, however, $\tilde{\theta}_{1,1}$ replaces $u_{3,11}$ etc. That is,

$$\begin{bmatrix} M_{11} \\ M_{22} \\ M_{21} \end{bmatrix} = D \begin{bmatrix} 1 & v & 0 \\ v & 1 & 0 \\ 0 & 0 & \frac{1-v}{2} \end{bmatrix} \begin{Bmatrix} \tilde{\theta}_{2,1} \\ -\tilde{\theta}_{1,2} \\ -(\tilde{\theta}_{1,1} - \tilde{\theta}_{2,2}) \end{Bmatrix} \quad \text{and,} \quad D = \frac{Et^3}{12(1-v^2)} \tag{8.305}$$

For shear forces,

$$\begin{Bmatrix} Q_1 \\ Q_2 \end{Bmatrix} = \alpha h G \begin{bmatrix} 1 & 0 \\ 0 & 1 \end{bmatrix} \begin{Bmatrix} u_{3,1} + \theta_2 \\ u_{3,2} - \theta_1 \end{Bmatrix} \tag{8.306}$$

It follows from eqns (8.304) and (8.305) that the bending and shear strain transformation matrices are given by

bending

$$\begin{Bmatrix} \tilde{\theta}_{2,1} \\ -\tilde{\theta}_{1,2} \\ \tilde{\theta}_{2,2} - \tilde{\theta}_{1,1} \end{Bmatrix} = - \begin{bmatrix} 0 & 0 & -N_{,1}^T \\ 0 & N_{,2}^T & 0 \\ 0 & N_{,1}^T & -N_{,2}^T \end{bmatrix} \{r_n\} = [a_B]\{r_n\} \tag{8.307}$$

shear

$$\begin{Bmatrix} u_{3,1} + \theta_2 \\ u_{3,2} - \theta_1 \end{Bmatrix} = \begin{bmatrix} N_{,2}^T & 0 & N^T \\ N_{,1}^T & -N^T & 0 \end{bmatrix} \{r_n\} = [a_s]\{r_n\} \tag{8.308}$$

It follows that the bending stiffness is calculated

$$[k_b] = \int_{area} [a_b]^T [k_b][a_b] dA \tag{8.309}$$

$$[k_s] = \int_{area} [a_s]^T [k_s][a_s] dA \tag{8.310}$$

The integration in eqns (8.304) and (8.305) can be carried out numerically, using eqn (1.213). It is seen that because rotations and displacements are interpolated independently, different node point schemes may be used for each (heretosis concept). Also, either selective or reduced integration can be used to improve the element performance with regard to shear locking. The phenomenon of shear locking will be discussed with reference to the simple element shown in Fig. 8.33 which is a degenerate form of the linear displacement plate element.

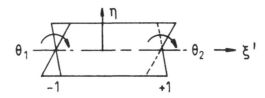

Fig. 8.33   Linear element rotations.

Excluding rigid–body displacements, the two rotations θ_1, θ_2 can be used as the only degrees of freedom of the element. The shape functions are

$$\{N_1\}^T = \{\frac{1}{2}(1 - \xi) \; \frac{1}{2}(1 + \xi)\}$$ (8.311)

Substituting in eqns (8.309) and (8.310) and noting

$$N^T_{1,x_1} = N^T_{1,\xi}\xi_{,x_1} = \frac{1}{L}[-1 \; 1]$$ (8.312)

it follows that

$$[k_b] = \frac{Et^3}{12L}\begin{bmatrix} 1 & -1 \\ -1 & 1 \end{bmatrix}$$ (8.313)

$$[k_s] = \frac{\alpha Glt}{6}\begin{bmatrix} 2 & 1 \\ 1 & 2 \end{bmatrix}$$ (8.314)

For constant bending distortions the rotations $\theta = \theta_1 = -\theta_2$ are the appropriate nodal displacements, and since the bending moment is constant the shear force should be zero. In the present approximation,

$$M_1 = \frac{Et^3}{12L}(2\theta) + \frac{\alpha GLt}{6}(\theta) = \frac{EI}{L}(2\theta) + \frac{\alpha t}{6}GL(\theta)$$ (8.315)

The first term of eqn (8.315) is the correct bending moment the second term is the extra moment caused by the extraneous shear strain. Write eqn (8.315) as;

$$M_1 = \frac{Et^3}{6L}\{1 + \frac{\alpha G}{E}\left(\frac{L}{t}\right)^2\}\theta$$ (8.316)

It is seen that the error in eqn (8.316) propagates as the square of the ratio (L/t); hence the reason for locking as the thickness decreases. The source of the error is easily understood by examining the deformation of the element, as shown in Fig. 8.34. Due to the kinematic assumptions used, the resulting deformations are as shown by the full lines in Fig. 8.34, whereas they should be as given by the dotted line for pure bending.

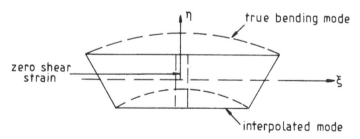

Fig. 8.34 Member deformation for constant moment.

The assumed deformations (solid lines in Fig. 8.34) obviously require a great deal more shear strain energy. It is possible to improve the efficiency of the element by a suitable choice of the integration order. From Fig. 8.33, it is seen that at $\xi = 0$, the shear strain is zero for constant bending modes. Therefore, since $\xi = 0$ is the position required for one–point integration, if adopted for $[k_s]$, in the ξ direction the extraneous shear energy for pure bending will be zero. Since the shear energy contribution to the bending mode is now zero, the effective deformation can now be thought of as being the true bending deformation. The quadratic displacement element suffers from a defect similar to that to the linear displacement element. The shear strains developed in this element vary quadratically, an unnecessarily high–order variation for a secondary effect. Again, the beam with quadratic variation of displacements is examined in the light of spurious shear strain energy. The beam element with three nodes is shown in Fig. 8.35.

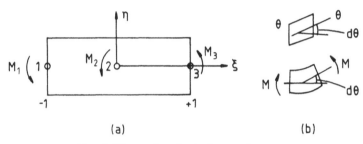

(a)   (b)

Fig. 8.35   Quadratic beam element.

The shape functions are given by

$$
\begin{bmatrix} N_1 \\ N_2 \\ N_3 \end{bmatrix} = \begin{bmatrix} -\dfrac{1}{2}\xi(1-\xi) \\ (1-\xi^2) \\ \dfrac{1}{2}\xi(1+\xi) \end{bmatrix}
\tag{8.317}
$$

For a linearly varying bending moment, the shear force distribution is constant. Therefore, the rotations due to shear are constant along the beam that is,

$$
\frac{dQ}{d\xi} = \frac{d\theta}{d\xi}\frac{6}{5GA} \quad \text{(for a rectangular section)}
\tag{8.318}
$$

If Q = constant, θ = constant = θ_s.

Similarly, for rotation,

$$
\frac{d\theta}{d\xi} = \frac{M}{EI} = -\frac{-\dfrac{1}{2}(1-\xi)M_1 + \dfrac{1}{2}(1+\xi)M_1}{EI}
$$

Therefore,

$$\theta = \frac{M_1\xi^2}{2EI} + C$$

Antisymmetry of deformation is maintained by taking the integration constant C:

$$C = -\frac{M_1}{6EI}$$

That is,

$$\theta = \frac{M_1\xi^2}{2EI} - \frac{M_1}{6EI} \qquad (8.319)$$

Thus, at

node 1 $\quad \theta_1 = \theta_b$
node 2 $\quad \theta_2 = -\theta_b/2$
node 3 $\quad \theta_3 = \theta_b$

The deflected shape is now as shown in Fig. 8.36.

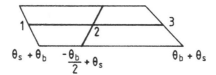

Fig. 8.36 Rotation modes.

Substitute these values into the first term of eqn (8.308):

$$\gamma_{13} = [-\frac{1}{2}\xi(1-\xi)\ (1-\xi^2)\ \frac{1}{2}\xi(1+\xi)] \begin{bmatrix} \theta_b + \theta_s \\ -\frac{\theta_b}{2} + \theta_s \\ \theta_b + \theta_s \end{bmatrix} = (-\frac{\theta_b}{2} + \theta_s) + \frac{3}{2}\theta_b\xi^2 \qquad (8.320)$$

Now, evidently, in this case, the correction value for γ_{13} is θ_s. Substituting this value of γ_{13} in eqn (8.320) gives the coordinates for which this holds true. That is,

$$\xi = \pm\frac{1}{\sqrt{3}} \qquad (8.321)$$

These are the coordinates for the two–point Gauss integration scheme. This implies that if the shear strain energy is intergrated at these points the energy computed by the integration is that for linear moments, so that the element represents this state correctly.

Zero energy modes associated with numerical intergration

For C_0 elements, Gaussian integration gives exact values for the integral of polynomials of degree $2(p-1)$ where p is the order of the complete polynomial present. Therefore, provided the integration is exact to the order $2(p-1)$, no loss of convergence will occur. The rate of convergence which would result if exact integration were used is preserved. The minimum integration formulae should be as follows:

order	linear	quadratic	cubic
n_i	1	2	3

Table 8.2

	Numerical integration	Estimated number of zero energy modes	Computed number of energy modes (eigenvalues)
Linear	bending 1 × 1 shear 1 × 1	4	4
	bending 2 × 2 shear 1 × 1		2
Quadratic serendipity	bending 2 × 2 shear 2 × 2	1	
	bending 3 × 3 shear 2 × 2	0	0
Quadratic Lagrange	bending 2 × 2 shear 2 × 2	4	4
	bending 3 × 3 shear 2 × 2	0	1
Heretosis	bending 3 × 3 shear 2 × 2	0	0
Serendipity	bending 3 × 3 shear 2 × 2	0	n.a.

However, it cannot be automatically assumed that stiffness matrices computed using these minimum integration rules will be non–singular. Numerical integration replaces the integrals by a weighted sum of independent linear relations between the nodal parameters $\{r\}$. If the number of unknowns in $\{r\}$ exceeds the number of independent relations supplied at all the integration points, then the stiffness matrix must be singular. That is, let

$$S = j - r - bk \tag{8.322}$$

where

j = the total number of available degrees of freedom

r = the number of rigid–body modes

b = the total number of integration points

k = the number of strains at the integration points used in the formulation of the stiffness matrix

If S is greater than zero, the stiffness matrix is singular. A more rigorous method of determining the singularity of the stiffness matrix is by determining the number of zero–valued eigenvalues. These zero values correspond to rigid–body motion or zero energy modes associated with the element mechanisms detected in a single element may propagate in a group of elements, but more probably they will disappear due to the incompatibility of the associated zero energy modes or the prescription of boundary conditions which suppress their formation. It is seen in Table 8.2 that there are at least three contenders for the quadratic element types of the isoparametric family, the difference between the Lagrange and serendipity elements being that in the former an internal node is required for use of the full Lagrange interpolation functions. These internal nodes require the use of static condensation to reduce the element stiffness matrix from 27×27 to 24×24, before incorporating it in the global system.

8.3.3   The heretosis concept

The heretosis element is simply the description which means the improvement in the characteristics exhibited in hybrids over those of the parents. The element has the same nodal configuration as the Lagrange element. However, now the the full Lagrange interpolation scheme is applied to the rotation terms (centre node included), but only the serendipity interpolation is applied to the transverse displacement terms (centre node excluded).

For the heretosis element shown in Table 8.2, let $\{N\}^T = \{N_1 \, N_2 \, N_3 \, \cdots \, N_8\}$, denote the serendipity, and $\{P\}^T = \{P_1 \, P_2 \, P_3 \, \cdots \, P_9\}$ the Lagrange shape functions. Then the element stiffness matrix is calculated as the sum of the bending and shear contributions. That is,

$$[k] = [k^b] + [k^s] \tag{8.323}$$

where

$$[k^b] = \int_{area} [B^b]^T [D^b][B^b]dA \quad \text{(bending stiffness)} \tag{8.324}$$

and

$$[k^s] = \int_{area} [B^s]^T [D^s][B^s]dA \quad \text{(shear stiffness)} \tag{8.325}$$

In eqns (8.324) and (8.325) the matrices $[B^b]$, $[B^s]$ are defined

$$[B^b] = [B_1^b \ B_2^b \ B_3^b \ \cdots \ B_9^b] \tag{8.326}$$

and

$$[B^s] = [B_1^s \ B_2^s \ B_3^s \ \cdots \ B_8^s] \tag{8.327}$$

The submatrices of eqns (8.326) and (8.327) can be found in eqns (8.307) and (8.308), so that

$$[B_a^b] = - \begin{bmatrix} 0 & 0 & P_{a,1}^T \\ 0 & P_{a,2}^T & 0 \\ 0 & P_{a,1}^T & P_{a,2}^T \end{bmatrix} \quad 1 \le a \le 8 \tag{8.328}$$

and

$$[B_9^b] = - \begin{bmatrix} 0 & P_{9,1}^T \\ P_{9,2}^T & 0 \\ P_{9,1}^T & P_{9,2}^T \end{bmatrix} \tag{8.329}$$

For the shear strain transformation matrices,

$$[B_a^s] = \begin{bmatrix} N_{a,1}^T & 0 & P_a^T \\ N_{a,2}^T & -P_a^T & 0 \end{bmatrix} \quad 1 \le a \le 8 \tag{8.330}$$

and

$$[B_9^s] = \begin{bmatrix} 0 & P_9^T \\ -P_9^T & 0 \end{bmatrix} \tag{8.331}$$

The $[D^b]$ and $[D^s]$ matrices have been given previously in eqns (8.305) and (8.306).

8.4 SOME COMPARATIVE STUDIES ON PLATE BENDING ELEMENTS

8.4.1 Scope of studies

It is impossible to include all test results on convergence studies for all elements, simply because the scope is too broad. However, a cross section of results will be given for selected elements falling into various catagories. From these the reader should be able to classify the results of any other particular element for which data is available.

8.4.2 The Morley constant moment element and the DKL(1) element

It will be remembered in the discussion on the DKL(1) element that its stiffness matrix is identical with the constant moment element. Hence, results will be the same. The results taken from [48] are for a square plate with either simply supported or clamped edges. In Tables 8.3 and 8.4, results are compared with the series solution, and the non–conforming (Bazeley–Irons) linear moment element. It is seen from the results that in terms of the accuracy of the deflection calculations the linear moment element clearly outperforms the constant moment element. For the cases shown, the bending moments in the constant moment element appear to be a lower bound on the solution, whereas in the linear moment case there is a pronounced saw–tooth effect in the second derivatives (moment values). These observations lead to the conclusion that care must be exercised in the interpretation of the results of the moment fields of both these elements. It would appear that for the constant moment case, moments should be taken at the triangle centroids, and, similarly, for the Bazeley–Irons element they should be sampled at the midpoints of the sides.

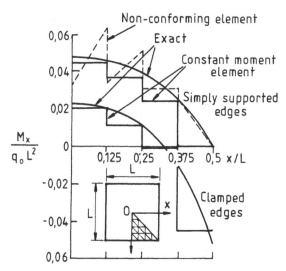

Fig. 8.37   Distribution of M_x along centre line of
square plate under uniformly distributed load.

Table 8.3

	Central deflection of square plate			
	Constant–moment element	Non–conforming element	Exact	Multiplier
Uniformly distributed load (simply supported)	0.00432 0.00412*	0.00405	0.00406	ql^4/D
Uniformly distributed load (clamped)	0.00170 0.00138*	0.00134	0.00126	ql^4/D
Concentrated load (simply supported)	0.01351 0.01219*	0.01165	0.01160	Pl^2/D
Concentrated load (clamped)	0.00776 0.00628*	0.00572	0.00560	Pl^2/D
* Finer elemental mesh where the side lengths are half those shown in Fig. 8.37				

It should be seen clearly from the results plotted in Fig. 8.38 that the shape of the triangular elements is of importance and that irregular shapes should be avoided at all costs. Because these elements are non–conforming their convergence on deflections should be from above as is illustrated in Table 8.3.

Table 8.4

	Centre of side M_n	Centre of plate M_x		Corner reaction M_y	Multiplier $\|2M_{xy}\|$
Moments in the square plate from the constant–moment element					
Uniformly distributed load		0.0447	0.0471	0.060	qL^2
(simply supported)		0.0471* (0.0479)	0.0477* (0.0479)	0.065* (0.065)	
Uniformly distributed load	−0.0443	0.0206	0.0235		
(clamped)	−0.0488* (−0.0513)	0.0224* (0.0231)	0.0231* (0.0231)		qL^2
Concentrated load				0.124	P
(simply supported)				0.122* (0.122)	
Concentrated load	−0.1065				P
(clamped)	−0.1184* (−0.1257)				
* Finer elemental mesh where the side lengths are half those in Fig. 8.37 The values in parentheses are exact.					

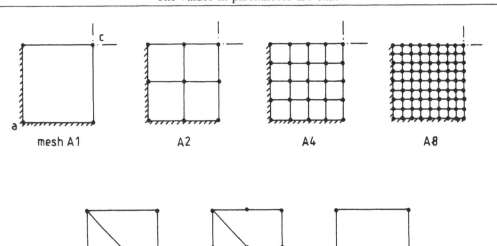

mesh A1   A2   A4   A8

BI9   LCCT12   Q19

Fig. 8.38   Mesh layouts for LCCT12 and Q19 studies.

8.4.3   The BI9, LCCT12 triangle and Q19 quadrilateral elements

These elements are shown in Figs. 8.10, 8.14 and 8.17 respectively. The LCCT12 triangular element is based on cubic interpolation functions (three per triangle Fig. 8.14(b)), and has 12–external degrees of freedom, two rotations and transverse displacement at the apices, and normal rotations at the mid–side nodes. The Q19 is composed of four of these LCCT12 triangles. Again, the square plate is studied, this time with simply supported edge conditions. The various mesh sizes are shown in Fig. 8.38. Each small square is taken as one Q19 element or two triangular elements in the case of the LCCT12. For comparison, the Bazeley–Irons element (BI9) has also been plotted. The values of the central deflections for the various mesh grids in Fig. 8.38 are given in Table 8.5.

Table 8.5

mesh	BI9	LCCT12	Q19
A1	4436(12)	3370(17)	3540(12)
A2	4600(27)	3954(43)	3930(27)
A4	4243(75)	4057(131)	4036(75)
A8	4125(243)		4062(243)
analytical solution	4062		
multiplier	$10^{-6}ql^4/D$		

Values of bending moments at the centre point (a), and twisting moments at the corner point (c) are shown in Table 8.6.

Table 8.6

mesh	Bending moment $M_x = M_y$ at centre (pt. c)			Twisting moment M_{xy} at corner (pt. a)		
	BI9	LCCT12	Q19	BI9	LCCT12	Q19
A1	8007	3115	4229	1991	2530	2202
A2	7595	4177	4441	2905	2851	2935
A4	7054	4621	4695	3149	3114	3166
A8	6896		4770	3222		3237
analytical solution	4789			3246		
multiplier	$10^{-5}ql^2$			$-10^{-5}ql^2$		

Finally, the bending moment M_x has been plotted along the centre line AB (Fig. 8.39) of the plate for both the BI9 and the Q19 elements.

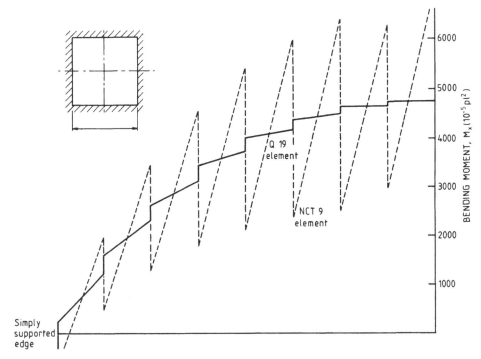

Fig. 8.39   Bending moments along axis of symmetry for the simply
supported square plate. (mesh A8, BI9 elements, Q19 elements).

It is seen from Table 8.5 that both the LCCT12 and Q19 elements, being C_1, compatible elements, converge from below, being too stiff. Their accuracy, both in the determination of deflections and in the moment fields within elements, is excellent. This last point is demonstrated clearly in Fig. 8.39 for the plot of the M_x moment. This accuracy is at the expense of internal degrees of freedom of the element, rather than of the total number of global degrees of freedom. Stiffness matrix generation is time consuming and could be lessened by the recognition of elements of the same dimensions, either in the data file by the user or internally by the computer software.

8.4.4   Rectangular plates

In this case a comparison of deflection calculations for simply supported plates for type I (12–dof.) elements is taken from [49] (Fig 8.40). It is seen that the incompatible 12–dof. element converges from above while the 16–dof. element, being C_1 compatible, converges from below. The grid sizes 1, 2, 4, in Fig. 8.40 are A1, A2 and A4 as given in Fig. 8.40. From Table 8.4, it is seen that the percentage errors in the central deflection for the LCCT12 and Q19 elements are as given in Table 8.7:

Table 8.7

Mesh	A2	A4
LCCT12	2.7	0.1
Q19	3.2	0.6

Comparing the values given in Table 8.7 with the data in Fig. 8.38, it is seen that both the LCCT12 and Q19 elements outperform the rectangular elements based on single polynomial expansions.

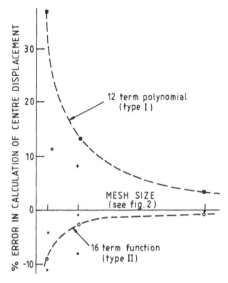

Fig. 8.40   Numerical comparisons–rectangular
plate element formulation.

8.4.5 The DKL element

As mentioned previously, the first member of the DKL family is the same as the Morley constant moment element, and these results have been given in Fig. 8.37 and Tables 8.3 and 8.4. It remains now to give the results for the second member of the family with 12 dof. (Fig. 8.25 and section 8.2.8 present the theory). It is considered that this element has considerable potential for applications in both plate and shell analyses, and so results are given in some detail.

The patch test

The patch test problem, as presented in [44], is used to test the validity of the element formulation. Four elements are used to model the rectangular plate shown in Fig. 8.41.

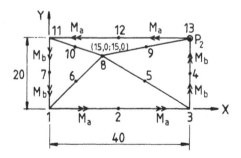

Fig. 8.41   Patch test problem.

The boundary conditions, material properties geometry and loading system shown in Fig. 8.41 will lead to a theoretical solution of $M_x = M_y = M_{xy} = 1.0$ everywhere in the plate. For the DKL element, moments can be evaluated at the corner mid–side and Loof nodes as well as at the centroid of the element. For this problem the moments M_x, M_y and M_{xy} were each computed to have a value of 1.0, correct to five significant figures at all these points.

Rectangular cantilever plate under twisting loads

According to [50], the critical test for a single quadrilateral element (or in the case of triangular elements, two elements), is the case of a cantilever plate under twist moments. Thus, twist moment results in differential bending. In this example, the differential bending is activated by using differential loads at the two free corners. The element meshes employed, the element properties and the plots of the deflection at a free corner against the increasing length L of the plate are illustrated in Fig 8.42. Since the element width is kept constant at unity, the length L is, effectively, the element aspect ratio. Results for the non–conforming elements BCIZ [74], HCT [147], DKT [119] and the hybrid element HSM, as obtained from ref. [49], are also plotted in Fig. 8.42. For the first mesh pattern the performance of the DKL element is excellent being indistinguishable from the benchmark curve obtained using 16 compatible 16–d.o.f. rectangular elements based on Kirchoff plate theory [49]. Results for the second mesh pattern are also very good, being closer to the benchmark curve than other elements.

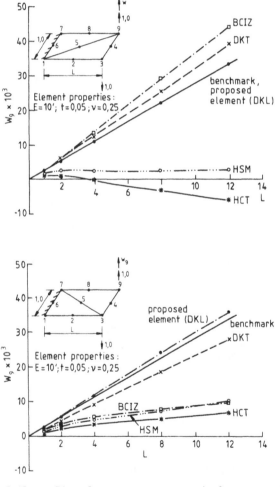

Fig. 8.42   Plot of w_9 verses aspect ratio for rectangular
cantilever plate subjected to differential loads.

Square plates

The convergence characteristics of the present DKL element are investigated by analysing a square plate with either simply supported or clamped edges subjected to uniform and concentrated loads. Due to symmetry, only one–quarter of the plate is analysed, and the meshes used are shown in Fig. 8.43 together with a table of symbols for the various triangular elements used for comparison.

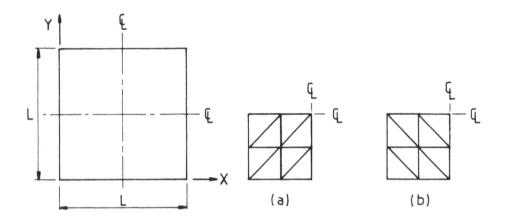

Legend

Element	ref.	symbol	mesh
DKL element	148	●	(b)
Discrete–Kirchoff (DKT)	86	×	(a)
Hybrid (HSM)	86	o	(a)
Non–conforming (BCIZ)	74	□	(a)
Clough (HCT)	147	v	(b)
Derivative–smoothed (A9)	75	▲	(a)

Fig. 8.43   Mesh orientations of square plate.

From Figs 8.44 and 8.47 it is observed that monotonic convergence to the exact value is exhibited by the present element for all the load and boundary support cases considered. Its performance under uniformly distributed loading is particularly good. However, its behaviour under concentrated loads is less satisfying, but it is still acceptable relative to the elements compared.

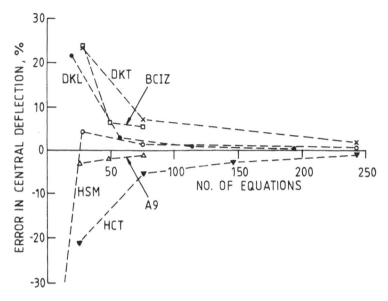

Fig. 8.44   Clamped square plate with uniformly distributed load–
convergence of central displacement.

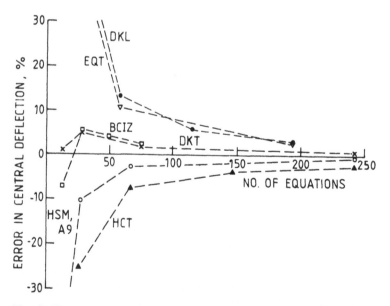

Fig. 8.45   Clamped square plate with concentrated central load–
convergence of central deflection.

The distribution of the moment M, along the plate centre line for a 3×3 mesh of the DKL element is plotted in Fig. 8.48. The computed stresses at the corner, mid–side and loof nodes exhibit a reasonably smooth distribution close to the exact solution. This performance is markedly better than that for the 4×4 mesh of the non–conforming element, also plotted on the same figure with its typical sawtooth distribution about the analytic curve.

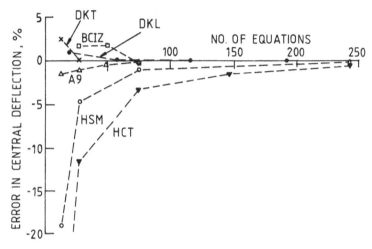

Fig. 8.46   Simply supported square plate with uniformly distributed load–convergence of central displacement.

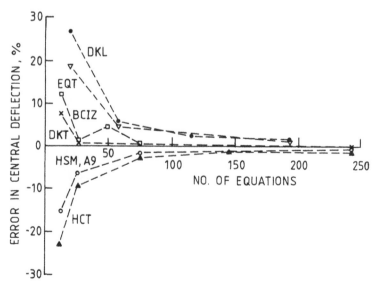

Fig. 8.47   Simply supported square plate with concentrated central load–convergence of central displacement.

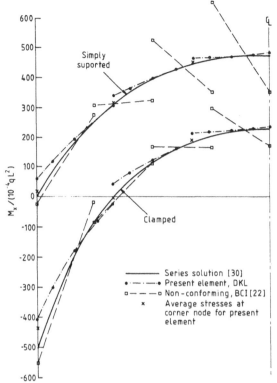

Fig. 8.48   Variation of M_x along centre
line for square plate under uniformly distributed load.

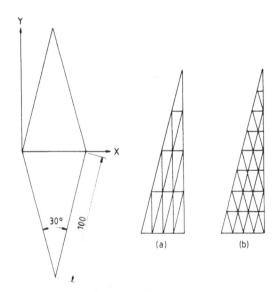

Fig. 8.49   Mesh orientation for acute skew plate.

Acute skew plate

The meshes used in analysing this problem and the results for the displacement and principal moments at the plate centre, are given in Fig. 8.49 and Table 8.8 respectively. This is a difficult problem to model due to the existence of singular moments at the obtuse corner. A thin plate solution, obtained analytically by Morley [51], is available. Results from both the mesh divisions are good, the centre displacement being out by less than 5%, and the principal moment at the plate center also being in error by less than 5% of Morley's results. The bending moments M_x and M_y along the plate centre lines are plotted in Fig. 8.50.

Table 8.8

mesh	Displacement, w_{max} $q_0 \dfrac{a^4}{D} \times 10^{-3}$	Principal moments at centre. $q_0 a^{-2} \times 10^{-2}$	
		M_{max}	M_{min}
(a)	0.426	1.97	1.15
(b)	0.422	1.96	1.11
exact	0.408	1.91	1.08

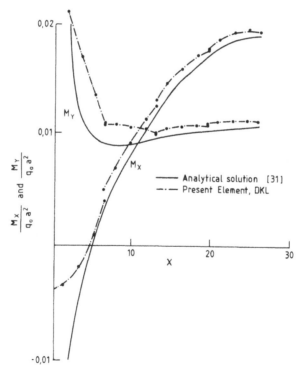

Fig. 8.50 Bending moments along center line of acute skew plate.

Clamped circular plate

The geometric data and mesh subdivisions used in this problem are illustrated in Fig. 8.51. The plate subjected to uniformly distributed load, has clamped edges. This problem is included to show the ability or otherwise of the straight sided element to model curved boundaries. in Table 8.9. The plate subjected to uniformly distributed load has clamped edges. This problem is included to show the ability or otherwise of the straight sided element to model curved boundaries. Results for the maximum displacements and bending moments at the plate boundary are tabulated in Fig. 8.51. The geometric idealization adopted will lead to stiffer solutions. As pointed out by Crisfield [52], it would be better to adopt 'best–fit' straight lines to the perimeter. Nevertheless, this method of putting all corner nodes on the boundary is simple. The results indicate that at least four elements are required to idealize the perimeter of the quarter plate, if reasonable results are to be obtained. That is, results which give displacements and stresses within 5% of their maximum value.

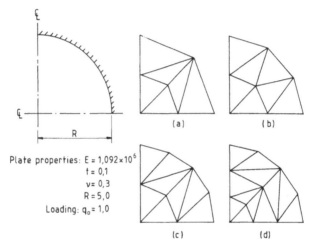

Plate properties: $E = 1,092 \times 10^6$
$t = 0,1$
$v = 0,3$
$R = 5,0$
Loading: $q_0 = 1,0$

Fig. 8.51   Mesh orientations and computed displacements and bending moments for clamped circular plate.

Table 8.9

	centre displacement	Bending moments at plate boundary	
		M_r	M_t
(a)	0.07789	−2.9095	−0.7896
(b)	0.08812	−3.0159	−0.8087
(c)	0.09232	−3.0293	−0.8314
(d)	0.09287	−3.1250	−0.9375

8.4.6 Isoparametric elements

Square plate

In this set of results, two isoparametric elements, IS08 and IS012, using serendipity functions have been compared with the Q19 element. The various meshes are shown in Fig. 8.52. In these meshes the nodal point density has been kept approximately constant along the sides. Such a comparison will probably favour the Q19 element with its large number of internal degrees of freedom.

The plate is 10 units square and 1 unit thick, and the two load cases considered are;

1. a concentrated load of 100 units;
2. a distributed load of 100 units over the whole plate.

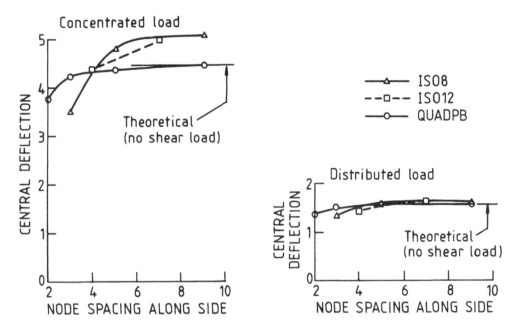

Fig. 8.52   Elements for square plates–isoparametric elements.

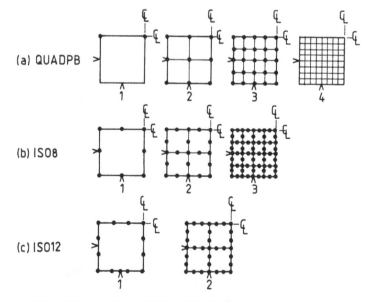

Fig. 8.53   ISO8, ISO12, Q19 deflection convergence.

The comparative studies of deflection convergence are shown in Fig. 8.53. It is seen that the isoparametric elements converge to higher deflection values than the Q19 element used herein because this latter element was formulated without transverse shear strains, whereas in the former they are automatically included.

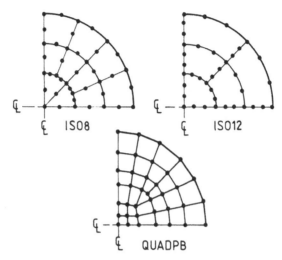

Fig. 8.54   Circular plate–mesh details.

Circular plate

One analysis for each type of element (IS08, IS012, and Q19) is presented in Fig. 8.55 for a circular plate 10 units in diameter and 1 unit thick. (Fig. 8.54) gives details of the meshes. Again, because the plate is moderately thick, transverse shear distortions can be expected to be significant. The plate is clamped at its outer circumference and carries a unit–distributed load.

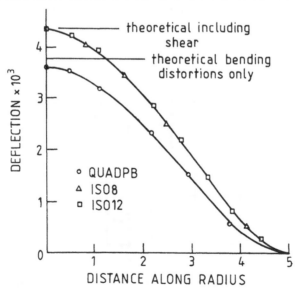

Fig. 8.55   Circular plate–transverse Deflection.

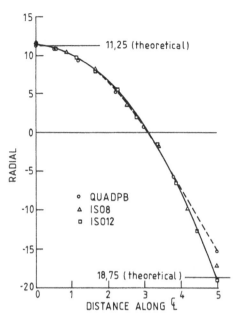

Fig. 8.56   Circular plate–radial extreme fibre stress.

In Fig. 8.56 the radial stress σ_r has been plotted, this time from averaged nodal point values. All element types give values which agree well with the analytic solution.

8.4.7   Heretosis, Lagrange and isoparametric elements

In this last study, three 8–node elements based on the isoparametric, Lagrange and heretosis concepts, as enunciated in section 8.2.2, are compared. Three plate types have been analysed;

1. square plate;
2. rhombic plate;
3. circular plate.

1. Square plate

The basic data for the square plate in these studies is as follows:

$l = 10$ length of side
$t = 0.1$ thickness
$\nu = 0.3$ Poisson's ratio
$E = 10.92 \times 10^5$ Young's modulus of elasticity

Figs 8.57 and 8.58 show the deflections for the various element configurations given in the same figures and are convergence studies for deflection. Results are normalized with respect to the central deflection series solution. From these results it is seen that the Lagrange and heretosis elements both perform better than the isoparametric element. In these former elements the direction of the convergence is not fixed, being either from above or below, or even oscillatory.

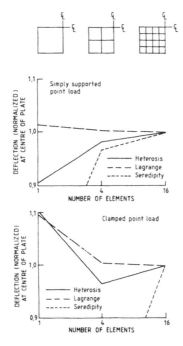

Fig. 8.57   Heretosis, Lagrange and isoparametric plate element deflections.

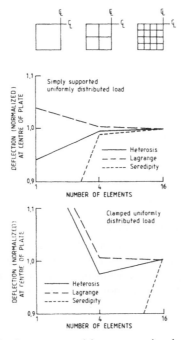

Fig. 8.58   Heretosis, Lagrange and isoparametric plate element deflections.

CHAPTER 9

Analysis of shells

9.1 INTRODUCTION

The analysis of shells from a purely computational point of view may be divided into:

1. membrane analysis;
2. analysis based on the interaction between membrane and flexural actions.

Of course, this division also applies to analysis based on the theory of elasticity in which various shell theories have been proposed for the inclusion of the bending deformations in the shell. In this theory, the membrane action is superficially simple, in that it entails the integration of the three equations of static equilibrium. For specific cases, e.g. spherical shells, hyperbolic paraboloids, under uniformly distributed loads, analytic solutions are available, and thus the straight application of the finite element method to membrane analysis has limited application. However, it does have special application to tensile fabric structures which are so thin that bending action is negligible and transverse stiffness is imparted to the shell surface by the tensile membrane stresses. This important application is discussed in some detail in section 9.2, and, follows from the study of tensile net structures given in Chapter 6. The formulation of finite elements for the analysis of shells (bending and membrane action), parallels the development of plate bending elements, and many of the problems found in that topic are exacerbated in the shell formulation. As with the plate bending elements, the derivation of effective shell elements has been limited by the requirement of C_1 continuity; that is, continuity in the displacement field derivatives, as well as in the displacement field itself, in the shell model adopting the Kirchoff assumptions to simplify the displacements across the element thickness. The fundamental difference between shell and plate action is that, in the former, there is coupling between the membrane and flexural actions as a result of the shell curvatures. The development of an effective curved–shell element is thus complicated by the necessity to incorporate the effects of the curvature into the element behaviour. The first application (and perhaps the most simplistic approach) of the finite element method to the analysis of shells consisted of replacing the curved shape with an assemblage of flat elements.

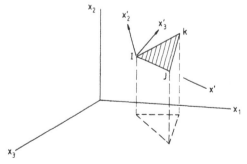

Fig. 9.1   Flat facet triangular element in 3–dimensional space.

For example, if a simple membrane analysis is required, the CST plane stress element can have its nodes located on the shell surface, and its in–plane stiffness is then transformed into the global coordinate system so that the equilibrium equations (three per node), may be solved at each node point. It should be noted that a curved surface can be approximated only by triangular elements, if the elements themselves are not to be curved or warped.

In Fig. 9.1, for nodes I, J, K the geometry and displacements are given:

1.   coordinates $(x_1, x_2, x_3)_{I,J,K}$;

2.   displacements $(u_1, u_2, u_3)_{I,K,J}$.

The in–plane geometry of the flat facet triangle is calculated, firstly, by taking a local global coordinate set parallel to x_1, x_2, x_3, through node I. In this system,

$$x_i = x_{0i} - x_{0Ii} \tag{9.1}$$

Then, let unit vector along $(I{-}J)$ be $\hat{e}'_1$ and its direction cosines l_{1i} relative to the global axes. Then from the Fig. 9.1, since $x_{Ii} \equiv 0$,

$$l_{1i} = \frac{x_{Ji}}{L_k} \quad (i - 1, 2, 3) \tag{9.2}$$

and the side length,

$$L_K = \sqrt{x_{Ji} x_{Ji}} \tag{9.3}$$

In these expressions, sum on the lower–case indices only. If $\hat{a}$ is the unit vector along the side $I{-}K$, by similar reasoning,

$$a_i = \frac{x_{Ki}}{L_J} \tag{9.4}$$

and

$$L_J = \sqrt{x_{Ki} x_{Ki}} \tag{9.5}$$

The unit vector normal to the plane $I{-}J{-}K$ will be $\hat{e}'_3$, with direction cosines, l_{3i}, and these are obtained from the cross product of $\hat{e}'_1$ and $\hat{a}$, normalizing the resulting vector. That is,

$$\tilde{k} = \hat{e}'_1 \times \hat{a} = \begin{vmatrix} i_1 & i_2 & i_3 \\ e_{11} & e_{12} & e_{13} \\ a_1 & a_2 & a_3 \end{vmatrix} \tag{9.6}$$

and

$$\hat{e}'_{3i} = \frac{k_i}{|k|} \quad ; \quad |k| = \sqrt{k_i k_i} \tag{9.7}$$

Finally, the unit vector $\hat{e}'_2$ is calculated as the cross product of $\hat{e}'_3$ and $\hat{e}'_1$,

$$\hat{e}'_2 = \hat{e}'_3 \times \hat{e}'_1 \tag{9.8}$$

Then the local coordinates of the triangle are given by

$$x'_i = l_{ij} x_j \tag{9.9}$$

or, in matrix notation,

$$\{x'\} = [L]\{x\} \tag{9.10}$$

This transformation will be found useful in obtaining triangle properties in the plane of the flat facet. The same transformation applies, of course, to the tangent plane coordinates at any point in a curved–shell element. The assemblage of elements produces the total stiffness relationship,

$$\{R\} = [K]\{r\} \tag{9.11}$$

It will be realized that problems arise with such membrane structure idealizations if flat sections occur such that the normal elastic stiffness at a node is zero. In general shell behaviour in which bending action occurs, it is possible to superpose the independent membrane and flexural (plate bending) behaviour with the appropriate spatial transformation to the global coordinate axes. Substantial difficulties exist in deriving the stiffness relationships of curved-shell elements, among which are

1. the choice of an appropriate shell theory which is consistent with the principles of compatibility and equilibrium;

2. the description of the element geometry (particularly its curvature);

3. the representation of rigid–body modes of behaviour by the assumed displacement fields;

4. retention of sufficient nodal variables to give the required degree of inter–element compatibility.

Derivation of a completely general, doubly curved–shell element requires an analytic description of the shell geometry as well as high–order representation of the displacements (see, for example, the Sheba element formulation Argyris [53]). Problems relating to the geometry representation of the shell surface are reduced somewhat by formulating the curved element in terms of a shallow shell theory. The necessary mathematical manipulations are then performed in a base reference plane, and it is sufficient to assume constant geometric curvature over the element. This, however, introduces geometric discontinuities into the approximation to the shell surface because adjacent elements are portions of different parabolic surfaces which will not match exactly. To ensure that shallow shell finite element approximations converge to the deep shell solution, it is essential that the shallowness assumption be enforced relative to a local base plane, rather than to the global horizontal plane. Regardless of whether or not the shallowness assumption is used, a proper description of the rigid–body modes in elements based on curvilinear shell theory is possible only with the inclusion of transcendental functions in the assumed displacement expressions, thus violating inter–element compatibility, or by specifying higher–order polynomials for the displacement fields, which leads to the introduction of additional nodal degrees of freedom, namely second–order derivatives. This complicates not only shell intersections but also skew boundaries in which the transformation of the stress–free boundary condition into these coordinates is not straightforward. In classical shell theory, assumptions are imposed on the three–dimensional equations of elasticity, and, these together with integration through the shell thickness reduce the problem from three to two dimensions. Due to the various assumptions and approximations that have been devised there are a variety of shell theories [54] to [58]. Thus, a shell theory combined with additional assumptions of geometry and the discretization of the field variables forms the basis of the curved elements in finite element approximations to shells. The isoparametric formulation, which has been so successfully applied to both two– and three– dimensional analysis (Chapter 7), obviously has the potential to provide an alternative approach to analysing shells. It was shown that the degenerated isoparametric concept was successful in the application to the analysis of plates (section 8.2.2). In historical sequence the degenerated isoparametric concept was actually first introduced by Ahmad [59] for the analysis of shells.

The three–dimensional field equations are discretized in terms of the mid–surface nodal variables from the outset. The constraint of straight normals is introduced, and the strain corresponding to stress perpendicular to the middle surface is neglected. The important distinction with shell theories is made, however, that normals while remaining straight are not constrained to remain normal to the mid–surface after the deformation. This allows the degenerated shell to experience shear deformations of the same type as those encountered in plate analysis (section 8.2.2 for a detailed discussion of this problem). It thus soon became apparent that unsatisfactory results are obtained when the fully integrated degenerated elements are applied in the thin–shell regime. The model is too stiff, and exhibits locking phenomena, and the the rate of convergence is too slow. Since thin shells are a more common occurrence than thin plates, the effect of locking is a serious problem which can destroy the solution through the build–up of computer rounding off errors. As with the isoparametric plate elements, shear locking is due to the inability of the assumed displacements effectively to model the zero shear strain condition in the limit as the thickness of the shell tends to zero. A distinguishing feature of the curved elements, as compared to the flat shell elements, is the presence of membrane–flexural coupling within each element. This coupling is achieved in the elements based on shell theory through the explicit presence of the curvature terms in the strain–displacement relationships, while for the degenerated element this is effected through the variation in the Jacobian. The presence of membrane flexural coupling, although desirable, again leads to the phenomena of membrane locking, in which a bending–dominated response is replaced with a membrane–dominated response. As will be shown later, this membrane locking can be alleviated in the quadratic isoparametric element by reduced integration. It is noted that the phenomenon is not present in the cubic (12 node) isoparametric element. A discussion on the relative merits of each of the above formulations for analysing these shells can also be found in the survey paper by Gallagher [49]. It is a fair comment, however, to suggest that no one particular approach is pre–eminent. However, it would appear that for non–linear analysis of shells where computational cost–effectiveness is paramount, there is considerable appeal in the flat facet formulation, due to its simplicity and, low cost of element stiffness matrix generation. The advantages in using an assemblage of flat facet elements to model a shell are

1.   we have simplicity and ease of formulation.

2.   The rigid–body motions are correctly represented.

3.   Convergence to the deep shell solution is obtained.

In this text the faceted shells will be discussed, and the advantages associated with the use of the DKL family (or similar element types) for the bending simulation outlined. Because the degenerated isoparametric shell elements appear to give the best compromise between geometric modelling of the curved shell on the one hand, and simplicity of formulation, on the other, these elements will also be given in some detail. First, however, membrane shell analysis and its application to the shape finding of tension membrane forms will be studied.

9.2 MEMBRANE ANALYSIS OF SHELLS

9.2.1   Introduction

The simple membrane analysis of shells, in which the finite element model is used to approximate the differential equations of equilibrium of the shell by discretizing the field variables at the node points, probably has very limited application, for a number of reasons. As mentioned in section 9.1, for regular shell shapes, such as spherical shells, paraboloids, etc., a wide variety of analytic solutions is already available, and it is most probably the deviation of the shell behaviour from the membrane solutions because of bending action produced by such effects as concentrated loads and restrained boundary conditions which are of primary interest. However, membrane shell analysis finds a very useful application in the shape finding of tension structures (similar to that used for cable net structures), and it is for this reason that the topic is given special consideration in this text. In the treatment given herein, only the CST (constant strain triangle) will be discussed for the following reasons:

1. The CST nodes can conform to the actual shell surface with the element being a flat facet of the surface.

2. The element formulation is particularly simple, and this allows an efficient application to non–linear analysis.

3. In non–linear analysis, the CST remains a planar element, whereas other elements such as the LST will assume parabolic (or higher–order polynomial shapes).

4. The geometric stiffness is particularly easy to incorporate, being derived from that of the linear cable elements in Chapter 6, for each of the stress fields taken parallel with the sides of the triangle.

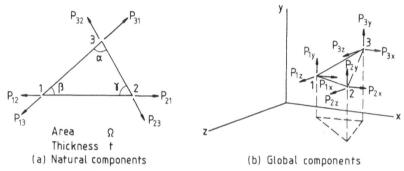

Fig. 9.2   CST triangular membrane element.

9.2.2   Elastic stiffness

The elastic stiffness of the CST is obtained using the natural mode method given in section 7.2.4. For convenience, the triangle (1–2–3) is shown in Fig. 9.2(a) in which the natural components $\{P_{32}, P_{13}, P_{21}\}$ are defined, and in Fig. 9.2(b), the corresponding three–dimensional global components are shown. From section 7.2.3 the (3×3) stiffness relationship between the natural forces $\{P_{32}, P_{13}, P_{21}\}$ and the elongations of the sides of the triangle $\Delta\{l_{32}, l_{13}, l_{21}\}$ is given (eqn (7.61)):

$$\{P_T\} = \Omega t\,[l_D]^{-1}[k_T][l_D]^{-1}\{\rho_T\} \qquad (9.12)$$

which is written

$$\{P_T\} = [k_N]\{\rho_T\} \qquad (9.13)$$

In eqn (9.12),

$$[k_T] = E\begin{bmatrix} 1 & \cos^2\alpha - v\sin^2\alpha & \cos^2\gamma - v\sin^2\gamma \\ \cos^2\alpha - v\sin^2\alpha & 1 & \cos^2\beta - v\sin^2\beta \\ \cos^2\gamma - v\sin^2\gamma & \cos^2\beta - v\sin^2\beta & 1 \end{bmatrix} \qquad (9.14)$$

and, using the lengths of the triangle sides l_{12} etc.,

$$[l_D] = \left| l_{32}, l_{13}, l_{21} \right| \tag{9.15}$$

If, in Fig. 9.2(b), the direction cosines of the sides $(J)\ J = 1, 2, 3$ are given by the vector, $\{C_J\}^T = \{C_{Jx}, C_{Jy}, C_{Jz}\}$, the components of the planar forces in Fig. 9.2(a) are transformed into their global components via the relationship,

$$\begin{Bmatrix} P_1 \\ P_2 \\ P_3 \end{Bmatrix} = \begin{bmatrix} 0 & C_2 & -C_3 \\ C_1 & 0 & C_3 \\ C_1 & -C_2 & 0 \end{bmatrix} \begin{Bmatrix} P_{32} \\ P_{13} \\ P_{21} \end{Bmatrix} \tag{9.16}$$

In eqn (9.16),

$$P_I = \begin{Bmatrix} P_{Ix} \\ P_{Iy} \\ P_{Iz} \end{Bmatrix} \tag{9.17}$$

Write eqn (9.16) as

$$\{P_{xyz}\} = [G]\{P_T\} \tag{9.18}$$

Application of contragredience gives the relationship between the corresponding displacements

$$\{\rho_T\} = [G]^T\{\rho_{xyz}\} \tag{9.19}$$

so that, in global components, the stiffness relationship is given, by

$$\{P_{xyz}\} = [G]^T[k_N][G]\{\rho_{xyz}\} \tag{9.20}$$

The (3×3) natural stiffness matrix is thus transformed into the (9×9) global stiffness:

$$[k_E] = [G]^T[k_n][G] \tag{9.21}$$

This matrix can then be used for the addition of the particular element's stiffness to the global stiffness equations of the whole membrane shell.

Example 9.1 *Membrane shell analysis for a dome*

A simple example of a dome of 100 in. (2540mm) radius is subjected to a uniform pressure of 100 lb/sq. ft. (4.778 kN/m^2). From radial symmetry, only a small slice of the dome need be considered, provided that the sides of the slice are allowed to roll on radial planes (Fig. 9.3).

The membrane stresses are given by

$$N_\phi = \frac{-ap}{(1 + \cos\phi)} \tag{i}$$

$$N_\theta = ap\left[\frac{1}{1 + \cos\phi} - \cos\phi \right] \tag{ii}$$

Stresses and displacements are plotted along line *AB*. The finite element stresses are averaged for elements (2–3), (6–7), etc. and plotted at the centroidal level along *AB*.

The results in Fig. 9.3(b) are compared with the solutions given in eqns (i) and (ii). In Fig. 9.9(c), x and y displacements have been plotted. These are of interest because they show, for example, that, if the x displacements are constrained at the springing line B, bending stresses will result in the shell.

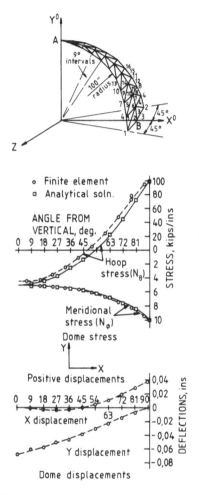

Fig. 9.3   Membrane stresses in spherical shell.

9.2.3   Geometric stiffness

The geometric stiffness of the element represents the effect which the displacements of the element have on the global force components while the member stresses remain constant. The geometric stiffness for the link element is given in Chapter 6, eqn (6.22), and the idea expressed there is extended to the generalized force fields of the CST in Chapter 7, eqn (7.70). To recapitulate, the geometric stiffness matrix for the link element I–J in Fig. 9.3(a) is defined (eqn (6.22)) as follows:

$$\Delta \begin{Bmatrix} P_I \\ P_J \end{Bmatrix} = \frac{P_N}{l} \begin{bmatrix} A & -A \\ -A & A \end{bmatrix} \begin{bmatrix} v_I \\ v_J \end{bmatrix} = [K_G] \begin{bmatrix} v_I \\ v_J \end{bmatrix} \tag{9.22}$$

The submatrix $[A]$ in eqn (9.22) is defined in terms of the direction cosines of the triangle sides:

$$[A] = [[I_3] - \{C\}\{C\}^T]]$$
(9.23)

in which the direction cosine matrix is defined

$$\{C\}^T = \{C_x \ C_y \ C_z\}$$
(9.24)

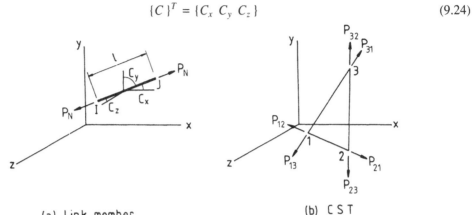

(a) Link member

(b) C S T

Fig. 9.4   Member forces–link element and CST

As was shown in section 7.2.4, the above relationship applies in turn for each of the sides (1–2), (2–3), (3–1) of the triangle, so that

$$[K_{Gij}] = \frac{P_{ji}}{l_k} \begin{bmatrix} [A_{ij}] & -[A_{ij}] \\ -[A_{ij}] & [A_{ij}] \end{bmatrix}$$
(9.25)

In this equation, $[A_{ij}]$ has been defined in eqn (9.23), with now the subscripts (ij) referring to (1–2) etc. For the whole triangle the displacements of the sides are obtained as follows:

$$\begin{Bmatrix} v_1 \\ v_2 \\ v_2 \\ v_3 \\ v_3 \\ v_1 \end{Bmatrix} = \begin{bmatrix} I_3 & 0 & 0 \\ 0 & I_3 & 0 \\ 0 & I_3 & 0 \\ 0 & 0 & I_3 \\ 0 & 0 & I_3 \\ I_3 & 0 & 0 \end{bmatrix} \begin{Bmatrix} v_1 \\ v_2 \\ v_3 \end{Bmatrix}$$
(9.26)

Write this equation

$$\{\tilde{v}\} = [M]\{r\}$$
(9.27)

Then, using contragredience,

$$[K_G] = [M]^T \begin{bmatrix} K_{G12} & K_{G23} & K_{G31} \end{bmatrix} [M]$$
(9.28)

As shown in eqn (7.70), this expression may also be written as the sum,

$$[K_G] = \sum_{i=1,3} [M_{ij}]^T [K_{Gij}][M_{ij}]$$
(9.29)

where, for example,

$$[M_{12}] = \begin{bmatrix} I_3 & 0 & 0 \\ 0 & I_3 & 0 \end{bmatrix}$$

with cyclic permutations on 1, 2, 3 for the other two matrices. It is also recalled from eqn (6.15) that the elastic stiffness matrix for the link member $I-J$ is given by

$$[K_E] = \frac{EA}{l_o} \begin{bmatrix} CC^T & -CC^T \\ -CC^T & CC^T \end{bmatrix} \tag{9.30}$$

Combining eqns (9.21), (9.28) or (9.22) and (9.30), the tangent stiffness matrix for the CST or the link member may be expressed as

$$[K_T] = [K_E] + [K_G] \tag{9.31}$$

This expression may then be used in the direct stiffness analysis of a structure composed of a combination of membrane and link members to determine the increment of displacements for a load increment. Of course, the process is an iterative one since $[K_T]$ depends on the displaced configuration. In section 6.8 a discussion was given on the methods for the non–linear analysis of framed structures. Therein the treatment was given to deal with those structures which exhibit geometric softening. In the present application the structures are usually geometric stiffening in the range of displacements considered, and this fact leads to stable iteration methods to obtain the equilibrium position. A brief resumé of such methods and some experience thereof is given herein.

9.2.4 Iteration schemes for tension structure analysis

Of the iterative schemes discussed, those based on the Newton–Raphson technique are most widely known, whereas the arc length method appears to be the most highly convergent, and the dynamic relaxation approach is a special technique which has the ability to find the correct equilibrium position even when the initial configuration is far from the correct location. Firstly variations on the Newton–Raphson method are given. The arc length method was discussed in Chapter 6.

Newton–Raphson method (NR)

In the NR method, at any iteration, j the force unbalance is given by

$$\{R_{uj}\} = \{R\} - \{R_{ij}\} \tag{9.32}$$

where $\{R\}$ is the applied load vector and $\{R_{ij}\}$ the internal member node force vector at iteration j. This vector is calculated from displacement increments at iteration $(j-1)$, and the element forces calculated using the updated geometry. Then the displacement increment is calculated:

$$\{\Delta r_j\} = [K_{Tj}]^{-1}\{R_{uj}\} \tag{9.33}$$

and the updated displacement vector is

$$\{r_{j+1}\} = \{r_j\} + \{\Delta r_j\} \tag{9.34}$$

A new member force vector is calculated from

$$\{R_{i(j+1)}\} = f(r_{j+1}) \tag{9.35}$$

and the updated tangent stiffness expressed as

$$[K_{T(j+1)}] = g(r_{(j+1)}) \tag{9.36}$$

The method suffers in convergence rate if $\{r_{j+1}\}$ is not a close approximation to the correct displacements, and then $[K_{T(j+1)}]$ will be inaccurate.

Modified Newton–Raphson (MNR)

In the MNR the inaccuracy with $[K_{Tj}]$ is recognized, and instead $[K_{T0}]$, the tangent stiffness at the equilibrium position before the current load or displacement step, is used throughout the iteration.

Newton secant method (NS)

Starting with the BFGS update [31], this method, introduced by Crisfield [60], is a modification of the Mathie and Strang vectorized version. The essence of the method is to project forward a better approximation to the displacement increment $\{\delta_j\}$ than that given in the NR method. In the NS method the jth displacement is given by

$$\{r_j\} = \{r_{j-1}\} + \{\delta_j\} \tag{9.37}$$

where $\{\delta_j\}$ is an accelerated iterative deflection increment calculated from the expression,

$$\{\delta_j\} = e_j\{\delta_{j-1}\} + f_j\{\delta_j^*\} \tag{9.38}$$

In eqn (9.38),

$$\{\delta_j^*\} = [K_{T0}]^{-1}\{R_{uj}\} \tag{9.39}$$

and e_j and f_j are scalars calculated as follows:

$$f_j = -\frac{a_j}{b_j} \tag{9.40}$$

and

$$a_j = \{\delta_{j-1}\}^T\{R_{u(j-1)}\} \; ; \; b_j = \{\delta_{j-1}\}^T\{R_{uj} - R_{u(j-1)}\} \tag{9.41}$$

Then e_j and c_j are defined by

$$e_j = f_j(1 - \frac{c_j}{b_j}) - 1 \tag{9.42}$$

and

$$c_j = \{\delta_j\}^T(\{R_{uj}\} - \{R_{u(j-1)}\}) \tag{9.43}$$

The method is based on the secant relationship obtained from Fig. 9.5 for a single degree of freedom system.

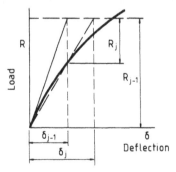

Fig. 9.5   Secant relationship.

From similar triangles in Fig. 9.5,

$$\frac{\delta_j}{\delta_{j-1}} = \frac{R_{j-1}}{R_{j-1} - R_j}$$

That is,

$$\delta_j (R_j - R_{j-1}) = -\delta_{j-1} R_{j-1}$$

This equation is obtained in vector form if both sides of eqn (9.38) are transposed and postmultiplied by $\{\gamma_i\} = \{R_{ui}\} - \{R_{u(i-1)}\}$.

Modified Newton secant method (MNS)

The displacements are again given by eqn (9.47) except that now $\{\delta_i\}$ is calculated by

$$\delta_i = g_j \delta_j^* \qquad (9.44)$$

where g_j is the scalar,

$$g_j = (f_i - 1)\frac{b_j}{c_j} \qquad (9.45)$$

It will be found in deriving the single parameter accelerator in eqn (9.45) that the MNS is not obtained directly from the BFGS formula.

The arc length method

The arc length method has been given in detail in section 6.8.3, eqns (6.205) *et seq.* A disadvantage of the arc length method is that the load increment factor corresponding to the equilibrium configuration is unknown at the outset of the iterations, as the initial load increment factor is different from the converged value. In the present application the equilibrium position corresponding to a particular load level or a particular set of support displacements is required. In such cases it may be better to proceed with one of the Newton methods, or, alternatively, make use of the arc length method over the major part of the required range converging to the prescribed values (loads or displacements) in the final step.

9.2.5 Convergence criteria

There are three basic classifications used to identify the various convergence criteria used for the nonlinear analysis of tension membrane structures. Because of the peculiar nature of tension membranes it is important that the most appropriate criterion be used.

9.2.5.1 Force criteria

These criteria are based on the comparison of the norms of the out–of–balance force vector $\{R_u\}$ and the external loads $\{R\}$. They are expressed

$$\|\{R_u\}\| \le f_c \|\{R\}\| \qquad (9.46)$$

In eqn (9.46) the convergence factor f_c is chosen of the order 10^{-2} to 10^{-4}, depending on the class of problem and also on the accuracy required. The Euclidean norm is taken as

$$\|\{R\}\| = (\{R\}^T \{R\})^{1/2} \qquad (9.47)$$

Care should be taken in the calculation of the inner product of $\{R\}$ to prevent overflow, and the load vectors may be normalized to prevent this. Other force norms may be chosen, for example, the maximum value of $\{R_u\}$ to be less than a factor of the maximum value of $\{R\}$. As these maxima may occur at different node points, it is difficult to see the justification for this approach.

9.2.5.2   Displacement criteria

A non–dimensional vector $\{E\}$ may be defined such that its ith term is given by

$$E_i = \frac{\delta r_i}{r_{i,ref}} \tag{9.48}$$

Here, if $(1/r_{i,ref})$ is set equal to K_{Tii}, the scaling then converts the criterion to an out–of–balance force vector. The possible displacement norms are

1. modified absolute norm:

$$\|E\|_1 = \frac{1}{n}\sum_{i=1}^{n}E_i \tag{9.49}$$

2. modified Euclidean norm:

$$\|E\|_2 = \frac{1}{n}(\{E\}^T\{E\})^{1/2} \tag{9.50}$$

3. maximum value norm:

$$\|E\|_3 = \max E_i \tag{9.51}$$

In applying any of the above criteria, the value of $R_{i,ref}$ is required. For simplicity it is convenient to compare the Euclidean norms of the present and the past iteration cycles of the displacements. That is,

$$\|\{\Delta r\}\|_{j+1} \le f_d\|\{\Delta r\}\|_j \tag{9.52}$$

Table 9.1

Structure	No. of load steps	Method			
		CPU 'n' (sec.-)	CPU 'n' (sec.-)	CPU 'n' (sec.-)	CPU 'n' (sec.-)
1. Square plate (10 × 10) mesh					
(a) point load at centre – 10 kN	9	71.36 8(+1)	154.96 42	148.35 41	264.05 34
(b) u.d.l. – 1.25 kN/m^2	10	74.66 9(+1)	201.52 56	177.61 51	259.25 32
2. Rectangular panel (10 × 20) mesh					
hydrostatic load – 10Kn/m^2	10	158.43 9(+1)	531.61 76	475.50 66	611.82 39
3. Circular dome (15 × 15) mesh					
non–conservative load – 10 kN/ m^2	10	99.62 9(+1)	257.97 62	235.45 58	347.57 35

If the norm used is simply the maximum expression in eqn (9.47), with displacements replacing forces, then care should now be taken to prevent underflow (such as the use of eqn (9.50)). The displacement norm criteria and, in particular, eqn (9.52) are useful in the membrane shape finding process because the smooth shape of the membrane is defined by displacements perpendicular to the membrane, whereas the force norm is determined largely by the out–of–balance in–plane stresses. In such cases the control on the force norm is not so critical, and using the displacement norm, in effect, relaxes this condition.

9.2.5.3   A comparison of non–linear solution techniques in shape finding

The general techniques involved in the shape finding process will be discussed further in section 9.3. In that section an additional technique for non–linear analysis, the dynamic relaxation method, will be introduced in connection with the location of geodesic string lines on the membrane surface. It is, however, a general technique and can be used for non–linear analysis of shells. In the present discussion several finite element meshes have been analysed (Table 9.1) with a number of load steps, starting from an initial position which is planar. In all these examples, the arc length method is clearly superior. It should be noted that the penalty incurred in reforming and reducing the tangent stiffness matrix in the NR method outweighs the increased accuracy obtained. In Table 9.2 a further comparison is made of the vector storage requirements for the various methods. In this it is seen that the NR and MNR require the least storage, with ARC being intermediate.

9.2.6   Shape finding procedures for membrane shells

In this section the discussion is centred on how the methods introduced in sections 9.2.4 and 9.2.5 are used in the practical application to the determination of membrane shapes. First the transverse equilibrium of a membrane subjected to a distributed load is examined to extract some general principles as to the types of shapes which are to be expected.

Table 9.2

Name	N-R	MNR	NSC	MNS	ARC
DISP	$\{r\}_{j+1}$, $\{r\}_i$	$\{r\}_{j+1}$, $\{r\}_j$	$\{r\}_{j+1}$, $\{r\}_j$	$\{r\}_{j+1}$, $\{r\}_j$	$\{r\}_{j+1}$
DISN	$\{\Delta r_u\}_j$	$\{\Delta r_u\}_j$	$\{\delta\}_j$, $\{\Delta r_u\}_j$	$\{\delta\}_j$, $\{\Delta r_u\}_j$	$\{\delta r\}_j$, $\{\Delta r_u\}_j$
DITT	-	-	-	-	$\{\Delta r_e\}$
STODIS	-	-	$\{\delta\}_{j-1}$	$\{\delta\}_{j-1}$	-
STOLOD	-	-	$\{R_u\}_{j-1}$	$\{R_u\}_{j-1}$	-

Legend

DISP	Incremental displacement vector
DISN	Iterative change in displacement vector
DITT	Tangential displacement vector
STODIS	Out–of–balance force vector from previous iteration
STOLOD	Out–of–balance force vector from previous iteration
j	Iteration number
N–R	Newton–Raphson method
MNR	Modified Newton–Raphson method
NSC	Newton secant method
MNS	Modified secant method
ARC	Arc–length method

General equilibrium principles of membrane shapes

An element of the membrane surface $(d\alpha, d\beta)$ bounded by the planes parallel to the XZ and YZ planes is shown in Fig. 9.6. As drawn, the Z direction has been taken approximately normal to the shell surface so that the load w_z also represents a pressure normal to the shell surface. The force components $(N_x, N_y, N_{yx} = N_{xy})$ are stress resultants per unit length of the shell surface in the shell tangent plane, and parallel to the X, Y, Y and X directions, respectively (Fig. 9.6).

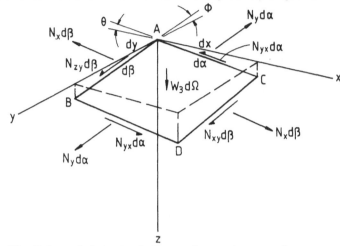

Fig. 9.6   Infinitesimal element in membrane surface.

In Fig. 9.6 the following quantities are defined:

$$p = \frac{\partial z}{\partial x} = \tan\theta \; ; \; q = \frac{\partial z}{\partial y} = \tan\phi \tag{i}$$

$$r = \frac{\partial p}{\partial x} \; ; \; s = \frac{\partial q}{\partial y} \; ; \; t = \frac{\partial p}{\partial y} = \frac{\partial q}{\partial x} \tag{ii}$$

Then the area $d\Omega$ is found by taking the cross product of the vectors $\vec{AB}$ and $\vec{AC}$, such that

$$\frac{d\Omega}{dxdy} = \sqrt{1 + p^2 + q^2} \tag{iii}$$

Define the stress resultants:

$$N_\alpha = N_x \frac{\cos\theta}{\cos\phi} \; ; \; N_\beta = N_y \frac{\cos\phi}{\cos\theta} \; ; \; N_{\alpha\beta} = N_{xy} \tag{iv}$$

and, for the distributed load,

$$w'_z = \frac{w_z d\Omega}{dxdy} \tag{v}$$

The equilibrium equations of the element are easily obtained by projecting the forces onto and at right angles to the XY plane and then writing their summations in the coordinate directions to be zero. In the present discussion only that in the Z direction is of interest, and, carrying out the above procedure it is found that,

$$N_\alpha r + N_\beta s + 2N_{\alpha\beta} t = -w'_z \tag{vi}$$

If the X and Y directions are chosen to be the principal directions in the surface, then $t = 0$, and the principal curvatures of the surface can be used to describe the shape which has been generated. There are two broad classifications of fabric structures, namely:

1. those stretched between supports (the tent type);

2. those supported by air pressure (the baloon type);

These may be further divided into:

1. air supported – low pressure;

2. air inflated – high pressure.

Returning to a discussion of eqn (vi), with $t = 0$ and $w'_z = 0$ (a membrane produced by support displacements), then with N_α and N_β both positive (tensile), it follows that for equilibrium r and s must have opposite signs, so that the membrane has an anticlastic shape. A hyperbolic paraboloid is a typical example. On the other hand, if w'_z is negative, (outwards pressure), then with r and s both positive, N_α and N_β can both be positive and the surface will be synclastic. A balloon is a typical example; a parachute another. These two simple ideas divide the membrane structures into the two types mentioned above. Of course, tensile stress is essential to the intregity of the shell shape because of the nature of the membrane fabric and its inability to carry any bending stress. Thus, any compressive strain will immediately cause the fabric to wrinkle in the zone where it occurs. In Fig. 9.7 some possible structural forms are shown, in which both regular and irregular plan forms have been covered.

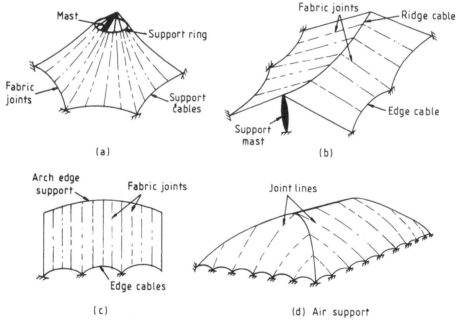

Fig. 9.7   Types of membrane structures.

It is seen from Fig. 9.7 that the support structural system (cables, beams, columns, support anchors), is an integral part of the whole design concept, which must be included when the total behaviour is being considered (e.g. dynamic wind or earthquake loading).

However, in the present context the interest is focused simply on the determination of the form of the membrane shell and its subsequent fabrication from strips of material of given width. From the various shapes illustrated in Fig. 9.7, it should be seen that the process of finding a suitable shape is no minor task. It will be approached herein as a problem of non–linear analysis of membrane shells, idealized by the CST finite elements with edge cables wherever necessary as part of the support structure. It should be noted that the non–linear analysis is simply a vehicle for forming the shape of the structure and does not represent a physical stretching of the material from its initial position.

Details of the shape finding process

From the discussions above, the design of membrane structures may be divided into three distinct stages, namely:

1.   shape finding for the initial equilibrium geometry;

2.   determination of the patterns from which the fabric strips are cut;

3.   evaluation of the structure performance under service loads (usually static wind loading).

As implied by the name, shape finding involves the selection of a suitable shape for the membrane shell, obtained in the present instance by the non–linear finite element analysis outlined in section 9.2.4. There is no unique solution to this statically indeterminate problem, as any of a number of shapes can be adopted for a particular shell plan form. The criteria which are applied to select admissible shapes are:

1.   They must conform to the form specified (plan, height, clearance, etc.).

2.   The stress levels in the membrane must be acceptable for the fabric properties.

3.   Sufficient prestress must be available to ensure adequate stiffness to carry wind loads.

As a result the designer has considerable scope to exercise judgement in the determination of the shape, and this makes shape finding subjective in nature. Once the initial equilibrium shape has been chosen, the response of the structure, at least to static loading, poses little difficulty if the material properties have been idealized. Realistic analysis at this stage which accounts for geometric non–linearity due to the large deformations encountered falls easily into the finite element solution techniques (section 9.2.4).

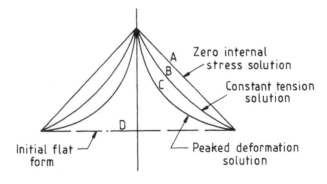

Fig. 9.8   Membrane shell cross sections.

Material non–linearity caused by the different material response in the warp and weft directions of the woven cloth is a far more demanding problem, and exact analysis is probably not yet available. In the construction phase the choice of cutting patterns for the manufacture of the fabric strips is also subjective in nature. Seams are important because the translucency of the fabric accentuates their presence, which, in turn, highlights, the geometric form of the shell. Thus, the aesthetics of the cutting pattern is a factor requiring careful examination. Fabrics of the membrane material come in rolls of constant width. For maximum efficiency of fabric use (least wastage), strips should be identified on the structure which when developed onto a plane have the straightest possible edges. This leads to the search for geodesic lines on the membrane surface. In comparison with traditional structural design, shape finding of membranes is unorthodox in that it is free from the consideration of material constitutive properties. Shapes can be chosen in a wide range between full geometric stiffness and zero elasticity and vice versa. The cross sections of three shape possibilities for a simple membrane over a circular plan are shown in Fig. 9.8. Any of the shapes in Fig. 9.8 may be obtained from the initial flat plan form D. A method of arriving at suitable shapes is as follows.

A finite element mesh is first established within the specified boundaries of the structure, projected onto a plane. This finite element mesh should, as far as possible, mirror the symmetries in the real structure (e.g. radial symmetry for the cone in Fig. 9.8). In the flat position, a uniform prestress is assigned. This is necessary because in the flat position the membrane has zero transverse stiffness. A fictitious elastic modulus is chosen to be artificially low so that the stress variations due to the elastic increments will be small when compared with the initially specified values. The elevations of the various support points are now produced by giving them displacement increments. As many as ten increments may be necessary. After each increment of specified displacements, equilibrium iterations are performed so that both compatibility of displacements and static equilibrium are satisfied. Since the critical attribute of the membrane (apart from structural adequacy) is its visual form, the equilibrated shape is examined on graphics output to ascertain that the desired surface form is being produced.

This is usually unnecessary until approximately half of the total specified displacement has been implemented. At each stage the analysis is continued, with the next increment of specified nodal displacements using the current configuration as the starting unstressed datum. In this way the accumulation of stresses or undesired elastic strains (curve C in Fig. 9.8) is avoided.

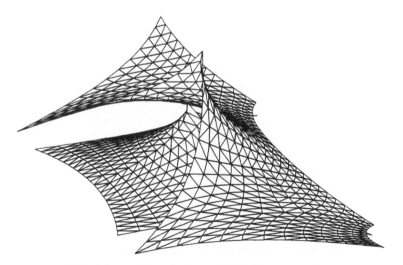

Fig. 9.9   Hyperbolic paraboloid type membrane.

The shape and stress distribution may be tailored in the later displacement steps by varying the assigned elastic modulus. A minimal surface is obtained by using zero elastic modulus for which strains are not accompanied by change in stress. Convergence of the analysis in this case will depend on the geometric stiffness derived from the level of prestress. Three shapes produced by the above procedures are shown in Figs 9.9, 9.10 and 9.11. The structures in Figs 9.9 and 9.10 are typical of those produced by the displacement of supports and the use of edge cables, whereas that in Fig. 9.11 is typical of a pneumatic structure– in this case for the cover of an Olympic sized–swimming pool.

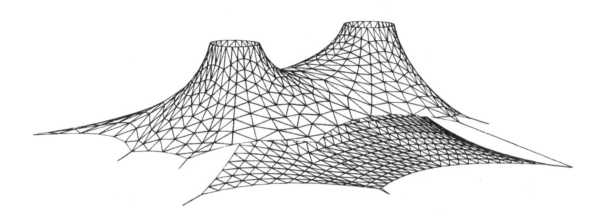

Fig. 9.10   Membrane with conical support rings.

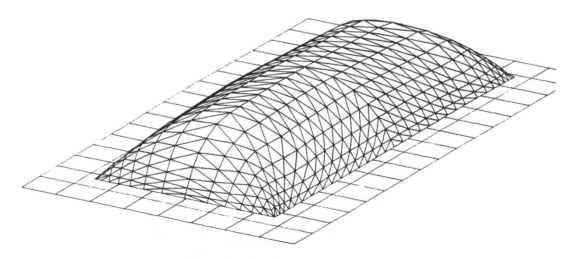

Fig. 9.11   Pneumatic membrane for pool cover.

9.2.7   Geodesic cutting patterning

Geodesics on the shell surface are curves of zero geodesic curvature, where the geodesic curvature at any point on the surface is defined as the orthogonal projection of the curvature vector onto the tangent plane at that point. When an enveloping developable surface of a geodesic is unrolled onto a plane, the geodesic becomes a straight line. Although tensile membrane structures generally do not have developable surfaces, the use of geodesics to define strip edges results, in practice, in strips which do not exhibit excessive bowing, thus economizing on material cut from a constant–width sheet. For a shell surface composed of a number of flat facet elements, a simple geometrical method is available for determining the geodesic line from a point such as A in the direction AB in Fig. 9.12. In Fig. 9.12, three triangular elements LNM, NOM, OPM are shown. The geodesic starts with the direction AB, and to determine the continuation BC, the angle MBC is chosen equal to ABN, and, similarly, for CD, angle BCD equals angle MCD. The reasoning is quite simple when the three triangles are flattened to be planar; the equalities above now mean that $ABCD$ is a straight line in the plane and hence is a geodesic on the surface. Although the above procedure is simple, it suffers, in practice, because it is usual in the design of the cutting patterns to define the end points of the fabric strips. Since the starting directon AB is unknown, to achieve a specified end point, the process above must be an interative one. An alternative procedure is to use the tangent concept that the geodesic on the surface is the line of shortest distance between two points which are connected by the geodesic.

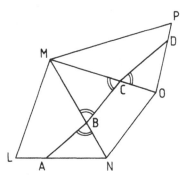

Fig. 9.12   A patch of surface elements.

If a cable is stretched on the surface and constant tension maintained, it will take up this shortest distance. Thus, consider a single such element with a prestress force F spanning between two fixed–end points. The element (or cable) lies on the surface, and the length L is composed of number of individual lengths L_i in each of the flat facets which contain the cable. Then the strain energy is given by

$$\bar{U} = \frac{1}{2}\sum F^2 L_i = \frac{1}{2}F^2\sum L_i \tag{9.53}$$

Because the reactions to the cable are at right angles to the surface, and the cable moves in the planes of the elements, this strain energy is also the potential energy of the system, which for equilibrium is a minimum. It follows that

$$\sum L_i = \text{a minimum}$$

and the cable on the frictionless surface will take up the shortest distance between the end points. Thus, if the cable is set up on the surface connecting the two end points, iteration such that the cable is allowed to slide over the surface, will converge to the geodesic as the equilibrium position. Since the cable has to slide over the surface the residual forces are modified so that only motion tangential to the surface occurs.

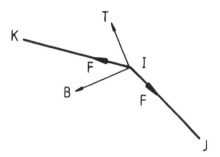

Fig. 9.13   Elements of a geodesic string on surface.

Fig. 9.13 shows two elements IJ and IK in the geodesic cable. To constrain the motion to be tangential to the surface, it is necessary to determine the tangent plane at I, which is perpendicular to the principal surface normal. The principal normal vector $\vec{N}$ can be determined by taking the cross product of the two vectors directed from a surface node along the two adjacent edges of that surface element. The principal normal will be the average of the two normals to the two surfaces containing IK and IJ. The direction of the tangent vector $\vec{T}$ to the cable at I is approximated by using the vector between the midpoints of the segments IJ and IK. The binormal is thus given by

$$\vec{B} = \vec{N} \times \vec{T}$$

The residual force at node I is calculated to be

$$|R_T| = \{B\}^T \{R\}$$

The global components of this force are

$$\{R_T\} = \{B\}\{B\}^T \{R\} \tag{9.54}$$

This force vector is now used in the DR method to iterate to the equilibrium position.

9.2.8   The dynamic relaxation (DR) method

The dynamic relaxation method (DR) is a pseudodynamic analysis which allows iteration of the statics situation to the correct equilibrium position. It is particularly well suited to the geodesic string determination in which an NR method may not converge because the initial position chosen is too far from the correct position. Thus, the equation of dynamic equilibrium is written

$$[M]\{\ddot{\delta}\} + [C]\{\dot{\delta}\} = P(t) - [K]\{\delta\} = R(t) \tag{9.55}$$

Because this is a pseudodynamic problem both the mass matrix and the damping matrix $[C]$ may be assumed to be diagonal. Eqn (9.55) is approximated at time t, by considering the values of the velocity V at times $t - \dfrac{\Delta t}{2} = t^-$, and $t + \dfrac{\Delta t}{2} = t^+$. Thus, the expressions for velocity and acceleration at time t are approximated

$$V_t = \dot{\delta}_t = \frac{V^- + V^+}{2} \; ; \quad \text{and} \quad \ddot{\delta}_t = \frac{V^+ - V^-}{\Delta t}$$

Substitution in eqn (9.55) gives

$$[M]\left\{\frac{V^- - V^+}{\Delta t}\right\} + [C]\left\{\frac{V^- + V^+}{2}\right\} = R(t)$$

Solving for V^+,

$$\{\frac{1}{\Delta t}[M] + \frac{1}{2}[C]\}V^+ = R(t) - \{\frac{-1}{\Delta t}[M] + \frac{1}{2}[C]\}V^- \qquad (9.56)$$

Now, since both $[M]$ and $[C]$ are diagonal matrices, eqn (9.56) may be written for the jth term, in the ith iteration,

$$V_{ij}^+ = \frac{\Delta t}{M_{ij}(1 + \frac{C\Delta t}{2})}R_{ij} + \left[\frac{1 - \frac{C\Delta t}{2}}{1 + \frac{C\Delta t}{2}}\right]V_{ij}^- \qquad (9.57)$$

Write this equation

$$V_{ij}^+ = A\frac{\Delta t}{M_{ij}}R_{ij} + BV_{ij}^- \qquad (9.58)$$

The definition of A and B follow directly from eqn (9.57). Using the velocity so obtained, the coordinate at node j at time $t + \Delta t$ is obtained as

$$x_{ij}^{t+\Delta t} = x_{ij}^t + \Delta t V_{ij}^+ \qquad (9.59)$$

In the DR method the response of the system is traced in small increments of time Δt until the motion is damped to the static equilibrium position. New node velocities are evaluated to give new updated coordinates. The residual force vector is then determined as the sum of the internal forces and any applied external forces. An alternativive approach to using the damping matrix $[C]$ is to use kinematic damping.

Kinematic damping

The concept of kinematic damping is based on the simple idea of a swinging pendulum. As the pendulum swings, there is an exchange between potential energy, and kinetic energy with the kinetic energy being a maximum when the system passes through the equilibrium position. This can be generalized to apply for the various modes in a multidegree of freedom system. In introducing kinetic damping into the DR analysis, viscous damping is dispensed with. The kinetic energy (KE) is traced for the free vibration of the system. At each increment after the nodal velocities have been determined, the kinetic energy is obtained as

$$\text{KE} = \frac{1}{2}\{V^+\}^T[M]\{V^+\} \qquad (9.60)$$

When a local peak in the KE is detected, the vector of nodal velocities is set to zero and the analysis restarted from the current configuration. This has the effect of extracting the KE associated with the mode which produces the current peak. Continuing, furthur peaks of decreasing magnitude are encountered a,nd the procedure is repeated until the energy in all modes is dissipated. The nodal mass components are chosen by

$$M_{ij} \geq \frac{1}{2}\Delta t^2 S_{ij} \qquad (9.61)$$

In eqn (9.61). S_{ij} equals the row sum of the absolute values of the stiffness components for the ith node in the jth direction. When the mass components are thus chosen, every d.o.f. in the structure will have similar stiffness–to–mass ratio, and consequently all periods of vibration are similar (or equal). This avoids the undesirable situation where the motion of each node is out of phase with all adjacent nodes, and so also allows the mass components to be chosen automatically. It is seen that the DR method has some advantage in that the stiffness matrix for the whole structure need not be assembled. When used in the determination of geodesic string lines, eqn (9.54) should be used to calculate the residual force component at the string nodes.

9.3 MEMBRANE AND BENDING SHELL ELEMENTS

9.3.1 Flat facet elements

The concept of using flat facet elements for analysing the behaviour of shell structures was suggested by Greene *et al.* [61] as early as 1960. Satisfactory bending stiffness matrices were first developed for the rectangular element. They were utilized in deriving the earliest successful flat shell elements for the analysis of arch dams and cylindrical roofs [62]. Gallagher *et al.* [63], by using an assumed stress field for the membrane action and a non–conforming 12–term polynomial for the transverse displacement, derived the first flat quadrilateral shell element. Another quadrilateral faceted shell element was formulated by Johnson [64] using the subdomain approach. Each of the four triangular elements consists of a superposition of a partly constrained LST with the HCT bending element. An improved version of this quadrilateral was later proposed by Yeh [65]. Both the rectangular and quadrilateral faceted shell elements are, however, restricted in use due to the need for all the four nodes to lie in the same plane. To model an arbitrary doubly curved shell with flat elements would require the use of triangular elements. Initial attempts at formulating faceted triangular shell elements by Clough and Tocher [66], Petyt [67] and Argyris [68] were not particularly satisfactory due to the lack of suitable triangular bending elements. This situation was rectified with the development of the HCT and BCIZ plate bending elements. These plate bending elements could then be combined with the plane stress elements to yield suitable triangular faceted shell elements. This was accomplished by Clough and Johnson [69] where the CST was used in conjunction with HCT bending element, and by Zienkiewicz *et al.* [70] with the non–conforming BCIZ plate bending element. The membrane approximation of the faceted shell proposed by Clough and Johnson [69] was improved on by Carr [71] by using the quadratic strain triangle with the resultant element having a total of 27 d.o.f. Another higher–order element with 27 d.o.f., which utilizes the LST to represent the membrane action, and a quartic plate bending element has been formulated by Chu and Schnobrich [72]. Razzaque [73] presented a faceted shell element where subcubic functions of the Zienkiewicz type [74] were used to describe the in–plane displacements. The derivative smoothed bending element [75] was employed to model the at mid–side nodes, exhibits good convergence characteristics. However, the generation of the shape functions for the faceted element is somewhat tedious. In an attempt to incorporate the rotational stiffeness about the normal to the shell, Olsen and Bearden [76] employed a 9–d.o.f. plane stress element with the in–plane rotation as a nodal parameter. This was combined with the BCIZ bending element to yield an 18–d.o.f. flat shell element. Care should be exercised when using this element, as convergence to the exact solution for the plane stress element is not guaranteed, due to the use of incomplete interpolation polynomials. The simplest formulation of a faceted shell element is due, (Dawe [77]), where the CST is combined with the Morley constant moment triangle [48] to yield a 12–d.o.f. constant stress shell element. This particular element was originally derived by Herrmann and Campbell [78] through the application of a mixed variational principle. Dungar *et al.* [79] developed a flat shell element using the hybrid stress model, by assuming a linear variation of the in–plane forces and a quadratic moment variation.

The in–plane displacements are assumed to vary linearly along an edge and cubically normal to it, while the transverse displacements possess a cubic variation. In another paper [80], the membrane approximation was improved on by using second–order polynomials for all interior stress fields. A similar type of element but with linearly varying equilibrium stress fields was presented by Yoshida [81]. The hybrid stress elements described utilize the in-plane rotation as a nodal parameter. The motivation for this, as is also the case with the displacement–based element presented by Olsen and Bearden [76], is to prevent a singularity arising in the global stiffness matrix when adjacent faceted elements meeting at a node are coplanar. Sander and Becker [82] have also suggested the use of a new family of triangular faceted shell elements based on a combination of compatible displacement elements for membrane action and equilibrium stress elements for the bending action. Recent impetus towards the non–linear analysis of shell structures has regenerated interest in the relatively simple and economical faceted formulation. The faceted Trump shell element with 18 d.o.f. formulated by Argyris et al. [83] through physical lumping ideas was proposed to solve non–linear plate and shell problems. Another element formulated with non–linear applications in mind is the triangular element presented by Horrigmoe [84]. A constant stress hybrid element was employed to represent the membrane action, while a hybrid stress model utilizing a modified form of Reissner's variational principle was used to derive the bending stiffness. A quadrilateral version of this hybrid flat shell element was also presented in [85]. The three node triangular discrete Kirchoff element formulated by Batoz et al. [86] was employed by Bathe and Ho [87] to represent the bending action in their flat shell element, with the CST modelling the membrane behaviour.

9.3.2   Curved elements based on a shell theory

9.3.2.1   Deep–shell theory

It is suggested by Morris [88] that the only suitable element for analysing deep shells satisfying the rigid–body, inter–element compatibility and satisfactory modelling of sensitive solutions, is one that is embedded in the curved–shell surface and formulated in terms of the surface coordinates. The formulation of such an element is complicated, and the existing literature on finite elements based on deep–shell theory is relatively sparse. Gallagher and Yang [89] presented a rectangular, doubly curved–shell element formulated using Norozhilov shell theory. Bilinear approximations were employed for the tangential displacements, while the 16–term polynomial initially proposed by Bogner et al. [11] for plate bending was used to model the normal displacements. Due to the low–order tangential displacements, the rigid body modes are not adequately approximated. A similar element was also proposed by Oden [90]. This doubly curved element was improved on by Greene et al. [91] and Gallagher [92] by using the same bicubic expressions employed for the normal displacements to approximate the tangential displacements, resulting in a 48–d.o.f. element. It is important that the shell theory employed satisfy the requirements of consistency, strain–free–rigid body displacements and coordinate invariance. Satisfactory compliance with these conditions is not achieved by the shell theory of Novozhilov, and the previously described, doubly curved rectangular elements were reformulated by Morris [88] using Naghdi's shell theory. The first successful derivation of a triangular–deep–shell element can be attributed to Argyris and Scharpf [53]. The Sheba element is formulated using the natural strain concept, and inter–element continuity is obtained, by the use of complete quintic polynomials for all three displacement components. Satisfaction of strain–free rigid–body motion is achieved by using the same displacement interpolation functions for the geometry of the surface. A 36–d.o.f. triangular element using cubic tangential displacements and an incomplete quintic polynomial for the normal displacements was proposed by Cowper et al. [93]. Inter–element continuity is satisfied by imposing a cubic variation of the normal rotations at the edges.

The centroidal displacements are statically condensed out, and different versions of the element based on Koiter–Sanders, Donnell–Vlasov or shallow–shell theory for the strain displacement relations are possible. A family of triangular and quadrilateral deep–shell elements have also been derived, using the subdomain approach, by Sander and Idelsohn [94]. The tangential displacements in each subregion are assumed to vary linearly, quadratically, or cubically and the normal displacement is modelled using a cubic polynomial. The elements are formulated so as to remain conforming even in the presence of a discontinuity in the radius of curvature. A consistent, first–order approximation, thin–shell theory is employed. Dupius and Goel [96], by expressing the strain tensor obtained from Koiter's thin–shell theory as a function of the Cartesian components, were able to formulate a doubly curved–shell element exactly satisfying the strain–free, rigid–body motion criterion. Two sets of rational functions, yielding a 54–d.o.f. and a 27–d.o.f. doubly curved, triangular element, respectively, were used to describe the Cartesian displacement components and the position of the shell above a datum line. Alternative variational principles have also been employed in formulating double curved, deep–shell elements. Thomas and Gallagher [97], through the use of the generalized potential energy principle, developed a 27–d.o.f. triangular element. Complete cubic polynomials were used for all the displacement components, and continuity of normal slopes was enforced at the element boundaries using Lagrange multipliers. A triangular element with 45 generalized d.o.f. was proposed by Gellert and Laursen [98] using a complementary energy expression and Naghdi's shell equations. A quadrilateral mixed element was proposed by Talaslides and Wunderlich [99] using a linear approximation for the geometry and field equations. The derivation of a non–linear, deep–shell element is computationally more difficult than the linear case. This, in part, is due to the difficulty in deriving a consistent non–linear shell theory. Nevertheless, a 36–d.o.f. non–linear, doubly curved quadrilateral element has been proposed by Matsui and Matsuoka [100]. The discrete Kirchoff approach is employed for the translational displacements, while the normal rotations are defined using biquadratic polynomials. Application of the Sheba element to geometrically non–linear problems is presented by Stein *et al.* [101], in which a moderate–rotation non–linear shell theory was used [102]. This particular large–displacement, moderate–rotation, non–linear shell theory was also employed by Kratzig *et al.* [103] to derive various doubly curved triangular elements. The accuracy obtained by using these non–linear, deep–shell elements is good; however, their practical application is complicated and, computationaly, rather expensive.

9.3.2.2 Shallow–shell theory

The formulation of curved–shell elements is simplified if the shallowness assumptions are used. One of the first shell elements so developed is the rectangular element proposed by Connor and Brebbia [104], using Reissner's shell theory specialised for the shallow Cartesian case. This element is modelled using bilinear tangential displacements, and a non–conforming, 12–term polynomial for the normal displacements. The same displacement assumptions were employed by Sabir and Ashwell [105], but using instead the analogy between doubly curved shells and plates on elastic foundations to obtain the strain displacement relations. These two elements do satisfy the rigid–body and compatibility requirements. Another rectangular planform element was presented by Yang [106], who used bicubic polynomials to approximate all three displacement components. An eigenvalue analysis reveals that all six rigid–body motions are present in the element. The same bicubic functions were also used by Jones [107] who used Marguerre shallow– shell theory, to formulate a curved quadrilateral element. Pecknold and Schnobrich [108] describe a parallelogram planform curved element, with bilinear tangential displacements and a spline interpolation function for the transverse displacements. The rigid–body modes are exactly included in the displacement functions, resulting in small inter–element discontinuities in the in–plane displacements. Doubly curved–shallow–shell elements of triangular planform are more versatile than the rectangular ones just described, due to their ability more accurately to model any given shape.

The first attempt to derive a triangular doubly curved element by Utku [109], using a shallow–shell theory admitting transverse shear deformations, was not particularly successful. Strickland and Loden [110], assuming linear tangential displacements and a non–conforming cubic variation due to Bazeley *et al.* [74] for the normal displacements, derived a 15–d.o.f. triangular element using Novozhilov's shallow–shell theory. Due to the low–order tangential displacements, this element does not satisfactorily reproduce the rigid–body modes. Bonnes *et al.* [111], employing Reissner's shell theory and the subdomain approach, presented a 36–.o.f. triangular element. On eliminating the mid–side nodes, a 27–d.o.f. version was obtained. A complete cubic polynomial was used to model the displacements in each subdomain. Cowper *et al.* [112], using Nozvozhilov's shallow–shell theory, with cubic tangential displacements and quintic normal displacements, derived a 36–d.o.f. triangular element. A series of triangular elements employing the displacement functions proposed by Bazeley *et al.* [75], Clough and Fellipa [40] and the Tuba formulation of Argyris *et al.* [113], for each displacement component was proposed by Vos [114]. Marguerre shallow–shell theory is used and a tensorial formulation employed in the stiffness generation. Dawe [115] proposed a very refined, doubly curved, triangular shell element using a constrained quintic for each displacement field. Koiter's shell theory, specialized for the shallow shell is used. The motivation for the use of such high–order in–plane displacements was to achieve a better approximation to the rigid–body modes. Mohr [116], using quadratic tangential displacements and a non–conforming quartic for the normal displacements, derived a 27–d.o.f. triangular element through the use of the Argyris natural–strain approach and the strain–displacement reactions of Koiter. An alternative approach to formulating displacement based curved elements is to use a linear shear theory of shells, and then impose discrete shear constraints to achieve a thin–shell solution. Various such triangular elements were formulated by Dhatt [117] using the subdomain approach. A quadratic variation is assumed for the normal rotations, and the tangential displacements are approximated by complete cubic polynomials. The various elements differ in whether a cubic or linear variation is used for the normal displacements, and whether the shear strain energy is included in the potential energy functional.

Another discrete Kirchoff element was proposed by Dhatt [118], using single–field representations, in which quadratic normal rotations and an incomplete cubic polynomial for the translational displacements are assumed. Linear tangential displacements with the rigid–body motions explicitly added, together with an incomplete 9 term–cubic variation of the normal displacements and quadratic normal rotations, are employed for the 15–d.o.f. triangular element presented in [119]. By assembling four of these 15–d.o.f. elements, and condensing out the internal d.o.f., a 20–d.o.f.–curved quadrilateral was obtained. Alternative variational principles have also been used to formulate shallow–shell elements. Tabarrok and Hoa [120], using a hybrid stress approach and Novozhilov shallow–shell theory, formulated both a triangular and rectangular planform, double–curved element. The field variables were assumed to vary linearly for the in–plane stress resultants and edge displacements, while the out–of–plane moments and edge displacements vary quadratically. A rectangular element with 6 d.o.f. per node, also formulated using a hybrid stress functional and Novozhilov shallow–shell theory, was proposed by Alayloiglu and Ali [121]. The tangential edge displacements vary linearly, while the normal displacements are approximated by a 16–term Hermitian polynomial. The resultant forces and moments were initially assumed to possess a cubic variation. After satisfying the equilibrium equations, those higher–order terms found to cause unnecessary duplication and cross–coupling were eliminated, to yield a 21–coefficient stress function. The assumed stress hybrid model entails the selection of an equilibrating stress field, which is not easily achieved for curved elements. Wolf [122], by using a functional suggested by Washizu [123], formulated a triangular shell element in which the stress field need not *a priori* satisfy the equilibrium conditions. Cubic in–plane stress resultants and quadratic moments were assumed. The in–plane edge displacements are linear with quadratic edge normal displacements.

Using a hybrid stress model which does not require *a priori* assumptions of equilibrating stress fields, Pian and Chen [124] proposed a 24–d.o.f. triangular and 32–d.o.f. quadrilateral semi–Loof type of element, using Marguerre shallow–shell theory. The stresses for the triangular and quadrilateral elements were described using 37 and 59 generalized parameters respectively. Unlike the displacement semi–Loof element proposed by Irons [125], the Loof nodes here are located at the third points along the edges, instead of the Gauss points. Mixed shallow–shell elements derived from the Hellinger–Reissner variational principle were first developed by Prato [126]. Marguerre's shallow–shell theory was used. Both a triangular and a quadrilateral element are derived, with linear interpolation functions employed for the kinematic and static field variables. The same approach was used by Connor and Will [127] to derive several triangular shell elements based on various combinations of linear and quadratic interpolation functions for the displacement and stress fields. Both three–node and six–node triangular elements were derived by Tahiani and Lachance [128] using the mixed method, with linear and quadratic interpolation, respectively, for the field variables.

9.3.3 Degenerated shell elements

The first degenerated shell element was proposed by Ahmad *et al.* [59]. Using the process described in that paper, any of the three–dimensional isoparametric elements can be degenerated to yield a corresponding shell element. As was previously discussed in relation to plate bending (section 8.3.2), this model is too stiff in the thin–shell regime, and it was observed by Ahmad *et al.* [59] that in such situations it is necessary to use the element with cubic interpolation functions. An improvement to the basic model was achieved by Zienkiewicz *et al.* [129] through the use of uniformly reduced integration and by Pawsey and Clough [130] through the use of selective reduced integration. A remarkable improvement in the performance of the eight–node parabolic element in thin–shell applications was observed. The eight–node parabolic element is reformulated by Lee and Pian [131] through the Hellinger–Reissner variational principle, where, in addition to the displacements, the strain field is also assumed to vary independently. This elements' performance was observed to be insensitive to the thickness of the shell without the need to use the uniformly or selective reduced integration techniques. In fact, the stiffness matrix of certain of the plane stress and plate bending elements presented in that study were found to be identical to the reduced integration element. Mixed formulations thus appear to provide an alternative means for reducing the severe constraints due to the condition of zero transverse shear strain in the thin–shell limit of the degenerated model.

The equivalence of some mixed models to reduced/selective integration displacement models is further established by Hughes *et al.* [132, 133] and Noor and Anderson [134]. The heretosis concept [135], discussed previously in relation to plate bending in Chapter 8, have also been generalized to derive shell elements and will be developed in the present chapter. Talha [136] noted that when a flat quadrilateral element with a particular local–global coordinate relation in the x–y plane is given a quadratic transverse deflection, one of the terms encountered is not present among the shape functions of the eight–node parabolic shell element. A central node with only a deflection normal to the mid–surface of the element is added to introduce the missing term. This internal node is then eliminated by constraining the integral of the boundary shear to be zero. Reduced integration is used and, unlike the normal parabolic element, it passes the patch test for quadrilaterals. One spurious mechanism is, however, associated with this element. The accuracy of the four-noded isoparametric shell element is very poor it is, in fact, non–convergent unless the thickness is larger than the other dimensions. By introducing modifications which relax the excessive constraints in the original element, MacNeal [137] was able to derive a 20–d.o.f., four–noded element which is both accurate and economical to compute. The modifications employed are the enforcement of curvature compatibility, a selective integration technique on the the membrane stiffness, a special reduced integration on shear stiffness and the augmentation of the transverse shear terms by the addition of a residual bending flexibility. These modifications were motivated by the results of a study on a beam element formulated using the isoparametric concept and an exact analysis. Another four node degenerated quadrilateral shell element but, with 6 d.o.f. at each node, was proposed by

Kanok–Nukulchai [138]. This element is a generalisation of the bilinear plate bending element of Hughes *et al.* [139]. Selective reduced integration was employed, with analytic integration performed over the thickness. Due to the use of 6 global d.o.f., a torsional stiffness is added to the element to prevent convergence problems when adjacent elements at a node become coplanar.

Wempner *et al.* [140], using the Hu–Washizu variational principle, which admits independent assumptions of stresses, strains and displacements, formulated another four–noded quadrilateral degenerated shell element. The displacements are bilinear, and the strains are described by the simplest possible expressions consistent with the criterion of homogeneous states of strain. This requires the identification of the 20 deformation modes associated with the element. The stresses are approximated following the form of the conjugate strains. Due to the orthogonality of the functions, the stress–strain and strain–displacement relations are uncoupled, and the final result of the formulation is thus a system of equations in terms of the displacement parameters only. The stiffness matrix is symmetrical and positive definite. The analysis of practical engineering shells is a difficult problem, and it is specially so in the presence of sharp corners and multiple junctions. A competitive thin–shell element should be able to model these without any additional complications. It should also be mixable with other triangular and quadrilateral elements while passing the patch test. The strain–free, rigid–body motion criterion must be satisfied, and it should not give rise to spurious low–energy modes. These desirable features described by Irons [125], epitomize the semi–Loof degenerated shell element. The element initially possesses 43 d.o.f., which are reduced to 32 d.o.f. with the introduction of 11 shear constraints. The shape function derivatives in terms of the 32 nodal parameters are then substituted into the strain expressions presented in [125]. This element has been found to provide accurate results even with relatively coarse meshes. A quadrilateral element with 16. d.o.f., similar to the semi–Loof element but based on lower–order interpolation functions, was proposed by Nagtegaal and Slater [141]. The element is integrated using a 2×2 Gauss rule with the modification that the shear stresses are constant throughout the element, to prevent it from being too stiff to in–plane bending. Due to judicious choice of the interpolation functions, the element converges well and does not exhibit spurious mechanisms.

9.3.4 Faceted shell elements

Apart from the obvious shortcoming of the inability to incorporate flexural membrane coupling within the element, flat facet elements have a number of deficiencies associated with their use. Thus, for example, most of the faceted elements presently developed possess only two in–plane corner rotational d.o.f. This is because the out–of–plane rotation (about an axis normal to the facet) is missing in available compatible, displacement–based, plane stress elements. Therefore, if the elements meeting at a node are coplanar, a singularity will arise in the global stiffness equations unless this d.o.f. is suppressed. This problem was overcome by Zienkiewicz *et al.* [70] using the artifice of applying a fictitious but consistent rotational stiffness to the out–of–plane rotation. The value of the stiffness added, however, depends on the precision of the computer and has no theoretical justification. If used, its influence should thus be kept to a mimimum. Clough and Johnston's [69] solution to the problem was to define a tangent plane to the normal (ξ_3) of the shell at a vertex. The local rotational d.o.f. are then transformed to a set of surface coordinates (ξ_1, ξ_2) in the principal directions. This will then serve as the base coordinates for the rotations, with the consequence that only two rotational d.o.f. are defined for each node in the global stiffness equations. Difficulties occur, however, with shell branches, as the normal here cannot readily be specified. In addition, the transformed local rotations will then have a significant component about the normal ξ_3, which cannot really be ignored, as is the case with this approach.

Olson and Beardon [76] and several hybrid elements [80], [81], [82] use a plane stress element with the in–plane rotation as a nodal parameter. This, however, introduces the added constraint that the included angle between the sides must remain constant throughout the deformation. As pointed out by Irons [7], serious errors are introduced if an extended patch of such coplanar elements is subjected to a horizontal tensile force. In modelling an arbitrary, doubly curved shell with flat elements, only triangular elements can be used. This leads to the CST and LST being possible contenders to model the membrane stress components.

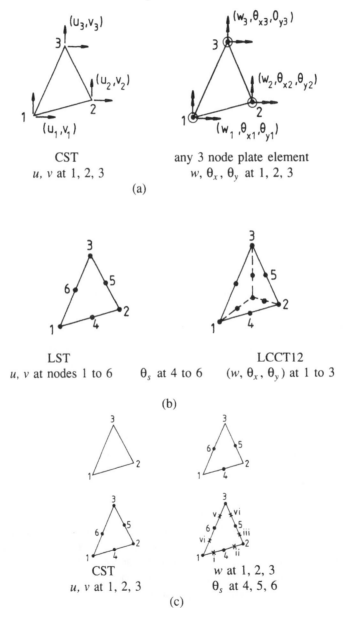

Fig. 9.14   Triangular element combinations.

Experience with the CST in plane stress situations indicates that it is a relatively poor performer. It is thus to be expected that a faceted shell element employing the CST will require fine mesh subdivision to analyse a shell with significant membrane action, regardless of the bending approximation. A rather common reason put forward to explain the relatively fine mesh required by most faceted shell elements to analyse the various shell examples in the literature is the approximation of the curved surface by the flat elements. Although it is difficult to generalize over a wide range of problems one may suspect that this is not so much the result of the inherent geometric approximation as a consequence of the poor membrane approximation. A further deficiency associated with displacement–based flat shell elements is the discontinuity of the element displacements in the shell assemblage. This is due to the different polynomial approximation orders of the displacement fields in available plane stress and plate bending elements, and thus continuity of the translational displacements is not maintained when the elements are assembled together, as the adjacent elements are not coplanar. This discontinuity between the in–plane and transverse displacements in most faceted shell elements does not appear to affect their behaviour in linear analysis. The existing facet elements are thus composed of a plane stress element plus a plate bending element. It is usually sound practice to combine four triangular elements into a single quadrilateral element. This has the two following major advantages:

1. The symmetry in the structure is retained in the finite element model.

2. Static condensation applied to the interior nodes of the quadrilateral element reduces the inter–element global degrees of freedom.

Possible element combinations are shown in Fig. 9.14. In Fig. 9.14, various combinations of plane stress and plate bending triangular elements are given. These may be grouped as follows:

1. CST or LST with global in–plane rotations at the apices;

2. LST with LCCT12 (a refined plate element which requires θ_s at mid–side nodes, as well as θ_x', θ_y' at apices)

3. CST or LST combined with the DKL family, in which case rotation nodes are embedded in the triangle sides. Rotation terms (θ_s) are defined along the triangle sides (either one rotation in the side with the CST, or two rotations in the side for the LST).

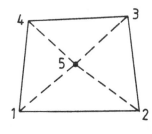

four CST
u,v at nodes 1 to 5

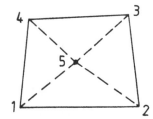

four three–node plate elements
$(, \theta_x, \theta_y)$ at nodes 1 to 5

(a)

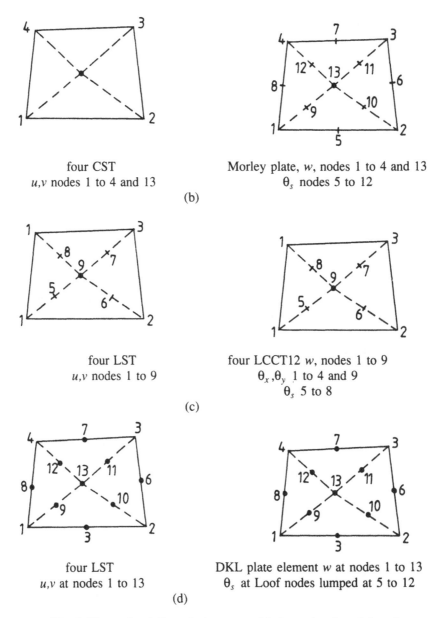

four CST
u,v nodes 1 to 4 and 13

Morley plate, *w*, nodes 1 to 4 and 13
θ_s nodes 5 to 12

(b)

four LST
u,v nodes 1 to 9

four LCCT12 *w*, nodes 1 to 9
θ_x, θ_y 1 to 4 and 9
θ_s 5 to 8

(c)

four LST
u,v at nodes 1 to 13

DKL plate element *w* at nodes 1 to 13
θ_s at Loof nodes lumped at 5 to 12

(d)

Fig. 9.15 Quadrilateral elements with four triangle subdomains.

In Fig. 9.15 various combinations of triangles into quadrilaterals are given as follows:

a. CST plus any three–node plate element
   Node 5, *u, v, w,* θ'_x, θ'_y condensed out
   Global d.o.f. *u, v, w,* θ_x, θ_y at nodes *1* to *4*
   Total external d.o.f. 20
   Internal d.o.f. 5

b. CST plus Morley plate element
   Nodes 10 to 13, θ_s condensed out
   Node 14, u, v, w condensed out
   Global d.o.f. u, v, w at nodes 1 to 4
   Local d.o.f. θ_s at nodes 6 to 9
   Total external d.o.f. 16
   Internal d.o.f. 7

c. LST plus LCCT12
   Nodes 5 to 8 θ_s condensed out
   Node 9 u, v, w, θ_x, θ_y condensed out
   (u, v, θ_s) constrained to be linear along external sides 1–2 etc.
   Total external d.o.f. 20
   Internal d.o.f. 17

d. LST plus DKL
   Nodes 9 to 12, u, v, w, θ_i, θ_{i+1} condensed out
   Node 13 u, v, w condensed out
   Total external d.o.f. 32
   Internal d.o.f. 23

If the membrane stresses vary significantly then (c) and (d) should perform better than (a) or (2) for a given mesh subdivision. Element (c) depends on its internal degrees of freedom to produce its flexibility, whereas element (d) retains the mid–side nodes on the external boundaries at the expense of the extra global degrees of freedom (and hence bandwidth in the solution of the equations). Comparisons of some results for various elements are given in examples 9.2 and 9.3.

Example 9.2 *Cylindrical shell roof*

The shell geometry is given in Fig. 9.16. This shell has been analysed by using many shell elements, and a summary of those plotted in Fig. 9.17 is given in Table 9.3. In Table 9.3, the DKL element, the two triangles combined as in Fig. 9.14(d), is given for comparison. The theory for the component parts of this element are given in Chapter 7 (membrane) and Chapter 8 (DKL plate bending).

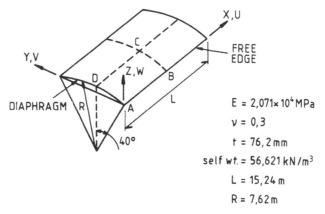

$E = 2{,}071 \times 10^4 \,\text{MPa}$

$v = 0{,}3$

$t = 76{,}2 \,\text{mm}$

self wt. $= 56{,}621 \,\text{kN/m}^3$

$L = 15{,}24 \,\text{m}$

$R = 7{,}62 \,\text{m}$

Fig. 9.16   Geometry of cylindrical shell roof.

Table 9.3

Shell elements used for cylindrical shell analysis			
Element	Ref.	d.o.f. / element	Symbol
DKL element (flat)	–	24	–
Carr (flat)	[71]	27	*
Clough and Johnson (flat)	[69]	15	×
Olson and Beardon (flat)	[76]	18	–
Bonnes and Dhatt (curved)	[111]	36	–
Brebbia and Connor (curved)	[104]	20	–
Cowper *et al.* (curved)	[112]	36	–
Dawe (curved)	[115]	54	
MacNeal (4–node, degenerated)	[137]	20	–
Nukulchai (4–node, degenerated)	[138]	24	–
Lagrange 9, (node full integration)	[145]	45	–
Lagrange 9, (node reduced integration)	[145]	45	–

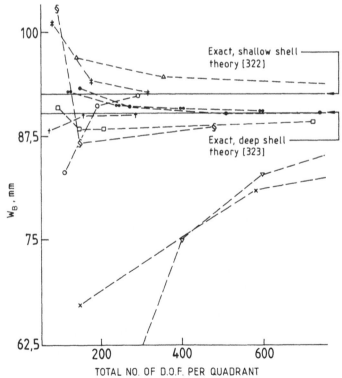

Fig. 9.17   Convergence of central deflection for cylindrical shell.

From Fig. 9.17 it is seen that elements converge to either the deep– or the shallow– shell solutions, but some elements appear to be excessively stiff. In Fig. 9.18 and 9.19, the plots are given for the longitudinal force N_x and bending moments M_x, M_y along the line CB shown in Fig. 9.16, in this case for the DKL element only.

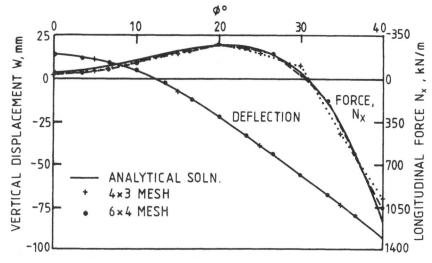

Fig. 9.18   Variation of vertical deflection and longitudinal force along CB.

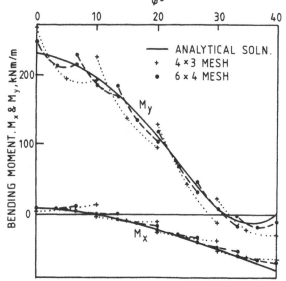

Fig. 9.19   Variation of bending moments along CB.

Example 9.3 *Pinched cylinder*

As a second example, the pinched cylinder shown in Fig. 9.20 is analysed. The various results are shown in Table 9.4. These results have been normalized with respect to the results presented by Dawe [115] for a (5×5) refined element analysis.

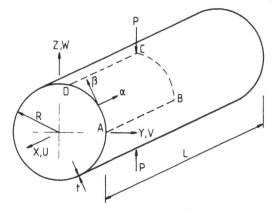

Fig. 9.20   Geometry of cylindrical shell under pinching loads.

Dimensions
$t =$ 2.3876 mm (0.094 in.)
$R =$ 125.806 mm (4.953 in.)
$L =$ 262.89 mm (10.35 in.)
$E =$ 72.4815 kN/m² (1.05×10^7 p.s.i.)
$v =$ 0.3125
$P =$ 0.4454 kN (100.34 lb)

The computation times for this analysis are available only for the supershell and Morley elements. For the results given for these two elements, the CPU times for the stiffness matrix generation and solution of equations are as follows:

	Stiffness generation	Equation solution	Active d.o.f.
Morley (10 × 10)	2.59	11.64	620
supershell (4 × 4)	5.21	8.04	271

It is seen from this comparison that the Morley element with the simple stiffness matrix generation plus the narrow bandwidth competes favourably with CPU time in this example.

Table 9.4

Finite element	Deflection under point load P (100 lb)
DKL (4 × 4)	0.9680
DKL (6 × 6)	0.9920
Cantin [143]	1.0030
Ashwell and Sabir [143] (8 × 8)	1.0010
Dawe [115] (5 × 5)	1.000)
Timoshenko [144]	0.9540
supershell (4 × 4)	0.9738
Morley (77)	1.0097

9.4 ISOPARAMETRIC DEGENERATED THREE–DIMENSIONAL SHELL ELEMENTS

9.4.1 Geometric definitions

The degenerated shell element family is shown in Fig. 9.21. The geometry of the shell is described by the node points 1 to 8 in the middle surface using the interpolation functions ϕ_1 to ϕ_N, given in section 1.4.4 (N = number of node points), and normals to this surface which are straight and when the element is deformed rotate about two axes in the middle surface. This is a similar assumption to that made in the analysis of plates using the isoparametric (heretosis) concept in section 8.2.2. The geometry of the shell may be described in either of two ways.

1. By defining the middle surface nodes and the shell thickness at these points. Assuming the geometric shape of the middle surface, normals may then be calculated from the cross product of vectors tangent to the parametric lines in the middle surface.

2. By using the node points on the inner and outer shell surface and calculating the mid–surface nodes as the mean between these nodes. In this case, the geometry lines through the thickness may not be perpendicular to the mid-surface.

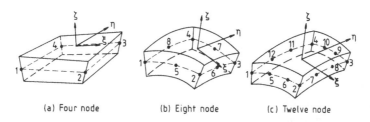

(a) Four node   (b) Eight node   (c) Twelve node

Fig. 9.21   Degenerated isoparametric shell element family.

This difference in the definition of shell geometry is shown in Fig. 9.22(a). It is seen that no .1 is normal to the mid–surface, but no. 2 is not. For all the elements in Fig. 9.21, the geometric definition within the element is given by

$$\{\tilde{X}\} = [\tilde{X}_1 \ \tilde{X}_2 \ \tilde{X}_3]^T = \sum_{i=1}^{m}\phi_i X_i + \sum_{i=1}^{m}\phi_i \frac{\zeta}{2}\tilde{n}_i \tag{9.62}$$

In eqn (9.63) the normal vector may be defined in terms of the unit normal and the shell thickness.

$$\tilde{n}_i = t_i \hat{n}_i \tag{9.63}$$

From Fig. 9.22(a), $\tilde{n}_i$ may not be normal to the mid–surface and is defined in eqn (9.64).

$$\tilde{n}_i = X_{i\,(top)} - X_{i\,(bottom)} \tag{9.64}$$

If the normals to the individual element mid–surfaces are used, the actual geometrical modelling is as shown in Fig. 9.23. Evidently, as the size of element decreases, this approximation decreases and, in addition, is not present if the 8– or 12– node quadratic and cubic elements are used to model shells defined by second–order curves. It is, of course, an approximation inherent to all flat facet modelling of curved shells.

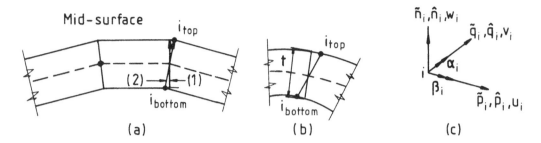

Fig. 9.22   Element idealization showing overlap of normals.

Returning to eqn (9.62) the derivatives of $\tilde{X}$ with respect to the (ξ, η, ζ) coordinates are given by

$$\frac{\partial \tilde{X}}{\partial \rho} = \sum_{i=1}^{m} \phi_{i,\rho} \tilde{X}_i + \sum_{i=1}^{m} \phi_{i,\rho} \frac{\zeta}{2} \tilde{n}_i \quad \rho = \xi, \eta \tag{9.65}$$

$$\frac{\partial \tilde{X}}{\partial \zeta} = \tilde{X}_{,\zeta} = \sum_{i=1}^{m} \phi_i \frac{\tilde{n}_i}{2} \tag{9.66}$$

In Fig. 9.22(c) various quantities are defined at node point i as follows:

{u, v, w}: local displacements
{α, β}: rotation about $\tilde{q}_i, \tilde{p}_i$ axes
{$r_{x_1}, r_{x_2}, r_{x_3}$}: global translation
{$\hat{p}, \hat{q}, \hat{n}$}: local orthogonal unit vectors
{$\tilde{p}, \tilde{q}, \tilde{n}$}: local orthogonal vectors

The assumption is now made that the effects of the node displacements and rotations can be interpolated independently. This is the same assumption made in the plate bending elements in section 8.2.2. Thus, the displacement vector at (ξ, η, ζ) is expressed in the standard isoparametric formulation:

$$\tilde{r} = \left\{ \begin{matrix} r_{x_1} \\ r_{x_2} \\ r_{x_3} \end{matrix} \right\} = \sum_{i=1}^{m} \phi_i \left\{ \begin{matrix} r_{iX_1} \\ r_{iX_2} \\ r_{iX_3} \end{matrix} \right\} + \sum_{i=1}^{m} \phi_i \zeta \frac{t_i}{2} [\hat{p}_i \ -\hat{q}_i] \left\{ \begin{matrix} \alpha_i \\ \beta_i \end{matrix} \right\} \tag{9.67}$$

If the heretosis concept is used for the quadratic element in Fig. 9.21(b), the centre node (9) is used with Lagrange interpolation of the nodal rotations, so that, now,

$$\tilde{r} = \sum_{i=1}^{8} \phi_i^S \tilde{r}_i + \sum_{i=1}^{9} \phi_i^L \zeta \frac{t_i}{2} [\hat{p}_i - \hat{q}_i] \left\{ \begin{matrix} \alpha_i \\ \beta_i \end{matrix} \right\} \tag{9.68}$$

In eqn (9.68), ϕ_i^S, ϕ_i^L are the serendipity and Lagrange shape functions, respectively. Displacement derivatives of eqn (9.68) with respect to (ξ, η, ζ) are hence given (for the heretosis element) by

$$\tilde{r}_{,\rho} = \sum_{i=1}^{8} \phi_{i,\rho}^{S} \tilde{r}_i + \sum_{i=1}^{9} \phi_{i,\rho}^{L} \zeta \frac{t_i}{2} [\hat{p}_i \ -\hat{q}_i] \begin{Bmatrix} \alpha_i \\ \beta_i \end{Bmatrix} \qquad \rho = \alpha, \beta \qquad (9.69)$$

and

$$\tilde{r}_{,\rho} = \sum_{i=1}^{9} \phi_i^{L} \frac{t_i}{2} [\hat{p}_i \ -\hat{q}_i] \begin{Bmatrix} \alpha_i \\ \beta_i \end{Bmatrix} \qquad (9.70)$$

The eqns (9.69), (9.70) can be grouped in matrix form as (for nodes 1 to 8)

$$\tilde{r}_{,\rho} = \left[[\bar{\phi}_{1,\rho}] \ [\bar{\phi}_{2,\rho}] \cdots [\bar{\phi}_{9,\rho}] \right] \begin{bmatrix} \{r_1\} \\ \{r_2\} \\ \cdot \\ \cdot \\ \cdot \\ \{r_9\} \end{bmatrix} = [\bar{\phi}_{,\rho}]^T \{r\} \qquad (9.71)$$

In eqn (9.71), the nodal displacement vectors are

$$\{r_i\} = \begin{Bmatrix} r_{iX_1} \\ r_{iX_2} \\ r_{iX_3} \\ \alpha_i \\ \beta_i \end{Bmatrix} \quad i = 1 \text{ to } 8; \quad \{r_9\} = \begin{Bmatrix} \alpha_9 \\ \beta_9 \end{Bmatrix} \qquad (9.72)$$

and the interpolation submatrices,

$$[\bar{\phi}_{i,\rho}] = \begin{bmatrix} \phi_{i,\rho}^{S} & 0 & 0 & \phi_{i,\rho}^{L} \zeta \frac{t_i}{2} p_{iX_1} & -\phi_{i,\rho}^{L} \zeta \frac{t_i}{2} q_{iX_1} \\ 0 & \phi_{i,\rho}^{S} & 0 & \phi_{i,\rho}^{L} \zeta \frac{t_i}{2} p_{iX_2} & -\phi_{i,\rho}^{L} \zeta \frac{t_i}{2} q_{iX_2} \\ 0 & 0 & \phi_{i,\rho}^{S} & \phi_{i,\rho}^{L} \zeta \frac{t_i}{2} p_{iX_3} & -\phi_{i,\rho}^{L} \zeta \frac{t_i}{2} q_{iX_3} \end{bmatrix} \qquad (9.73)$$

and, for node 9,

$$[\bar{\phi}_{9,\rho}] = \begin{bmatrix} \phi_{9,\rho}^{L} \zeta \frac{t_9}{2} p_{9X_1} & -\phi_{9,\rho}^{L} \zeta \frac{t_9}{2} q_{9X_1} \\ \phi_{9,\rho}^{L} \zeta \frac{t_9}{2} p_{9X_2} & -\phi_{9,\rho}^{L} \zeta \frac{t_9}{2} q_{9X_2} \\ \phi_{9,\rho}^{L} \zeta \frac{t_9}{2} p_{9X_3} & -\phi_{9,\rho}^{L} \zeta \frac{t_i}{2} q_{9X_3} \end{bmatrix} \qquad (9.74)$$

The displacements $\tilde{r}_\rho$ and $\tilde{r}_{,\rho}$ are expressed in the global coordinate system, and it is necessary to obtain these values and their derivatives in the local coordinates at each point in the element (and, of course, at the Gauss points for numerical evaluation of the element stiffness matrix). The directions of the local coordinates system are defined by the three unit vectors $\hat{T}_1, \hat{T}_2, \hat{T}_3$.

First the normal to the ξ, η parametric surface is given by

$$\tilde{n} = \tilde{X}_{,\xi} \times \tilde{X}_{,\eta} \tag{9.75}$$

in which the surface vectors are calculated in eqn (9.66). Thence,

$$\hat{T}_3 = \frac{\tilde{n}}{|\tilde{n}|} \tag{9.76}$$

The vector $\hat{T}_1$ is obtained by normalizing $\tilde{X}_{,\xi}$ such that

$$\hat{T}_1 = \frac{\tilde{X}_{,\xi}}{|\tilde{X}_{,\xi}|} \tag{9.77}$$

Finally, the cross product of $\hat{T}_1$ and $\hat{T}_3$ gives $\hat{T}_2$ as

$$\hat{T}_2 = \hat{T}_3 \times \hat{T}_2 \tag{9.78}$$

The transformation between global coordinates $\{\tilde{X}\}$ and local coordinates $\{\tilde{x}\}$ is given,

$$\{\tilde{x}\} = [T]^T \{\tilde{X}\} \tag{9.79}$$

where the columns of $[T]$ are defined in eqns (9.76), to (9.78),

$$[T] = [T_1 \ T_2 \ T_3] \tag{9.80}$$

9.4.2 Derivation of the strain transformation matrix

Displacement derivative transformation

The usual isoparametric derivative transformation between (ξ, η, ζ) and (X_1, X_2, X_3) coordinates is used. That is,

$$\frac{\partial r_{X_i}}{\partial \xi} = \frac{\partial r_{X_i}}{\partial X_j} \frac{\partial X_j}{\partial \xi} \quad \text{sum 1 to 3 on } j \tag{9.81}$$

This equation may be written in matrix form,

$$\begin{bmatrix} \{r_{,\xi}\}^T \\ \{r_{,\eta}\}^T \\ \{r_{,\zeta}\}^T \end{bmatrix} = \begin{bmatrix} \{X_{,\xi}\}^T \\ \{X_{,\eta}\}^T \\ \{X_{,\zeta}\}^T \end{bmatrix} \begin{bmatrix} \{r_{X_1}\}^T \\ \{r_{X_2}\}^T \\ \{r_{X_3}\}^T \end{bmatrix} \tag{9.82}$$

or, in shorthand notation,

$$[r_{,\xi\eta\zeta}] = [J] [r_{X_1 X_2 X_3}] \tag{9.83}$$

The matrix $[J]$ is the Jacobian of the transformation, and the inverse relation gives $\{X_1, X_2, X_3\}$ derivatives:

$$[r_{X_1 X_2 X_3}] = [J]^{-1}[r_{,\xi\eta\zeta}] \tag{9.84}$$

The Jacobian is written:

$$[J] = \sum_{i=1}^{8} \begin{bmatrix} \phi_{i,\xi}^S & \phi_{i,\xi}^S \dfrac{\zeta}{2} \\[2mm] \phi_{i,\eta}^S & \phi_{i,\eta}^S \dfrac{\zeta}{2} \\[2mm] 0 & \dfrac{\phi_i^S}{2} \end{bmatrix} \begin{bmatrix} X_i^T \\[2mm] t_i \hat{n}_i^T \end{bmatrix} \tag{9.85}$$

Write eqn (9.84) as

$$[\phi_X] = [J]^{-1}[\phi'] \tag{9.86}$$

To transform from global to local displacement derivatives, the tensor transformation in eqn (1.48) is used:

$$[\phi_x] = [T]^T [\phi_X][T] = [T]^T [J]^{-1}[\phi'][T] \tag{9.87}$$

$$[\phi_X] = \begin{bmatrix} u_{,x_1} & v_{,x_1} & w_{,x_1} \\[1mm] u_{,x_2} & v_{,x_2} & w_{,x_2} \\[1mm] u_{,x_3} & v_{,x_3} & w_{,x_3} \end{bmatrix} \tag{9.88}$$

The tensor transformation in eqn (9.87) may be cast in vector form as follows. Let

$$[T]^T [J]^{-1} = [A] = \begin{bmatrix} \tilde{a}_1^T \\[1mm] \tilde{a}_2^T \\[1mm] \tilde{a}_3^T \end{bmatrix} \tag{9.89}$$

and

$$[T] = \begin{bmatrix} t_{11} & t_{12} & t_{13} \\ t_{21} & t_{22} & t_{23} \\ t_{31} & t_{32} & t_{33} \end{bmatrix}$$

The eqn (9.87) is now rearranged in vector formas

$$\{\delta_\rho\} = \begin{bmatrix} t_{11}[A] & t_{21}[A] & t_{31}[A] \\ t_{12}[A] & t_{22}[A] & t_{32}[A] \\ t_{13}[A] & t_{23}[A] & t_{33}[A] \end{bmatrix} \{r_{,\xi\eta\zeta}\} \tag{9.90}$$

where, now, the vectors in eqn (9.90) are defined by

$$\{\delta_\rho\}^T = [u_{,x_1}\ u_{,x_2}\ u_{,x_3}\ v_{,x_1}\ v_{,x_2}\ v_{,x_3}\ w_{,x_1}\ w_{,x_2}\ w_{,x_3}]$$

and

$$\{r_{,\xi\eta\zeta}\}^T = [r_{X_1,\xi}\ r_{X_1,\eta}\ r_{X_1,\zeta}\ r_{X_2,\xi}\ r_{X_2,\eta}\ r_{X_2,\zeta}\ r_{X_3,\xi}\ r_{X_3,\eta}\ r_{X_3,\zeta}]$$

From eqns (9.71) to (9.74) the vector of displacement derivatives $\{r_{,\xi\eta\zeta}\}$ is expressed as

$$\{r_{,\xi\eta\zeta}\} = \begin{bmatrix} \hat{b}_1 & 0 & 0 & p_{1X_1}\tilde{c}_1 & -q_{1X_1}\tilde{c}_1 & \cdot & p_{9X_1}\tilde{c}_9 & -q_{9X_1}\tilde{c}_9 \\ 0 & \hat{b}_1 & 0 & p_{1X_2}\tilde{c}_1 & -q_{1X_2}\tilde{c}_1 & (2\text{--}8) & p_{9X_2}\tilde{c}_9 & -q_{9X_2}\tilde{c}_9 \\ 0 & 0 & \hat{b}_1 & p_{1X_3}\tilde{c}_1 & -q_{1X_3}\tilde{c}_1 & \cdot & p_{9X_3}\tilde{c}_9 & -q_{9X_3}\tilde{c}_9 \end{bmatrix} \tag{9.91}$$

In this equation,

$$\tilde{b}_j = \begin{Bmatrix} \phi_{j,\xi}^S \\ \phi_{j,\eta}^S \\ 0 \end{Bmatrix} \; ; \; \tilde{c}_i = \frac{t_i}{2} \begin{Bmatrix} \zeta\phi_{i,\xi}^L \\ \zeta\phi_{i,\eta}^L \\ \phi_i^L \end{Bmatrix} \; ; \; i = 1,9, \; j = 1,8 \tag{9.92}$$

In the element stiffness matrix formulation, the transformation from nodal point displacements to integration point strains is required (eqn (3.16)). That is,

$$\{\varepsilon\} = [a]\{r\} \tag{9.93}$$

Here the strain terms are,

$$\{\varepsilon_x\} = \begin{Bmatrix} \varepsilon_{x_1} \\ \varepsilon_{x_2} \\ \varepsilon_{x_1x_2} \\ \varepsilon_{x_2x_3} \\ \varepsilon_{x_3x_1} \end{Bmatrix} = \begin{Bmatrix} u_{,x_1} \\ v_{,x_2} \\ u_{,x_2} + v_{,x_1} \\ v_{,x_3} + w_{,x_2} \\ w_{,x_1} + u_{,x_3} \end{Bmatrix} \tag{9.94}$$

$$\{\sigma\}^T = [\sigma_{x_1} \; \sigma_{x_2} \; \sigma_{x_1x_2} \; \sigma_{x_2x_3} \; \sigma_{x_3x_1}] \tag{9.95}$$

and the constitutive equation, is then,

$$\{\sigma_x\} = [D]\{\varepsilon_x\} \tag{9.96}$$

The elastic constitutive matrix is

$$[D] = \frac{E}{1-v^2} \begin{bmatrix} 1 & v & 0 & 0 & 0 \\ v & 1 & 0 & 0 & 0 \\ 0 & 0 & \dfrac{1-v}{2} & 0 & 0 \\ 0 & 0 & 0 & \dfrac{1-v}{2.4} & 0 \\ 0 & 0 & 0 & 0 & \dfrac{1-v}{2.4} \end{bmatrix} = \begin{bmatrix} [D_M]_{3\times3} & [0]_{3\times2} \\ [0]_{2\times3} & [D_S]_{2\times2} \end{bmatrix} \tag{9.97}$$

For the general material write this equation

$$[D] = \begin{bmatrix} d_{11} & d_{12} & 0 & 0 & 0 \\ d_{21} & d_{22} & 0 & 0 & 0 \\ 0 & 0 & d_{33} & 0 & 0 \\ 0 & 0 & 0 & d_{44} & 0 \\ 0 & 0 & 0 & 0 & d_{55} \end{bmatrix} \tag{9.98}$$

In this expression it is seen that the plane stress condition σ_{x_3} has been assumed, and the factor 1/1.2 has been used on the out–of–plane shear strains because of their assumed parabolic distribution.

Strain displacement relationship

It is now possible, from eqns (9.78) and (9.81) etc., to form the strain displacement matrix,

$$\{\varepsilon_x\} = [B]\{r\} = [[B_1]_{(5\times5)}[B_2]_{(5\times5)} \cdots [B_9]_{(5\times5)}]\{r\}_{(42\times1)} \tag{9.99}$$

The $[B_i]$ submatrices are defined by

$$[B_i] = \begin{bmatrix} t_{11}f_1 & t_{21}f_1 & t_{31}f_1 & e_{11}g_1 & e_{21}g_1 & e_{21}g_1 \\ t_{12}f_2 & t_{22}f_2 & t_{32}f_2 & e_{12}g_2 & e_{22}g_2 & e_{22}g_2 \\ t_{11}f_2+t_{12}f_1 & t_{21}f_2+t_{22}f_1 & t_{31}f_2+t_{32}f_1 & e_{11}g_2+e_{12}g_1 & e_{21}g_2+e_{22}g_1 & e_{21}g_2+e_{22}g_1 \\ t_{12}f_3+t_{13}f_2 & t_{22}f_3+t_{23}f_2 & t_{32}f_3+t_{33}f_2 & e_{12}g_3+e_{13}g_2 & e_{22}g_3+e_{23}g_2 & e_{22}g_3+e_{23}g_2 \\ t_{13}f_1+t_{11}f_3 & t_{23}f_1+t_{21}f_3 & t_{33}f_1+t_{31}f_3 & e_{13}g_1+e_{11}g_3 & e_{23}g_1+e_{21}g_3 & e_{23}g_1+e_{21}g_3 \end{bmatrix} \tag{9.100}$$

Where $i = 1$ to 8, and if $i = 9$ only the last two columns are required. In eqn (9.100), the various terms are obtained from

$$f_i = \tilde{a}_j^T \tilde{b}_i; \quad g_j = \tilde{a}_j^T \tilde{c}_i; \quad e_{1j} = \hat{T}_j^T \hat{p}_i; \quad e_{2j} = \hat{T}_j^T \hat{q}_i \tag{9.101}$$

Element stiffness matrix

Having obtained the strain transformation matrix, the element stiffness matrix can be calculated by the isoparametric integration:

$$[K]_e = \int_{vol} [B]^T[D][B]dv = \int_{-1}^{+1}\int_{-1}^{+1}\int_{-1}^{+1} [B]^T[D][B]detJd\xi\, d\eta\, d\zeta \tag{9.102}$$

For integration through the thickness ζ, two–point Gauss integration is used. The integration with respect to ξ and η will be discussed further in the following. For the heterosis formulation, the element stiffness matrix thus generated has dimensions (42×42), and the interior node rotations may be condensed out to give the final size (40×40). The strain displacement matrix may be partitioned to reflect the contribution from membrane, rotation and shear effects. Thus, write $[B_i]$ in the partitioned form of eqn (9.103):

$$[B_i]_{5\times5} = \begin{bmatrix} [B_{MT}]_{3\times3} & [B_{MR}]_{3\times2} \\ [B_{RM}]_{2\times3} & [B_{SR}]_{2\times2} \end{bmatrix} \tag{9.103}$$

where the submatrices in eqn (9.103) are defined as follows

　　$[B_{MT}]$: relates the membrane strain due to translation
　　$[B_{MR}]$: relates the strain due to rotation
　　$[B_{ST}]$: relates the shear strain due to translation
　　$[B_{SR}]$: relates the shear strain due to rotation

Using this partitioned form, the stiffness matrix can be generated from

$$[k_{ij}] = \int_{vol} \begin{bmatrix} B_{MT} & B_{MR} \\ B_{ST} & B_{SR} \end{bmatrix}_i^T \begin{bmatrix} [D_M] & 0 \\ 0 & [D_S] \end{bmatrix} \begin{bmatrix} B_{MT} & B_{MR} \\ B_{ST} & B_{SR} \end{bmatrix}_j dv$$

Carrying out the multiplication of the submatrices in this equation,

$$[k_{ij}] = \int_{vol} \begin{bmatrix} B_{MT_i}^T D_M B_{MT_j} + B_{ST_i}^T D_S B_{ST_j} & B_{MT_i}^T D_M B_{MR_j} + B_{ST_i}^T D_S B_{SR_j} \\ B_{MR_i}^T D_M B_{MT_j} + B_{SR_i}^T D_S B_{ST_i} & B_{MR_i}^T D_M B_{MR_j} + B_{SR_i}^T D_S B_{SR_i} \end{bmatrix} dv \qquad (9.104)$$

where $i \leq 8$ and $j \leq 8$. If $i \leq 8$ and $j = 9$, only the last column of the j block in eqn (9.104) is formed, whereas for $i = j = 9$, only the last partitioned block is required in both i and j. The volume integral in eqn (9.104) will be calculated numerically as

$$[k]_e = \sum_{i=1}^{m_1} \sum_{j=1}^{m_2} \sum_{k=1}^{m_3} w_i w_j w_k \{ [B(\xi,\eta,\zeta)]^T [D] [B(\xi,\eta,\zeta)] detJ \}_{ijk} \qquad (9.105)$$

It was shown in section 8.2.2. that for the plate element, reduced (2×2) integration was necessary to minimize the effect of excessive shear stiffness when the plate becomes thin. This situation will also occur in the shell elements. In addition, membrane locking will occur for the eight–node element, and a discussion is given of this problem and how selective integration will reduce this effect.

9.4.3 Membrane locking and its elimination

For the flat parabola shown in Fig. 9.23 under the uniform bending moment M, it is easily shown that if the difference between the member length ds, and it's horizontal projection dx, is ignored then the displacement of the middle surface may be calculated as

$$u = -\frac{2M}{3EI} \frac{r}{l^2} x^3 \qquad (i)$$

and

$$v = \frac{M}{2EI} x^2 \qquad (ii)$$

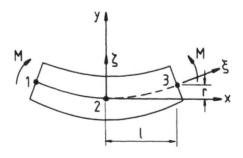

Fig. 9.23   Shallow arch element.

The quadratic element cannot represent the cubic power of x and hence cannot represent this simple constant moment state. For example, under the moment field shown, let $(+a, -a)$ be the displacements of nodes 1 and 3 in the x direction. Then,

$$u_\xi = \frac{1}{2}\xi(\xi - 1)a + \frac{1}{2}\xi(\xi + 1)(-a) = -\xi a$$

When $x = 1$,

$$u_l = -a = -\frac{2M}{3EI}rl \quad \text{from equation (i)}$$

and, since $\xi = \frac{x}{l}$, it follows that

$$u_\xi = -\frac{2M}{3EI}rx \tag{iii}$$

Similarly for v_ξ. In this case (b, b) are the displacements of 1 and 3 in the y direction

$$v_\xi = \{\frac{1}{2}\xi(\xi - 1) + \frac{1}{2}\xi(\xi + 1)\}b = \xi^2 b$$

With $b = \frac{Ml^2}{2EI}$, it follows that

$$v_\xi = \frac{Mx^2}{2EI} \tag{iv}$$

It is seen that the finite element models v_ξ correctly but not u_ξ. The effect of this assumption on the flat element can be ascertained by considering the transformation of derivatives in Fig. 9.21 in which the angle α is small, so that $\cos \alpha \approx 1$, $\sin \alpha \approx \alpha$.

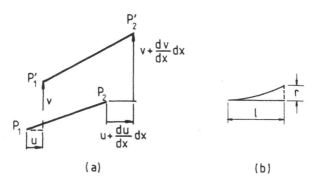

(a)   (b)

Fig. 9.24   Deformation of flat arch element.

For the flat parabola in Fig. 9.24(b),

$$y = r(\frac{x}{l})^2; \quad \frac{dy}{dx} = \frac{2rx}{l^2}$$

For the transformation matrix, $P_1 P_2$ to $P'_1 P'_2$,

$$\begin{bmatrix} \cos \alpha & \sin \alpha \\ -\sin \alpha & \cos \alpha \end{bmatrix} \approx \begin{bmatrix} 1 & \alpha \\ -\alpha & 1 \end{bmatrix}$$

Thence the strain in the direction $P'_1 P'_2$:

$$\begin{Bmatrix} \dfrac{\partial u'}{\partial x'} \\[2mm] \dfrac{\partial v'}{\partial x'} \end{Bmatrix} = \begin{bmatrix} 1 & \alpha \\ -\alpha & 1 \end{bmatrix} \begin{Bmatrix} \dfrac{\partial u}{\partial x} \\[2mm] \dfrac{\partial v}{\partial x} \end{Bmatrix}$$

Therefore,

$$\frac{\partial u'}{\partial x'} = \frac{\partial u}{\partial x} + \alpha \frac{\partial v}{\partial x} = \frac{\partial u}{\partial x} + \frac{dy}{dx} \frac{\partial v}{\partial x} \tag{v}$$

Substitute the correct relations eqns (i) and (ii), in this expression, such that

$$\frac{\partial u'}{\partial x'} = -\frac{2M}{EI} \frac{rx^2}{l^2} + \frac{2rx}{l^2} \frac{Mx}{EI} = 0$$

and so no membrane strain occurs. However, if the finite element expressions, eqns (iii) and (iv), are used,

$$\frac{\partial u'}{\partial x'} = -\frac{2M}{3EI} r + \frac{2rx}{l^2} \frac{Mx}{EI}$$

$$\varepsilon_{x'x'} = \frac{2Mr}{EI} \{ -\frac{1}{3} + \frac{x^2}{l^2} \} \tag{vi}$$

As the size of the element decreases, this term decreases because $r/l \rightarrow 0$. It is seen that $\varepsilon_{x'x'}$ is zero, however, when the term in braces in eqn (vi) is zero. That is, when

$$-\frac{1}{3} + \frac{x^2}{l^2} = 0$$

Solving for x gives the values,

$$\xi = \frac{x}{l} = \pm \frac{1}{\sqrt{3}} \tag{9.106}$$

These values are, of course the (2×2) Gauss integration point coordinates found to be suitable for removing the spurious shear strain in the isoparametric plate bending elements.

Reduced and selective integration

For the quadratic element, a minimum of (3×3) integration points in the (ξ, η, ζ) surface is necessary for a complete integration of the stiffness matrix. As the geometric variation in the ζ direction is linear, a two–point integration is sufficient through the thickness. Therefore, to evaluate the element exactly $(3 \times 3 \times 2)$ integration points in the ξ, η, ζ directions, respectively, are required. However, this produces the spurious shear and membrane stiffness which, when the length/thickness ratio is large, can destroy the solution. The remedy is to introduce either reduced or selectively reduced integration techniques. Since the two–point integration in the ζ direction is considered a minimum, modification to the integration order applies to the reduction in the ξ and η directions only. Four schemes for reduced integration are suggested and their results discussed. These are as follows.

URI (uniform reduced integration)

A uniform reduced integration of order (2×2) is applied to the complete stiffness matrix. Besides improved convergence, this scheme provides the incentive of a reduction in computational cost in stiffness matrix generation.

SRIA (selective reduced integration A)

This scheme involves using either two or three integration points in each of the ξ and η directions to evaluate the contribution to the element–strain energy. The integration order is given in Table 9.5.

Table 9.5

Selective and reduced integration order	
Energy term	Integration order ($\xi \times \eta$)
$\varepsilon_{x_1} d_{11} \varepsilon_{x_1}$	2×3
$\varepsilon_{x_2} d_{22} \varepsilon_{x_2}$	3×2
$\varepsilon_{x_1} d_{12} \varepsilon_{x_2}$	2×2
$\varepsilon_{x_1 x_2} d_{33} \varepsilon_{x_1 x_2}$	2×2
$\varepsilon_{x_2 x_3} d_{44} \varepsilon_{x_2 x_3}$	3×2
$\varepsilon_{x_1 x_3} d_{55} \varepsilon_{x_1 x_3}$	2×3

The implementation of this scheme requires formulation of the element stiffness matrix $[k_{ij}]$ in eqn (9.105) in three seperate parts. Let $[B_M]$ ($M = 1$ to 5) be the row vector of the strain displacement matrix $[B_l]$ ($l = 1$ to 9) in eqn (9.103). Hence, for (2×2) order, i and $j = 1$ to 9:

$$[k_{ij}] = \int_{vol} [B_1^T \; B_2^T \; B_3^T]_i \begin{bmatrix} d_{12} & 0 & 0 \\ 0 & d_{12} & 0 \\ 0 & 0 & d_{33} \end{bmatrix} \begin{bmatrix} B_1 \\ B_2 \\ B_3 \end{bmatrix}_j dv \qquad (9.107)$$

For 3×2 order,

$$[k_{ij}] = \int_{vol} [B_2^T \; B_4^T] \begin{bmatrix} d_{22} & 0 \\ 0 & d_{44} \end{bmatrix} \begin{bmatrix} B_2 \\ B_4 \end{bmatrix}_j dv \qquad (9.108)$$

and, for the (2×3) order:

$$[k_{ij}] = \int_{vol} [B_1^T \; B_5^T] \begin{bmatrix} d_{11} & 0 \\ 0 & d_{55} \end{bmatrix} \begin{bmatrix} B_1 \\ B_5 \end{bmatrix}_j dv \qquad (9.109)$$

Evidently, this scheme requires considerable computer time in the element stiffness matrix evaluation.

SRIB (selective reduced integration B)

In this scheme the reduced integration is applied to the shear energy as well as the membrane effects. Referring to eqn (9.104), the integration order of the partitioned matrices is as follows:

$$\begin{bmatrix} (2 \times 2) + (2 \times 2) & (2 \times 2) + (2 \times 2) \\ (2 \times 2) + (2 \times 2) & (3 \times 3) + (2 \times 2) \end{bmatrix}$$

The use of the (3×3) order is to ensure that the bending contribution is not underintegrated.

UNI3 (full integration)

Uniform (3×3) integration order is included to compare its accuracy with the above schemes.

9.4.4 Equivalent nodal forces

To conform to the element nodal displacement vector given in eqn (9.71), the element nodal force vector is defined as

$$\{R\}^T = [\{R_1\}^T \ \{R_2\}^T \ \cdots \ \{R_9\}^T]$$ (9.110)

where the individual submatrices are

$$\{R_i\}^T = [R_{ix_1} \ R_{ix_2} \ R_{ix_3} \ M_{i\alpha} \ M_{i\beta}] \quad i = 1 \text{ to } 8$$ (9.111a)

and

$$\{R_9\} = [M_{9\alpha} \ M_{9\beta}]$$ (9.111b)

For the body–force vector,

$$\{\tilde{p}\}^T = [\rho_{x_1} \ \rho_{x_2} \ \rho_{x_3}]$$

being the density vector in the global directions; the equivalent nodal force vector $\{R_b\}$ can be obtained via contragredience. Thence,

$$\{R_b\} = \int_{vol} [\phi]^T \{\bar{p}\} dv = \int_{-1}^{+1}\int_{-1}^{+1}\int_{-1}^{+1} [\phi]\{\bar{p}\} detJ \ d\xi \ d\eta \ d\zeta$$ (9.112)

where $[\phi]$ is the shape function matrix (eqn (9.67)). For a surface traction of uniform pressure $\{\tilde{p}\} = [p_{x_1} \ p_{x_2} \ p_{x_3}]$, the equivalent element nodal force vector is evaluated from eqn (9.112). That is,

$$\{R_S\} = \int_A [\phi]^T \{\tilde{P}_u\} = \int_{-1}^{+1}\int_{-1}^{+1} [\phi]^T \{\tilde{p}_u\} |\tilde{n}| d\xi \ d\eta$$ (9.113)

where $\tilde{n}$ is the normal vector evaluated at the Gauss points, and $dA = |\tilde{p} \times \tilde{q}| = |\tilde{n}| d\xi \ d\eta$. For a normal pressure $\tilde{p}_n$, the pressure vector p_u in eqn (9.113) is replaced by $p_n \ \hat{n}$, where $\hat{n}$ is the unit normal vector evaluated at the Gauss points. Thus, the equivalent nodal force vector $\{R_n\}$ is expressed,

$$\{R_n\} = \int_A [\phi]^T \{p_n\} \hat{n} dA = \int_{-1}^{+1}\int_{-1}^{+1} \{\phi\}^T p_n \hat{n} |\tilde{n}| d\xi \ d\eta$$ (9.114)

If the pressure is hydrostatic, then it's interpolated value is given

$$p_h = \{\phi\}\{p\}$$

In this equation $\{p\}$ are the nodal values and p_h is calculated at the Gauss points, so that from eqn (9.114) the expression for the noadal forces becomes

$$\{R_h\} = \int_{-1}^{+1}\int_{-1}^{+1} (\{\phi^L\}^T \{p\})[\phi]^T \hat{n} |\tilde{n}| d\xi \ d\eta$$ (9.115)

Evaluation of element stresses

From eqns (9.94) and (9.97) the element stresses are obtained from

$$\{\sigma_x\} = [D][B]\{r\}$$ (9.116)

where the strain displacement matrix $[B]$ is evaluated at the top ($\zeta = 1$) and middle ($\zeta = 0$) surfaces of the shell.

The stresses $\{\sigma\}_{mid}$ correspond to the membrane action, whereas the bending stresses $\{\sigma\}_{bending}$ are calculated from the difference of the top surface and mid–surface values. That is,

$$\{\sigma\}_{bending} = \{\sigma\}_{top} - \{\sigma\}_{mid}$$

For the quadratic element, the optimum position for calculating the stresses is at the 2×2 Gauss points $\xi = \eta = \pm\frac{1}{\sqrt{3}}$. At these points the displacement derivatives (and hence the stresses) acquire the same degree of accuracy as the nodal displacements. The nodal stresses may be then obtained by extrapolation, using a local least square smoothing technique. Thus, let the stresses at the 2×2 Gauss points be

$$[\sigma_I \ \sigma_{II} \ \sigma_{III} \ \sigma_{IV}]^T$$

Considering the two–dimensional ξ, η interpolation, the variation of stresses is written

$$\sigma(\xi, \eta) = [1 \ \xi \ \eta \ \xi\eta]\begin{bmatrix} a_1 \\ a_2 \\ a_3 \\ a_4 \end{bmatrix}$$

Substituting for $(\xi, \eta = \pm\frac{1}{\sqrt{3}})$ and inverting gives the $\{a\}$ coefficients as

$$\begin{Bmatrix} a_1 \\ a_2 \\ a_3 \\ a_4 \end{Bmatrix} = \frac{1}{4}\begin{bmatrix} 1 & 1 & 1 & 1 \\ -\sqrt{3} & \sqrt{3} & \sqrt{3} & -\sqrt{3} \\ -\sqrt{3} & -\sqrt{3} & \sqrt{3} & \sqrt{3} \\ 3 & -3 & 3 & -3 \end{bmatrix}\begin{Bmatrix} \sigma_I \\ \sigma_{II} \\ \sigma_{III} \\ \sigma_{IV} \end{Bmatrix} \qquad (9.117)$$

The element nodal stresses $[\sigma_1 \ \sigma_2 \ \sigma_3 \ \sigma_4]^T$ are found by substituting $(\xi, \eta) = \pm 1$, in eqn (9.115), so that using, eqn (9.116),

$$\begin{Bmatrix} \sigma_1 \\ \sigma_2 \\ \sigma_3 \\ \sigma_4 \end{Bmatrix} = \begin{bmatrix} 1+\frac{\sqrt{3}}{2} & -\frac{1}{2} & 1-\frac{\sqrt{3}}{2} & -\frac{1}{2} \\ \cdot & 1+\frac{\sqrt{3}}{2} & -\frac{1}{2} & 1-\frac{\sqrt{3}}{2} \\ symm & \cdot & 1+\frac{\sqrt{3}}{2} & -\frac{1}{2} \\ \cdot & \cdot & \cdot & 1+\frac{\sqrt{3}}{2} \end{bmatrix}\begin{Bmatrix} \sigma_I \\ \sigma_{II} \\ \sigma_{III} \\ \sigma_{IV} \end{Bmatrix} \qquad (9.118)$$

The nodal stresses obtained in this way are not unique between adjacent elements. To overcome this, a simple averaging is required. In order to show the influence of the various integration schemes (URI, SRIA, SRIB, UNI3), two examples are given. The first is a cylindrical shell shown in Fig. 9.16 with uniform loading and the second a pinched cylinder with symmetrical point loading. The first example combines bending with membrane action, whereas the second experiences a stress singularity at the load point.

Example 9.4 *Cylindrical shell*

The dimensions of the shell are shown in Fig. 9.16. The deflection is plotted for the points A and B shown in Fig. 9.16, for various mesh sizes. It is seen from the results plotted in Fig. 9.25 (point B) and Fig. 9.26 (point C) that results converge to the deep–shell solution [144] and that UNI3 exhibits a very slow convergence rate. The CPU ($\times$ sec.) required for stiffness matrix formulation and assembly are shown for the various integration schemes. It is seen from this figure that SRIA is the least efficient and that in comparing results from Figs 9.25 to 9.27, for the cylindrical shell, the conclusion may be drawn that the URI (2×2) integration scheme gives the best combination of accuracy and computational cost–effectiveness.

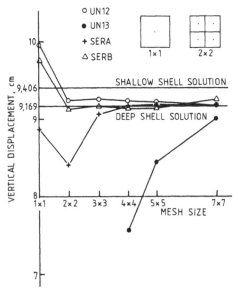

Fig. 9.25   Cylindrical shell – vertical displacement of point C (Fig. 9.16).

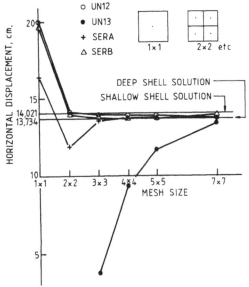

Fig. 9.26   Cylindrical shell – horizontal displacement of point B (Fig. 9.16).

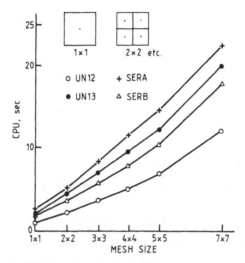

Fig. 9.27   Cylindrical shell – comparison of efficiencies of various integration schemes.

[1] Argyris, J.H. (1965) *Continua and Discontinua.* Proc. 1st Conference on Matrix Methods in Structural Mechanics. Wright–Patterson Air Force Base, Ohio.

[2] Pian, T.H. (1964) Derivation of element stiffness matrices by assumed stress distributions. *A.I.A.A. Journal.* Vol. 2, Pp. 1333–6.

[3] Argyris, J.H. and Kelsey, S. (1969) *Energy Theorems and Structural Analysis.* Butterworth, London.

[4] Weatherburn, C.E. (1944) *Advanced Vector Analysis.* G. Bell & Sons, London.

[5] Stewart, G.W. (1973) *Introduction to Matrix Computations.* Academic Press, New York.

[6] Wilson, E.L. (1978) *CAL87: Computer Analysis Language for the Static and Dynamic Analysis of Structural Systems.* Report No. UC SESM 79–1, Dept. of Civil Engineering. University of California, Berkeley.

[7] Irons, B. and Ahmad, S. (1980) *Techniques of Finite Elements.* Ellis Horwood Series in Engineering Science, Chichester, England.

[8] Abramowitz, M. and Stegun, I.A (ed) (1965) *Handbook of Mathematical Functions.* Dover Publications, Inc., New York.

[9] Meek, J.L. (1971) *Matrix Structural Analysis.* McGraw–Hill. New York.

[10] Prager, W. (1961) *Introduction to Mechanics of Continua.* Ginn and Co. Boston.

[11] Bogner, F.K., Fox, R.L. and Schmit, L.A. (1965) The generation of interelement compatible stiffness and mass matrices by the use of interpolation formulas. *Proc. 1st Conference on Matrix Methods in Structural Mechanics.* Wright–Patterson Air Force Base, Ohio. Pp. 397–444.

[12] Prezemieniecki, J.S. (1968) Theory of Matrix Structural Analysis. McGraw–Hill. New York.

[13] Livesley, R.K. (1964) *Matrix Methods of Structural Analysis.* Pergamon Press and Macmillan Co., New York.

[14] Popov, E. (1965) *Introduction to Mechanics of Solids.* Prentice–Hall, Englewood Cliffs, NJ.

[15] Denke, P. (1954) A matrix method of structural analysis. *Proc. of 2nd US National Congress of Applied Mechanics.* Pp. 445–51.

[16] Lawo, M., and Thierauf, G. (1980) *Stabtragwerke, Matrizenmethoden der Statik und Dynamik, Teil 1: Statik.* Friedr. Vieweg & Sohn, Braunschweig/Weisbaden.

[17] Argyris, J.H., Kelsey, S. and Kamel, H. (1964) Matrix methods of structural analysis. – A Précis. in: *Matrix Methods of Structural Analysis.* Prof. B. Fraeije de Veubeke ed. Pp. 81–98 Macmillan Co., New York.

[18] Timoshenko, S. and Goodier, J.N. (1970) *Theory of Elasticity.* 3rd edn. McGraw Hill, New York.

[19] Meek, J.L. and Swannell, P. (1977) The lateral torsional buckling problem reviewed from virtual displacement principles. *Institution of Engineers Aust. Civil Eng. Transactions CE19 (2),* Pp. 153–61.

[20] Vlasov, V.Z. (1961) *Thin Walled Elastic Beams* 2nd edn., Translation by the Israel Program for Scientific Translations, for NSF and the USA Dept. of Commerce Office of Technical Services, Washington, DC.

[21] Leonhardt, F. and Schlaich, J. (1972) Structural design of roofs over the sports arenas for the 1972 Olympic Games: some problems of prestressed cable net structures. *The Structural Engineer.* Vol. 50, No. 3. Pp. 113–19

[22] Jennings, A. (1968) Frame analysis including changes of geometry. *Journal of Structures Division, ASCE.* Vol. 94, ST3, Pp. 627–44.

[23] Argyris, J. and Scharpf, D. (1972) Large deflection analysis of prestressed networks. *Journal of Structures Division, ASCE.* Vol. 98, ST3 Pp. 633–53.

[24] Brown, P.D (1978) Design and analysis of tension structures. M. Eng. Sc. thesis, University of Queensland.

[25] Poskitt, F. (1967) Numerical solution of non–linear structures. *Journal of Structures Division, ASCE.* Vol. 93, ST3, Pp. 69–94.

[26] Fung, Y.C. (1965) *Foundations of Solid Mechanics.* Prentice–Hall, Inc., Englewood Cliffs, NJ.

[27] Timoshenko, S.P. and Gere, J.M. (1961) *Theory of Elastic Stability.* McGraw–Hill, New York.

[28] Powell, G. and Klinger, (1970) Elastic lateral buckling of steel beams. *Journal of Structures Division, ASCE.* Vol. 96, Pp. 1919–32.

[29] Korn, A. (1970) Bounding of frame buckling loads. *Journal of Structures Division, ASCE.* Vol. 96, Pp. 1639–55.

[30] Oran, C. (1973) Tangent Stiffness of space frames. *Journal of Structures Division, ASCE.* Vol. 99, Pp. 987–1001.

[31] Matthies, H. and Strang, G. (1979) The solution of nonlinear finite element equations. *International Journal of Numerical Methods in Engineering.* Vol. 14, Pp. 1613–26.

[32] Bergan, P.G. and Soreide, T.H. (1977) Solution of large displacement and instability problems using the current stiffness parameter. In: *Proc. Conf. on Finite elements in Nonlinear Solid and Structural Mechanics. Geilo, Norway.* Bergan, P.G. ed. Tapir (Technical University of Norway Press) Trondheim, Pp. 747–669.

[33] Sharifi, P. and Popov, E.P. (1971) Non–linear buckling analysis of sandwich arches. *Journal of the Engineering Mechanics Division, ASCE.* Vol. 97, Pp. 1397–1411.

[34] Batoz, J.L. and Dhatt, G. (1979) Incremental displacement algorithms for non–linear problems. *International Journal of Numerical Methods in Engineering.* Vol. 14, Pp. 1262–67.

[35] Riks, E. (1979) An incremental approach to the solution of snapping and buckling problems. *International Journal of Solids and Structures.* Vol. 15, Pp. 529–51.

[36] Ramm, E. (1981) Strategies for tracing the non–linear response near limit points. In: *Non–linear Finite Element Analysis in Structural Mechanics.* Wunderlich, W.,Stein, E. and Bathe, K.J. eds. Springer–Verlag, Berlin, Pp. 63–89.

[37] Crisfield, M. (1981) A fast incremental/iterative solution procedure that handles snap–through. *Computers and Structures.* Vol. 13, Pp. 55–62.

[38] Noor, A.K. and Peters, J.M. (1983) Instability analysis of space trusses. *Computer Methods in Applied Mechanics and Engineering.* Vol. 40, Pp. 199–218.

[39] Argyris, J.H., *et al.* (1979) Finite element method –the natural approach. *Computer Methods in Appied Mechanics and Engineering* Vol.7 No.18, Pp. 1–106.

[40] Clough, R. and Felippa, C., (1968) A refined quadrilateral element for the analysis of plate bending. *Proc. 2nd Conference on Matrix Methods in Structural Mechanics.* Wright–Patterson Air Force Base, Ohio. Pp. 399–440.

[41] Cooke, R.D. (1972) Two hybrid elements for analysis of thick, thin and sandwich plates. *International Journal of Numerical Methods in Engineering.* Vol. 5, Pp. 277–88.

[42] Anderheggen, E. (1968) A finite element bending equilibrium analysis. *Journal of the Engineering Mechanics Division, ASCE.* Vol. 95, EM4 Pp. 841–58.

[43] Herrmann, L.R. (1968) Finite element bending analysis of plates. *Journal of the Engineering Mechanics Division, ASCE.* EM–5, Pp. 13–25.

[44] Irons, B.M. (1973) Comment on a higher order conforming rectangular plate element. by Gopalacharyulu, S. *International Journal of Numerical Methods in Engineering.* Vol. 6, Pp. 308–9.

[45] Melosh, R.J. (1961) Stiffness matrix for analysis of thin plates in bending. *Journal of Aero. Science.* Vol. 28, no. 3 Pp. 34–42.

[46] Dawe, D.J. (1967) On assumed displacements for the rectangular plate bending element. *Journal of the Aeronautical Society.* Vol. 71, Pp. 722.

[47] Holand, I. and Bell, K. (eds) (1969) Finite element methods in stress analysis. Tapir (Technical University of Norway Press), Trondheim.

[48] Morley, L.S.D. (1971) The constant–moment plate bending element. *Journal of Strain Analysis.* Vol. 6, Pp. 20–24.

[49] Gallagher, R.H. (1969) Analysis of plate and shell structures. *Proc. Conference on Appication of Finite Element Methods in Civil Engineering.* Vanderbilt University, Pp. 155–205.

[50] Robinson, J. and Haggenmacher, G. (1979) LORA – an accurate four node stress plate bending element. *International Journal of Numerical Methods in Engineering.* Vol. 14, Pp. 296–306.

[51] Morley, L.S.D. (1963) *Skew Plates and Structures.* Pergamon Press, Oxford, England.

[52] Crisfield, M.A. A four–noded thin plate bending element using shear constraints – a modified version of Lyons' element. *Computer Methods in Appied Mechanics and Engineering.* Vol. 38, Pp. 93–120.

[53] Argyris, J.H. and Scharpf, D.W. (1968) The Sheba family of shell elements for the matrix displacemnt method. *Journal of the Royal Aeronautical Society.* Vol. 72, Pp. 873–83.

[54] Reissner, E. (1960) On some problems in shell theory. *Proc. 1st Symposium on Naval Structural Mechanics.* Pergamon Press, Inc., New York Pp. 74–114.

[55] Koiter, W.T. (1960) A consistent first approximation in the general theory of thin elastic shells. in: Koiter, W.T. (ed.), *Theory of Thin Elastic Shells.* North Holland, Amsterdam. Pp. 12–33.

[56] Naghdi, P.M. (1963) Foundations of elastic shell theory. in: Sneddon, A.H. (ed.), *Progress in Solid Mechanics.* Vol. 4, North–Holland, Amsterdam. Pp. 1–90.

[57] Novozhilov, V.V. (1964) *The Theory of Thin Shells.* 2nd edn. Noordhoff. Groningen.

[58] Kraus, H. (1967) *Thin Elastic Shells.* John Wiley, New York.

[59] Ahmad, S., Irons, B.M. and Zienkiewicz, O.C. (1970) Analysis of thick and thin shell structures by curved finite elements. *International Journal of Numerical Methods in Engineering.* Vol. 2, Pp. 419–51.

[60] Crisfield, M.A. (1979) A faster modified Newton–Raphson iteration. *Computer Methods in Appied Mechanics and Engineering.* Vol. 20, Pp. 267–78.

[61] Greene, B.E., Strome, D.R. and Weikel, R.C. (1961) Appication of the stiffness method to the analysis of shell structures. *Proc. Aviation Conference ASME.* Los Angeles Pp. 354.

[62] Zienkiewicz, O.C., and Cheung, Y.K. (1965) Finite element method of analysis of arch dam shells and comparison with finite difference procedures. *Proc. Symposium on Theory of Arch Dams. Pergamon Press, New York.*

[63] Gallagher, R.H., Gellatly, R.A., Padlog, J. and Mallett, R.H. (1967) A discrete element procedure for thin shell instability analysis. *Journal of American Institute of Aeronautics and Astronautics.* Vol. 5, Pp. 138–45

[64] Johnson, C.P. (1967) The Analysis of Thin Shells by a Finite Element Procedure. PhD thesis, Rept. No. SESM 67–22 University of California, Berkeley.

[65] Yeh, C. (1970) Large Deflection Dynamic Analysis of Thin Shells Using the Finite Element Method. PhD thesis, Rept. No. SESM 70–18, University of California, Berkeley.

[66] Clough, R.W. and Tocher, J.L. (1965) Analysis of thin arch dams by the finite element method. In: Rydzewski, J.R. (ed.), *Theory of Arch Dams.* Pergamon Press, Oxford. Pp. 107–21.

[67] Petyt, M. (1965) *The Application of Finite Element Techniques to Plate and Shell Problems.* Institute of Sound and Vibration, University of Southampton, ISVR Rept. No. 120.

[68] Argyris, J.H. (1965) Matrix displacement analysis of anisotropic shells by triangular elements. *Journal of the Royal Aeronautical Society, Vol. 69,* Pp. 801–05.

[69] Clough, R.W. and Johnson, C.P. (1968) A finite element approximation for the analysis of thin shells. *International Journal of Solids and Structures.* Vol. 4, Pp. 43–60.

[70] Zienkiewicz, O.C., Parekh, C.J. and King, I.P. (1968) Arch dam analysis by a linear finite element shell solution program. *Proc. of the Symposium on Arch Dams ICE.* London, ed. T.L. Dennis. London. Pp. 19–22.

[71] Carr, A.J. (1967) A Refined Finite Element Analysis of Thin Shell Structures Including Dynamic Loadings. Rept. No. SESM 67–9, University of California, Berkeley.

[72] Chu, T.C. and Schnobrich, W.C (1972) Finite element analysis of translational shells. *Computers and Structures.* Vol. 2, Pp. 197–222.

[73] Razzaque, A. (1972) Finite Element Analysis of Plates and Shells. PhD thesis. University of Wales, Swansea.

[74] Bazeley, G.P., Cheung, Y.K., Irons, B.M. and Zienkiewicz, O.C. (1965) Triangular elements in plate bending – conforming and non–conforming solutions. *Proc. 1st Conference on Matrix Methods in Structural Mechanics.* Wright–Patterson Air Force Base, Ohio. Pp. 547–76.

[75] Razzaque, A. (1973) Program for triangular bending elements with derivative smoothing. *International Journal of Numerical Methods in Engineering.* Vol. 6, Pp. 331–43.

[76] Olson, M.D. and Bearden, T.W. (1979) The simple flat triangular shell element revisited. *International Journal of Numerical Methods in Engineering.* Vol. 14, Pp. 51–68.

[77] Dawe, D.J. (1972) Shell analysis using a facet element. *Journal of Strain Analysis.* Vol. 7, Pp. 266–70.

[78] Herrmann, L.R. and Campbell, D.M. (1968) A finite element analysis for thin shells. *Journal of the American Institute of Aeronautics and Astronautics.* Vol. 6, Pp. 1842–847

[79] Dungar, R., Severn, R.T. and Taylor, P.R. (1967) Vibration of plate and shell structures using triangular finite elements. *Journal of Strain Analysis.* Vol. 2, Pp. 73–83.

[80]   Dungar, R. and Severn, R.T. (1969) Triangular finite elements of variable thickness and their application to plate and shell problems. *Journal of Strain Analysis.* Vol. 4, Pp. 10–21.

[81]   Yoshida, Y. (1974) A hybrid stress element for thin shell analysis. *Proc. of the International Conference on Finite Element Methods in Engineering.* University of New South Wales, Australia, Pp. 271–84.

[82]   Sander, G. and Becker, P. (1975) Deliquent finite elements for shell idealization. *Proc. World Congress on Finite Element Methods in Structural Mechanics,* Vol. 2, Bournemouth, England, Pp. 1–31.

[83]   Argyris, J.H., Dunne, P.C., Malejannakis, G.A. and Schelke, E. (1977) A simple triangular facet shell element with application to linear and nonlinear equilibrium and elastic stability problems. *Computer Methods in Appied Mechanics and Engineering.* Vol. 10, Pp. 371–403 and Vol 11, Pp. 97–131.

[84]   Horrigmoe, G. and Bergan, P.G. (1978) Non–linear analysis of free–form shells by flat finite elements. *Computer Methods in Appied Mechanics and Engineering.* Vol. 16, Pp. 11–35.

[85]   Horrigmoe, G. (1977) *Finite Element Instability Analysis of Free–Form Shells.* Division of Structural Mechanics, University of Trondheim, Norway.

[86]   Batoz, J.L., Bathe, K.J. and Ho, L.W. (1980) A study of three–node triangular plate bending elements. *International Journal of Numerical Methods in Engineering.* Vol. 15, Pp. 1771–812.

[87]   Bathe, K.J. and Ho, L.W. (1981) A simple and effective element for analysis of general shell structures. *Computers and Structures.* Vol. 13, Pp. 673–81.

[88]   Morris, A.J. (1973) A deficiency in current finite elements for thin shell applications. *International Journal of Solids and Structures.* Vol. 9, Pp. 331–46.

[89]   Gallagher, R.H. and Yang, T.Y. (1968) Elastic instability predictions for doubly curved shells. *Proc. 2nd Conference on Matrix Methods in Structural Mechanics.* Wright–Patterson Air Force Base, Ohio. Pp. 711–39.

[90]   Oden, J.T. (1968) Calculation of stiffness matrices for finite elements of thin shells of arbitrary shape. *Journal of American Institute of Aeronautics and Astronautics,* Vol. 6, Pp. 969–72.

[91]   Greene, B.E., Jones, R.E., McLay, R.W. and Strome, D.R. (1968) Dynamic analysis of shells using doubly–curved finite elements. *Proc. 2nd Conference on Matrix Methods in Structural Machanics,* Wright–Patterson Air Force Base, Ohio. Pp. 185–212.

[92]   Gallagher, R.H., Lain, S. and Mau, S.T. (1971) A procedure for finite element plate and shell pre– and post–buckling analysis. *Proc. 3rd Conference on Matrix Methods in Structural Mechanics.* Wright–Patterson Air Force Base, Ohio. Pp. 857–79.

[93]   Cowper, G.R., Lindberg, G.M. and Olson, M.D. (1971) Comparison of two high–precision triangular finite elements for arbitrary deep shells. *Proc. 3rd Conference on Matrix Methods in Structural Mechanics.* Wright–Patterson Air Force Base, Ohio. Pp. 277–304.

[94]   Sander, G. and Idelhson, S. (1983) A family of conforming finite elements for deep shell analysis. *International Journal of Numerical Methods in Engineering.* Vol. 18, Pp. 363–80.

[95]   Budiansky, B. and Sander, J.L. (1963) On the 'best' first order linear shell theory. *Progress in Solid Mechanics.* Prager Anniversary Volume. Macmillan, New York. Pp. 129–40.

[96] Dupius, G. and Goel, J.J. (1970) A curved element for thin elastic shells. *International Journal of Solids and Structures.* Vol. 6, Pp. 1413–428.

[97] Thomas, G.R. and Gallagher, R.H. (1975) *A Triangular Thin Shell Finite Element: Linear Analysis.* Rept. No. NASA CR–2482, Cornell University, Ithaca, N.Y.

[98] Gellert, M. and Laursen, M.E. (1977) A new high–precision stress finite element for analysis of shell structures. *International Journal of Solids and Structures.* Vol. 13, Pp. 683–97.

[99] Talaslides, D. and Wunderlich, W. (1979) Static and dynamic analysis of Kirchoff shells based on a mixed finite element formulation. *Computers and Structures.* Vol. 10, Pp. 239–49.

[100] Matsui, T. and Matsuoka, O. (1976) A new finite element scheme for instability analysis of thin shells. *International Journal of Numerical Methods in Engineering.* Vol. 10, Pp. 145–70.

[101] Stein, E., Berg, A. and Wagner, W. (1982) Different levels of non–linear shell theory in finite element stability analysis. In: Ramm, E. (ed.), *Buckling of Shells.* Proc. State–of–Art Colloquium, Springer–Verlag, Berlin. Pp. 91–136.

[102] Simmonds, J.C. and Danielson, D.A. (1972) Nonlinear shell theory with finite rotations and stress function vectors. *Journal of Appied Mechanics.* Vol. 39, Pp. 1085.

[103] Kratzig, W.B., Basar, Y. and Wittek, U. (1982) Nonlinear behaviour and elastic stability of shells– theoretical concepts – numerical computations – results. in: Ramm, E. (ed) *Buckling of Shells.* Proc. State–of–the Art Colloquium. Springer–Verlag, Berlin. Pp. 19–56.

[104] Connor, J.J. and Brebia, C. (1967) Stiffness matrix for shallow rectangular shell element. *Journal of the Engineering Mechanics Division, ASCE.* Vol. 93, Pp. 43–65.

[105] Sabir, A.B. and Ashwell, D.G. (1969) A stiffness matrix for shallow shell finite elements. *International Journal of Mechanical Sciences.* Vol. 11, Pp. 269–79.

[106] Yang, T.Y. (1973) High order rectangular shallow shell finite elements. *Journal of the Engineering Mechanics Division, ASCE.* Vol. 99, Pp. 157–81.

[107] Jones, R.F. (1973) Shell and plate analysis by finite element. *Journal of the Structural Division, ASCE.* Vol. 99, Pp. 889–902.

[108] Pecknold, D.A. and Schnobrich, W.C. (1969) Finite element analysis of skewed shallow shells. *Journal of the Structural Division, ASCE.* Vol. 95, Pp. 715–44.

[109] Utku, S. (1967) Stiffness matrices for thin triangular elements of nonzero Gaussian curvature. *Journal of the American Institute of Aeronautics and Astronautics.* Vol. 5, Pp. 1659–67.

[110] Strickland, G. and Loden, W. (1968) A doubly curved triangular shell element. *Proc. 2nd Conference on Matrix Methods in Structural Mechanics.* Wright–Patterson Air Force Base, Ohio. Pp. 641–66.

[111] Bonnes, G., Dhatt, G., Girounx, Y.M. and Robichaud, L.P.A. (1968) Curved triangular lements for the analysis of shells. Proc. 2nd Conference on Matrix Methods in Structural Mechanics. Wright–Patterson Air Force Base, Ohio. Pp. 617–39.

[112] Cowper, G.R., Lindberg, G.M. and Olson, M.D. (1970) A shallow shell finite element of triangular shape. *International Journal of Solids and Structures.* Vol. 6, Pp. 1133–56.

[113] Argyris, J.H., Fried, I. and Scharpf, D.W. (1968) Tuba family of plate elements for the matrix displacement method. *Journal of the Royal Aeronautical Society.* Vol. 72, Pp. 701–9.

[114] Vos, R.G. (1972) Generalization of plate finite elements to shells. *Journal of the Engineering Mechanics Division, ASCE.* Vol. 98, Pp. 385–400.

[115] Dawe, D.J. (1975) High order triangular finite element for shell analysies *International Journal of Solids and Structures.* Vol. 11, Pp. 1097–10.

[116] Mohr, G.A. (1980) Numerically integrated triangular element for doubly curved thin shells. *Computers and Structures.* Vol. 11, Pp. 565–71.

[117] Dhatt, G. (1969) Numerical analysis of thin shells by curved triangular elements based on discrete Kirchoff hypothesis. *Proc. ASCE. Symposium on Appication of F.E.M. in Civil Engineering. Tennessee.* Pp. 255–78.

[118] Dhatt, G. (1970) An efficient triangular shell element. *Journal of American Institute of Aeronautics and Astronautics.* Vol. 8, Pp. 2100–2.

[119] Batoz, J.L. and Dhatt, G. (1972) Development of two simple shell elements. *Journal of the American Institute of Aeronautics and Astronautics.* Vol. 10, Pp. 237–8.

[120] Tabarrok, B. and Hoa, V.S. (1974) Thermal stress analysis of plates and shallow shells by hybrid finite element method. *Journal of Strain Analysis.* Vol. 9, Pp. 152–8.

[121] Alaylioglu, H. and Ali, R. (1977) A hybrid stress doubly curved shell finite element. *Computers and Structures.* Vol. 7, Pp. 477–80.

[122] Wolf, J.P. (1975) Alternate hybrid stress finite element models. *International Journal of Numerical Methods in Engineering.* Vol. 9, Pp. 601–15.

[123] Washizu, K. (1971) Some remarks on basic theory for finite element method. in: Gallagher, R.H. *et al.* (eds.), *Recent Advances in Matrix Methods of Structural Analysis and Design.* University of Alabama Press, Pp. 25–48.

[124] Pian, T.H.H. and Chen, D.P. (1982) Alternative ways for formulation of hybrid stress elements. *International Journal of Numerical Methods in Engineering.* Vol. 18, Pp. 1679–84.

[125] Irons, B.M. (1976) The SemiLoof shell element. in: Ashwell, D.G. and Gallagher, R.H. (eds.), *Finite Elements for Thin Shells and Curved Members.* John Wiley, New York. Pp. 197–222.

[126] Prato, C.A. (1969] Shell finite element method via Reissner's principle. *International Journal of Solids and Structures.* Vol. 5, Pp. 1119–33.

[127] Connor, J.J. and Will, G.T. (1971) A mixed finite element shallow shell formulation. in: Gallagher, R.H. *et al.* (eds.), *Recent Advances in Matrix Methods of Structural Analysis and Design.* University of Alabama Press, Pp. 105–37.

[128] Tahiani, C. and Lachance, L. (1975) Linear and non–linear analysis of thin shallow shells by mixed finite elements. *Computers and Structures.* Vol. 5, Pp. 167–177.

[129] Zienkiewicz, O.C., Taylor, R.L. and Too, J.M. (1971) Reduced integration technique in general analysis of plates and shells. *International Journal of Numerical Methods in Engineering.* Vol. 3, Pp. 275–90.

[130] Pawsey, S.F. and Clough, R.W. (1971) Improved numerical integration of thick shell finite elements. *International Journal of Numerical Methods in Engineering.* Vol. 3, Pp. 575–86.

[131] Lee, S.W. and Pian, T.H.H. (1978) Improvement of plate and shell finite elements by mixed formulations. *Journal of American Institute of Aeronautics and Astronautics.* Vol. 16, Pp. 29–34.

[132] Malkus, D.S. and Hughes, T.J.R. (1978) Mixed finite element methods–reduced and selective integration techniques: a unification of concepts. *Computer Methods in Appied Mechanics and Engineering.* Vol. 15, Pp. 63–81.

[133] Hughes, T.J.R., and Cohen, M. and Haroun, M. (1978) Reduced and selective integration technique in the finite element analysis of plates. *Nuclear Engineering and Design.* Vol. 46, Pp. 203–22.

[134] Noor, A.K. and Anderson, C.M. (1982) Mixed models and reduced/selective integration displacement models for nonlinear shell analysis. *International Journal of Numerical Methods in Engineering.* Vol. 18, Pp. 1429–54.

[135] Hughes, T.J.R. and Liu, W.K. (1981) Nonlinear finite element analysis of shells: Part 1. Three dimensional shells. *Computer Methods in Applied Mechanics and Engineering.* Vol. 26, Pp. 331–62.

[136] Talha, M.A. (1979) A theoretically improved and easily implemented version of the Ahmad thick shell element. *International Journal of Numerical Methods in Engineering.* Vol. 14, Pp. 125–42.

[137] MacNeal, R.H. (1978) A simple quadrilateral shell element. *Computers and Structures.* Vol. 8, Pp. 175–83.

[138] Kanok–Nukulchai, W. (1979) A simple and efficient finite element for general shell analysis. *International Journal of Numerical Analysis in Engineering.* Vol. 14, Pp. 179–200.

[139] Hughes, T.J.R., Taylor, R.L. and Kanok–Nukulchai, W. (1977) A simple and efficient finite element for plate bending. *International Journal of Numerical Methods in Engineering.* Vol. 11, Pp. 1529–43.

[140] Wempner, G., Talaslidis, D. and Hwang, C.M. (1982) A simple and efficient approximation of shells via finite quadrilateral elements. *Journal of Appied Mechanics.* Vol. 49, Pp. 115–20.

[141] Nagtegaal, J.C. and Slater, J.G. (1981) A simple incompressible thin shell element based on discrete Kirchoff theory. in: Hughes, T.J.R *et al.* (eds.) *Nonlinear Finite Element Analysis of Plates and Shells.* Proc. ASME Winter Annual Meeting. AMD, Vol. 48, Washington, DC.

[142] Cantin, G. and Clough, R.W. (1968) A curved cylindrical shell finite element. *Journal of the American Institute of Aeronautics and Astronautics.*

[143] Ashwell, D.G. and Sabir, A.B. (1972) A new cylindrical shell finite element based on simple independent strain functions. *International Journal of Mechanical Sciences.* Vol. 14, Pp. 171–83.

[144] Timoshenko, S.P. and Kreiger, S.W. (1959) *Theory of Plates and Shells.* 2nd edn, McGraw–Hill, New York.

[145] Parisch, H. (1979) A critical survey of the 9 node degenerated shell element with special emphasis on thin shell applications and reduced integration. *Computer Methods in Appied Mechanics and Engineering.* Vol. 20, Pp. 323–50.

[146] Scordelis, A.C. and Lo, K.S. (1964) Computer analysis of cylindrical shells *Journal of American Concrete Institute.* Vol. 61, Pp. 539–60.

[147] Clough, R.W. and Tocher, J.L. (1965) Plate element stiffness matrices for plate bending. *Proc. 1st Conference on Matrix Methods in Structural Mechanics,* Wright–Patterson Air Force Base, Ohio. Pp. 515–46

[148] Meek, J.L. and Tan, H.S. (1985) Flat facet plate elements based on the discrete Kirchoff hypothesis. *Proc. 2nd Int. Conference on Computer Aided Analysis and Design in Civil Engineering.* University of Roorkee, India Pp. III–1, III–6.

[149] Frish–Fay, R. (1962) *Flexible Bars.* Butterworth. London.

INDEX

Printed and bound by CPI Group (UK) Ltd, Croydon, CR0 4YY

22/10/2024

01777634-0007